109.95
70E

Foams: Fundamentals and Applications in the Petroleum Industry

ADVANCES IN CHEMISTRY SERIES 242

Foams: Fundamentals and Applications in the Petroleum Industry

Laurier L. Schramm, EDITOR
Petroleum Recovery Institute

American Chemical Society, Washington, DC 1994

Library of Congress Cataloging-in-Publication Data

Foams: fundamentals and applications in the petroleum industry / Laurier L. Schramm, editor.

p. cm.—(Advances in Chemistry Series, ISSN 0065–2393; 242).

Includes bibliographical references and index.

ISBN 0–8412–2719–5

1. Foam—Industrial applications. 2. Petroleum engineering.

I. Schramm, Laurier Lincoln. II. Series.

QD1.A355 no. 242.
[TP690.65]
540 s—dc20
[622'.3382]
94–28454
CIP

The paper used in this publication meets the minimum requirements of American National Standard for Information Sciences—Permanence of Paper for Printed Library Materials, ANSI Z39.48–1984.

Copyright © 1994

American Chemical Society

All Rights Reserved. The appearance of the code at the bottom of the first page of each chapter in this volume indicates the copyright owner's consent that reprographic copies of the chapter may be made for personal or internal use or for the personal or internal use of specific clients. This consent is given on the condition, however, that the copier pay the stated per-copy fee through the Copyright Clearance Center, Inc., 27 Congress Street, Salem, MA 01970, for copying beyond that permitted by Sections 107 or 108 of the U.S. Copyright Law. This consent does not extend to copying or transmission by any means—graphic or electronic—for any other purpose, such as for general distribution, for advertising or promotional purposes, for creating a new collective work, for resale, or for information storage and retrieval systems. The copying fee for each chapter is indicated in the code at the bottom of the first page of the chapter.

The citation of trade names and/or names of manufacturers in this publication is not to be construed as an endorsement or as approval by ACS of the commercial products or services referenced herein; nor should the mere reference herein to any drawing, specification, chemical process, or other data be regarded as a license or as a conveyance of any right or permission to the holder, reader, or any other person or corporation, to manufacture, reproduce, use, or sell any patented invention or copyrighted work that may in any way be related thereto. Registered names, trademarks, etc., used in this publication, even without specific indication thereof, are not to be considered unprotected by law.

PRINTED IN THE UNITED STATES OF AMERICA

1994 Advisory Board

Advances in Chemistry Series

M. Joan Comstock, *Series Editor*

Robert J. Alaimo
Procter & Gamble Pharmaceuticals

Mark Arnold
University of Iowa

David Baker
University of Tennessee

Arindam Bose
Pfizer Central Research

Robert F. Brady, Jr.
Naval Research Laboratory

Margaret A. Cavanaugh
National Science Foundation

Arthur B. Ellis
University of Wisconsin at Madison

Dennis W. Hess
Lehigh University

Hiroshi Ito
IBM Almaden Research Center

Madeleine M. Joullie
University of Pennsylvania

Lawrence P. Klemann
Nabisco Foods Group

Gretchen S. Kohl
Dow-Corning Corporation

Bonnie Lawlor
Institute for Scientific Information

Douglas R. Lloyd
The University of Texas at Austin

Cynthia A. Maryanoff
R. W. Johnson Pharmaceutical Research Institute

Julius J. Menn
Western Cotton Research Laboratory, U.S. Department of Agriculture

Roger A. Minear
University of Illinois at Urbana–Champaign

Vincent Pecoraro
University of Michigan

Marshall Phillips
Delmont Laboratories

George W. Roberts
North Carolina State University

A. Truman Schwartz
Macalaster College

John R. Shapley
University of Illinois at Urbana–Champaign

L. Somasundaram
DuPont

Michael D. Taylor
Parke-Davis Pharmaceutical Research

Peter Willett
University of Sheffield (England)

FOREWORD

The ADVANCES IN CHEMISTRY SERIES was founded in 1949 by the American Chemical Society as an outlet for symposia and collections of data in special areas of topical interest that could not be accommodated in the Society's journals. It provides a medium for symposia that would otherwise be fragmented because their papers would be distributed among several journals or not published at all.

Papers are reviewed critically according to ACS editorial standards and receive the careful attention and processing characteristic of ACS publications. Volumes in the ADVANCES IN CHEMISTRY SERIES maintain the integrity of the symposia on which they are based; however, verbatim reproductions of previously published papers are not accepted. Papers may include reports of research as well as reviews, because symposia may embrace both types of presentation.

ABOUT THE EDITOR

LAURIER L. SCHRAMM is a senior staff research scientist and group leader for process sweep improvement at the Petroleum Recovery Institute and adjunct associate professor of chemistry at the University of Calgary, where he lectures in applied colloid and interface chemistry. Dr. Schramm received his B. Sc. (Hons.) in chemistry from Carleton University in 1976 and Ph.D. in physical and colloid chemistry in 1980 from Dalhousie University, where he studied as a Killam and NRC Scholar. From 1980 to 1988 he held research positions with Syncrude Canada Ltd. in its Edmonton Research Centre.

His research interests have included many aspects of colloid and interface science applied to the petroleum industry, including research into mechanisms of processes for the improved recovery of light, heavy, or bituminous crude oils, such as in situ foam, polymer or surfactant flooding, and surface hot water flotation from oil sands. These mostly experimental investigations have involved the formation and stability of dispersions (foams, emulsions, and suspensions) and their flow properties, electrokinetic properties, interfacial properties, phase attachments, and the reactions and interactions of surfactants in solution.

Dr. Schramm is a Fellow of the Chemical Institute of Canada (including serving on the Local Section Executive) and is a member of the American Chemical Society and the International Association of Colloid and Interface Scientists. He has written more than 60 scientific publications and patents. This is his third ACS book, following *Emulsions: Fundamentals and Applications in the Petroleum Industry* and *The Language of Colloid and Interface Science.*

CONTENTS

Preface xi

FOAM FUNDAMENTALS

1. **Foams: Basic Principles** 3
   Laurier L. Schramm and Fred Wassmuth

2. **Mechanisms of Aqueous Foam Stability and Antifoaming Action with and without Oil: A Thin-Film Approach** 47
   D. T. Wasan, K. Koczo, and A. D. Nikolov

3. **Fundamentals of Foam Transport in Porous Media** 115
   A. R. Kovscek and C. J. Radke

4. **Foam Sensitivity to Crude Oil in Porous Media** 165
   Laurier L. Schramm

ENHANCING OIL RECOVERY FROM POROUS MEDIA TESTING FOAMS

5. **CO_2 Foams in Enhanced Oil Recovery** 201
   John P. Heller

6. **Steam-Foams for Heavy Oil and Bitumen Recovery** 235
   E. E. Isaacs, J. Ivory, and M. K. Green

7. **Adsorption of Foam-Forming Surfactants for Hydrocarbon-Miscible Flooding at High Salinities** 259
   Karin Mannhardt and Jerry J. Novosad

NEAR-WELL AND OILWELL APPLICATIONS OF FOAM

8. **Gas-Blocking Foams** 319
   Jan Erik Hanssen and Mariann Dalland

9. **Foams for Well Stimulation** 355
   David J. Chambers

10. **Role of Nonpolar Foams in Production of Heavy Oils** 405
    Brij B. Maini and Hemanta Sarma

FOAMS IN SURFACE FACILITIES

11. **Bituminous Froths in the Hot-Water Flotation Process** 423
    Robert C. Shaw, Jan Czarnccki, Laurier L. Schramm, and Dave Axelson

12. **Antifoaming and Defoaming in Refineries** 461
    V. E. Lewis and W. F. Minyard

GLOSSARY AND INDEXES

Glossary 487

Author Index 536

Affiliation Index 536

Subject Index 536

PREFACE

FOAMS CAN BE FOUND in almost every part of the petroleum production and refining process, from deep in the producing reservoirs, through oilwell drilling, stimulation, and production, to downstream vessels in the refining process. In these cases the foams may occur naturally or by design and may be desirable or undesirable. In all cases the presence and nature of the foam can determine both the economic and technical successes of the industrial process concerned.

This book provides an introduction to the nature and occurrence, properties and uses, and formation and breaking of foams in the petroleum industry. It is aimed at scientists and engineers who may encounter foams or apply foams, whether in process design, petroleum production, or research and development. Primarily the focus is on the introduction and subsequent application of foam principles, and includes attention to practical foam problems. Books available up to now are either principally theoretical (such as the colloid chemistry texts) or focus on foams in general (like Bikerman's classic books). A significant gap in this coverage concerns foams in the petroleum industry, a topic that is not only of great practical importance but also presents many problems of fundamental interest.

In this book a wide range of authors' expertise and experiences have been brought together to yield the first book on foams that focuses on the uses and occurrences of foams in the petroleum industry[1]. This broad range has allowed for a variety of foams and applications to be highlighted, foams that are bulk or lamellar, aqueous or nonaqueous, and flowing or static. It also covers the occurrences of foams in a wide variety of situations: in porous media, well-bores, flotation vessels, and process plants.

To provide an introduction to the science and engineering of foams in the petroleum industry, the book does not assume a knowledge of colloid chemistry, the initial emphasis being placed on a review of the basic concepts important to understanding foams. As such, it is hoped that the book will also be of interest to senior undergraduate and graduate stu-

---

[1]For those interested in foams for mobility control, I also recommend reading *Surfactant-Based Mobility Control*, Smith, D. H., Ed.; ACS Symposium Series No. 373, American Chemical Society: Washington, DC, 1988.

dents in science and engineering, because these topics are not normally part of university curricula.

The focus of the book is practical rather than theoretical. In a systematic progression, beginning with the fundamental principles in bulk foams, the reader is soon introduced to the case of lamellar foams being generated and flowing in porous media, followed by commercial foam applications and treatments. The first four chapters deal with foam fundamentals, including bulk foam stability and antifoaming, foams in porous media, and foam sensitivity to oil in porous media. Armed with the necessary tools, the reader is next introduced to some exemplary petroleum industry applications of foams. Chapters 5–7 cover the application of foams to improving oil recovery from porous media, including the uses of CO_2, steam, and hydrocarbon foams. Chapters 8–10 present some examples of the application of foams to oil well and near-well petroleum production problems. Chapters 11 and 12 give some examples of foams in surface processes: froth flotation in oil sands processing, and antifoaming and defoaming in refineries. Finally, a comprehensive and fully cross-referenced glossary of foam terminology is included.

Overall, the book illustrates how to understand, make, and use desirable foams and how to approach breaking, or preventing, undesirable foams. It also serves as a companion volume to *Emulsions: Fundamentals and Applications in the Petroleum Industry*[2].

Acknowledgments

I thank all the authors who contributed considerable time and effort to their chapters. This book was made possible through the support of my family, Ann Marie, Katherine, and Victoria, who gave me the time needed for the organization, research, and writing. I am also very grateful to Conrad Ayasse for his consistent encouragement and support. Throughout the preparation of this book many valuable suggestions were made by the reviewers of individual chapters and by the staff of ACS Books, particularly Cheryl Shanks, Stephanie Patton, Margaret Brown, and Colleen P. Stamm.

LAURIER L. SCHRAMM
Petroleum Recovery Institute
Calgary, Alberta
Canada T2L 2A6

August 31, 1993

[2] Schramm, L. L., Ed; Advances in Chemistry Series No. 231, American Chemical Society: Washington, DC, 1992.

Foam Fundamentals

Foams: Basic Principles

Laurier L. Schramm and Fred Wassmuth

Petroleum Recovery Institute 100, 3512 33rd Street, N.W., Calgary, Alberta T2L 2A6, Canada

> *This chapter provides an introduction to the occurrence, properties, and importance of foams as they relate to the petroleum industry. The fundamental principles of colloid science may be applied in different ways to stabilize or destabilize foams. This application has practical importance because a desirable foam that must be stabilized at one stage of an oil production process, may be undesirable in another stage and necessitate a defoaming strategy. By emphasizing the definition of important terms, the importance of interfacial properties of foam making and stability is demonstrated.*

Importance of Foams

If a gas and a liquid are mixed together in a container, and then shaken, examination will reveal that the gas phase has become a collection of bubbles that are dispersed in the liquid: A foam has been formed (Figure 1). Foams have long been of great practical interest because of their widespread occurrence in everyday life. Some important kinds of familiar foams are listed in Table I. In addition to their wide occurrence, foams have important properties that may be desirable in a formulated product, such as fire-extinguishing foam, or undesirable, such as a foam in an industrial distillation tower.

Petroleum occurrences of foams may not be as familiar but have a similarly widespread, long-standing, and important occurrence in industry. Foams may be applied or encountered at all stages in the petroleum recovery and processing industry (oil well drilling, reservoir injection, oil

Table I. Some Examples of Foams in Everyday Experience

Group	Product
Foods	Champagne, soda heads
	Beer head
	Whipped cream
	Meringue
Detergency	Manual dishwashing suds
	Machine dishwashing suds
	Commercial bottle-cleaning process foam
	Machine clothes-washing suds
Personal Care Products	Shaving cream
	Hair shampoo suds
	Contraceptive foams
	Bubble bath foam
Process Industries	Foam blankets on electroplating baths
	Sewage treatment effluent foams
	Mineral or oil flotation froths
	Foam fractionation
	Pulping black liquor foam
Other	Fire extinguishing foams
	Explosion suppressing foam blankets
	Fumigant, insecticide, and herbicide blankets

well production, and process plant foams). This chapter is intended to provide an introduction to the basic principles involved in the occurrence, "making", and "breaking" of foams in the petroleum industry. Subsequent chapters in this volume will go into specific areas of occurrence in greater detail.

As suggested in Table II, petroleum industry foams may be desirable or undesirable. For example, one kind of oil well drilling-fluid (or "mud") is foam-based. Here, a stable foam is used to lubricate the cutting bit and to carry cuttings up to the surface. Drilling with a foam drilling-fluid also allows lower pressures to be applied to the formation, which is important when drilling into low-pressure reservoirs. This foam is obviously desirable, and great care goes into its proper preparation. Other foams that are desirable near well-bores include fracturing foams and acidizing foams.

It may happen that a foam that is desirable in one part of the oil production process may be undesirable at the next stage. For example, in the oil fields, an in situ foam that is purposely created in a reservoir to increase viscosity (and thereby improve volumetric sweep efficiency as part of an oil recovery process) may present a handling problem when produced.

Table II. Some Examples of Foams in the Petroleum Industry

Type	Occurrence
Undesirable foams	Producing oil well and well-head foams
	Oil flotation process froth
	Distillation and fractionation tower foams
	Fuel oil and jet fuel tank (truck) foams
Desirable foams	Foam drilling fluid
	Foam fracturing fluid
	Foam acidizing fluid
	Blocking and diverting foams
	Gas-mobility control foams

Foams may contain not just gas and liquid, such as water, but solid particles, and even oil. In the large Canadian oil–sands mining and processing operations, bitumen is separated from the sand matrix in large tumblers, as an emulsion of oil dispersed in water, and then further separated from the tumbler slurry by a flotation process. The product of the flotation process is bituminous froth, a foam that may be either air and water dispersed in the oil (primary flotation), or air and oil dispersed in water (secondary flotation). In either case, the froths must be "broken" and deaerated before the bitumen can be upgraded to synthetic crude oil.

Finally, many kinds of foams pose difficult problems wherever they may occur. In surface emulsion treaters (e.g., oil–water separators) and in refineries (e.g., distillation towers), the occurrence of foams is generally undesirable, and any such foams will have to be "broken".

The same basic principles of colloid science that govern the nature, stability, and properties of foams apply to all of the previously mentioned petroleum industry foam applications and problems. The widespread importance of foams in general and scientific interest in their formation, stability, and properties have precipitated a wealth of published literature on the subject. The present chapter provides an introduction, and is intended to complement the other chapters dealing with petroleum industry foams in this book. A good starting point for further basic information is the classic text, J. J. Bikerman's *Foams: Theory and Industrial Applications* (1) and several other books on foams (2–4). Most comprehensive colloid chemistry texts contain introductory chapters on foams (5–7), but some of the chapters in specialist monographs (8–13) give a much more detailed treatment of advances in specific foam-related areas. With regard to the occurrence of other colloidal systems in the petroleum industry, a recent book describes the principles and occurrences of emulsions in the petroleum industry (14).

Foams as Colloidal Systems

Definition and Classification of Foams. Colloidal species of any kind (bubbles, particles, or droplets), as they are usually defined, have at least one dimension between 1 and 1000 nm. Foams are a special kind of colloidal dispersion: one in which a gas is dispersed in a continuous liquid phase. The dispersed phase is sometimes referred to as the internal (disperse) phase, and the continuous phase as the external phase. In practical occurrences of foams, the bubble sizes usually exceed the size limit given, as may the thin liquid-film thicknesses. Table II lists some simple examples of petroleum industry foam types. Solid foams, dispersions of gas in a solid, will not in general be covered in this chapter. A glossary of frequently encountered foam terms in the science and engineering of petroleum industry foams is given at the end of this volume.

A two-dimensional slice of a general foam system is depicted in Figure 1. The general foam structure is contained on the bottom by the bulk liquid and on the upper side by a second bulk phase, in this case, gas. Within the magnified region, the various parts of the foam structure are clarified. The gas phase is separated from the thin liquid-film, by a two-dimensional interface. In reality, a sharp dividing surface does not exist between gas and liquid properties. Dictated by mathematical convenience, the physical behavior of this interfacial region is approximated by a two-dimensional surface phase (the Gibbs surface). For the purposes of this book, a lamella is defined as the region that encompasses the thin film, the two interfaces on either side of the thin film, and part of the junction to other lamellae. The connection of three lamellae, at an angle of 120°,

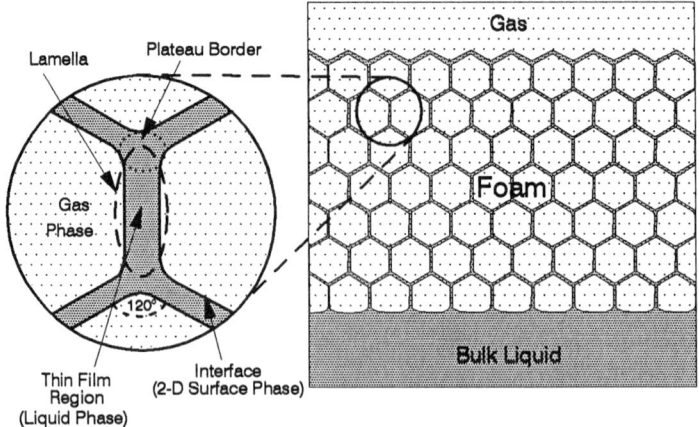

Figure 1. A generalized foam system.

is referred to as the Plateau border. Because Figure 1 represents only a two-dimensional slice, the Plateau border extends perpendicularly, out of the page. The relevance of the definitions, given here, will become clear in the following paragraphs.

"Making" Thin Films and Foams. Much of this chapter is concerned with foam properties and stability, and as a practical matter one frequently has to contend with already formed foams. Nevertheless, a few comments on how foams are made is appropriate. The "breaking" of foams will be discussed later.

A foam structure can always be formed in a liquid if bubbles of gas are injected faster than the liquid between bubbles can drain away. Even though the bubbles coalesce as soon as the liquid between them has drained away, a temporary dispersion is formed. An example would be the foam formed when bubbles are vigorously blown into a viscous oil. Such a foam, comprising spherical, well-separated bubbles, is referred to as a wet foam, or kugelschaum. Wet foams in which the liquid lamellae have thicknesses on the same scale as the bubble sizes are sometimes referred to as "gas emulsions". Here, the distinction of whether this is a foam or not relates to stability. But it is complicated by the fact that, as for other types of colloidal dispersions, no foams are thermodynamically stable. Eventually they all collapse.

In pure liquids, gas bubbles will rise up and separate, more or less according to Stokes' law. When two or more bubbles come together, coalescence occurs very rapidly, without detectable flattening of the interface between them; that is, there is no thin-film persistence. The adsorption of surfactant at the gas–liquid interface promotes thin-film stability between the bubbles and lends a certain persistence to the foam structure. Here, when two bubbles of gas approach, the liquid-film thins down to a persistent lamella instead of rupturing at the point of closest approach. In carefully controlled environments, it has been possible to make surfactant-stabilized, static bubbles and films with lifetimes on the order of months and years (*3*).

Arrangement of the Phases. In a persistent foam, the spherical bubbles become transformed into foam cells, polyhedra separated by almost flat liquid-films. Such foams are referred to as dry foams, or polyederschaum. The polyhedra are almost, but not quite, regular dodecahedra. Figure 1 illustrates the two-dimensional structure of such a foam. These arrangements of films, which come together at equal angles, result from the surface tensions, or contracting forces, along the liquid-films. The bubbles in a foam arrange themselves into polyhedra such that, along the border of a lamella, three lamellae always come together at angles of

120°; the border where they meet is termed a Plateau border. In three dimensions, four Plateau borders meet at a point at the tetrahedral angle, approximately 109°. Observations of dynamic foams show that any time more than three films come together, a rearrangement immediately takes place to restore junctions of only three films along lamella borders.

Interfacial Properties and Foam Stability

Most foams that have any significant persistence contain gas, liquid, and a foaming agent. The foaming agent may comprise one or more of the following: surfactants, macromolecules, or finely divided solids. The foaming agent is needed to reduce surface tension, and thereby aid in the formation of the increased interfacial area with a minimum of mechanical energy input, and it may be needed to form a protective film at the bubble surfaces that acts to prevent coalescence with other bubbles. These aspects will be discussed later; the resulting foam may well have considerable stability as a metastable dispersion.

Colloidal species can come together in very different ways. In the definition of foam stability, one considers stability against two different processes: film thinning and coalescence (film rupturing). In film thinning, two or more bubbles approach closely together. The liquid-films that separate them thin, but the bubbles do not actually touch each other, and there is no change in total surface area. In coalescence, two or more bubbles fuse together to form a single, larger bubble. In foam terminology, the thin liquid-film(s) rupture and reduce the total surface area. Because foams are thermodynamically unstable, the term stable is used to mean relatively stable in a kinetic sense. It is important to distinguish the degree of change and the time scale.

The stability of a foam is determined by a number of factors involving both bulk solution and interfacial properties:

- gravity drainage
- capillary suction
- surface elasticity
- viscosity (bulk and surface)
- electric double-layer repulsion
- dispersion force attraction
- steric repulsion

To see how these phenomena operate, surfactants, surface tension, and force balances will be introduced next.

Surfactants and Surface Tension. In two-phase colloidal systems, a thin, intermediate region or boundary, known as the interface, lies between the dispersed and continuous phases. Interfacial properties are very important in foams because the gas bubbles have a large surface area, and even a modest surface energy per unit area can become a considerable total surface energy. An example is a foam prepared by dispersion of gas bubbles into water. For a constant gas-volume fraction, the total surface area produced increases as the bubble size that is produced decreases. Because a free-energy is associated with surface area, this increases as well with decreasing bubble size. Energy has to be added to the system to achieve the dispersion of small bubbles. If sufficient energy cannot be provided through mechanical energy input, then another alternative is to use surfactant chemistry to lower the interfacial free-energy, or interfacial tension. If a small quantity of a surfactant is added to the water, possibly a few tenths of a percent, then the surface tension is significantly lowered. This decrease in turn would lower the amount of mechanical energy needed for foam formation.

The terms surface tension and surfactant were just introduced without explanation. If the molecules in a liquid are considered, the attractive van der Waals forces between molecules are felt equally by all molecules except those in the interfacial region. This imbalance pulls the molecules of the interfacial region toward the interior of the liquid. The contracting force at the surface is known as the surface tension. Because the surface has a tendency to contract spontaneously in order to minimize the surface area, droplets of liquid and bubbles of gas tend to adopt a spherical shape: This shape reduces the total surface free energy. For two immiscible liquids, a similar situation applies, except that it may not be so immediately obvious how the interface will tend to curve. An imbalance of intermolecular force will result in an interfacial tension, and the interface will adopt a configuration that minimizes the interfacial free energy.

From thermodynamic principles, surface tension has units of energy per unit area, and this illustrates the fact that area expansion of the surface requires energy. Physically, surface tension is viewed as the sum of the contracting forces that act parallel to the surface or interface. This latter viewpoint defines surface tension (γ°), or interfacial tension (γ), as the contracting force per unit length (l) around a surface. The classic illustration, Figure 2, demonstrates that the work required to expand the surface against contracting forces is equal to the increase in surface free energy (dG) that accompanies this expansion,

$$\text{work} = \gamma^\circ l \, dx = \gamma^\circ \, dA$$

$$dG = \gamma^\circ \, dA$$

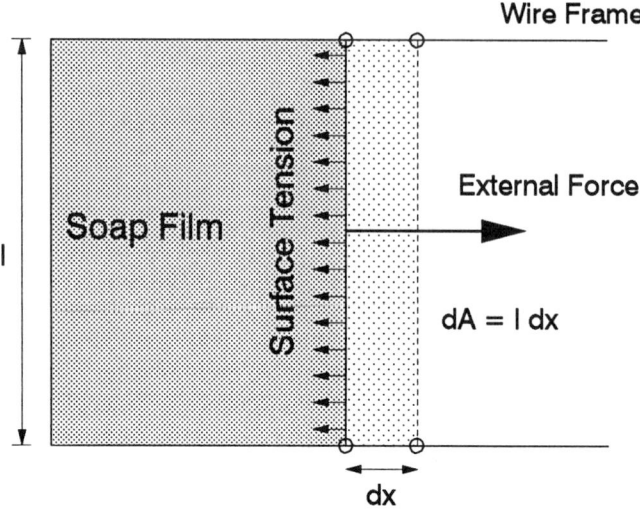

Figure 2. Surface tension: a surface-contracting force per unit length and a surface energy per unit area.

where the slide wire is moved a distance dx and the increase in area dA equals l times dx. These relationships show why surface tension may either be expressed as force per unit length or energy per unit area.

The lamella in a foam contains two gas–liquid interfaces separated by a layer of fluid thin-film, and each interface has a surface tension. For this reason, the term film tension is sometimes used, the film tension being equal to twice the surface tension of the surfaces. The film tension is twice the surface tension of the surfaces, but this is not necessarily the same as twice the surface tension of the bulk solution. In fact, the surface tension of a fluid-film surface is similar to that of the bulk solution when the fluid-film is thick, but departs from the bulk solution value as the fluid-film thins.

Many methods are available for the measurement of surface and interfacial tensions. Details of these experimental techniques and their limitations are available in several good reviews (*15–17*). Some methods that are used frequently in foam work are the du Nouy ring, Wilhelmy plate, drop weight or volume, pendant drop, and the maximum bubble-pressure method. In all cases, when solutions, rather than pure liquids, are involved, appreciable changes can take place with time at the surfaces and interfaces.

Equation of Young–Laplace. Interfacial tension causes a pressure difference to exist across a curved surface, the pressure being greater on the concave side (i.e., on the inside of a bubble). An interface between a gas phase G, in a bubble, and a liquid phase L, that surrounds the bubble, will have pressures P_G and P_L. For spherical bubbles of radius R in a wet foam,

$$\Delta P = P_G - P_L \tag{1}$$

$$\Delta P = 2\gamma^\circ / R$$

so that ΔP varies with the radius, R. This is the Young–Laplace equation (7). It shows that when $P_G > P_L$, the pressure inside a bubble exceeds that outside. If the bubbles were of more complex geometry, then the two principal radii of curvature R_1 and R_2 would be used,

$$\Delta P = P_G - P_L \tag{2}$$

$$\Delta P = \gamma^\circ (1/R_1 + 1/R_2)$$

Figure 3 shows an additional pressure variation between the Plateau borders (P_B), where the radius of curvature is relatively small (R_{1B}), and in the more laminar part of the lamella (P_A), where the radius of curvature is relatively large (R_{1A}). In the Figure 3 illustration, the principal radii of curvature have been assumed to be equal at a given location in the lamella ($R_{1A} = R_{2A}$ and $R_{1B} = R_{2B}$).

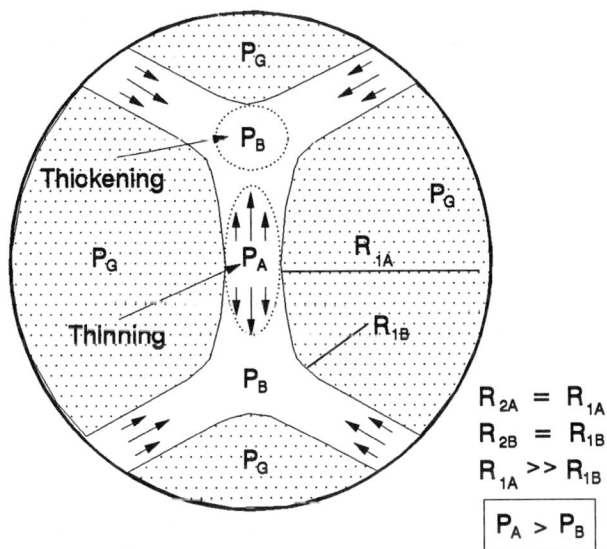

Figure 3. Pressure differences across curved surfaces in a foam lamella.

The Young–Laplace equation forms the basis for some important methods for measuring surface and interfacial tensions, such as the pendant and sessile drop methods, the spinning drop method, and the maximum bubble-pressure method (*15–18*). Liquid flow, in response to the pressure difference expressed by equations 1 or 2, is known as Laplace flow or capillary flow.

Surfactants. Some compounds, like short-chain fatty acids, are amphiphilic or amphipathic; that is, they have one part that has an affinity for the nonpolar media (the nonpolar hydrocarbon chain), and one part that has an affinity for polar media, that is, water (the polar group). The most energetically favorable orientation for these molecules is at surfaces or interfaces so that each part of the molecule can reside in the fluid for which it has the greatest affinity (Figure 4). These molecules that form oriented monolayers at interfaces show surface activity and are termed surfactants. As there will be a balance between adsorption and desorption (due to thermal motions), the interfacial condition requires some time to establish. Because of this time requirement, surface activity should be considered a dynamic phenomenon. This condition can be seen by measuring surface tension versus time for a freshly formed surface.

A consequence of surfactant adsorption at an interface is that it provides an expanding force acting against the normal interfacial tension.

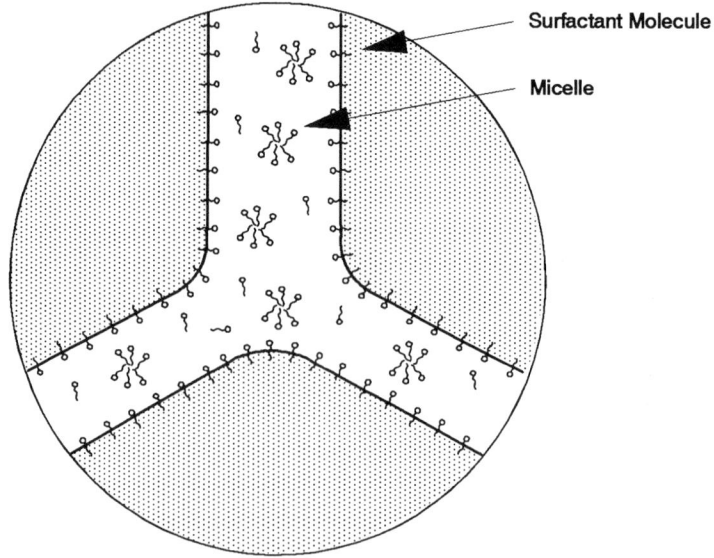

Figure 4. Surfactant associations in a foam lamella. The size of the surfactant molecules compared to the lamella has been exaggerated for the purposes of illustration.

Thus surfactants tend to lower interfacial tension. Gibbs has described the lowering of surface free energy due to surfactant adsorption in terms of thermodynamics. The general Gibbs adsorption equation for a binary, isothermal system that contains excess electrolyte is

$$\Gamma_s = -(1/RT)(d\gamma/d \ln C_S) \qquad (3)$$

where Γ_S is the surface excess of surfactant (mol/cm^2), R is the gas constant, T is absolute temperature, C_S is the solution concentration of the surfactant (M), and γ may be either surface or interfacial tension (mN/m). This equation can be applied to dilute surfactant solutions in which the surface curvature is not great and in which the adsorbed film can be considered to be a monolayer. The packing density of surfactant in a monolayer at the interface can be calculated as follows. According to equation 3, the surface excess in a tightly packed monolayer is related to the slope of the linear portion of a plot of surface tension versus the logarithm of solution concentration (Figure 5). From this relationship, the area per adsorbed molecule (A_S) can be calculated from

$$A_S = 1/(N_A \Gamma_s) \qquad (4)$$

where N_A is Avogadro's number. Numerous examples are given by Rosen (19).

When surfactants concentrate in an adsorbed monolayer at a surface, the interfacial film may provide a stabilizing influence in thin-films and foams because they can both lower interfacial tension and increase the interfacial viscosity. Increased interfacial viscosity provides a mechanical resistance to film thinning and rupturing.

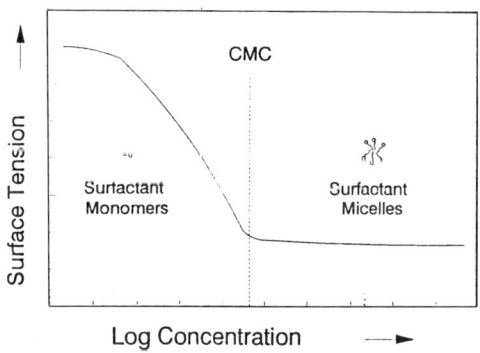

Figure 5. *Illustration of the association behavior of surfactants in solution, showing the critical micelle concentration (CMC). (Reproduced from reference 14. Copyright 1992 American Chemical Society.)*

Classification of Surfactants. Surfactants are classified according to the nature of the polar (hydrophilic) part of the molecule, as illustrated in Table III. In-depth discussions of surfactant structure and chemistry can be found in references 19–21. A good discussion of the chemistry of commercially available surfactants is given by Myers (*11*).

In aqueous solution, dilute concentrations of surfactant act much as normal electrolytes, but at higher concentrations very different behavior results. This behavior (illustrated in Figures 4 and 5) is explained in terms of the formation of organized aggregates of large numbers of molecules called micelles, in which the lipophilic parts of the surfactants associate in the interior of the aggregate and leave the hydrophilic parts to face the aqueous medium. The concentration at which micelle formation becomes significant is called the critical micelle concentration (CMC). The CMC is a property of the surfactant and several other factors, because formation of micelles is opposed by thermal and electrostatic forces. A lower CMC is produced by increasing the molecular mass of the lipophilic part of the molecule, lowering the temperature (usually), and adding electrolyte (usually). Surfactant molecular masses range from a few hundreds up to several thousands of g/mol.

Some typical molar CMC values for low electrolyte concentrations at room temperature are

Typical Molar CMC Values

Surfactants	*CMC (M)*
Nonionics	10^{-5}–10^{-4}
Anionics	10^{-3}–10^{-2}
Amphoterics	10^{-3}–10^{-1}

The solubilities of micelle-forming surfactants show a strong increase above a certain temperature, termed the Krafft point (T_k). This increased solubility is explained by the fact that the single surfactant molecule has limited solubility, whereas the micelles are very soluble. As shown in Figure 6, at temperatures below the Krafft point, the solubility of the surfactant is too low for micellization, and solubility alone determines the surfactant–monomer concentration. As temperature increases, the solubility increases until at T_k the CMC is reached. At this temperature, a relatively large amount of surfactant can be dispersed in micelles, and solubility increases greatly. Above the Krafft point, maximum reduction in surface or interfacial tension occurs at the CMC, because now the CMC determines the surfactant–monomer concentration. Krafft points for a number of surfactants are listed in reference 11.

Table III. Surfactant Classifications

Class	Examples	Structures
Anionic	Na stearate	$CH_3(CH_2)_{16}COO^-Na^+$
	Na dodecyl sulfate	$CH_3(CH_2)_{11}SO_4^-Na^+$
	Na dodecyl benzenesulfonate	$CH_3(CH_2)_{11}C_6H_4SO_3^-Na^+$
Cationic	Laurylamine hydrochloride	$CH_3(CH_2)_{11}NH_3^+Cl^-$
	Cetyltrimethylammonium bromide	$CH_3(CH_2)_{15}N^+(CH_3)_3Br^-$
Nonionic	Poly(oxyethylene) alcohol	$C_nH_{2n+1}(OCH_2CH_2)_mOH$
	Alkylphenol ethoxylate	$C_9H_{19}-C_6H_4-(OCH_2CH_2)_nOH$
Zwitterionic	Lauramidopropylbetaine	$C_{11}H_{23}CONH(CH_2)_3N^+(CH_3)_2CH_2COO^-$
	Cocoamido-2-hydroxypropylsulfobetaine	$C_nH_{2n+1}CONH(CH_2)_3N^+(CH_3)_2CH_2CH(OH)CH_2SO_3^-$

SOURCE: Reproduced from reference 14. Copyright 1992 American Chemical Society.

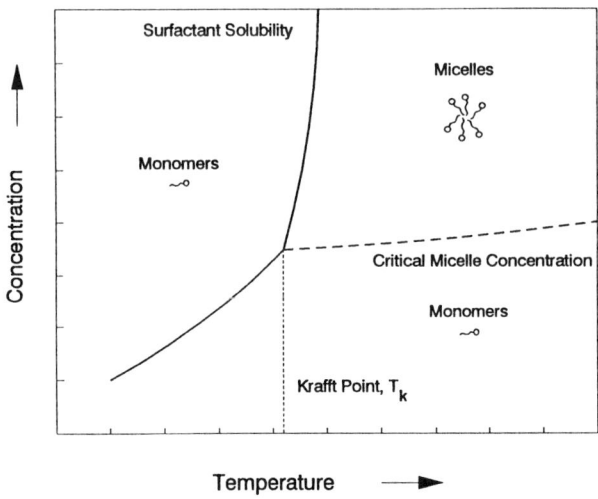

Figure 6. The solubility–micellization behavior of surfactants in solution. (Adapted from reference 14. Copyright 1992 American Chemical Society.)

Nonionic surfactants do not exhibit Krafft points. Rather, the solubility of nonionic surfactants decreases with increasing temperature, and the surfactants begin to lose their surface-active properties above a transition temperature referred to as the cloud point. This loss occurs because above the cloud point, a separate surfactant-rich phase of swollen micelles separates.

Gravity and Laplace Capillary Suction. Immediately after foam generation, there will always be a tendency for liquid to drain due to the force of gravity. This liquid will drain by flowing downward through the existing liquid-films, the interior of the lamellae. Eventually the gas bubbles will no longer be even approximately spherical (kugelschaum), and relatively planar lamellae will separate polyhedral-shaped bubbles (polyederschaum). At this point, the capillary forces will become competitive with the forces of gravity. Figure 3 shows that at the Plateau borders the gas–liquid interface is quite curved, and this curve generates a low-pressure region in the Plateau area (as shown by the Young–Laplace Equation). Because the interface is flat along the thin-film region, a higher pressure resides here. This pressure difference forces liquid to flow toward the Plateau borders and causes thinning of the films and motion in the foam. Unchecked, this thinning process will lead to film rupture and cause foam collapse. But a restoring force can be present through the Marangoni effect.

Surface Elasticity. A foam film must be somewhat elastic in order to be able to withstand deformations without rupturing. The surface-chemical explanation for film elasticity comes from Marangoni and Gibbs (*12*). If a surfactant-stabilized film undergoes a sudden expansion, then immediately the expanded portion of the film must have a lower degree of surfactant adsorption than unexpanded portions because the surface area has increased (Figure 7). This expansion causes an increased local surface tension that provides increased resistance to further expansions. Unchecked, further thinning would ultimately lead to film rupture. A local rise in surface tension produces immediate contraction of the surface. The surface is coupled by viscous forces to the underlying liquid layers. Thus, the contraction of the surface induces liquid flow, in the thin-film, from the low-tension region to the high-tension region. The transport of bulk liquid due to surface tension gradients is termed the Marangoni effect and provides the resisting force to film thinning.

This kind of resisting force exists only until the surfactant adsorption equilibrium is reestablished in the film, a process that may take place within seconds or over a period of hours. In thick films, this step can take place quite quickly; however, in thin films there may not be enough surfactant in the extended surface region to reestablish the equilibrium quickly, and diffusion from other parts of the film is required. The restoration processes are the movement of surfactant along the interface from a region of low surface tension to one of high surface tension and the movement of surfactant from the thin-film into the now-depleted surface region. Thus the Gibbs–Marangoni effect provides a force to counteract film rupture, but is probably significant mainly for either rapid deforma-

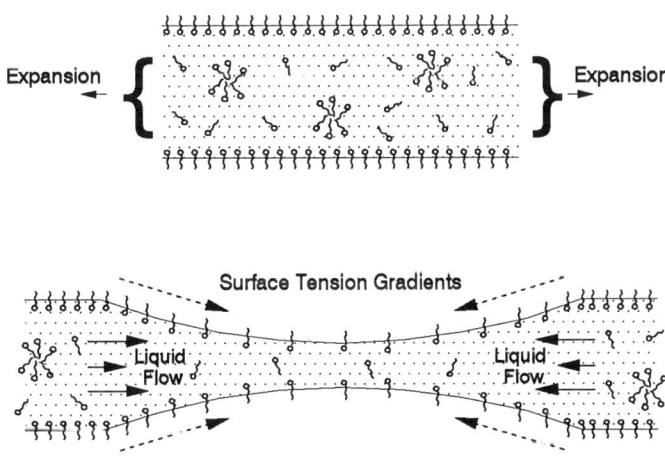

Figure 7. The origin of surface elasticity.

tions or for stabilizing very thin films. The effect can, in principle, be studied either by direct elasticity measurements, or by dynamic surface tension measurements coupled with some information about diffusion rates.

Many surfactant solutions show dynamic surface tension behavior. That is, some time is required to establish the equilibrium surface tension. If the surface area of the solution is suddenly increased or decreased (locally), then the adsorbed surfactant layer at the interface would require some time to restore its equilibrium surface concentration by diffusion of surfactant from or to the bulk liquid. In the meantime, the original adsorbed surfactant layer is either expanded or contracted; because surface tension gradients are now in effect, Gibbs–Marangoni forces arise and act in opposition to the initial disturbance. The dissipation of surface tension gradients to achieve equilibrium embodies the interface with a finite elasticity. This fact explains why some substances that lower surface tension do not stabilize foams (6): They do not have the required rate of approach to equilibrium after a surface expansion or contraction. In other words, they do not have the requisite surface elasticity.

At equilibrium, the surface elasticity, or surface dilatational elasticity, E_G, is defined (6, 7) by

$$E_G = \frac{d\gamma^\circ}{d \ln A} \quad (5)$$

where γ° is the surface tension and A is the geometric area of the surface. Surface elasticity is related to the compressibility of the surface film, K, by $K = 1/E_G$. For a foam lamella, there are two such surfaces, and the elasticity becomes

$$E_G = \frac{2 d\gamma^\circ}{d \ln A} \quad (6)$$

The subscript G specifies elasticity determined from isothermal equilibrium measurements, such as for the spreading pressure–area method, which is a thermodynamic property and is termed the Gibbs surface elasticity, E_G. E_G occurs in very thin films where the number of molecules is so low that the surfactant cannot restore the equilibrium surface concentration after deformation, as illustrated in Figure 8.

The elasticity that is determined from nonequilibrium dynamic measurements depends upon the stresses applied to a particular system, is generally larger in magnitude, and is termed the Marangoni surface elasticity, E_M (6, 22) (equation 12). The time-dependent Marangoni elasticity is il-

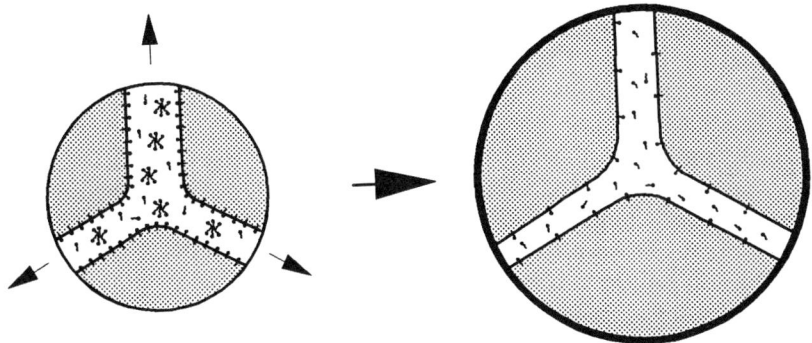

Figure 8. The Gibbs surface elasticity.

lustrated in Figure 9. For foams, this dynamic property is of most interest. Surface elasticity measures the resistance against creation of surface tension gradients and the rate at which such gradients disappear once the system is again left to itself (23).

Malysa et al. (22) found a correlation between E_M and the "frothability" of 1-octanol and 1-octanoic acid solutions. Huang et al. (24) determined E_M for a series of alpha-olefin sulfonates, and found that the elasticity increased with increasing hydrocarbon chain length. Their results also

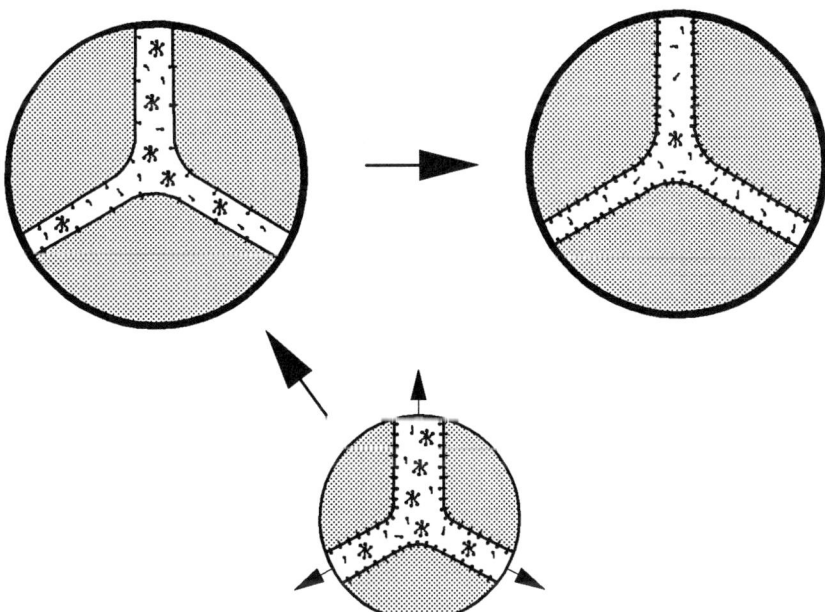

Figure 9. Illustration of the Marangoni surface elasticity.

suggested a correlation between E_M and each of (1) thin-film drainage times, measured in a single-film drainage apparatus; (2) static foam half-lives; (3) dynamic foam heights; and (4) reduction of gas permeability in core-floods with foam. These correlations all indicated that high surface elasticity promoted high foam stability; however, each comparison was based on only three data points and only one high ($\simeq$1%) concentration; a systematic study remains to be done, especially for surfactants at high concentrations. Lucassen-Reynders (23) illustrated the effects of surfactant concentration on the Gibbs elasticity. For low bulk surfactant concentrations, the Gibbs elasticity increases with an increase in surfactant concentration until a maximum in elasticity is reached. From this point forward, any further addition of surfactant decreases the Gibbs elasticity. Surfactant concentrations above the CMC lie well beyond this maximum elasticity region. Lucassen-Reynders (23) cautioned that no direct relationship exists between elasticity and foam stability. Additional factors, such as film thickness and adsorption behavior, have to be taken into account.

The first surface elasticity measurements appear to have been the Gibbs elasticity measurements made by Mysels et al. (25) for thin-films that are supported by glass frames. Their results indicated that for a number of surfactant systems E_G (film) $\simeq$10 mN/m, but the rigid films produced by the addition of a suitable alcohol could increase the elasticity of dodecyl sulfate to E_G (film) $\simeq$ 100 mN/m. Lucassen-Reynders (23) reported a similar range of values of E_M, from near-zero to about 70 mN/m, and suggested that adding simple electrolyte into an anionic surfactant solution could increase the elasticity by factors of 2 to 3.

The Marangoni elasticity can be determined experimentally from dynamic surface tension measurements that involve known surface area changes. One such technique is the maximum bubble-pressure method (MBPM), which has been used to determine elasticities in this manner (24, 26). In the MBPM, the rates of bubble formation at submerged capillaries are varied. This amounts to changing dA/A because approximately equal bubble areas are produced at the maximum bubble pressure condition at all rates. Although such measurements include some contribution from surface dilational viscosity (23, 27), the result will be referred to simply as surface elasticity in this work.

The first requirements for foam formation are thus surface tension lowering and surface elasticity. A greater elasticity tends to produce more-stable bubbles. But if the restoring force contributed by surface elasticity is not of sufficient magnitude, then persistent foams may not be formed because of the overwhelming effects of the gravitational and capillary forces. These foams are termed evanescent foams. Important properties that determine the stability will include bubble size, liquid viscosity, and density difference between gas and liquid. More-stable foams may require additional stabilizing mechanisms.

Surface Rheology. Surface rheology deals with the functional relationships that link the dynamic behavior of a surface to the stress that is placed on the surface. The complex nature of these relationships is often expressed in the form of a surface stress tensor, $\mathbf{P}^S$. Both elastic and viscous resistances oppose the expansion and deformation of surface films. The isotropic (diagonal) components of this stress tensor describe the dilatational behavior of the surface element. The components that are off-diagonal relate the resistance to changing the shape of the surface element to the applied shear stresses. Equation 7 demonstrates the general form of the surface stress tensor.

$$P^S = [\gamma + (\kappa^S - \eta^S)\nabla \cdot v^S] I^S + \eta^S[\nabla v^S \cdot I^S + I^S \cdot (\nabla v^S)^T] \quad (7)$$

where v^S is surface velocity, I^S or $(I - nn)$ is surface projection tensor, n is unit vector normal to the surface, I is the unit tensor, T is the transpose operator, and ∇ is the gradient operator.

Previously introduced, the thermodynamic surface tension γ represents the elastic resistance to surface dilation. Furthermore, two types of viscosities are defined within the interface, a dilational viscosity and a shear viscosity. For a surfactant monolayer, the surface shear viscosity η^S is analogous to the three-dimensional shear viscosity: the rate of yielding of a layer of fluid due to an applied shear stress. The phenomenological coefficient κ^S represents the surface dilational viscosity, and expresses the magnitude of the viscous forces during a rate expansion of a surface element. Figures 10a and 10b illustrate the difference between the two surface viscosities.

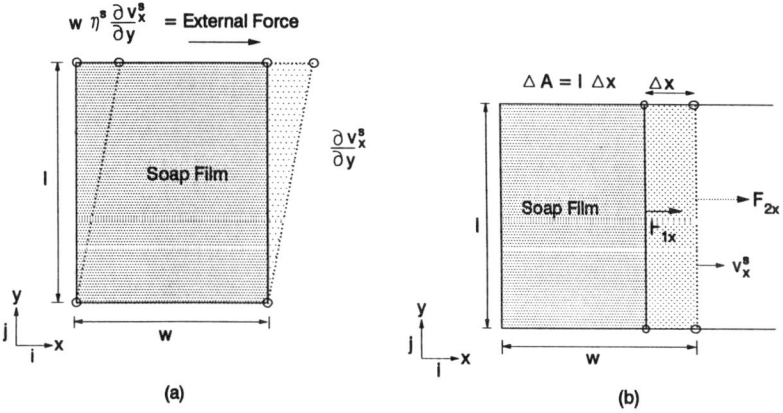

Figure 10. The concepts of surface shear and surface dilational viscosities: (a) deformation of a soap film and (b) dilation of a soap film.

To demonstrate the details of the surface stress tensor, $\mathbf{P}^S$, a simplified surface geometry is presumed. For a soap film that is stretched on a wire frame parallel to the x–y plane, the surface stress tensor, confined to this geometry (Figure 10b), reduces to

$$\mathbf{P}^S = \begin{bmatrix} \left[\gamma + \left(\kappa^S - \eta^S\right)\right]\left(\dfrac{\partial}{\partial x}v^S_x + \dfrac{\partial}{\partial y}v^S_y\right)\mathbf{ii} & \eta^S\left(\dfrac{\partial}{\partial x}v^S_x + \dfrac{\partial}{\partial y}v^S_x\right)\mathbf{ij} \\[1em] \eta^S\left(\dfrac{\partial}{\partial x}v^S_y + \dfrac{\partial}{\partial y}v^S_y\right)\mathbf{ji} & \gamma + (\kappa^S - \eta^S)\left(\dfrac{\partial}{\partial x}v^S_x + \dfrac{\partial}{\partial x}v^S_y\right)\mathbf{jj} \end{bmatrix} \quad (8)$$

where i is a unit vector in the x-direction and j is a unit vector in the y-direction. In this example, three of the four sides of the wire frame are rigidly fixed. The movable side, initially at rest (position 1), moves forward (increasing x-direction) through position 2 at a particular velocity v^S_x. Because both sides, perpendicular to the y-direction, are fixed, the velocity v^S_y is set to zero. Under these conditions the force experienced on the movable bar at position 1, F_{1x}, is equal to integrating the stress tensor along the length of the bar.

$$F_{1x} = \gamma_1 l \quad (9)$$

A similar integration results in the force experienced at position 2, F_{1x}. This force consists of an elastic and viscous contribution.

$$F_{2x} = \gamma_2 l + (\kappa^S - \eta^S)\frac{\partial}{\partial x}v^S_x l = \gamma_2 l + (\kappa^S - \eta^S)\frac{v^S_x}{w}l \quad (10)$$

$$F_{2x} = \gamma_2 l + (\kappa^S - \eta^S)\frac{dA}{w\,dt}$$

Accordingly, the dynamic tension difference $\Delta\gamma_d$ between position 1 and 2 is given by

$$\Delta\gamma_d = \frac{F_{2x} - F_{1x}}{l} = \gamma_2 - \gamma_1 + (\kappa^s - \eta^s)\frac{d\ln A}{dt} \quad (11)$$

If the areal change remains small, then the difference in thermodynamic tension can be approximated by using the surface dilational elasticity E_M.

$$\Delta\gamma_d = E_M \Delta \ln A + (\kappa^s - \eta^s)\frac{d \ln A}{dt} \qquad (12)$$

Unlike in three dimensions, where liquids are often considered incompressible, a surfactant monolayer can be expanded or compressed over a wide area range. Thus, the dynamic surface tension experienced during a rate-dependent surface expansion is the resultant of the surface dilatational viscosity, the surface shear viscosity, and elastic forces. Often, the contributions of shear or dilational viscosities are neglected during stress measurements of surface expansions. Isolating interfacial viscosity effects is rather difficult. The interface is connected to the substrate on either side of it, and so are the interfacial viscosities coupled to the bulk viscosities. Therefore, it becomes laborious to determine purely interfacial viscosities without the influence of the surroundings.

As bubbles in a foam approach each other, the thinning of the films between the bubbles, and their resistance to rupture, are thought to be of great importance to the ultimate stability of the foam. Thus, a high interfacial viscosity can promote foam stability by lowering the film drainage rate and retarding the rate of bubble coalescence (28). Fast-draining films may reach their equilibrium film thickness in a matter of seconds or minutes because of low surface viscosity, but slow-draining films may require hours because of their high surface viscosity. Bulk viscosity and surface viscosity do not normally contribute a direct stabilizing force to a foam film, but rather act as resistances to the thinning and rupturing processes. The bulk viscosity will most influence the thinning of thick films, and the surface viscosity will be dominant during the thinning of thin films.

To the extent that viscosity and surface viscosity influence foam stability, one would predict that stability would vary according to the effect of temperature on the viscosity. Thus some petroleum industry processes exhibit serious foaming problems at low process temperatures that disappear at higher temperatures. Ross and Morrison (6) cited some examples of petroleum foams that become markedly less stable above a narrow temperature range that may be an interfacial analogue of a melting point.

The presence of mixed surfactant adsorption seems to be a factor in obtaining films with very viscous surfaces (12). For example, in some cases, the addition of a small amount of nonionic surfactant to a solution of anionic surfactant can enhance foam stability because of the formation of a viscous surface layer, which is possibly a liquid-crystalline surface phase in equilibrium with a bulk isotropic solution phase (6, 8). In general, some very stable foams can be formed from systems in which a liquid-crystal phase is present at lamella surfaces and in equilibrium with an isotropic interior liquid. If only the liquid-crystal phase is present, stable foams are not produced. In this connection, foam phase diagrams may be used to delineate compositions that will produce stable foams (6, 8).

Adamson (7) illustrated some techniques for measuring surface shear viscosity. Further details on the principles, measurement, and applications to foam stability of interfacial viscosity are reviewed in Chapter 2 of this volume. Most experimental studies deal with the bulk and surface viscosities of bulk solution rather than the rheology of films themselves.

Surface Potential. *Electric Double Layer.* Because the interfaces on each side of the thin-film are equivalent, any interfacial charge will be equally carried on each side of the film. If a foam film is stabilized by ionic surfactants, then their presence at the interfaces will induce a repulsive force that opposes the thinning process. The magnitude of the force will depend on the charge density and the film thickness.

Having a charged interface influences the distribution of nearby ions in a polar medium. Ions of opposite charge (counterions) are attracted to the surface, while those of like charge (coions) are repelled. An electric double layer, which is diffuse because of mixing caused by thermal motion, is thus formed. The electric double layer (EDL) consists of the charged surface and a neutralizing excess of counterions over coions, distributed near the surface. The EDL can be viewed as being composed of two layers:

- an inner layer that may include adsorbed ions
- a diffuse layer where ions are distributed according to the influence of electrical forces and thermal motion

Taking the surface electric potential to be ψ°, and applying the Gouy–Chapman approximation, the electric potential ψ at a distance x from the surface is approximately:

$$\psi = \psi^\circ \exp(-\kappa x) \qquad (13)$$

The surface charge density is given as, $\sigma^\circ = \epsilon \kappa \psi^\circ$, where ϵ is the permittivity; thus ψ° depends on surface charge density and the solution ionic composition (through κ). $1/\kappa$ is called the double-layer thickness and for water at 25 °C is given by

$$\kappa = 3.288\sqrt{I} \quad (\text{nm}^{-1}) \qquad (14)$$

where I is the ionic strength, given by $I = (1/2) \Sigma_i c_i z_i^2$. Where c_i is the concentration of ions i, and z_i is the charge number of ions i. For a 1-1 electrolyte, $1/\kappa = 1$ nm for $I = 10^{-1}$ M, and $1/\kappa = 10$ nm for $I = 10^{-3}$ M.

In fact, an inner layer exists because ions are not really point charges, and an ion can approach a surface only to the extent allowed by its hydration sphere. The Stern model incorporates a layer of specifically adsorbed ions bounded by a plane, the Stern plane (Figure 11). In this case, the potential changes from ψ° at the surface, to $\psi(\delta)$ at the Stern plane, to $\psi = 0$ in bulk solution.

An indirect way to obtain information about the potential at foam lamella interfaces is by bubble electrophoresis, in which an electric field is applied to a sample causing charged bubbles to move toward an oppositely charged electrode. The electrophoretic mobility is the measured electrophoretic velocity divided by the electric field gradient at the location where the velocity was measured. These results can be interpreted in terms of the electric potential at the plane of shear, also known as the zeta potential, using well-known equations available in the literature (29–31). Because the exact location of the shear plane is generally not known, the zeta potential is usually taken to be approximately equal to the potential at the Stern plane (Figure 11):

$$\zeta \approx \psi(\delta) \qquad (15)$$

Where δ is the distance from the surface to the Stern plane.

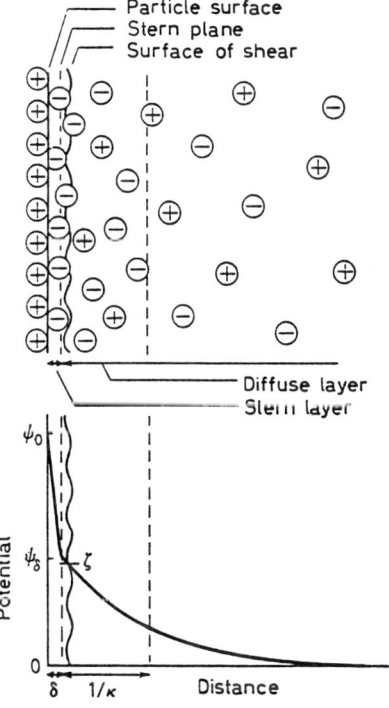

Figure 11. A simplified illustration of the surface and zeta potentials for a charged foam lamella surface. (Reproduced from reference 57. Copyright 1975 Butterworth.)

Good descriptions of practical experimental techniques in conventional electrophoresis can be found in references 29, 30, and 32. For the most part, these techniques are applied to suspensions and emulsions, rather than foams. In bubble microelectrophoresis, the dispersed bubbles are viewed under a microscope, and their electrophoretic velocity is measured taking the horizontal component of motion, because bubbles rapidly float upwards in the electrophoresis cells (*33, 34*). A variation on this technique is the spinning cylinder method, in which a bubble is held in a cylindrical cell that is spinning about its long axis. An electric field is applied, and the electrophoretic mobility is determined (*2, 35*). Other electrokinetic techniques, such as the measurement of sedimentation potential (*36*) have been used as well.

Several studies (e.g., Huddleston and Smith in reference 2 p 163) have shown that gas bubbles in water of neutral pH tend to adopt a negative surface charge even in the absence of surfactants.

Repulsive Forces. In the simplest example of foam stability, gas bubbles and the liquid-films between them would be stabilized entirely by the repulsive forces created when two charged interfaces approach each other and their electric double layers overlap. The repulsive energy, V_R, for the double layers at each interface in the thin-film is given approximately as

$$V_R = V_1 \exp(-\kappa t) \tag{16}$$

where t is the film thickness. The constant of proportionality V_1 depends on the particular system, and includes the term $\psi(\delta)$ (*37*).

A third important force at very small separation distances where the atomic electron clouds overlap causes a strong repulsion, called Born repulsion. Therefore, in extremely thin films such as the Newton black films, Born repulsion becomes important as an additional repulsive force.

Dispersion Forces. van der Waals postulated that neutral molecules exert forces of attraction on each other that are caused by electrical interactions between three types of dipolar configurations. The attraction results from the orientation of dipoles that may be (1) two permanent dipoles, (2) dipole–induced dipole, or (3) induced dipole–induced dipole (also called London dispersion forces). Except for quite polar materials, the London dispersion forces are the most significant of the three. For molecules, the force varies inversely with the sixth power of the intermolecular distance.

For a liquid-film in a foam, the dispersion forces can be approximated by adding up the attractions between all interdroplet pairs of molecules. When added this way, the dispersion force in a liquid-film decays less rapidly as a function of separation distance than is the case for individual molecules, and the attractive energy V_A can be approximated by

$$V_A = -\frac{V_2}{t^2} \qquad (17)$$

where t is the film thickness and V_2 is a constant for the particular system and contains the effective Hamaker constant (37).

Total Interaction Energy: The DLVO Theory. Derjaguin and Landau (*38*), and independently Verwey and Overbeek (*39*), developed a quantitative theory for the stability of lyophobic colloids, now known as the DLVO theory. The DLVO theory accounts for the energy changes that take place when two charged species approach each other and involves estimating the energies of attraction and repulsion, versus distance, and adding them together to yield the total interaction energy V. In our example of a liquid-film that separates two bubbles in a foam, and where only the electrical and van der Waals forces are considered,

$$V = V_R + V_A = [V_1 \exp(-\kappa t)] - \frac{V_1}{t^2} \qquad (18)$$

V_R decreases exponentially with increasing separation distance, and has a range about equal to κ^{-1}, as V_A decreases inversely with the square of increasing separation distance. Figure 12 shows an example of a total interaction energy curve for a thin liquid-film stabilized by the presence of ionic surfactant. Either the attractive van der Waals forces or the repulsive electric double-layer forces can predominate at different film thicknesses.

In the example shown, attractive forces dominate at large film thicknesses. As the thickness decreases, the attraction increases, but eventually the repulsive forces become significant, so that a minimum in the curve may occur. This is called the secondary minimum and may be thought of as a thickness in which a metastable state exists, that of the common black film. As the film thickness decreases further, repulsive forces increase but eventually the attractive forces dominate again, and a much stronger minimum may occur in the curve. This is referred to as the

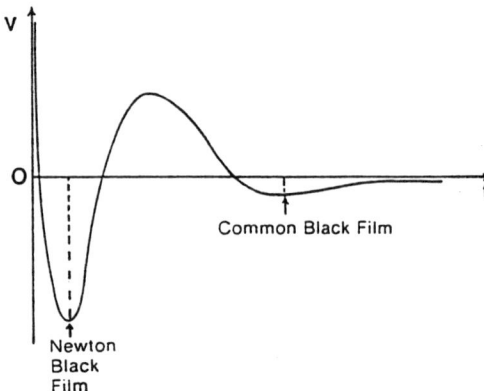

Figure 12. The total interaction potential energy, V, between the surfaces of a foam lamella as a function of the lamella thickness, t. (Reproduced from reference 3. Copyright 1978 Advanced Educational Toys.)

primary minimum; the film is now in a state of greater stability than before, that of the Newton black film. At smaller film thicknesses, repulsive forces dominate once more because of Born repulsion.

Disjoining Pressure. When the two interfaces that bind a foam lamella are electrically charged, the interacting diffuse double layers exert a hydrostatic pressure that acts to keep the interfaces apart. In thin lamellae (film thicknesses on the order of a few hundred nanometers), the electrostatic, dispersion, and steric forces all may be significant, and the disjoining pressure concept is frequently employed. The disjoining pressure represents the net pressure difference between the gas phase (bubbles) and the bulk liquid from which the lamellae extend (*38*), and is the total of electrical, dispersion, and steric forces (per unit area) that operate across the lamellae (perpendicular to the interfaces). The disjoining pressure, π, may be expressed by taking the derivative of the interaction potential with respect to the film thickness:

$$\pi(t) = -\frac{dV}{dt} \qquad (19)$$

The relationship between π and V is illustrated in Figure 13. The metastable minima in V correspond to the condition $\pi = 0$.

Figure 14 illustrates how the disjoining pressure can be determined. A film is formed, and liquid is withdrawn from it through a slit in the ver-

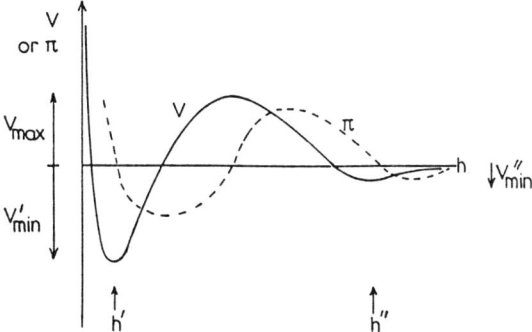

Figure 13. Comparison of the total interaction energy, V (—) and the disjoining pressure π (---) in a foam lamella as a function of film thickness, h. (Adapted from reference 58. Copyright 1984 Academic Press).

tical wall of the apparatus. This step produces a free liquid-film (a), whose surface is under the same pressure, $P = P_N$, as the bulk gas phase. The free liquid-film is in mechanical equilibrium with the adjacent Plateau borders, or wetting menisci. The Plateau borders at b, away from the transition zone c contain liquid under pressure P_0. The disjoining pressure is then given as $\pi(h) = P_N - P_0$, where h is the film thickness. Thus, measurements of π as a function of h may be performed (38). To the extent that the disjoining pressure arises from electrostatic forces, there will be an obvious influence of electrolyte concentration. For very thin films (<1000 Å), the disjoining pressure is very important. But for thicker films and wetter foams, the factors discussed previously are more important.

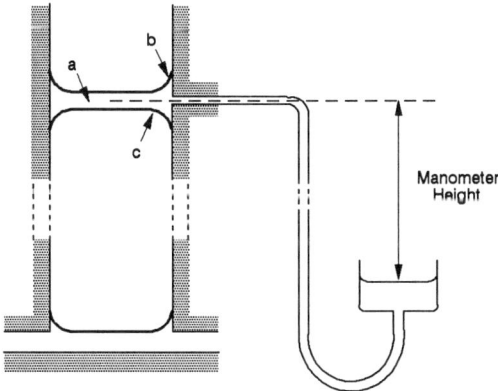

Figure 14. Schematic of a disjoining pressure apparatus. (Adapted from reference 38. Copyright 1987 Consultants Bureau.)

Other Influences on Stability. Where there exists a bubble size distribution, there will exist pressure gradients between bubbles of different size in a foam. Concentration gradients may form across lamellae that promote gas diffusion between bubbles. This effect will cause larger bubbles to grow at the expense of smaller bubbles, and it becomes a mechanism for foam degeneration.

Bulk Properties of Foams

Experimental Assessment of Foam Stability. Usually foam stability has been tested by one of three methods (*4, 6, 13*): (1) the lifetime of single bubbles; (2) the steady-state (dynamic) foam volume under given conditions of gas flow, shaking, or shearing; or (3) the rate of collapse of a (static) column of foam generated as described.

The first method is quite difficult to reproduce because of the strong influence on the results that small contaminations or vibrations can have. The latter two are also difficult to reproduce because the foam generation and collapse is not always uniform, yet these methods are very commonly used. The dynamic foam tests are most suitable for evanescent foams because their lifetimes are transient. For more stable foams, the static foam tests may be used.

In a typical dynamic foam test, foam is generated by flowing gas through a porous orifice into a test solution, as shown in Figure 15. The steady-state foam volume maintained under constant gas flow into the column is then measured. There are many variations of this kind of test (*4, 40*), including an ASTM standard test (*41*). This technique is frequently used to assess the stability of evanescent foams.

For more stable foams, a static foam test may be employed. Again, there are many variations of this kind of test including at least four ASTM standard test methods (*42–45*). ASTM method D 1173–53 (*42*) involves filling a pipet with a given volume of foaming solution; the solution is allowed to fall a specified distance into a separate volume of the same solution that is contained in a vessel (Figure 16). The volume of foam that is produced immediately upon draining of the pipet (initial foam volume) is measured, as is the decay in foam volume by some elapsed time period. This is sometimes referred to as the Ross–Miles test. Data obtained using this method on numerous surfactant solutions were provided in reference 19. In an alternative standard static test, air sparging is used to initially generate the foam (*43*). In another static method, a reproducible foam is generated, and its decay is monitored in a sealed chamber by means of pressure changes caused by the collapsing foam. Such pressure changes can be related to changes in interfacial area present in the foams (*46*).

Static foam tests can even measure for specific shear rate regimes:

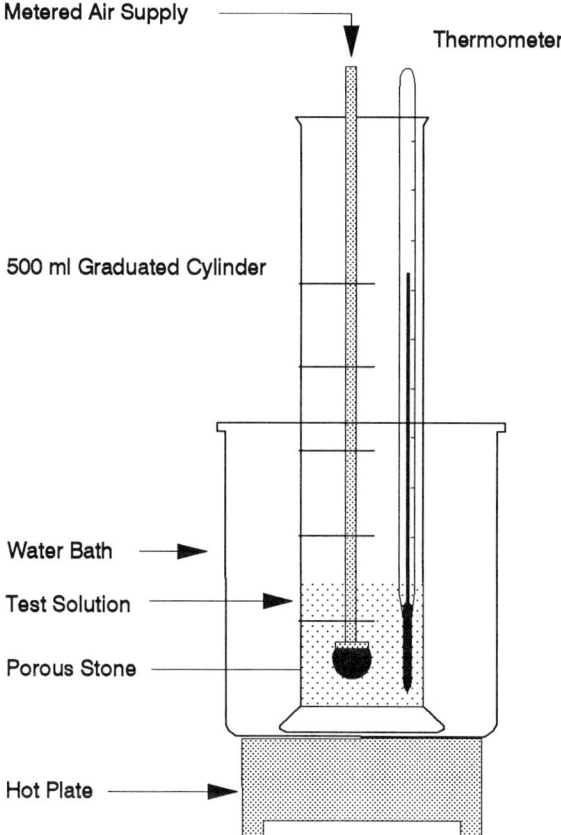

Figure 15. A typical dynamic foam stability test apparatus in which foam is generated by flowing gas through a diffuser. Not drawn to scale. (Adapted from reference 41. Copyright 1988 American Society for Testing and Materials.)

Bottle shaking can be used to generate foam under conditions of relatively low shear (44), and a blender can be used (at about 8000 rpm) to generate foam under conditions of very high shear (45).

In either the dynamic or static foam tests, but especially the static tests, many changes in a foam may occur over time, including gas diffusion and changing bubble size distribution.

Density. In calculating the density of a foam (ρ_F), one can usually ignore the mass of gas involved, so that

$$\rho_F = \frac{m_L}{V_F} \tag{20}$$

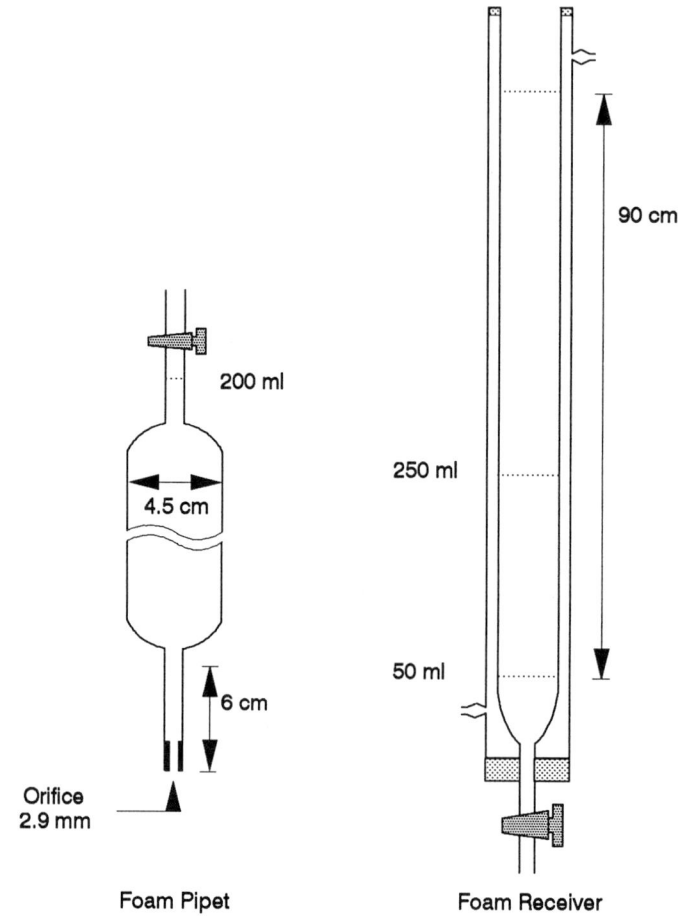

Figure 16. A typical static foam stability test apparatus in which surfactant solution falls into a volumetric receiver. Not drawn to scale. (Adapted from reference 42. Copyright 1988 American Society for Testing and Materials.)

where m_L is the mass of liquid in the foam, and V_F is the total volume of the foam. For foams made from a liquid of density ρ_L and volume V_L:

$$\frac{\rho_L}{\rho_F} = \frac{V_F}{V_L} \tag{21}$$

Equation 21 forms the basis for some further terminology. The factor V_F/V_L is termed the expansion factor (or ratio) of a foam. A reformulation of the factor ρ_L/ρ_F yields the so-called increase of volume upon foaming, or foaming power as $100[(\rho_L/\rho_F) -1]$. Here, the term in square brackets is the ratio of gas to liquid volume in the foam.

The densities of quite stable foams can be measured by pycnometer, or by calculation from known volumes of initial liquid and final foam, from equation 21. In some cases, conductivity measurements can be used.

Foam densities typically vary from about 0.02 to about 0.5 g/mL (8). However, it should be borne in mind that bulk foams are not necessarily homogeneous and usually exhibit a distribution of densities throughout the vertical direction due to gravity-induced drainage.

Rheological Properties. The rheological properties of a foam can be very important: High viscosity may be part of the reason that a foam is troublesome, a resistance to flow that must be dealt with, or a desirable property for which a foam is formulated. The simplest description applies to Newtonian behavior in laminar flow. The coefficient of viscosity, η, is given in terms of the shear stress, τ, and shear rate, $\dot{\gamma}$, by

$$\tau = \eta\dot{\gamma} \tag{22}$$

where η has units of milliPascal–seconds. Many colloidal dispersions, including the more concentrated foams, do not obey the Newtonian equation. For non-Newtonian fluids the coefficient of viscosity is not a constant, but is a function of the shear rate; thus

$$\tau = \eta(\dot{\gamma})\ \dot{\gamma} \tag{23}$$

A convenient way to summarize the flow properties of fluids is by plotting flow curves of shear stress versus shear rate (τ versus $\dot{\gamma}$). These curves can be categorized into several rheological classifications. Foams are frequently pseudoplastic; that is, as shear rate increases, viscosity decreases. This is also termed shear-thinning. Persistent foams (polyederschaum) usually exhibit a yield stress (τ_Y), that is, the shear rate (flow) remains zero until a threshold shear stress is reached, then pseudoplastic or Newtonian flow begins. An example would be a foam for which the stress due to gravity is insufficient to cause the foam to flow, but the application of additional mechanical shear does cause flow (Figure 17).

Pseudoplastic flow that is time dependent is termed thixotropic. That is, viscosity decreases at constant applied shear rate, and in a flow curve, hysteresis occurs. Several other rheological classifications are covered in the Glossary of this book. Even viscosity itself is represented in many ways, as shown in Table IV.

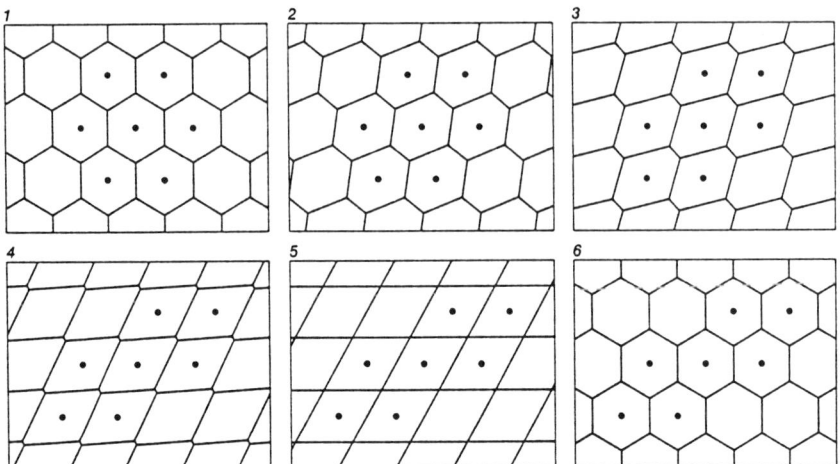

Figure 17. Shearing stress applied to a bulk foam causes distortion (frames 2–4) and eventually bulk flow once the yield stress is exceeded (frames 5 and 6). Dots in a constellation of cells show how the relative positions of those cells have changed. (Reproduced from reference 50. Copyright 1986 Scientific American).

Some very useful descriptions of standard experimental techniques have been given by Whorlow (47) and others (48, 49). With due care, these may be applied to foams; for example, measurements may be made with a foam sample placed in the annulus between two concentric cylinders (4). The shear stress is calculated from the measured torque required to maintain a given rotational velocity of one cylinder with respect to the other. Knowing the geometry, the effective shear rate can be calculated from the rotational velocity. One reason for the relative lack of such rheological data for foams, compared with that for other colloidal systems, is the difficulty associated with performing the measurements in these systems. In attempting to conduct a measurement, a number of changes may occur in the sample chamber, making the measurements irreproducible and not representative of the original foam, including

- creaming of the bubbles, which causes a nonuniform distribution within the chamber, or even removal of all bubbles into an upper phase away from the region in which measurements are made
- centrifugal separation of gas and liquid phases, making the foam radially inhomogeneous, and possibly collapsing the foam
- shear-induced coalescence or finer dispersion of bubbles, changing the properties of the sample

Table IV. Glossary of Bulk Viscosity Terms

Term	Symbol	Explanation
Absolute viscosity	η	$\eta = \tau/\dot\gamma$ and can be traced to fundamental units independent of the type of instrument.
Apparent viscosity	η_{APP}	$\eta_{APP} = \tau/\dot\gamma$ but as determined for a nonNewtonian fluid, usually by a method suitable only for Newtonian fluids.
Differential viscosity	η_D	$\eta_D = d\tau/d\dot\gamma$
Specific increase in viscosity	η_{SP}	$\eta_{SP} = \eta_{Rel} - 1$
Intrinsic viscosity	$[\eta]$	$[\eta] = \lim_{c \to 0} \lim_{\dot\gamma \to 0} \eta_{SP}/C$ $= \lim_{c \to 0} \lim_{\dot\gamma \to 0} (1/C) \ln \eta_{Rel}$
Reduced viscosity	η_{Red}	$\eta_{Red} = \eta_{SP}/C$
Relative viscosity	η_{Rel}	$\eta_{Rel} = \eta/\eta_0$
Inherent viscosity	η_{Inh}	$\eta_{Inh} = C^{-1} \ln (\eta/\eta_0)$

ABBREVIATIONS: C is the solute or dispersed-phase concentration and η_0 is the viscosity of the pure solvent or dispersion medium.
SOURCE: Adapted from reference 14. Copyright 1992 American Chemical Society.

Because wet foams contain approximately spherical bubbles, their viscosities can be estimated by the same means that are used to predict emulsion viscosities (*14*). In this case, the foam viscosity is described in terms of the viscosity of the continuous phase (η_0) and the amount of dispersed gas through an empirical extension of Einstein's equation for a dilute dispersion of spheres:

$$\eta = \eta_0 (1 + \alpha_0\phi + \alpha_1\phi^2 + \alpha_2\phi^3 + \cdots) \quad (24)$$

where η_0 is the liquid viscosity, ϕ is the gas phase volume fraction, and the αs are empirical constants. Such equations assume Newtonian behavior, or at least apply to the Newtonian region of a flow curve, and they usually apply if the bubbles are not too large and if there are no strong electrostatic interactions.

In dry foams, where the internal phase has a high volume fraction, the foam viscosity increases strongly, because of bubble "crowding" or structural viscosity, becomes non-Newtonian, and frequently exhibits a yield stress. The maximum volume fraction possible for an internal phase made up of uniform, incompressible spheres is 74%, but because the gas bubbles are very deformable and compressible, foams with an internal volume fraction of 99% or more are easy to make. As mentioned earlier, the structure of such a foam consists of irregular polyhedra with a maximum of three lamella meeting at the Plateau borders. Figure 17 shows an illustration of foam polyhedra that are deformed in the presence of applied shear (50). Following the cells marked with dots shows that no flow occurs until there is so much distortion that four films meet at Plateau borders. This situation is energetically unstable and also marks the point at which the yield stress is reached. The foam layers will shift to relieve the stress and attain an energetically more favorable structure. By restoring the stable three film contacts, flow results.

Optical Properties. Between the time a fluid film is first formed until it thins to an equilibrium black film, the thickness can be measured by techniques based on the interference of light. For monochromatic light, the intensity of light, I_R, reflected at angle θ from a fluid film compared to the maximum intensity yielded by constructive interference, I_0, enables the determination of the film thickness, t, as

$$t = \frac{\lambda}{2\pi\mu \cos \theta} \sin^{-1} \left(\frac{I_R}{I_0} \right)^{1/2} \tag{25}$$

where μ is the refractive index and λ is the wavelength.

If white light is used, then interference will produce colored fringes in the film. These fringes can be used to determine the film thickness and also its variation over the surface of the film, so they can be used to experimentally study film drainage and thinning processes. Some examples were given by Isenberg (3). If the thickness of the film is much less than that of the illuminating light, then the film will appear black (5–30 nm). At larger thicknesses, the shorter wavelength components of the illuminating light will begin to interfere constructively, and colored fringes will appear, with colors due to longer wavelength components being added as the thickness increases further (30–120 nm). When these liquid-films naturally thin, they may reach a metastable film thickness (in the secondary minimum), typically about 30 nm; the colored fringes disappear; and the film is termed a common black film. With further thinning, they may reach a more stable thickness (in the primary minimum), typically about 5 nm, in which case the film is termed a Newton black film.

For bulk foams, some researchers have fitted the intensity of light (I_T) passing through a sample of foam to an equation of the form:

$$I_0 / I_T = an_F + b \tag{26}$$

where I_0 is the intensity of incident light, n_F is the number of foam lamellae per unit path length, and a and b are empirical constants. Equation 26 can be used as the basis for a method of measuring foam decay.

Bubble Sizes. Despite the fact that the bubbles in persistent foams are polyhedral and not spherical, it is nevertheless conventional to refer to the "diameters" of gas bubbles in foams as if they were spherical. It was stated previously that colloidal droplets are between about 10^{-3} and 1 μm in diameter while allowing that in practice, foam bubbles are often larger. In fact, foam bubbles usually have diameters greater than 10 μm and may be larger than 1000 μm. Foam stability is not necessarily a function of drop size, although there may be an optimum size for an individual foam type. It is common but almost always inappropriate to characterize a foam in terms of a given bubble size because there is inevitably a size distribution. This size distribution is usually represented by a histogram of sizes, or, if there are sufficient data, a distribution function.

Some foams that have a drop-size distribution that is heavily weighted toward the smaller sizes will represent the most stable foam. In such cases, changes in the size distribution curve with time yield a measure of the stability of the foams. The bubble size distribution also has an important influence on the viscosity. For bubbles that interact electrostatically or sterically, foam viscosity will be higher when bubbles are smaller (for a given foam quality). This condition results because the increased interfacial area and thinner films increase the resistance to flow. The viscosity will also be higher when the bubble sizes are relatively homogeneous, that is, when the bubble size distribution is narrow rather than wide (also for given foam quality).

Electrical Properties. Bikerman (4) described studies of the specific conductivity of individual foam lamellae. For bulk foams, the specific conductivity is proportional to the volume fraction of liquid in the foam:

$$\frac{\kappa_L}{\kappa_F} = \frac{f\rho_L}{\rho_F} \tag{27}$$

where f is also a function of ρ_L/ρ_F. The proportionality constant f typically varies from about 2.0 to 2.5, although in the limit of $\rho_L/\rho_F = 1$ then $f = 1$ also because in this limit foam no longer exists at all.

Influence of Additional Phases

The presence of a third phase can promote or impair foam stability, and in some cases, even prevent foaming. As mentioned previously, stable foams can be formed from mixtures of an isotropic liquid with a liquid-crystal phase: The foam lamellae become covered with layers of liquid-crystal; the foam stability is increased through surface viscosity. Foam stability can also be affected by the presence of other dissolved species, an additional liquid phase such as oil in an aqueous foam, or fine solids. In these cases, whether the effect is one of stabilizing or destabilizing depends on several factors. First, it depends on whether or not the third-phase species have a strong affinity for the liquid phase, and therefore whether they tend to accumulate at the gas–liquid interface. Second, once accumulated, any effect they may have on the interfacial properties is important.

Antifoaming. Some agents will act to reduce the foam stability of a system (termed foam "breakers" or defoamers), and others can prevent foam formation in the first place (foam preventatives and foam inhibitors). There are many such agents. Kerner (51) described several hundred different formulations for foam inhibitors and foam "breakers". In all cases, the cause of the reduced foam stability can be traced to changes in the nature of the interface, but the changes can be of various kinds.

The addition to a foaming system of any soluble substance that can become incorporated into the interface may decrease dynamic foam stability if the substance does any combination of the following: increase surface tension, decrease surface elasticity, decrease surface viscosity, or decrease surface potential. Such effects may be caused by a cosolubilization effect in the interface or by a partial or even complete replacement of the original surfactants in the interface. Some branched, reasonably high molecular mass alcohols can be used for this purpose. Not being very soluble in water, they tend to be adsorbed at the gas–liquid interface and displace foam-promoting surfactant and "break" or inhibit foam. Alternatively, a foam can be destroyed by adding a chemical that actually reacts with the foam-promoting agent(s).

Foams may also be destroyed or inhibited by the addition of certain insoluble substances. The next two sections deal with the addition of a second liquid phase or of a solid phase into a foam.

Effect of a Second Liquid Phase. For illustration, we will use the example of an oil phase that is introduced into an aqueous foam. However, the same considerations apply to an aqueous phase being introduced into an oil-continuous foam.

When a drop of oil comes into contact with the gas–liquid interface, the oil may form a bead on the surface or it may spread and form a film. If the oil has a strong affinity for the new phase, it will seek to maximize its contact (interfacial area) and form a film. A liquid with much weaker affinity may form into a bead. Petroleum emulsions have been used to prevent the formation of foams or to destroy foams already generated in various industrial processes (52). Such a destabilizer may act by penetrating the interface and either remaining as a lens or spreading over the interface.

If an insoluble agent is dispersed in the liquid interior of foam lamellae, then, from thermodynamics, it can penetrate or "enter" the gas–liquid interface if the entering coefficient, E, is positive (originally defined (53) as a rupture coefficient, R). E is defined, for unit surface areas, as

$$E = \gamma^\circ_F + \gamma_{OF} - \gamma^\circ_O \tag{28}$$

When oil enters, a new gas–oil interface is created, while some gas–aqueous and aqueous–oil interfacial areas are eliminated, and $E = -\Delta G_E$ for this process. If E is positive, the oil will be drawn up into the lamellar region between bubbles and will breach the aqueous–gas interface to form a lens. With oil present in this manner, the interfacial film can lose its foam stabilizing capability, although Ross (54) pointed out that entering does not always destabilize a foam. If $E < 0$, then the oil should be ejected from the lamellar surface region and not destabilize the foam by this mechanism. Whereas the foregoing description comes from thermodynamics, a dynamic description has been given by Wasan and co-workers, in terms of pseudoemulsion film thinning, which is described further in Chapters 2 and 4 of this book.

Once the oil is present at the interface, and it bridges the lamellar liquid and the gas phase, it would be predicted (from thermodynamics) to spread spontaneously over a foam if its "spreading coefficient", S, is positive (55). For an oil–foam system S is given, for unit surface areas, by

$$S = \gamma^\circ_F - \gamma_{OF} - \gamma^\circ_O \tag{29}$$

When oil spreads, a new gas–oil and aqueous–oil interface is created, some gas–aqueous interfaces are eliminated, and $S = -\Delta G_S$ for this process. This condition is usually sufficient for defoaming.

Depending upon the degree of entering and "spreading" of an insoluble agent, several consequences for foam stability can result. If the agent enters the gas–liquid interface, it may be able to bridge adjacent bubbles. This bridging will cause lamella ruptures because the lamellae will be less cohesive and will probably not have the conditions of low surface tension and the Marangoni effect. Even if the agent cannot bridge adjacent bubbles, as long as it can enter the interface, it may also be able to spread over the lamella surfaces, replacing the liquid medium as the phase in contact with the gas. In this case, because the new interface will normally not have the capability of stabilizing the lamellae, a reduction in foam stability will occur. In this way, the rapid spreading of drops of oil of low surface tension over lamellae ruptures them by providing weak spots (13). Poly(dimethylsiloxane)s are frequently used as practical antifoaming agents because they are insoluble in aqueous media (and some oils), have low surface tension, and are not overly volatile. They are usually formulated as emulsions for aqueous foam inhibiting so that they will readily become mixed with the aqueous phase of the foam.

More detailed aspects of these processes are discussed in Chapter 4 of this book, in the context of foam–oil interactions in porous media. Because the interfacial tensions mentioned can change with time after an initial spreading of defoamer, it follows that the parameters E and S are also time-dependent and that some defoamers are effective only for a limited amount of time.

Effect of Solids. The presence of dispersed particles can increase or decrease aqueous foam stability. One mechanism for the stability enhancement is the bulk viscosity enhancement that results from having a stable dispersion of particles present in the solution. A second stabilizing mechanism is operative if the particles are not completely water-wetted. In this case, particles would tend to collect at the interfaces in the foam where they may add to the mechanical stability of the lamellae.

Because there can be degrees of wetting of particles at an interface, another quantity is needed. The contact angle, θ, in an oil–water–solid system is defined as the angle, measured through the aqueous phase, that is formed at the junction of the three phases. Whereas interfacial tension is defined for the boundary between two phases, the contact angle is defined for a three-phase junction. If the interfacial forces that act along the perimeter of the drop are represented by the interfacial tensions, then an equilibrium force balance can be written as

$$\gamma_{W/O} \cos \theta = \gamma_{S/O} - \gamma_{S/W} \tag{30}$$

where the subscripts refer to water (W), oil (O), and solid (S). This is Young's equation. The solid is completely water wetted if $\theta = 0$, and only partially wetted otherwise. This equation is frequently used to describe wetting phenomena, so two practical points are important. In theory, complete nonwetting by water would mean that $\theta = 180°$, but this is not seen in practice. Also, values of $\theta < 90°$ are often considered to represent "water-wetting", and values of $\theta > 90°$ are considered to represent "non-water-wetting". This rather arbitrary assignment is based on correlation with visual appearance of drops on surfaces.

A foam-stabilizing mechanism is operative if the particles are not completely water-wetted. In this case, particles would tend to collect at the interfaces in the foam where they may add to the mechanical stability of the lamellae. On the other hand, quite hydrophobic particles may actually act to destabilize foam. Thus intermediate contact angles (between about 40° and 70°) appear to be optimum for solid-stabilized foams (8).

Summary

The stability of a foam is determined through the interplay of a number of factors that involve bulk solution, interfacial properties, and also external forces. We have summarized some of the effects on foam stability of gravity drainage, capillary suction, surface elasticity, viscosity, electric double-layer repulsion, dispersion force attraction, and steric repulsion. Foams are such complex systems that Lucassen (56) has stated that "any attempt to understand their properties in terms of a 'simple' all-embracing theory is doomed to failure." Nevertheless, we have attempted to provide an introduction to the occurrence, properties, and importance of foams as they relate to the petroleum industry. More detailed aspects are taken up in the subsequent chapters of this book.

List of Symbols

a	empirical constant
A	surface area
A_S	area per adsorbed surfactant molecule
b	empirical constant
c_i	concentration of ions i
C_S	surfactant concentration in solution
CMC	critical micelle concentration
E	entering coefficient
E_G	Gibbs surface elasticity

E_M	Marangoni surface elasticity
f	proportionality constant
$F_{1x, 2x}$	force in the x direction at positions 1 or 2
G	Gibbs free energy
i	unit vector in the x-direction
I	unit tensor
I	solution ionic strength
I_0, I_R, I_T	intensities of incident, reflected and transmitted light
I^S	surface projection tensor
j	unit vector in the y-direction
m_L	mass of liquid
MBPM	maximum bubble pressure method
n	unit vector normal to the surface
n_F	number of foam lamellae per unit distance
N_A	Avogadro's number
p_A, p_B	pressures at different locations within a lamella
p_G, p_L	pressures on each side of an interface
Δp	pressure difference across an interface
P^S	surface stress tensor
R	radius of a curved surface or interface, also used as the gas constant
R_1, R_2	principal radii of curvature of a surface or interface
S	spreading coefficient
t	fluid film thickness
T	absolute temperature
T_k	Krafft point
T	transpose operator
v^S	surface velocity
V_1, V_2	constants in potential energy equations
V_A	attractive potential energy
V_F	volume of foam
V_L	volume of liquid
V_R	repulsive potential energy
V	total potential energy
x	distance from a surface or interface
z_i	charge number of ions i

Greek

$\alpha_0, \alpha_1, \alpha_2$	empirical constants
Γ_S	surface excess concentration of surfactant
γ	interfacial tension

$\dot{\gamma}$	shear rate
γ_d	dynamic surface tension
γ°	surface tension
δ	distance from a surface to the Stern plane
ς	zeta potential
η	viscosity
η^S	surface shear viscosity
θ	contact angle
κ	Debye length (inverse of the double layer thickness)
κ_F, κ_L	specific conductivities of foam and liquid phases
κ^S	surface dilational viscosity
λ	wavelength
μ	refractive index
μ_E	electrophoretic mobility
π	disjoining pressure
ρ_F	foam density
ρ_L	liquid density
σ°	surface charge density
τ	shear stress
ϕ	volume fraction
ψ	electric potential
ψ°	surface electric potential
∇	gradient operator

Acknowledgments

We are grateful to the Alberta Oil Sands Technology and Research Authority (AOSTRA) for an industrial post-doctoral fellowship held by Fred Wassmuth.

References

1. Bikerman, J. J. *Foams; Theory and Industrial Applications;* Reinhold: New York, 1953.
2. *Foams;* Akers, R. J., Ed.; Academic: Orlando, FL, 1976.
3. Isenberg, C. *The Science of Soap Films and Soap Bubbles;* Tieto: Clevedon, England, 1978.
4. Bikerman, J. J. *Foams;* Springer-Verlag: New York, 1973.
5. Osipow, L. I. *Surface Chemistry Theory and Industrial Applications;* Reinhold: New York, 1962.
6. Ross, S.; Morrison, I. D. *Colloidal Systems and Interfaces;* Wiley: New York, 1988.

7. Adamson, A. W. *Physical Chemistry of Surfaces,* 4th ed.; Wiley-Interscience: New York, 1982.
8. Ross, S. In *Kirk–Othmer Encyclopedia of Chemical Technology,* 3rd ed.; Wiley: New York, 1980; Vol. 11, pp 127–145.
9. *Surfactants;* Tadros, Th. F., Ed.; Academic Press: London, 1984.
10. *Interfacial Phenomena in Petroleum Recovery;* Morrow, N. R., Ed.; Surfactant Science Series 36; Dekker: New York, 1991.
11. Myers, D. *Surfactant Science and Technology;* VCH: New York, 1988.
12. Clunie, J. S.; Goodman, J. F.; Ingram, B. T. In *Surface and Colloid Science;* Matijevic, E., Ed.; Wiley: New York, 1971; Vol. 3, pp 167–239.
13. Kitchener, J. A. In *Recent Progress in Surface Science;* Danielli, J. F.; Pankhurst, K. G. A.; Riddiford, A. C., Eds.; Academic Press: Orlando, FL, 1964; Vol. 1, pp 51-93.
14. *Emulsions: Fundamentals and Applications in the Petroleum Industry;* Schramm, L. L., Ed.; ACS Advances in Chemistry 231; American Chemical Society: Washington, DC, 1992.
15. Harkins, W. D.; Alexander, A. E. In *Physical Methods of Organic Chemistry;* Weissberger, A., Ed.; Interscience: New York, 1959; pp 757–814.
16. Padday, J. F. In *Surface and Colloid Science;* Matijevic, E., Ed.; Wiley-Interscience: New York, 1969; Vol. 1, pp 101–149.
17. Miller, C. A.; Neogi, P. *Interfacial Phenomena Equilibrium and Dynamic Effects;* Dekker: New York, 1985.
18. Cayias, J. L.; Schechter, R. S.; Wade, W. H. In *Adsorption at Interfaces;* Mittal, K. L., Ed.; ACS Symposium Series 8; American Chemical Society: Washington, DC, 1975; pp 234–247.
19. Rosen, M. J. *Surfactants and Interfacial Phenomena,* 2nd ed.; Wiley: New York, 1989.
20. *Industrial Applications of Surfactants;* Karsa, D. R., Ed.; Royal Society of Chemistry: London, 1987.
21. *Anionic Surfactants Physical Chemistry of Surfactant Action;* Lucassen-Reynders, E. H., Ed.; Dekker: New York, 1981.
22. Malysa, K.; Lunkenheimer, K.; Miller, R.; Hartenstein, C. *Colloids Surf.* **1981,** *3,* 329–338.
23. Lucassen-Reynders, E. H. In *Anionic Surfactants Physical Chemistry of Surfactant Action;* Lucassen-Reynders, E. H., Ed.; Dekker: New York, 1981; pp 173–216.
24. Huang, D. D. W.; Nikolov, A.; Wasan, D. T. *Langmuir* **1986,** *2,* 672–677.
25. Mysels, K. J.; Cox, M. C.; Skewis, J. D. *J. Phys. Chem.* **1961,** *65,* 1107–1111.
26. Schramm, L. L.; Green, W. H. F. *Colloid Polym. Sci.* **1992,** *270,* 694–706.
27. Edwards, D. A.; Wasan, D. T. In *Surfactants in Chemical Process Engineering;* Wasan, D. T.; Ginn, M. E.; Shah, D. O., Eds.; Dekker: New York, 1988; pp 1–28.
28. Joly, M. In *Recent Progress in Surface Science;* Danielli, J. F.; Pankhurst, K. G. A.; Riddiford, A. C., Eds.; Academic: Orlando, FL, 1964; Vol. 1, pp 1–50.
29. Hunter, R. J. *Zeta Potential in Colloid Science;* Academic: Orlando, FL, 1981.
30. James, A. M. In *Surface and Colloid Science;* Good, R. J.; Stromberg, R. R., Eds.; Plenum: New York, 1979; Vol. 11, pp 121–186.
31. O'Brien, R. W.; White, L. R. *J. Chem. Soc. Faraday 2* **1978,** *74,* 1607–1626.

32. Riddick, T. M. *Control of Stability through Zeta Potential;* Zeta Meter Inc.: New York, 1968.
33. Okada, K.; Akagi, Y. *J. Chem. Eng. Jpn.* **1987**, *20(1)*, 11–15.
34. Yoon, R.-H.; Yordan, J. L. *J. Colloid Interface Sci.* **1986**, *113(2)*, 430–438.
35. Whybrew, W. E.; Kinzer, G. D.; Gunn, R. *J. Geophys. Res.* **1952**, *57*, 459–471.
36. Usui, S.; Sasaki, H. *J. Colloid Interface Sci.* **1978**, *65(1)*, 36–45.
37. Hiemenz, P. C. *Principles of Colloid and Surface Chemistry*, 2nd ed.; Dekker: New York, 1986.
38. Derjaguin, B. V.; Churaev, N. V.; Miller, V. M. *Surface Forces;* Consultants Bureau: New York, 1987.
39. Verwey, E. J. W.; Overbeek, J. Th. G. *Theory of the Stability of Lyophobic Colloids;* Elsevier: New York, 1948.
40. Ross, S.; Suzin, Y. *Langmuir* **1985**, *1*, 145–149.
41. *ASTM Designation D 1881–86;* American Society for Testing and Materials: Philadelphia, PA, 1988.
42. *ASTM Designation D 1173–53,* Reapproved 1986; American Society for Testing and Materials: Philadelphia, PA, 1988.
43. *ASTM Designation D 892–74,* Reapproved **1984**, *American Society for Testing and Materials: Philadelphia, PA, 1988.*
44. *ASTM Designation D 3601–88;* American Society for Testing and Materials: Philadelphia, PA, 1988.
45. *ASTM Designation D 3519–88;* American Society for Testing and Materials: Philadelphia, PA, 1988.
46. Nishioka, G.; Ross, S. *J. Colloid Interface Sci.* **1981**, *81*, 1–7.
47. Whorlow, R. W. *Rheological Techniques;* Wiley: New York, 1980.
48. Van Wazer, J. R.; Lyons, J. W.; Kim, K. Y.; Colwell, R. E. *Viscosity and Flow Measurement;* Wiley: New York, 1963.
49. Fredrickson, A. G. *Principles and Applications of Rheology;* Prentice-Hall: Englewood Cliffs, NJ, 1964.
50. Aubert, J. H.; Kraynik, A. M.; Rand, P. B. *Sci. Amer.* **1986**, *254(5)*, 74–82.
51. Kerner, H. T. *Foam Control Agents;* Noyes Data Corp.: Park Ridge, NJ, 1976.
52. Currie, C. C. In *Foams;* Bikerman, J. J., Ed.; Reinhold: New York, 1953; pp 297–329.
53. Robinson, J. W.; Woods, W. W. *J. Soc. Chem. Ind.* **1948**, *67*, 361.
54. Ross, S. *J. Phys. Colloid Chem.* **1950**, *54*, 429–436.
55. Harkins, W. D. *J. Chem. Phys.* **1941**, *9*, 552.
56. Lucassen, J. In *Anionic Surfactants Physical Chemistry of Surfactant Action;* Lucassen-Reynders, E. H., Ed.; Dekker: New York, 1981; pp 217–265.
57. Shaw, D. J. *Introduction to Colloid and Surface Chemistry*, 3rd Ed.; Butterworth: London, 1981.
58. Vincent, B. In *Surfactants;* Tadros, Th. F., Ed., Academic: Orlando, FL, 1984; pp 175–196.

RECEIVED for review October 14, 1992. ACCEPTED revised manuscript March 15, 1993.

2

Mechanisms of Aqueous Foam Stability and Antifoaming Action with and without Oil

A Thin-Film Approach

D. T. Wasan, K. Koczo, and A. D. Nikolov

Department of Chemical Engineering, Illinois Institute of Technology, Chicago, IL 60616

Mechanisms of stability of foams with and without oil are presented in this chapter based on the structure and stability of the thin liquid films (foam lamellae). The drainage and stability of single-foam lamellae and bulk foams depend on the surfactant concentration. At low surfactant concentrations, the drainage of the single-foam lamellae is governed by the surface tension gradient; at high concentrations, micellar layering inside the film governs the stability. Solubilized oil reduces the micellar structuring effect and accelerates the film drainage and thus destabilizes the foam. The foam stability in the presence of emulsified oil (three-phase foam) is controlled by the stability of the pseudoemulsion film; that is, the water film, between an oil drop trapped inside the Gibbs–Plateau border and the gas phase. If this film is stable, then the foam is stable. If the pseudoemulsion film is unstable, then the oil acts as an antifoam. A new understanding of the role played by the pseudoemulsion film in antifoaming action is discussed.

IN MOST APPLICATIONS IN THE PETROLEUM INDUSTRY, such as in steam-foam flooding, aqueous foams are used to control flow resistance, and this capability makes them attractive for mobility control for improving oil recovery. When the surfactant solution comes in contact with oil in the porous medium, oil is emulsified, and Figure 1 shows the presence of emulsified oil droplets inside of a capillary network. The oil has

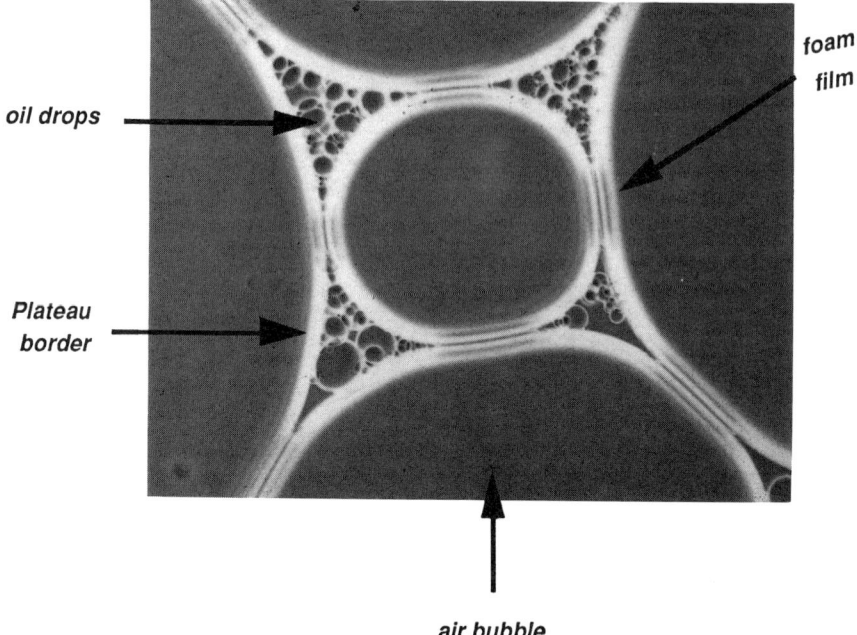

Figure 1. Photomicrograph of a two-dimensional foam formed from 4 wt% Enordet AE1215-30 (ethoxylated alcohol, a nonionic surfactant) solution containing emulsified octane in which the oil drops have drained from the foam films into the Plateau borders.

drained from the thinning foam film and is trapped inside the Gibbs–Plateau borders. Under the capillary pressure inside the Gibbs–Plateau border, the three phases (gas, water, and oil) come into contact, and three types of aqueous films are formed (Figure 2): foam film between the bubbles; emulsion film between the oil drops; and an asymmetrical, oil–water–gas film between the oil drops and the gas phase (bubbles), the so-called "pseudoemulsion film" (1). The foam life and the effect of oil on the foam stability are controlled by the structure and stability of these films as summarized in Figure 3.

In the first part of the chapter, the mechanisms of foam film drainage and bulk foam stability at both low and high concentrations are discussed. We show that at low surfactant concentrations, the surface rheological properties of the adsorbed surfactant layers, such as interfacial tension gradients, control the film stability. Surface rheological properties are also shown to possess a direct significance to the rheology of foams. At high surfactant concentrations, much above the critical micelle concentration (CMC), the formation of micellar layering or ordered microstructure formation inside the film governs the foam film drainage and stability.

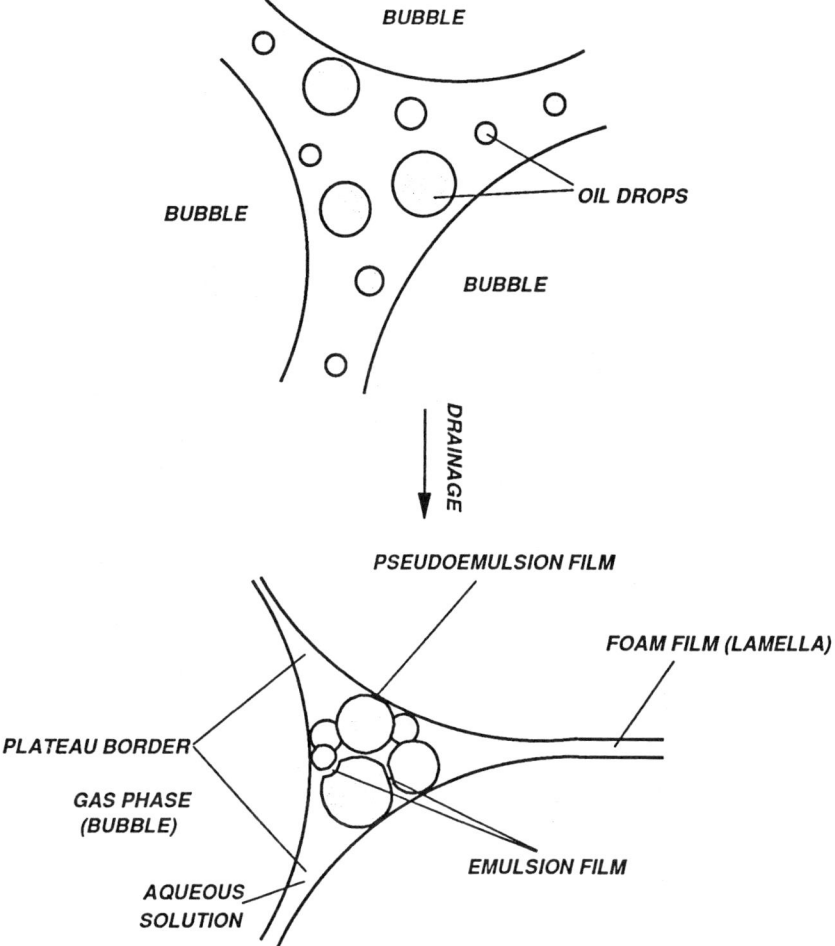

Figure 2. Aqueous foam, pseudoemulsion and emulsion films, and Plateau borders containing oil drops.

Factors such as micellar concentration, micelle size, chain lengths, film area, electrolyte concentration, temperature, presence of solubilized oil on the micellar film structuring, and thus foam stability are highlighted.

The interactions between foams and emulsified oil drops are discussed in the second part of this chapter. In the presence of emulsified oil, the mechanisms of foam stability are more complex than without oil. The mechanism of foam stability in the presence of oil drops is shown to be determined by the stability of the pseudoemulsion film. When the pseudoemulsion film is stable, the oil drops enhance the foam stability; when the film is unstable, the oil acts as an antifoam (defoamer). In

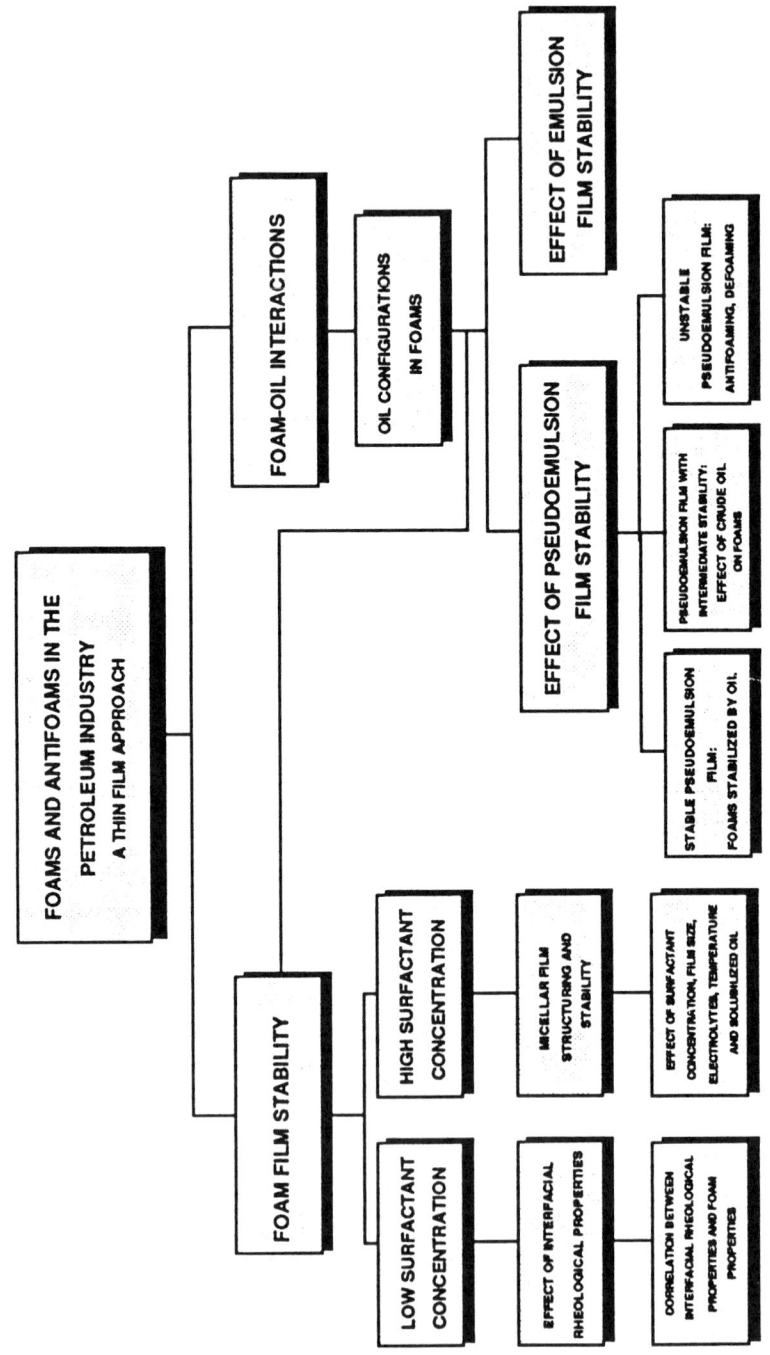

Figure 3. Chapter overview.

several practical systems, such as in foam-enhanced oil processes, the pseudoemulsion film may display intermediate stability as the emulsified oil can have both destabilizing (at low oil contents) and stabilizing effects. We have observed a direct correlation between pseudoemulsion film stability and crude oil recovery efficiency in porous media. The stability of emulsion films between oil drops inside three-phase foams also has a direct influence on foam stability. We review the various antifoaming mechanisms when nonpolar oil, hydrophobic solid particles, or mixtures of insoluble oil and hydrophobic particles are used as antifoaming agents and present a new understanding of the role played by the pseudoemulsion film in antifoaming action.

Foam Film Stability

Mechanisms of Single-Foam Film Stability. Soap bubbles and soap films have been the focus of scientific interest since the days of Hooke and Newton (2–9). The stability and structure of foams are determined primarily by the relative rate of coalescence of the dispersed gas bubbles (10). The process of coalescence in foams is controlled by the thinning and rupture of the foam films separating the air bubbles. Experimental observations suggest that the lifetime (stability) of foam films is determined primarily by the thinning time rather than by the rupture time. Hence, if the approaching bubbles have equal size, the process of coalescence can be split into three stages:

1. formation of a thick lamella

2. thinning of the lamella to a film. By forming spots, the film may reduce its thickness by a single thickness transition when dark spots form in the thinning film at low surfactant concentration (below the CMC) or by multiple thickness transitions at concentrations several times higher than the CMC. Before the CMC, the thick foam film at a thickness corresponding to common black film (with about 10–100-nm thickness)

3. ruptures, according to Vrij (11), due to an unbounded increase in the surface corrugations with time. However, very thin Newton black film ruptures because of nucleation (hole formation) according to Derjaguin and Gutop (12) and Derjaguin and Prokhorov (13).

Stages (1) and (3) for small size film occur very fast so that the lifetime of the intervening film is essentially given by stage (2).

The driving force for the drainage of a thick film is the capillary pressure (suction at the Plateau borders). The thinning rate and stability of the thin lamella are governed by the hydrodynamic and thermodynamic interactions between the two film surfaces. At the first stage of film thinning, at thicknesses >100 nm, the hydrodynamic interactions, which are greatly influenced by the deformation and mobility of the surfaces, dominate. When the film has thinned to <100 nm, thermodynamic interactions begin to dominate. The thermodynamic properties of thin liquid films are different from those of the bulk surfactant solutions. These films possess an excess chemical potential that is manifested as an excess pressure. Derjaguin and Kusakov (14) coined the term "disjoining pressure" to characterize this excess pressure.

Generally, the disjoining pressure consists of the electrostatic repulsion forces between the two overlapping surface double layers, the attractive van der Waals forces among all the molecules of the film, the steric forces due to steric hindrance in closely packed monolayers and, in the presence of micelles, the structural forces.

The main stages in the formation and evolution of the thin liquid film between two equal size approaching foam bubbles as shown in Figure 4 can be summarized as:

a. When two equal bubbles approach each other, the result is hydrodynamic interaction, and a thick lamella is formed.

b. Deformation of the bubble surfaces leads to a bell-shaped formation that is called a "dimple".

c. The dimple gradually disappears, and a plane-parallel film of radius, R, is formed. The flat film drains under the combined actions of suction from Plateau borders and the disjoining pressure, Π. The mechanism of subsequent thinning of the film depends on the surfactant concentration.

d. At low surfactant concentrations (i.e., below the CMC), when $d\Pi/dh > 0$, (where h is the liquid film thickness) the lamella favors the growth of corrugations at the film surfaces, and at a critical thickness, h_{cr}, either the film ruptures or a jump transition in thickness occurs, leading to a stable or metastable structure. This process of transition to stable or metastable state is known as "black spot formation" because at these thicknesses the film appears grey or black.

e. The black spots increase in size and cover the whole film.

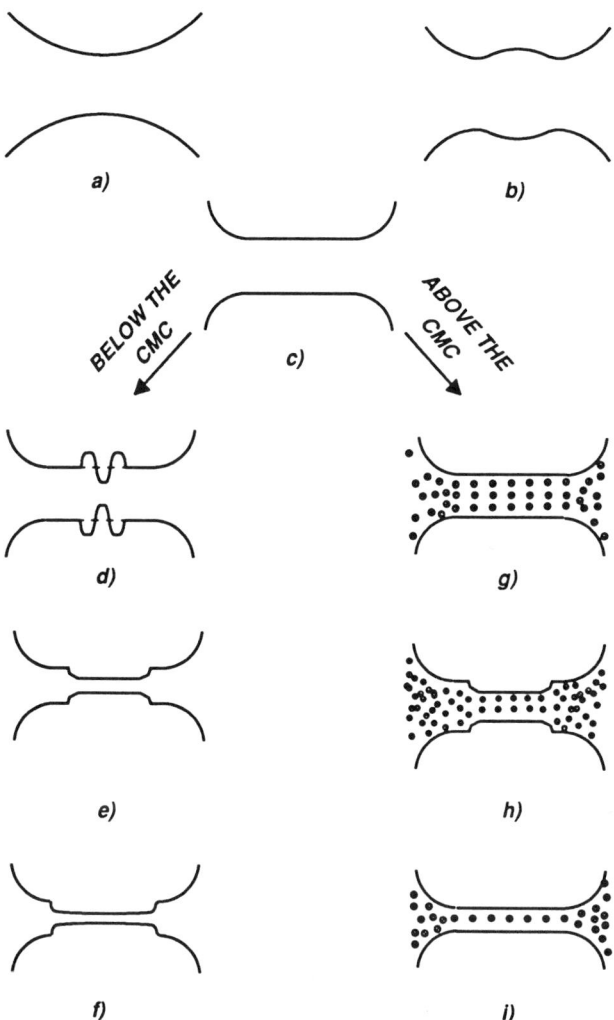

Figure 4. Main stages in the evolution of a thin liquid film; a–i are described in the text.

f. The formation of an equilibrium film whose lifetime can be virtually unlimited and is dependent upon the magnitude of the capillary pressure.

g. At high surfactant concentrations (i.e., several times higher than CMC), when the structural component of the disjoining pressure, Π_{st}, is positive, a long-range colloid, crystal-like structure is formed because of the internal layering of micelles inside the film (15, 16).

h. The thinning film exhibits a number of metastable states, and its thickness changes in a stepwise fashion; the stratification depends on the micellar concentration and film size.

i. The film attains an equilibrium state with no more stepwise changes, and the resulting film is generally stable and thick, and contains micelles.

The drainage characteristics and surface rheology of films and foams with low surfactant concentrations will be discussed in the following sections. Structural forces become dominant in a thinning liquid film (15, 16) when the surfactant concentrations are several times higher than the CMC. The properties of these films and foams will be discussed in subsequent sections.

Foam Stability at Low Surfactant Concentrations: Role of Surface Rheological Properties. At surfactant concentrations below or around the CMC, the adsorption of the surface-active molecules on the film surfaces and the properties of the adsorbed layers control the drainage and stability of the film and the foam.

The approach of two bubbles under the capillary pressure acting normal to the surfaces causes liquid to be squeezed out of the thinning film into the adjoining Plateau borders. This liquid flow results in the convective flux of surfactant in the sublayer (Figure 5). Therefore, the surfactant concentration at the surface is increased in the direction of the flow, and

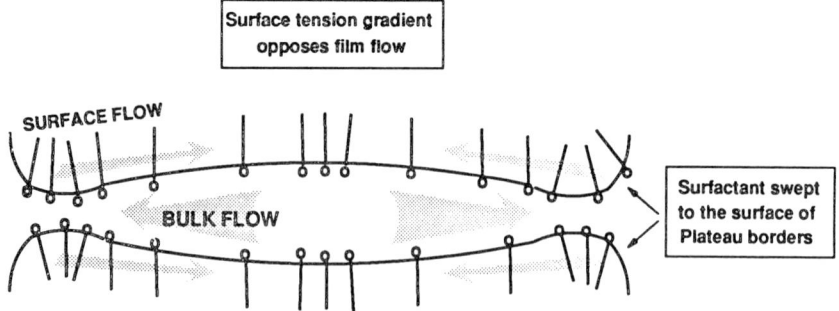

Figure 5. Gibbs–Marangoni effect in the thin-film drainage process. Surfactant is swept to the Plateau borders by flow in the film and droplet phases, and thereby create surface concentration gradients that engender surface tension gradients.

the result is decreased surface tension. This surface tension gradient along the surfaces produces a force opposite to liquid flow. This phenomenon is known as the Gibbs–Marangoni effect.

Reynolds (17) was the first to formulate the rate of film drainage (U_{RE}) between two flat and rigid surfaces:

$$U_{RE} = -\frac{dh}{dt} = \frac{8h^3}{3\mu R_f^2}\Delta P_c \tag{1}$$

where h is half of the film thickness, t is time, μ is viscosity of the film liquid, R_f is film radius, and ΔP_c is capillary pressure causing drainage.

The time required for the film to drain from an initial thickness, h_i, to final thickness, h_f, is given by:

$$t = \frac{3\mu R_f^2}{16\Delta P_c}\left[\frac{1}{h_f^2} - \frac{1}{h_i^2}\right] \tag{2}$$

As pointed out by Ivanov and Dimitrov (18), Zapryanov et al. (19) and Wasan and Malhotra (20), Reynolds' equation 1 represents a most conservative prediction; it underestimates the velocity of thinning and hence overestimates the film drainage time. Both theoretical and experimental research have shown that drainage between two foam film surfaces is generally much more rapid partially because of a fluidic mobility within the boundary surfaces of the film.

The surface tension gradient in the thinning film, which is created by the efflux of liquid from the film and the sweeping of surfactant along the film surfaces to the Plateau borders (Figure 5), can be characterized by the dimensionless elasticity number, E_s, which is defined for one surface-active component (21) by

$$E_s = E_o\frac{R_f}{\mu D} \tag{3}$$

where μ is bulk viscosity of the liquid, D is diffusivity of the surfactant, and E_o is the Gibbs elasticity:

$$E_o = -\left(\frac{d\sigma}{d\ln\rho^s}\right)_o \tag{4}$$

where ρ_s is surface density of the surfactant, and σ is surface tension.

The Boussinesq number (Bo) measures the ratio of interfacial and bulk viscous effects in the draining film:

$$\mathrm{Bo} = \frac{\mu^s + \kappa^s}{\mu R_f} \tag{5}$$

where μ^s is surface-shear viscosity, and κ^s is surface dilatational viscosity.

Figure 6 shows the interfacial mobility, defined as the velocity ratio of the rate of film drainage to the rate of film drainage between two flat and rigid surfaces, U/U_{RE}, as a function of film thickness and E_s. In the presence of surfactant, the surface tension gradient, characterized by the elasticity number, opposes film drainage, and the adsorbed layer can create immobile and more stable film surfaces if E_s is high enough. At high values of tension gradient; that is, high E_s, bulk and surface diffusion and surfactant adsorption cannot counterbalance the surface tension gradient (the Gibbs–Marangoni effect) and, hence, the velocity of thinning (or the drainage time) is essentially given by Reynolds' equation. However, for small values of E_s, the thinning or approach velocity is several times greater than Reynolds' velocity.

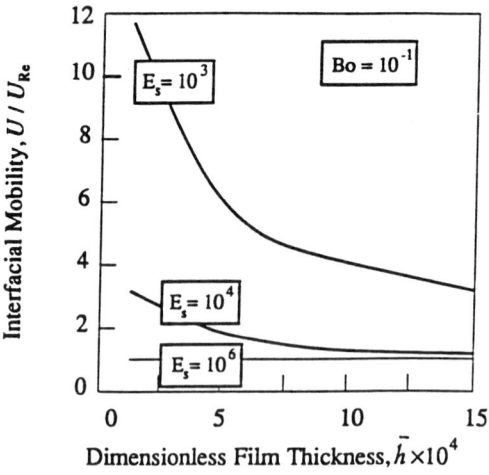

Figure 6. Interfacial mobility, or dimensionless drainage velocity, versus dimensionless film thickness, at three values of the dimensionless interfacial elasticity number. (Reproduced with permission from reference 21. Copyright 1991 Butterworth–Heinemann.)

The dilatational modulus (or the dilatational elasticity) gives the surface tension variation of a liquid surface with respect to the unit fraction area change, and it is a measure of the ability of the surface to adjust its surface tension to an instantaneous stress. Figure 7 shows the dilatational modulus of α-olefin sulfonate (AOS) solutions with various chain lengths ($C_{12}AOS$, $C_{14}AOS$, and $C_{16}AOS$) as a function of the rate of surface dilatation (the rate of surface dilatation is characterized by the bubble frequency) (22). $C_{16}AOS$ possesses the highest elasticity, $C_{14}AOS$ has a much lower, and $C_{12}AOS$ has a negligible elasticity. As the dilatational modulus measures the ability of the surface to develop surface tension gradient, this property characterizes the film and foam stability. The elasticity of the film is proportional to the dilatational modulus of the solution from which the film was formed. The film drainage time as well as the dilatational modulus of these surfactant solutions are plotted in Figure 8. These properties are indeed correlated: Films from solutions with higher

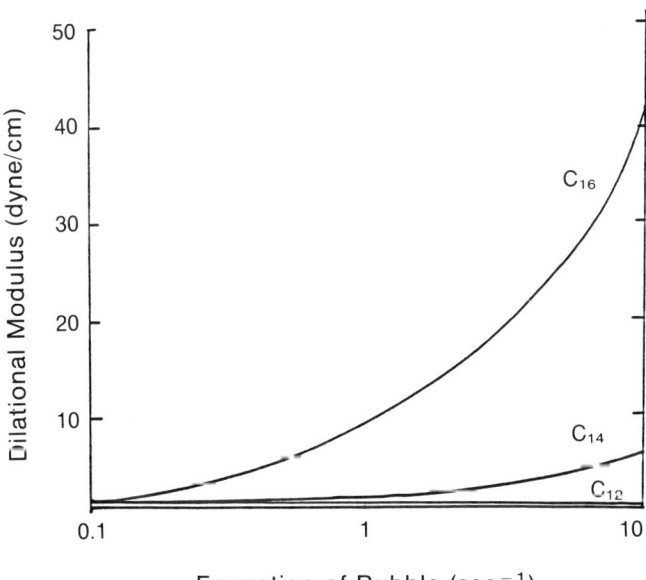

Figure 7. Dilatational modulus of $C_{12}AOS$, $C_{14}AOS$, and $C_{16}AOS$ solutions at 3.16×10^{-2} M + 1 % NaCl concentration as a function of the bubbling frequency. (Reproduced from reference 22 Copyright 1986 American Chemical Society.)

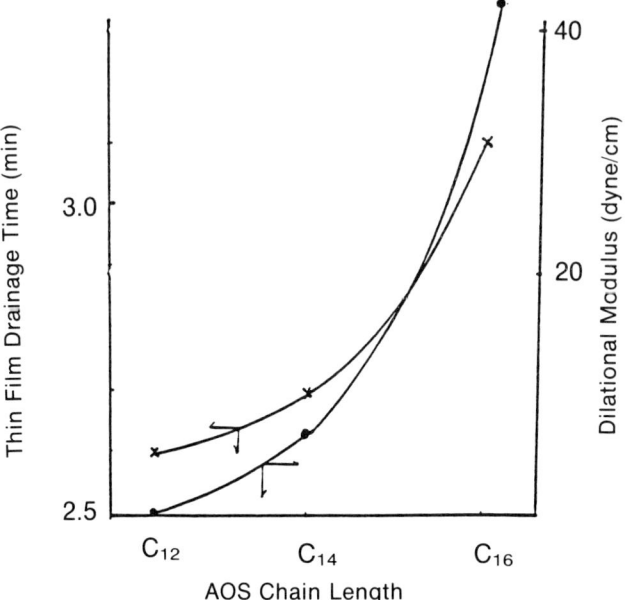

Figure 8. Thin-film drainage time (×) and dilatational modulus (●) at 10 s^{-1} bubbling frequency of α-olefin sulfonate solutions as a function of surfactant chain length. (Reproduced from reference 22. Copyright 1986 American Chemical Society.)

elasticity have higher drainage time, and both properties increase with the α-olefin sulfonate chain length (22). (Although the concentration of the α-olefin sulfonates was the same, the CMC values are different.)

During film drainage, the surfactant monolayer may undergo dilating and shearing deformations that also produce surface stresses, and the drainage also depends on these rheological properties of the film interfaces. The combined effects of surface tension gradient and interfacial viscosity on the film drainage time are depicted in Figure 9. An increase in surface viscosity, which is characterized by increasing Bo, results in decreased surface mobility and, hence, higher drainage time. If, however, the surface tension gradient; that is, E_s, is high, then the film drainage will be slow even at low values of surface viscosity (i.e., Bo < 10) (21). This behavior was the case in α-olefin sulfonate solutions, which possess low shear viscosity (22): The shear viscosities of the AOS solutions were 1.1×10^{-4} g/s for the C_{12}, 1.3×10^{-4} g/s for the C_{14}, and 1.7×10^{-4} g/s for the C_{16} at 25 °C and a concentration of 3.16×10^{-2}/M + 1 wt% NaCl. Hence, in the film drainage of these solutions the Gibbs–Marangoni effect, not the surface viscosity appears to play a major role.

Several authors have analyzed the drainage of flat thin films and have

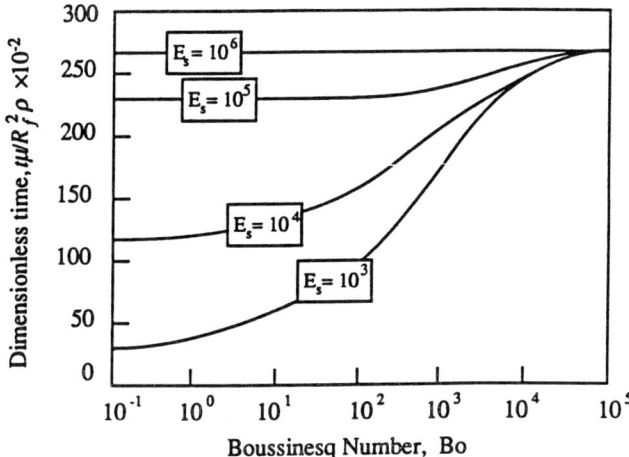

Figure 9. Dimensionless drainage time for the film to drain from a dimensionless thickness, h_i, to the thickness, h_f versus Boussinesq number, at various values of the dimensionless interfacial elasticity number. (Reproduced with permission from reference 21. Copyright 1991 Butterworth–Heinemann.)

taken into account the surface viscosity (*18*), bulk and interfacial diffusion (*23*), shear and dilatational viscosities (*19, 24*), and bulk and interfacial mass transfer (*25*). For example, Malhotra and Wasan (*26*) developed a generalized model that accounts for the effect of the mobility of the surfaces on film-thinning phenomena by considering the kinetics of adsorption–desorption of surfactants, surface and bulk diffusion, surface rheological properties, and flow in both film and bulk phases. The generalized model predicted results that were in fair agreement with the experimental data (*26*).

Jain and Ruckenstein (*27*) and Gumerman and Homsy (*28*) reported that surface rheological properties may also considerably stabilize a thin film by imparting a rigidity to liquid-film surfaces. The differences between estimated rupture times for films with mobile surfaces and immobile surfaces may also be very high.

According to Malhotra and Wasan (*29*), the previous theoretical models, which were based only on the flow in the film phase, do not take into account the flow in the Plateau borders. When the flow in the Plateau borders was taken into account (*29*), the model results were in much better agreement with the observed dependence of drainage times on the film radius.

The theoretical findings of thin-film drainage models clearly suggest the important role that surface viscosities and elasticities play in foam stability. Indeed, correlations between the surface-shear viscosity, surface dilatational elasticity, and foam stability have been reported by several investigators (30–32).

The surface dilatational elasticity has frequently been either directly or indirectly implicated as the most important dynamic surface property for creating stable foam film surfaces (33). Figure 10 shows that the foam stability, defined as foam lifetime, and foaminess, defined as the initial foam height obtained in a foam test of α-olefin sulfonates, is a function of the dilatational elasticity, as measured by Huang et al. (22). The increase in the dilatational elasticity, which increases with the chain length of the surfactant (Figure 7), results in an increase in both foaminess and foam stability. Thus, the interfacial rheological properties show similar correlations with the foam stability as with the film drainage time (Figure 8).

Foam Rheology and Surface Rheological Properties. Surface rheological properties such as surface viscosity and surface tension gradients possess a direct significance to the macroscopic rheology of foams. This significance may be attributed both to the presence of surfactants adsorbed to the surfaces within foams and their large specific surface

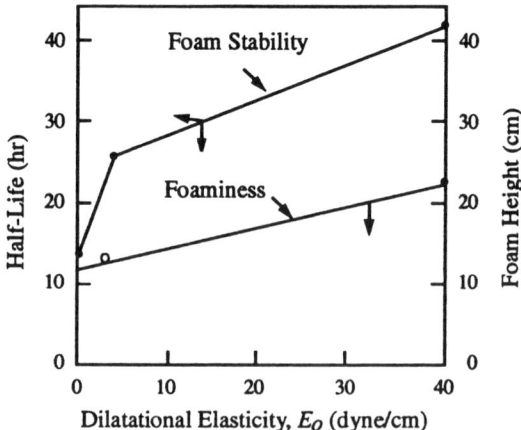

Figure 10. Foam half-life measurements of foam stability and foam height measurements of foaminess versus dilatational elasticity for aqueous foams stabilized by α-olefin sulfonates. (Reproduced from reference 22. Copyright 1986 American Chemical Society.)

area. Most theoretical studies of foam rheology avoided an explicit consideration of surface rheology (34, 35). Edwards et al. (36) were the first to consider the relationship between the macroscopic foam rheological behavior and surface dilatational viscosity and surface tension gradient as well as thin foam film parameters such as disjoining pressure. Their analysis considers the rheological response of foam material to shear and dilatational foam deformations by employing various two- and three-dimensional, spatially periodic foam models (21, 36, 37). In a pure dilation of a spatially periodic foam, bubbles expand or contract uniformly with adjacent surfaces of neighboring bubbles approaching to or receding from each other in the course of a global motion. The observance of a bulk foam viscosity is due to microscale shearing flow of viscous fluid through thin liquid films and Plateau borders, as well as to surface viscous effects. Practically, the dilatational deformation of foam material arises not only because of the compressible nature of the dispersed gas phase, but also on account of the geometrical deformability of bubble surfaces and foam films within the global foam structure.

The rheological behavior of the foam depends on the liquid–gas ratio (38). For monodisperse, spatially periodic foams possessing a finite foam film contact angle, and relatively low disperse-phase volume fraction (just beyond that of touching spheres), the dilatational viscosity of the foam depends primarily upon interfacial stresses owing to the large surface-to-volume ratio of the foam and is localized within the Plateau border zones of the local foam structure (36). Interfacial viscosities were most important for these wet foams. The leading-order dilatational viscosity, K, of a wet foam (composed of periodic rhomboidal dodecahedron unit cells) was calculated as a function of surface dilatational viscosity, κ^s, surface (σ) and film tension (σ_f), contact angle, θ, and film thickness, (h), by Edwards et al. (36) as follows:

$$K = \frac{4\kappa^s}{3h} \left\{ \frac{\dfrac{\sigma_f}{2\sigma} - \cos\theta}{1 + \dfrac{\cos\theta \tan^2\theta}{\dfrac{3}{2\pi} - \tan^2\theta}} \right\} + \cdots \quad (6)$$

where

$$\cos\theta = \frac{\sigma_f}{\sigma} - \frac{h\Pi(h)}{2\sigma} \quad (7)$$

and $\Pi(h)$ is the disjoining pressure.

Equation 6 clearly shows that the foam-dilatational viscosity is directly proportional to surface viscosity and inversely to foam film thickness. The terms in the parenthesis (in eq 6) represent the disjoining pressure contribution and the Plateau border curvature radius.

Unlike the wet foam calculation where primary viscous stresses are localized within Plateau border regions and derived from interfacial viscous properties, the total viscous stress for a dry foam (i.e., dispersed-phase volume fraction approaching 1) is distributed throughout the thin liquid films. The leading-order dilatational viscosity of a dry foam composed of a spatially periodic array of tetrakaidecahedron (*39*) bubbles was given by Edwards and Wasan (*40*) as follows:

$$K = \mu \frac{4(4 + \sqrt{2})}{3\epsilon\sqrt{2}}(1 - N_\sigma + N_\sigma^2) + 0(1) \tag{8}$$

where ϵ is the relative foam film size

$$\epsilon = \frac{h}{f} \quad \text{and} \quad N_\sigma = \frac{1 + 1/2 E_o Pe_s}{1 + 1/3 E_o Pe_s} \tag{9}$$

f is the characteristic foam cell dimensions, E_o is the Gibbs elasticity, Pe_s is the surface Peclet number, and N_σ is the capillary number (*21*).

The foam-dilatational viscosity, K, arises because of two primary mechanisms (*37*): (1) viscous flow within the thin films, and (2) interfacial tension gradients acting along the foam bubble surfaces. The effect of interfacial tension gradients is to increase the foam viscosity as they impede flow near the surfaces of the thin foam films by contributing to a larger film stress. As in the wet foam (eq 6), the foam dilatational viscosity for a dry foam, K, is inversely proportional to film thickness as well (eq 9).

Figure 11 displays the dependence of the foam dilatational viscosity upon the rate of foam expansion for $N_\sigma = 1$ (*21*). The foam viscosity, K, increases for foam expansion, as the foam films thin and velocity gradients occurring over the thickness of the film increase in magnitude, whereas, for opposite reasons, the foam viscosity, K, decreases as the foam is compressed.

Flow resistance of foams is much higher than that of the liquid or gas phases alone. This characteristic is utilized for foam-based mobility control in enhanced oil recovery (*41–46*). The gas mobility in porous medium and the dilatational modulus of α-olefin sulfonates are plotted in Figure 12 (*22*). The displacement efficiency of the foam in the porous medium is higher if the blocking efficiency of the foam is higher. In the systems

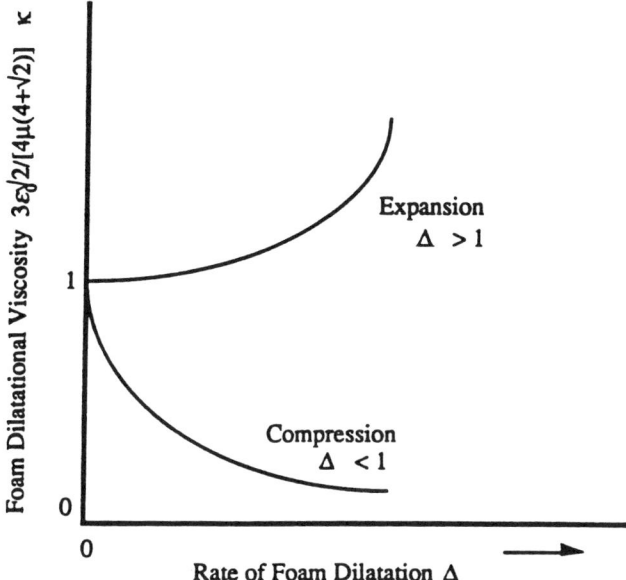

Figure 11. Dependence of the foam dilatational viscosity upon rate of dilatation for $N_\sigma = 1$. (Reproduced from reference 22. Copyright 1986 American Chemical Society.)

investigated, a direct correlation is seen between the elasticity and the gas blocking performance of foam (i.e., higher dilatational modulus results in lower gas mobility.) The relationship between the two parameters is nonlinear because of the complex nature of the flow in porous medium. Modeling of flow in porous medium will be discussed in more detail in Chapter 3.

Foam Film Stability at High Surfactant Concentrations: Micellar Stratification Phenomenon. As discussed in the preceding sections, the presence of adsorbed surfactant layers stabilizes the foam films by controlling the film surface mobility and, thereby, the flow through the lamellae and their Plateau borders. At a concentration several times higher than the CMC, the effect of surface rheological properties is less pronounced and plays a secondary role in the film stability. However, foam stability increases with the surfactant concentration even far beyond CMC (7). The reason for this increase is another stabilizing mechanism: ordered microstructure formation (stratification) in the draining foam films. In industrial systems, the concentration of surfactant solutions is mostly higher than the CMC, thus this second mechanism is very important from a practical point of view. The phenomenon of microstruc-

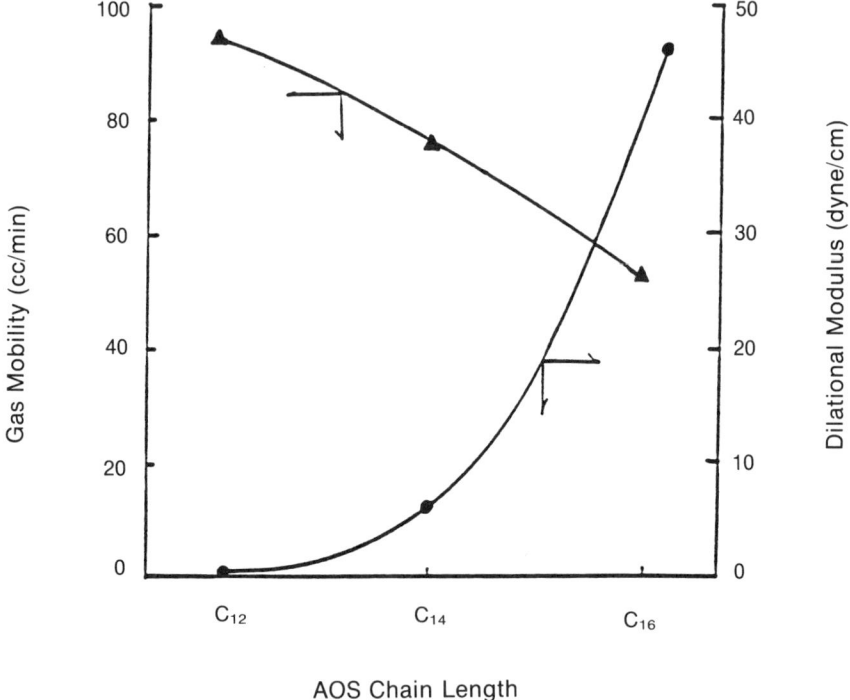

Figure 12. Correlation of gas mobility (▲) with dilatational modulus (●) with varying α-olefin sulfonate (AOS) chain length. (Reproduced with permission from reference 21. Copyright 1991 Butterworth–Heinemann.)

turing in thin liquid films and foams, its importance, and the effect of several parameters such as surfactant concentration, film size, electrolyte concentration, temperature, and the presence of solubilized oil are discussed in the sections that follow. At very high surfactant concentrations (greater than 5–10%, for instance), ordering of micelles not only inside liquid films but also inside the bulk phase takes place (47). However, in the petroleum industry applications, surfactant concentrations are generally well below this concentration range, and we will not discuss these systems here.

At high surfactant concentrations (i.e., much above the CMC), thin foam films were observed (48–53) to become thinner in a stepwise fashion; that is, thinning foam films formed from micellar surfactant solutions exhibit a number of metastable states before attaining an equilibrium film thickness. This process can be followed in Figure 13, which shows a photocurrent (film thickness)–time interferogram of a horizontal flat film

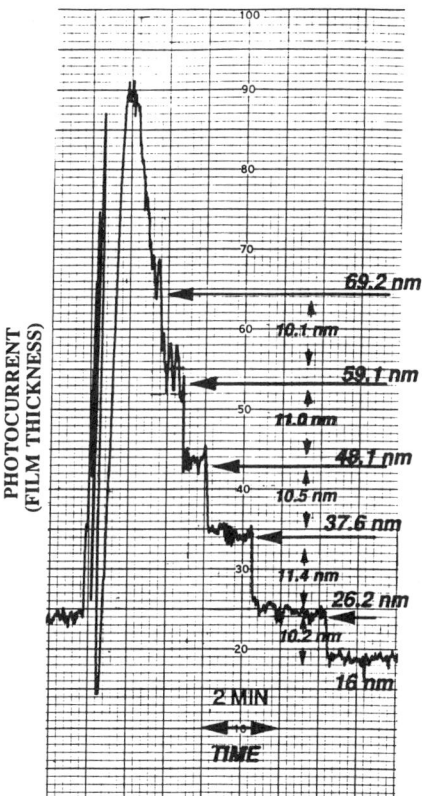

Figure 13. Interferogram of film formed from solution of nonionic detergent (Enordet AE1215-30, 0.052 mol/L). As the film thins, less light is reflected. Formation of metastable states of uniform thickness is revealed by "steps". The height of the step corresponds to the thickness of film. The width of the steps is proportional to the lifetimes of respective metastable states. The vertical distance between steps corresponds to micelle diameter, about 10 nm. (Reproduced with permission from reference 54. Copyright 1990 Steinkopff Verlag Darmstadt.)

formed from the micellar solution of a nonionic surfactant (ethoxylated alcohol) (54). As soon as the film is formed, it starts to thin. After it is thinner than 104 nm; that is, 1040 Å (the last interferential maximum corresponding to the applied monochromatic 546-nm light reflected from the film), the film thickness changes in steps (i.e., stratification). The film rests for a few seconds in a metastable, uniformly thick state. Then, dark spots with smaller thickness than the remaining part of the film appear and gradually increase in size (Figure 14). The spots cover the entire film

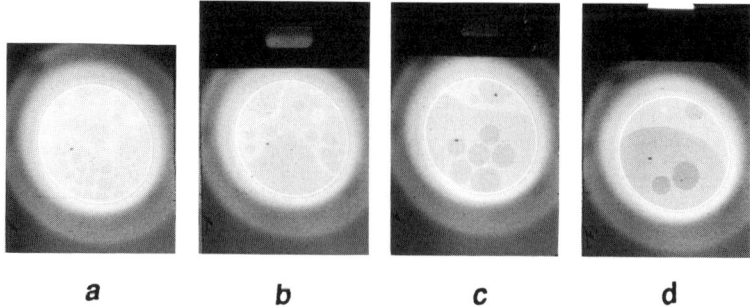

Figure 14. Stratification of horizontal foam film from 0.1 mol/L sodium dodecyl sulfate surfactant solution.

and the film "rests" for a time in a new metastable state. Then, even darker spots appear, and after their expansion, a subsequent metastable state ensues. This process continues until the film finally reaches a stable state with no more stepwise changes. The metastable state of the film appears in the interferogram as a step-wise width in proportion to the lifetime of the respective state. The height of the steps is nearly equal, as shown in Figure 13 (about 10.6 nm), and it corresponds to the micelle diameter, about 10 nm.

The same phenomenon was observed for ionic surfactant solutions (*16, 55*), such as α-olefin sulfonates (*1*), but here the height of the steps was close to the effective micellar diameter, which includes the electric double layer around the ionic micelles. Furthermore, foam films formed from concentrated monodisperse suspensions of polystyrene latexes (*16*) or silica particles (*56*) stratify in a similar way.

The phenomenon of stepwise thinning was observed not only in horizontal, but in large, vertical films and foams as well (*56*). Figure 15 shows foam films formed from a latex suspension containing 150-nm particles in a vertical frame (2.5-cm diameter). A series of stripes of different, uniform colors can be observed at the upper part of the frame. The different colors are due to interference of the common (polychromatic) light reflected by the surface of the different, uniform thickness stripes. The boundaries between the stripes are very sharp, a consequence of the stepwise profile of the film in this region, and the liquid meniscus below the film appears as a region with gradually changing colors. When observed in reflected light, the top stripes have the following, sharply distinguished colors: black, white, yellow, blue, red, and green–yellow. Following the stripes, a sequence of diffuse, alternate green and red stripes indicates the gradual change in film thickness where the order–disorder transition region is observed.

Similar, sharply defined stripes were found with vertical films from

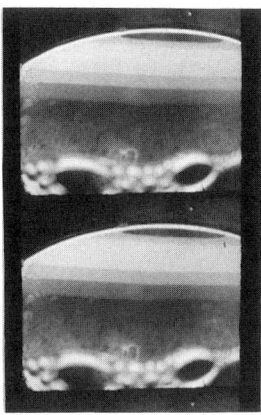

Figure 15. Color interference bands in vertical foam film formed from a latex suspension with particle diameters of 150 nm.

micellar solutions of nonionic surfactant (e.g., ethoxylated alcohol with 30 ethoxy groups and 12–15 carbon chains) with a micellar diameter of about 10 nm (56). However, all stripes were very grey though with different intensities because the diameter of the micelles is small compared to the wavelength of the visible light.

Nikolov et al. (57) were the first to explain the stratification phenomenon as a layer-by-layer thinning of ordered structures of micelles or colloidal particles inside the film. According to the colloid crystal-like model first proposed by them, the different stripes in the stratifying (horizontal or vertical) films contain different numbers of micelle (or particle) layers. The micelles interact via screened electrostatic repulsion forming an ordered structure because of the restricted volume of the film. The model permitted, for the first time, calculation of the structural contribution to the disjoining pressure of the film that arises from the presence of micellar structure within the films. From the disjoining pressure isotherms, the film's excess free energy was calculated as shown in Figure 16. The curves exhibit minima that correspond to the metastable state ($n = 1$, 2, 3, ...) and to the final stable state ($n = 0$) of the film. A stepwise film thickness transition can be interpreted as a transition from a given metastable state to the next one. The experimental values of the film thickness were in good agreement with the ones calculated from the theoretical model.

All the experimental and theoretical data for stratifying films show that stratification is a universal phenomenon and is due to the formation of a long-range colloid crystal-like structure within the foam film and a layer-by-layer thinning of such an ordered structure. This ordering occurs because highly charged Brownian particles interact via repulsive forces and

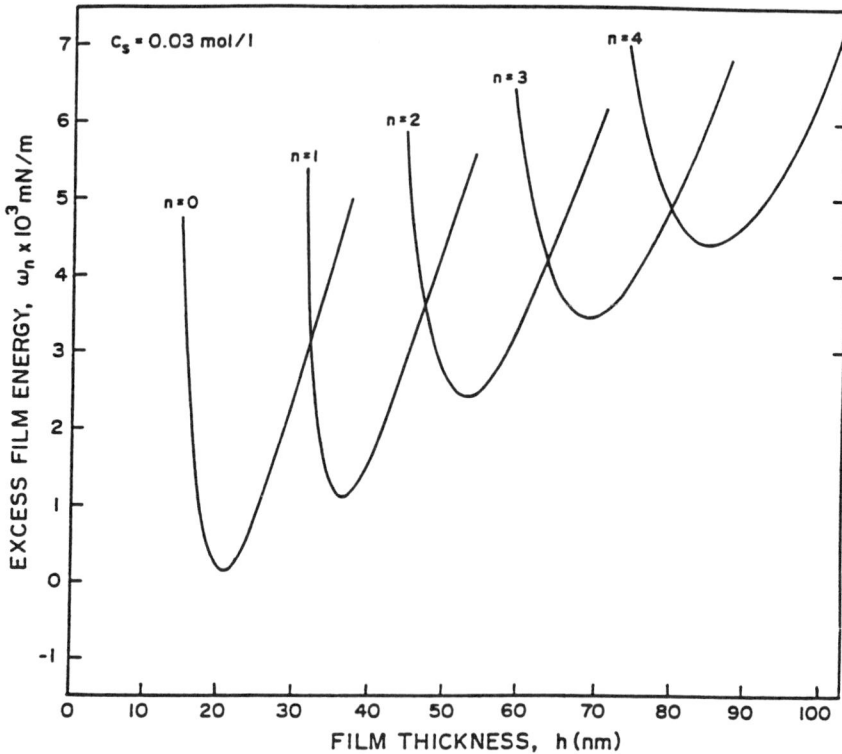

Figure 16. Calculated isotherms of the excess energy for unit area of the film, $\omega_n(h)$, at a surfactant concentration of 0.03 mol/L and at different micellar layers inside the film. (Reproduced with permission from reference 57. Copyright 1989 Academic.)

are forced into the restricted volume of the film. The classical DLVO theory of colloid stability (58, 59) (an acronym for a theory of the stability of colloidal dispersions developed independently by B. Derjaguin and L. D. Landau on one hand, and by E. J. W. Verwey and J. Th. G. Overbeek on the other), which explained order in such systems as a balance of van der Waals attraction forces and electrostatic forces, cannot be used here because the interparticle distances are too large for the van der Waals forces to be significant to balance the repulsive forces.

The formation of long-range, ordered structures inside thin foam films and the fact that this mechanism inhibits the film drainage have many implications of both fundamental and practical significance. The dynamic process of stratification or multilayer microstructuring in submicrometer thin liquid foam films can serve as an important tool for probing the long-range structural or interaction forces in colloidal dispersions.

For example, the rheology of such dispersions containing stratifying films will be quite different. The bulk viscosity of the stratifying foam film was much higher than that of the pure solvent (water) (60).

The formation of ordered microstructure in thin liquid films offers a new mechanism for the stabilization of foams. As a proof of microstructuring in real foams, Figure 17 shows a photograph of an aqueous foam system stabilized because of the stratification in the foam bubble lamellae. The practical importance of the film microstructuring is that the lifetimes of foams with stratifying films are much longer.

Effects of Micellar Concentration and Size. The effect of important technological parameters on the film stratification phenomenon is discussed next.

Figure 18 shows a plot of the effective volume fraction of the micelles as a function of the stepwise thickness transitions for anionic micellar solutions of sodium dodecyl sulfate (with a mean micellar diameter of 4.8 nm) and nonionic micellar solutions of ethoxylated alcohol with 30 ethoxy groups and 12–15 carbon atoms (with a mean micellar diameter of 10 nm) (61). These curves show the effect of both the effective concentration and the size of micelles on film thickness transitions. (The curves of the different systems can be compared because the mechanism of film mi-

Figure 17. Aqueous foam stabilized due to the stratification in the foam bubble lamellae (0.1 mol/L sodium dodecyl sulfate solution).

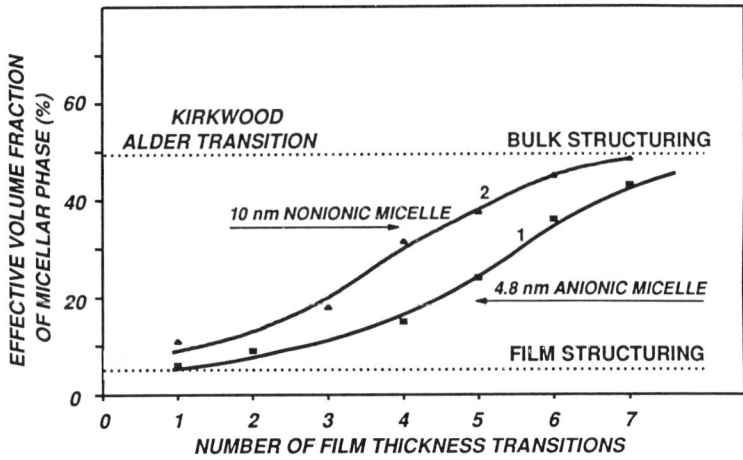

Figure 18. Phase diagram of the film disorder–order structure transition. Key: curve 1, in the presence of anionic micelles of sodium dodecyl sulfate; and curve 2, in the presence of nonionic micelles of Enordet AE1215-30 (ethoxylated alcohol with 30 ethoxy groups).

crostructuring is basically the same.) The smaller-sized micelles induce more thickness transitions than the larger ones at the same effective volume fraction. With the smaller, anionic micelles the film structuring process (i.e., order–disorder transitions) begins at a lower effective micellar volume percent (5 vol%) than with the larger, nonionic micelles (11 vol%). At those micellar concentrations, the intermicellar distance is about 2–3 times the micelle radius. This observation shows that an order–disorder transition can occur at a film thickness of several particle diameters and at much lower concentration than 50 vol%, which was the concentration theoretically predicted by Kirkwood (62) and Alder and Wainwright (63) for colloid crystal structure transition in bulk phase. This variation leads to the conclusion that the particles may form a loose structure inside the thinning film; however, this structure contains many vacancies (dislocations).

Kralchevsky et al. (64) described the film transition from one metastable state to the next one by a vacancy mechanism. The driving force behind the thinning process is the difference between the chemical potential of the micelles inside the film and inside the adjoining meniscus (Plateau border). Because the micelles diffuse, vacancies appear in their places (initially created at the film periphery) and move throughout the entire film area. For example, micelles with 19-nm diameter can pass through the film with a diameter of about 6×10^{-2} cm in about 2×10^{-2} s. The film-thickness transitions (by spot formation) usually occur when there is a sufficiently large number of vacancies and, as a result, a spot is formed and gradually increases its area. The vacancies created at the film periphery diffuse and "condense" at the spot and thus increase its area. By increasing the concentration of micelles, the number of vacancies decreases, and thus the thickness transition at the same number of layers will be slower, as was also observed experimentally (*61*). Moreover, at high concentration of micelles, film structuring occurs at high film thickness and, depending on the film radius, the film can be several micelle layers thick.

Nikolov et al. (*1, 16*) investigated the effect of surfactant chain length with alkyl sulfates and α-olefin sulfonates. They found that the number of stepwise transitions increases with the chain length. For example, with C_{16}-α-olefin sulfonate ($C_{16}AOS$) two step transitions were observed, but with C_{12}-α-olefin sulfonate ($C_{12}AOS$) only a one-step transition was observed (*1*). The effect can be explained by the lower CMC of the longer surfactant molecules. At the same concentration, more micelles are present in the solution of the longer chain surfactants, and the result is easier micellar ordering.

Effect of Foam Film Area. Figure 19 shows the film-stability diagram of a horizontal flat film in the presence of 19-nm silica hydrosol representing the effect of particle or micelle concentration and film diameter (*61*). From the practical point of view, the film stability can be improved by increasing the particle concentration (or concentration of micelles) or by decreasing the film diameter. With a 6×10^{-2}-cm film diameter, for example, a stable film with a thickness of 100 nm containing three particle layers inside could be formed. By increasing the film diameter to 10^{-1} cm, three more transitions were observed with no layers in the final film. In total, six film-thickness transitions were seen, which is the same number as that found in large macroscopic film. Foam films stabilized by micelles showed similar behavior (*65*). The effect of film area can also be explained by the vacancy mechanism (*64*). An obvious way of increasing the film area in a foam system is to increase the bubble size.

Effect of Electrolytes. Adding electrolytes to the surfactant solution, the number of stepwise transitions decreases, and the process becomes irregular (some of the steps passed over), especially with ionic surfactants in which the repulsive force between the micelles (or particles) is electrostatic (*16*). Moreover, above a threshold salt concentration, no stepwise transition occurs; that is, the electrolytes prevent ordering in the thinning film. In the C_{16}-α-olefin sulfonate solution (3.16×10^{-2} mol/L), for example, two transitions were observed without added electrolyte (*1*). At one-to-one electrolyte concentrations higher than 1 wt%, the film thinned by a single-step transition. Addition of salt inhibits the ordering of ionic mi-

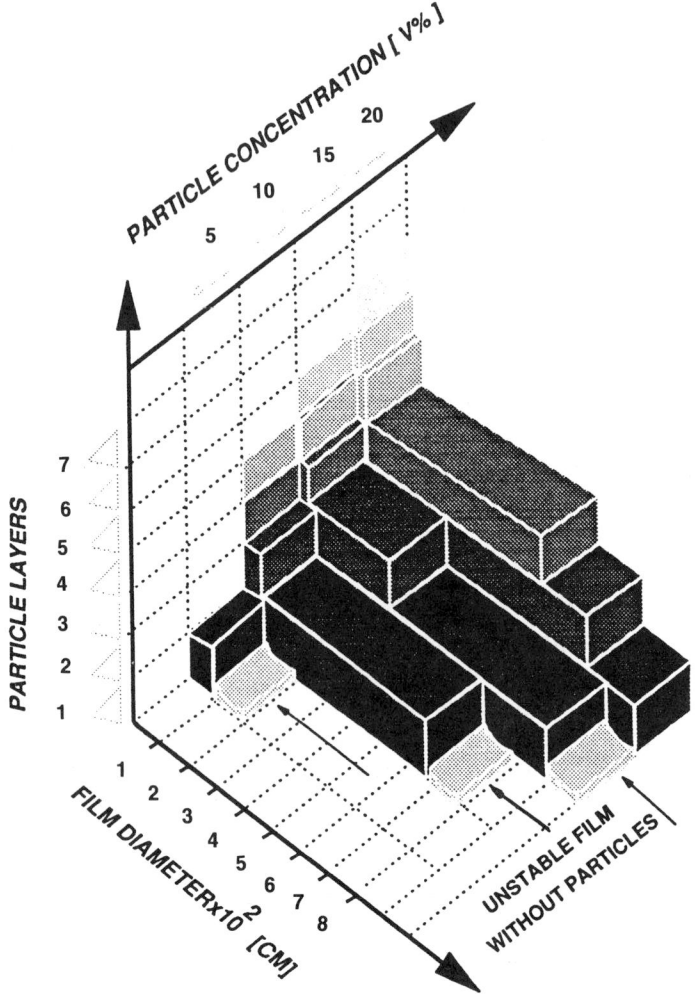

Figure 19. Three-dimensional film thickness–stability diagram of stratifying horizontal microscopic film in the presence of 19-nm silica particles; number of particle layers versus particle concentration and film diameter.

celles because the added ions screen the electric double layer around the micelles, and the result is a decreased effective volume fraction of micelles. The threshold concentration of added NaCl for order–disorder increases with the micellar volume fraction *(16)*.

Effect of Temperature. In the application of foams in the petroleum industry, the temperature is a very important parameter. The effect of temperature was studied with nonionic surfactants in which the intermicellar repulsion is the result of steric forces (54). A decrease of temperature similar to the effect of film area can prevent the occurrence of the last few stepwise transitions, and the stratification stops at a larger thickness; higher film stability results. In the Enordet AE1215-30 ethoxylated alcohol, for example, at 26 °C, two transition steps with a final thickness of 49 nm were observed; as opposed to the six transitions and 14-nm final thickness at 32 °C. At higher temperatures, the rate of stratification increased; that is, the drainage time decreased with the temperature, as shown in Figure 20. The surfactant with lower degree of ethoxylation is more sensitive to the temperature. Above 35 °C, the stepwise transitions become irregular, and at even higher temperatures, no transitions occur. Near the cloud point of the surfactant (but below it), the film ruptures without reaching a stable final thickness.

In summary, the formation of colloid crystal structure and the corresponding positive structural component of the disjoining pressure in-

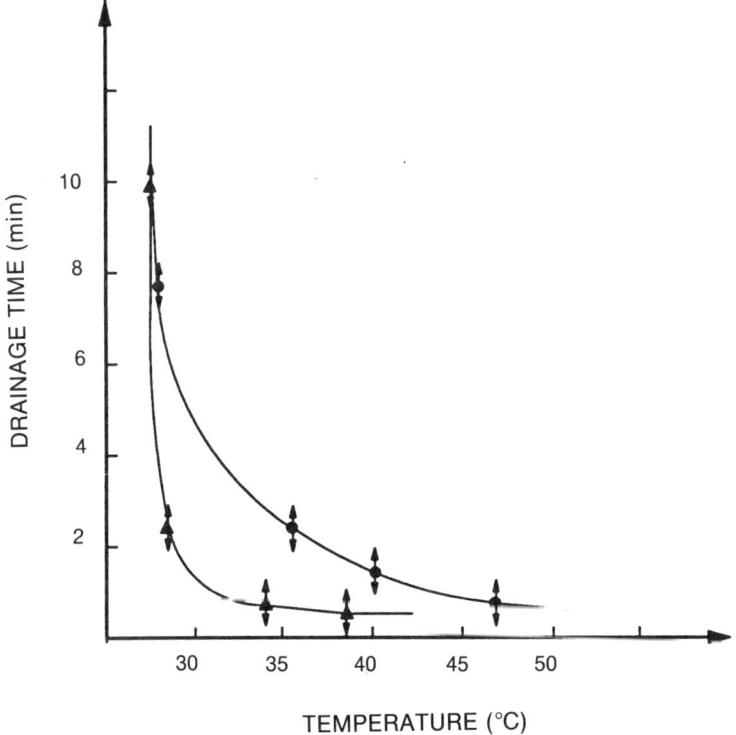

Figure 20. Effect of temperature on the drainage time of horizontal foam films from ethoxylated alcohols: 4 wt% Enordet AE1215-30 (●) and 2 wt% Enordet AE1215-9.4 (▲), with 12–15 C-atom chains and 30 and 9.4 ethoxy groups, respectively (54). Film radius was 0.3 mm.

creases film and foam stability. The film, and thus foam, stability can be improved by increasing the micellar concentration, decreasing the individual film area (for example by decreasing the bubble size), decreasing the electrolyte concentration, or lowering the temperature.

Effect of Solubilized Oil. Figure 21 shows the interferogram of a foam film formed from the nonionic Enordet AE1215-30 (ethoxylated

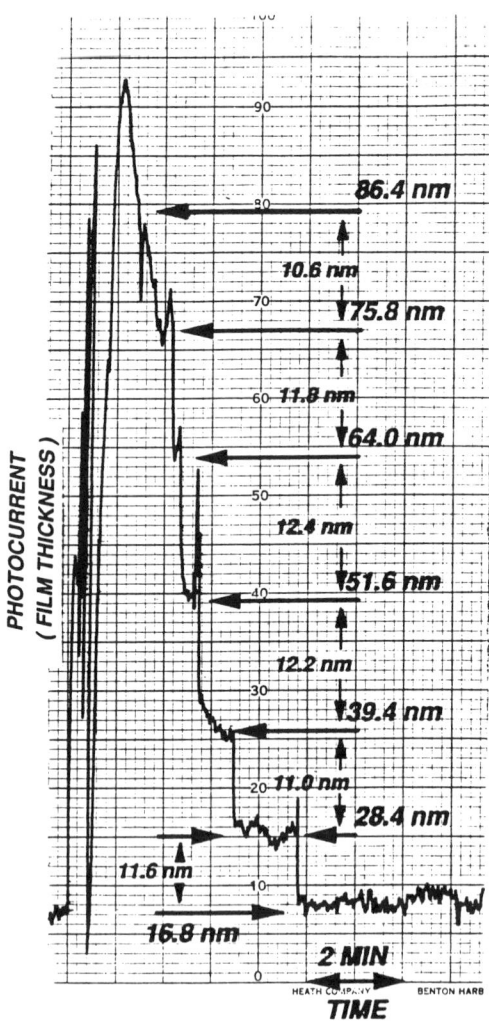

Figure 21. Interferogram of thinning of the Enordet AE1215-30 film at surfactant concentration 5.2×10^{-2} mol/L. The solution is preequilibrated with n-decane. (Reproduced with permission from reference 54. Copyright 1990 Steinkopff Verlag Darmstadt.)

alcohol) surfactant solution in the presence of solubilized n-decane (*54*). Comparing Figures 13 and 21, the number of stepwise transitions increases from five to six, but the film drainage time decreases from 6 to 4.3 min in the presence of solubilized oil. These phenomena can be explained by the effect of oil solubilization on the size and polydispersity of micelles and on the intermicellar interactions (*66, 67*). Nilsson (*68*) and Lobo et al. (*66*) found that the solubilization of decane leads to an increase in the micellar aggregation number and a reduction in the second virial coefficient (increased attraction between the micelles) of nonionic surfactants. Oil solubilization also increases the polydispersity of micelles (*54*). As already discussed, the number of film-thickness transitions is higher at higher micellar volume fractions. Therefore, oil solubilization is expected to result in an increased number of step transitions. At the same time, however, higher micellar attraction and polydispersity lead to the formation of a loose, less-packed colloid crystal-like structure inside the thinning film that results in faster stepwise transitions, less stability of the colloid structure, and lower foam stability in the presence of solubilized oil.

Foam–Oil Interactions

Aqueous foams containing emulsified oils have a wide variety of practical applications. The effect of oil on foam stability is of primary importance because the oil can stabilize or destabilize the aqueous foam. Emulsification of some crude oils can lead to a decrease in foam stability; thus the application of foams for mobility control in enhanced oil recovery (EOR) is strongly affected by the foam–crude oil interactions in the porous medium.

Oil Configurations in Foams. In the presence of oil, the mechanisms of foam stability are more complex than without oil. Solubilized oil decreases the stability by accelerating the stepwise foam film thinning, as shown in the previous section. The effect of emulsified oil on foams is closely connected with the configuration of oil relative to the aqueous and gas phases. This configuration can be, in most cases, one of the following (Figure 22):

a. Oil drop without interaction with the solution surface is generally the initial oil configuration inside the foam.

b. If the oil drop interacts with the solution surface, the drop gets deformed and separated by a pseudoemulsion film from the gas phase.

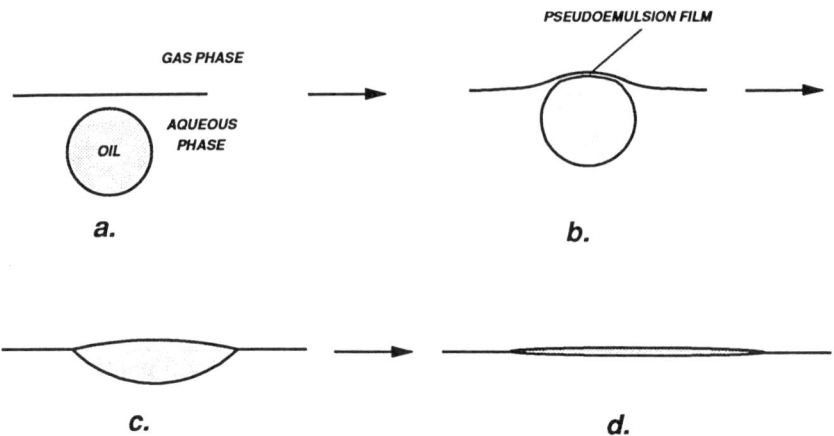

Figure 22. Configurations of oil at the gas–aqueous interface. Key: a, oil drop inside the solution; b, oil drop at the surface separated by a pseudoemulsion film from the gas phase; c, oil lens; and d, spread oil layer at the solution surface.

c. If the pseudoemulsion film ruptures, the oil drop enters the surface, and an oil lens is formed at the solution surface.

d. A spread oil layer or film can form from the lens on the solution surface.

Nikolov et al. (1) observed the configuration of crude oil drops at the surface of α-olefin sulfonate solutions at 1 wt% electrolyte and found different oil configurations depending on the surfactant chain length. Figure 23 shows movie pictures on the configuration of crude oil at the surface of (a) $C_{12}AOS$, (b) $C_{14}AOS$, and (c) $C_{16}AOS$ solutions. The pictures were taken by differential interferometry (DI), in which the original, optical microscopic image is split into two images that interfere and produce characteristic interference patterns. Three different oil configurations could be observed. The first configuration is the thick pseudoemulsion film, which produces the double white spots as seen in Figure 23b and 23c. This configuration was very common for $C_{16}AOS$, less frequent for $C_{14}AOS$, and least frequent for $C_{12}AOS$. The second oil configuration was a thin pseudoemulsion film, and the third is the oil lens. The thin pseudoemulsion film and the oil lens produce similar interference patterns with parallel streaks in the middle. The lens configurations (the large interference patterns in Figure 23a and 23b) differ from the thin pseudoemulsion film in Figure 23c in the distance of the parallel streaks. In the lenses, the streaks are denser because of the higher curvature of the lens than in the picture of the pseudoemulsion film.

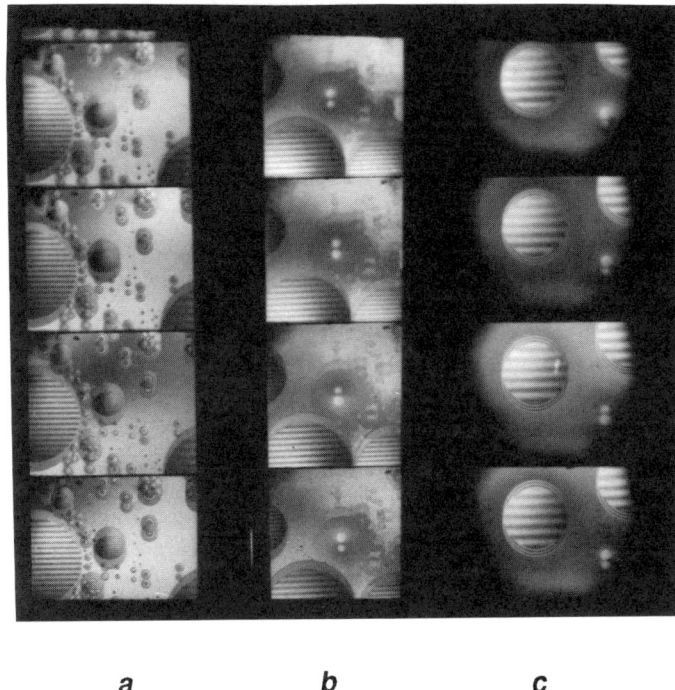

| a | b | c |

Figure 23. Differential interferometric (DI) photographs of Salem crude oil droplets at foam surface. Key: a, $C_{12}AOS$; b, $C_{14}AOS$; and c, $C_{16}AOS$ (1).

The observation that the oil preferred the pseudoemulsion film configuration in the $C_{16}AOS$ solution indicated that this film is stable in this system, which prevented the entering of oil (i.e., lens formation) (Figure 22b and 22c). However, in the $C_{12}AOS$ solution in which the lens configuration was preferred, the pseudoemulsion film was unstable.

For the oil to break the foam, the drops have to enter the aqueous surface and spread on the aqueous surface (i.e., to change from configuration a to c or d in Figure 22). Ross and McBain (69) postulated that the oil breaks the foam film by spreading on both sides of the foam film, thereby driving out the original film liquid and leaving an oil film that is unstable and breaks easily. Prins (70) developed a model of foam rupture by oil spreading. As a criterion for this process, Ross (71) used the spreading coefficient, S, defined by Harkins (72) as the change of surface free energy from configuration a to d (Figure 22):

$$S = \sigma_{W/A} - \sigma_{W/O} - \sigma_{O/A} \tag{10}$$

where $\sigma_{W/A}$ and $\sigma_{O/A}$ are the surface tensions of the aqueous and the oil phases, respectively, and $\sigma_{W/O}$ is the oil–water interfacial tension.

Robinson and Woods (73) used a similar, thermodynamic approach, but they found better correlation between the defoaming action and the entering coefficient (E); that is, the change from configuration a to c (Figure 22) defined as

$$E = \sigma_{W/A} + \sigma_{W/O} - \sigma_{O/A} \tag{11}$$

These theories were tested by several authors, and many exceptions were found (69, 73–75). Thus, in foam-flooding systems for enhanced oil recovery, Schramm and Novosad (76) and Manlowe and Radke (77) did not find a correlation between E, S, and the effect of oil on foam stability. Hence, the classical spreading–entering theory is not consistent with many experimental findings.

Nikolov et al. (1) were the first to identify the importance of the pseudoemulsion film; that is, configuration b in Figure 22, in the foam–oil interactions. To observe the behavior of emulsified oil (macroemulsion) within the foam system, the foam containing the oil was placed between two glass sheets and observed through the microscope. In Figure 24, the process of coalescence between a small air bubble (white circular object entering the movie frame from the upper middle portion of the frame) and an oil lens on the large bubble surface (the thick, dark edge of the large white object in the upper portion of the movie frame is the oil lens on the large bubble surface) in the foam system containing $C_{12}AOS$ solution is illustrated. After a certain thickness, the pseudoemulsion film formed between the oil lens and the air bubble surface ruptures, and the oil phase spreads on the surface forming an oil "bridge" between small and large bubble surfaces. After this the oil bridge ruptured, and the bubbles coalesced. As seen, the crude oil had mostly lens configuration on the surface of $C_{12}AOS$ solution, because the pseudoemulsion film was unstable for this system (1, 78).

In contrast, Figure 25 shows frames with the $C_{16}AOS$ solution. The oil drops drain from the films into the Plateau borders without entering or spreading, and the foam does not break. This observation was also in accordance with the observation that the typical oil configuration on the surface of $C_{16}AOS$ solution (Figure 23) was stable (thick or thin) pseudoemulsion film. These experiments clearly showed that the foam stability in the presence of emulsified oil is controlled by the stability of the pseudoemulsion film.

Krugljakov (79) suggested that the asymmetrical aqueous film between an antifoam drop and the air phase (i.e., the pseudoemulsion film) should be unstable enough for the drop to break the foam. He

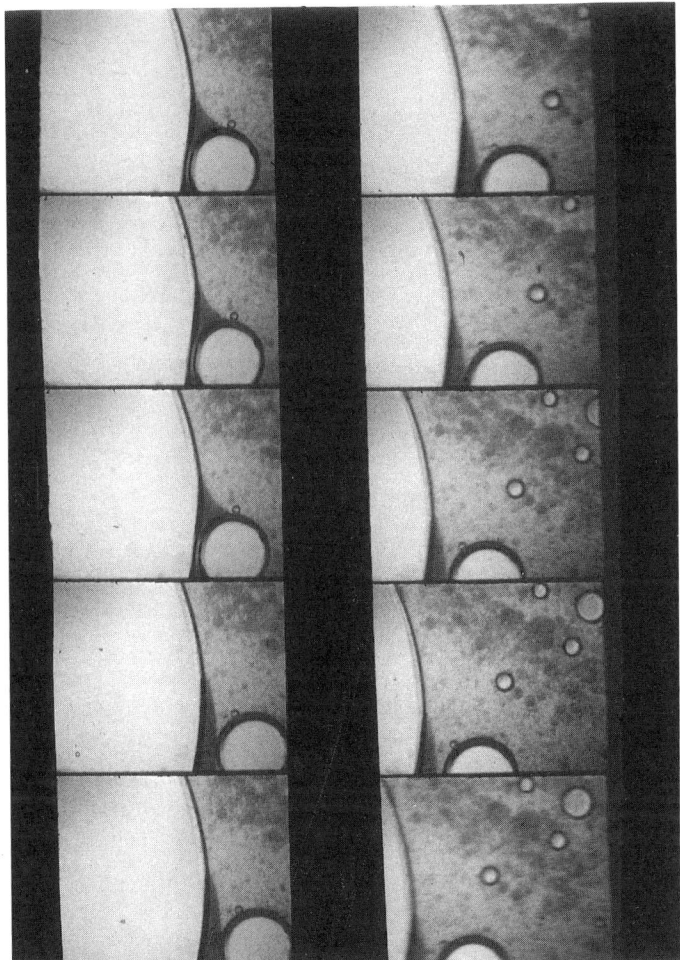

Figure 24. Spreading of Salem crude oil between two bubble surfaces for C_{12} α-olefin sulfonate with 1 wt% NaCl (1).

pointed out that the stability of this film depends on the surfactant concentration. At low surfactant concentrations, the pseudoemulsion film can be unstable when there is insufficient surfactant adsorption at the oil–water and air–water interfaces. At higher concentrations, however, this film can become stable.

Schramm and Novosad (76), Manlowe and Radke (77), and Hanssen and Dalland (80) also concluded that the pseudoemulsion film stability is a controlling factor in the stability of three-phase foams within porous media.

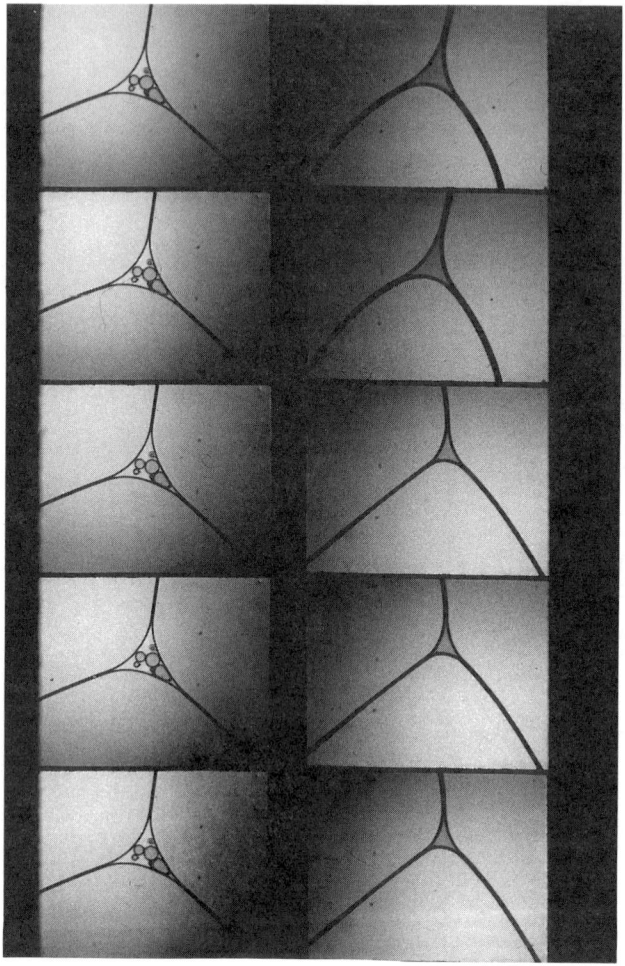

Figure 25. Two-dimensional foam drainage for C_{16} α-olefin sulfonate with 1 wt% NaCl (1).

The classical spreading–entering theories do not predict the behavior of oil drops, because they do not take into account the role of the pseudoemulsion film as an intermediate step (Figure 22b) between the oil drop (a) and oil lens (c) or spread oil layer (d) configurations. If this film is stable, it provides an energy barrier that prevents the coalescence of bubbles. If this film is unstable; that is, the barrier is small, oil drops can enter and spread on the bubble surfaces, and the oil can act as defoamer.

The theories of foam–oil interactions will also be discussed in Chapter 4.

Structure and Stability of the Pseudoemulsion Film. The structure and stability of the pseudoemulsion film depend on several parameters such as surfactant concentration, electrolyte concentration, film size, and capillary pressure (i.e., drop size).

The behavior of the pseudoemulsion film is controlled by mechanisms similar to foam films: the interfacial rheological properties of the surfactant molecules at low surfactant concentrations and micellar ordering at high surfactant concentrations (i.e., much above the CMC). Not much above the CMC, both of these mechanisms can play a role in film thinning and stability. The stability of a thin pseudoemulsion film also depends on the van der Waals interactions between the phases at the two sides of the film, that is between air and oil, acting across the aqueous film. In a water pseudoemulsion film the Hamaker constant is generally negative, the van der Waals interactions are repulsive and stabilize the film.

Figure 26 shows the dynamic interfacial tension of Enordet AE1215-30 ethoxylated alcohol, and $C_{12}AOS$ and $C_{16}AOS$ solutions against two different oil phases (n-octane and n-dodecane) (1). For all these measurements, the surfactant concentration was kept constant at 3.16×10^{-3} mol/L (which is about 1 wt% and is above the CMC). The electrolyte concentration was 1 wt%. No stratification was observed. The film stability in the absence of micellar layering and the observed film-thickness transition from common to Newton black film are controlled by the capillary pressure (drop size). The effect of interfacial tension gradient on the three-phase foam stability was studied by measuring the change of the interfacial tension with the bubble frequency. Figure 26 shows that for both $C_{12}AOS$ and $C_{16}AOS$, the change of interfacial tension is higher with n-dodecane than with n-octane.

The oil phases have the opposite effect with Enordet AE1215-30 solution. Moreover, in the $C_{12}AOS$ system, the dynamic interfacial tension changes less than in the $C_{16}AOS$ system. A correlation was found between the stability of foams from these systems and the change of the dynamic interfacial tension (1): Higher stability foams were obtained by $C_{16}AOS$ in the presence of n-dodecane and Enordet AE1215-30 in the presence of n-octane, and lower stability foams formed from $C_{12}AOS$ in the presence of n-octane and n-dodecane and from Enordet AE1215-30 in the presence of n-dodecane; that is, higher change in dynamic interfacial tension resulted in higher foam stability. This correlation confirms that the Gibbs–Marangoni effect, as characterized by the change of dynamic interfacial tension, plays a role in the stability of pseudoemulsion film, similar to that of the foam film.

In the presence of micelles, solubilization of oil into the micelles of the solution takes place. The stability of aqueous pseudoemulsion film will be lower (faster film-thinning) in the presence of solubilized oil, similar to that shown in the stability of foam films, because oil solubilization

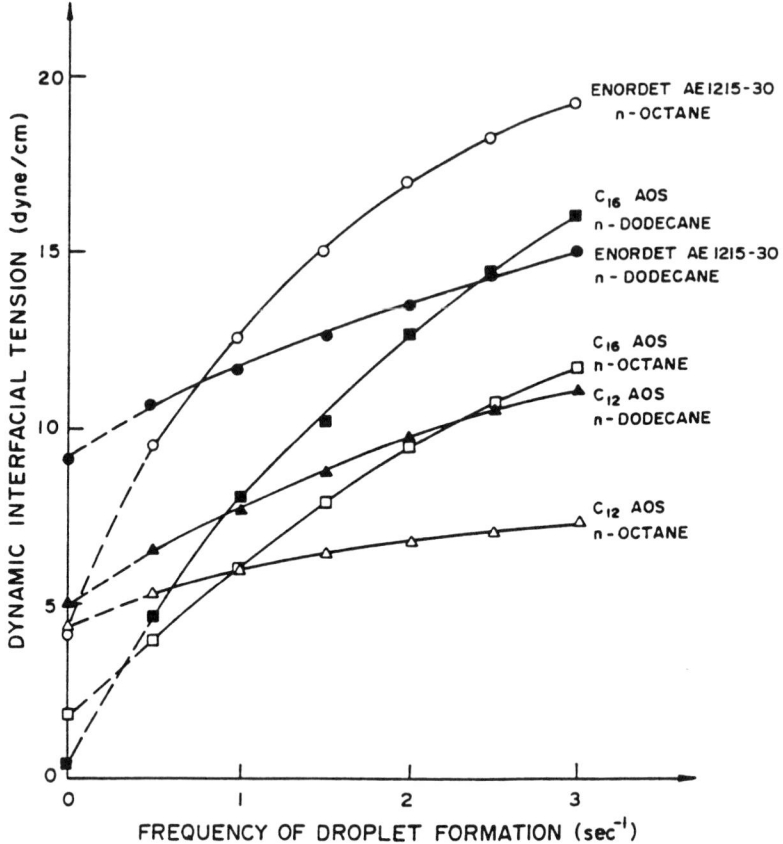

Figure 26. Dynamic interfacial tension vs. droplet frequency of various surfactant solution–oil systems (1).

increases the polydispersity and aggregation number of the surfactant micelles and the intermicellar attractions (66, 67).

Lobo and Wasan (81) observed the drainage and stability of pseudoemulsion films from nonionic surfactant solutions (Enordet AE1215-30 ethoxylated alcohol) at concentrations much above the CMC. They observed that, for a 4 wt% surfactant system, the film thinned stepwise by stratification (Figure 27), in a fashion similar to the foam films from micellar solutions (Figure 14). Three thickness transitions were observed (81) at 4 wt% concentration with *n*-octane as oil, which was the same number of steps as observed by Nikolov et al. (54) in foam films at the same concentration. This observation on the micellar layering in the pseudoemulsion film confirms, again, the universality of the stratification phenomenon.

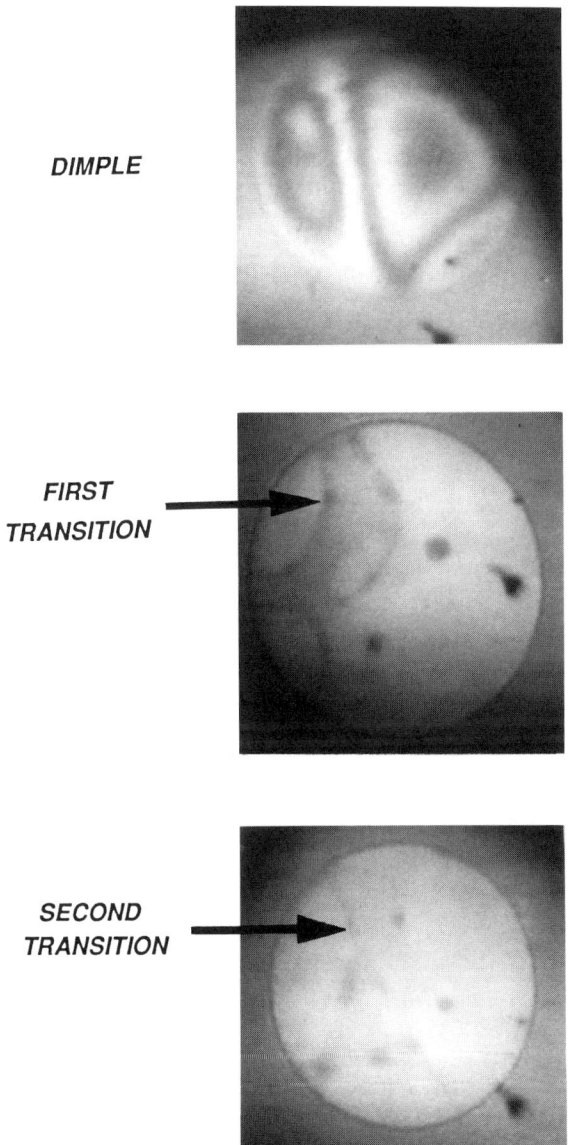

Figure 27. Stratification in a pseudoemulsion film formed from 4 wt% Enordet AE1215-30 solution between n-*octane and air.*

The pseudoemulsion film in Figure 27 gets brighter during the step-wise transitions, as opposed to the foam film where the film gets darker and the thinnest film is black, as shown in Figure 14. The reason for this variation is that the interference characteristics of the two types of films are different. In foam film, the incident light rays encounter an optically denser medium at one surface and a rarer medium at the other surface. In pseudoemulsion film, however, the incident rays encounter an optically denser medium at both surfaces (air–water and water–oil), which results in constructive interference in thin films, thus bright film.

These observations suggest that an ordered structure forms in the pseudoemulsion film, which initially contained three layers of micelles (for a 4 wt% surfactant). For the solution containing 1 wt% of surfactant, only one such transition was observed.

Lobo and Wasan (*81*) also measured the contact angle between the pseudoemulsion film and the meniscus and calculated the excess film energy as a function of the oil drop size, as shown in Figure 28. The excess energy is lower at high surfactant concentrations, a condition that implies thicker, more stable film, because it contains more micellar layers even at high capillary pressure (i.e., smaller size droplets). The excess film energy increases with the increasing drop size, and this observation can be explained by the vacancy mechanism of film stratification. The number of

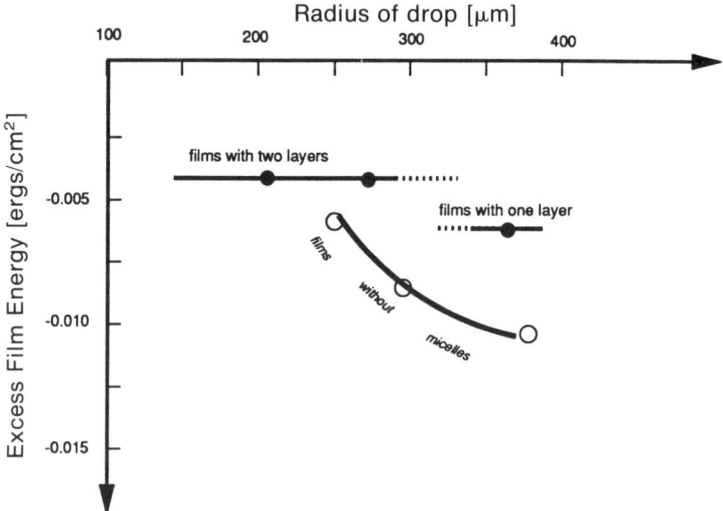

Figure 28. Variation of the pseudoemulsion film interaction energy with the oil drop radius. The solutions were 4 wt% (●) and 1 wt% (○) Enordet AE1215-30 solution, and the oil was n-octane (Reproduced from reference 81. Copyright 1993 American Chemical Society).

vacancies increases with the increase in film size (i.e., drop size), and this increase in film size enhances the film transition to lower thicknesses. When the (individual) film area decreases with decreasing drop size, the capillary pressure increases simultaneously. The increasing capillary pressure and the decreasing film area have two opposing effects on the film drainage. However, the stabilizing effect of the micellar layering overwhelms the effect of capillary pressure, as seen with the 4 wt% surfactant micellar system. The excess film energy for the pseudoemulsion film for the 1 wt% surfactant system depends strongly on the drop size (i.e., capillary pressure).

In applying the properties of a single pseudoemulsion film to actual foam–oil interactions, the relative film stability is relevant. Let us consider a freshly formed foam containing emulsified oil. After stopping the foam generation, the emulsion starts draining from the "static" foam because of gravity. Experiments show (82) that the oil drops generally drain out of the foam films (without getting trapped inside the thinning film) into the adjoining Plateau borders, as seen in Figure 1. During drainage, the oil drops get compressed into each other and into the bubble surfaces and form pseudoemulsion and emulsion films (Figure 2). If the foam films are stable, the lifetime of the three-phase foam containing emulsified oil is determined primarily by the stability of the pseudoemulsion films. When the pseudoemulsion films are stable, the oil stabilizes the foam; when they are relatively unstable, the oil will act as antifoam, as shown in Figures 25 and 24, respectively. When the stability of the pseudoemulsion films is intermediate, as in several crude oil–foam systems, the stabilizing and destabilizing effects compete. These mechanisms are discussed in the following sections.

Influence of Stable Pseudoemulsion Film on Foam Stability. In Figure 29, the foam lifetime is plotted as a function of the emulsified oil volume fraction for the foams of 4 wt% sodium dodecylbenzenesulfonate (Siponate DS-10) solution with decane in which the pseudoemulsion films are stable (82). The foam stability drops after the solution is equilibrated with the oil, then increases with increasing oil content. The curves demonstrate the two independent effects of the oil: the destabilizing effect of the solubilized oil, which is constant after equilibration; and the stabilizing effect of the emulsified oil. The measured stability is the sum of these two effects, and it can be higher or lower than the stability of the oil-free system. The stabilizing effect is stronger if the drops are larger, and it can overcompensate for the negative effect of the solubilized oil. The nonionic, Enordet AE1215-30 solution showed similar behavior with octane.

The mechanism of foam stability in these systems can be explained by the hydrodynamics of foam drainage (82). Because of the buoyancy of the

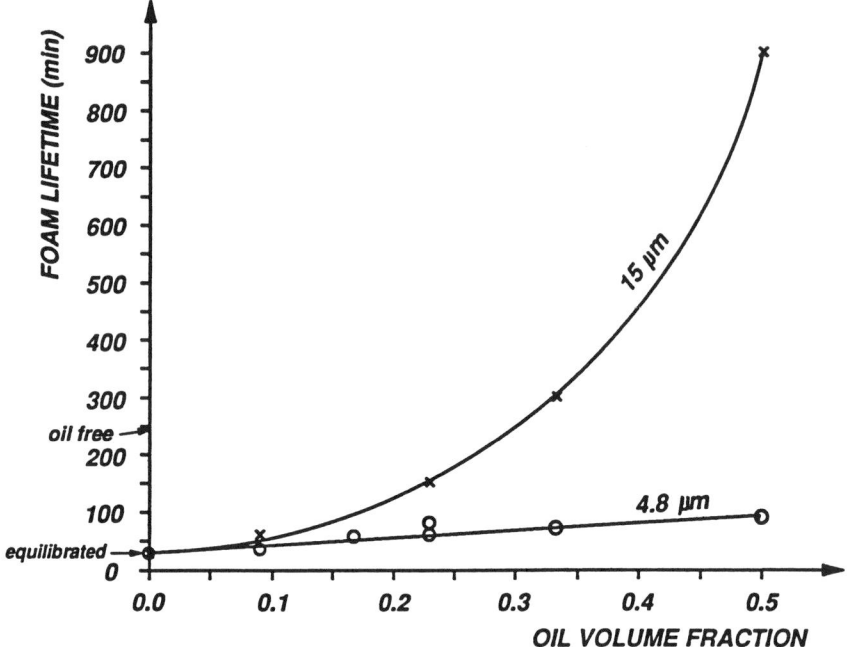

Figure 29. Foam lifetime as a function of initial emulsified decane volume fraction for two different drop sizes (15 and 4.8 µm) for foams formed from 4 wt% sodium dodecylbenzenesulfonate (Siponate DS-10). (Reproduced with permission from reference 82. Copyright 1992 Academic.)

oil drops and because of resistance to their movement within the restricted volume of the Plateau borders, the oil drops in the Plateau borders drain slower than the surrounding aqueous phase, and this difference in drainage rate results in an increase in the oil fraction within the Plateau borders. As the concentration of the oil drops increases, they get trapped in the Plateau borders because of the highly viscous nature of the emulsions (Figure 1). When trapped, the shrinking Plateau borders squeeze the oil drops into a close-packed configuration. This process results in a decrease in the area available for the flow of the aqueous liquid from the foam films through the Plateau borders, thus slowing down the entire foam drainage. Another effect of the trapped oil drops is that in the presence of macroemulsions, the Plateau borders bordering the foam cells are relatively thick. Thicker Plateau borders result in a lower capillary pressure difference (because of the smaller curvature) between the foam films and the Plateau borders. Because the capillary pressure is the driving force for the film liquid to drain, the change in the thickness of the Plateau borders is an additional factor that slows down the film drainage. Koczo et al. (82) measured the capillary pressure within the Plateau borders and found that, in the presence of oil, the capillary pressure was, indeed, reduced.

Figure 30 shows the effect of drop size on the foam lifetime for two oil volume fractions (9.1 and 33 %) in the same system. The foam stability goes through a minimum. The increase of stability with drop size can be explained by the mechanism of oil accumulation. The smaller drops accumulate in the Plateau borders to a lesser extent because of their size and lower buoyancy force; thus they have less resistance to movement within the Plateau borders. As a result, they are less likely to get trapped

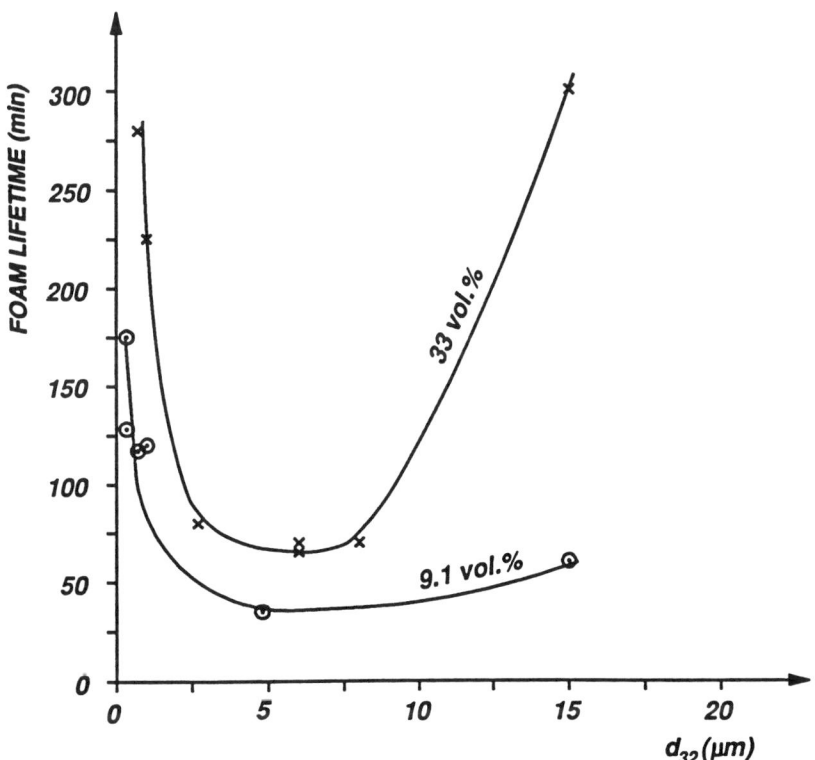

Figure 30. Foam lifetime as a function of the Sauter mean drop diameter for two initial emulsified decane volume fractions (33 and 9.1 vol%) for foams from 4 wt% sodium dodecylbenzenesulfonate (Siponate DS-10). (Reproduced with permission from reference 82. Copyright 1992 Academic.)

within the Plateau borders. This mechanism explains why with increasing volume fraction, increased stability is not as pronounced if the drops are smaller (Figure 29). As the drop size decreases (as in the left side of the curves in Figure 30), the accumulation of oil decreases. However, the viscosity of emulsions steeply increases with decreasing drop size below the 1–2-μm range because the interaction between the oil drops becomes significant. As a result, the liquid drainage is much slower, and thus the foam stability is much higher in the presence of very fine emulsions.

For different types of oils, the stability is a function of the oil phase and the accompanying buoyancy forces that govern the drainage rate, oil accumulation, and consequently, the foam stability (*82*).

Influence of Unstable Pseudoemulsion Film: Antifoaming and Defoaming Action.

When the pseudoemulsion films are unstable in the three-phase foam, the oil generally destabilizes the foam. Destabilizing of foam has important applications as antifoaming or defoaming in several industrial processes (*83, 84*). Foam inhibitors are also used in the petroleum industry, such as in gas–oil separation, in natural gas processing, and in distillation processes (*85*).

It is important to distinguish between the two, basic methods of foam destabilization: (1) defoaming, when the foam breaking agents are added to the foam to destroy it and; (2) antifoaming, when the foam inhibiting agents are mixed, generally in dispersed form, into the liquid prior to foam formation.

The mechanism of the two processes can be very different, although the distinction between them is very vague in practice (*83*). In defoaming, the fresh oil spreads and forms lenses on the foam surfaces, as the foam and the defoamers interact, (i.e., the oil configuration shown in Figure 22c and 22d). The oil spreading process, by defoaming, can play an important role in foam breaking. This important role was found by Aveyard et al. (*86*), who studied the effect of liquid alkenes on the foams of various surfactant solutions. When oil made contact with both surfaces of a foam film, the oil spread as a duplex film on both film sides, but the stability hardly decreased. If, however, the oil could spread on only one side of the film, the film stability strongly decreased. Aveyard et al. (*86*) explained this finding by the effect of spreading oil on the surface tension of the solution. Oil spreading as a duplex film resulted in about 8-mN/m surface tension reduction of the studied cetyltrimethylammonium bromide solution. Thus, when the oil spread on only one film side, there was a significant difference between the tensions of the two film sides, which destabilized the foam film. Similar tests with another solution, in which oil spreading was accompanied only by a slight surface tension reduction, did not show spreading oil to have a film breaking effect. Similarly, one-sided spreading can occur when a defoamer oil makes contact with foam: the oil can spread first on only one, the outer, surface of the foam lamellae, because the foam is a closed structure.

In antifoaming, the oil drops (or particles) get mixed into the solution, and thus, their configuration can be drop and drop with pseudoemulsion film, not only lens and spread oil (Figures 22a–22d).

Next, the mechanisms of antifoaming action will be discussed primarily, concentrating on the antifoaming of aqueous surfactant solutions by heterogeneous agents.

Not only oil but also solid particles are used as heterogeneous antifoaming agents. Three types of antifoaming agents can be distinguished:

1. Nonpolar oil with very low solubility in the aqueous phase. Generally silicone oils are used.

2. Hydrophobic solid particles. Typically hydrophobic, amorphous silica or hydrophobic polymers are used with 0.2–30-μm particle size. Hydrophobic silica is produced by covering the surface of hydrophilic silica with a nonpolar, molecular layer from silicone oil or paraffinic hydrocarbon (83).

3. Mixture of insoluble oil and hydrophobic particles. The oil phase can be silicone or hydrocarbon oil, the solid phase is any of class 2. The solid content of the mixture is 1–20%. These mixed-type antifoams are very effective, even at very low concentrations (10–1000 ppm); thus they are widely applied for the inhibition of aqueous foams (83–85, 87, 88).

Table I shows the initial foam height, h_f^o, and the foam half-life, $\tau_{1/2}$, in the presence of different types of antifoams. The oil alone has low, foam-breaking efficiency (high initial foam height and half-life), the hydrophobic particles alone are more effective, and the mixed type of agent is much better antifoam. The synergism between the oil and the hydrophobic solids is a general phenomenon. It generally occurs independently from the nature of the nonpolar oil (silicone or hydrocarbon oil) and the nature of the surfactant solution (anionic, cationic, or nonionic) (86, 88).

Antifoaming Mechanisms. Several theories can be found in the literature on the mechanisms of antifoaming by the three types of antifoams.

The spreading mechanism by oil (69–71), characterized by the spreading coefficient (equation 10), can be important in defoaming action, but seems to be negligible when the oil is applied as antifoam. Garrett (87) suggested that oil spreading is not a necessary condition of antifoaming action by oils.

Table I. Antifoaming Efficiency of Different Types of Antifoaming Agents Measured by Foam Stability Tests

Antifoaming Agent	Concentration (ppm)	h_f^o (cm)	$\tau_{1/2}$ (min)
–	0	20	600–800
DMPS-5X silicone oil	2000	18	350
T-500, hydrophobic silica	200	13	30
T-500 + DMPS-5X (1:40)	200	9.2	1.5

NOTE: DMPS-5X is poly(dimethylsiloxane) (silicone) oil with 50-cSt viscosity; T-500 and Tullanox-500 (Tulco Co.) are hydrophobic silica with 0.2-μm particle size. The surfactant solution is 0.06 M SDS. h_f^o is the initial foam height, and $\tau_{1/2}$ is the foam half-life.

In most antifoaming mechanisms that were suggested in the literature, the last step of foam breaking by oil drops is the formation of an oil bridge in the foam film, as was shown in Figure 24. According to Ross and McBain (*69*), an oil bridge forms when the spreading oil squeezes the aqueous liquid out of the film. Others (*89, 90*) suggested that the lens, which forms after the oil drop enters the air–water surface inside the foam film (with or without the formation of duplex oil film), enters the opposite film surface on further film (foam) drainage, and results in the bridge configuration (Figure 31). The film then breaks because the capillary forces thin out the film around the bridge and, finally, the film pinches off from the oil (*88, 91*).

Analogous bridging mechanisms were suggested for antifoaming by hydrophobic particles (*92–94*) (Figure 32a) or by the mixture of particles and oil (*89*). Garrett (*92*) and Frye and Berg (*95*) pointed out that the wettability of the particle by the solution controls the effect of solid particles on foam stability. For hydrophobic particles, the curvature of the film at the particle is convex (Figure 32a); thus the capillary pressure thins the film and pushes it off the particle. (In these theories, it is assumed that the second film curvature along the three-phase contact line is small.) If, however, the particle is hydrophilic, the film has a concave meniscus; thus, the capillary pressure has an opposite direction (Figure 32b). For effective foam-breaking action, the (receding) contact angle (Θ_{AW}) should exceed a critical value. In surfactant solutions, the contact angles are not much higher than 90° even with very hydrophobic solids such as poly(tetrafluoroethylene) (*88*). However, if the particles are rough with edges, which is common in the commercial antifoams, they can rupture foam films even if $\Theta_{AW} < 90°$ (*88, 95*). As a result, rough hydrophobic particles are more effective antifoams than smooth particles (*92, 95*). The contact angle hysteresis phenomenon may also affect the mechanism.

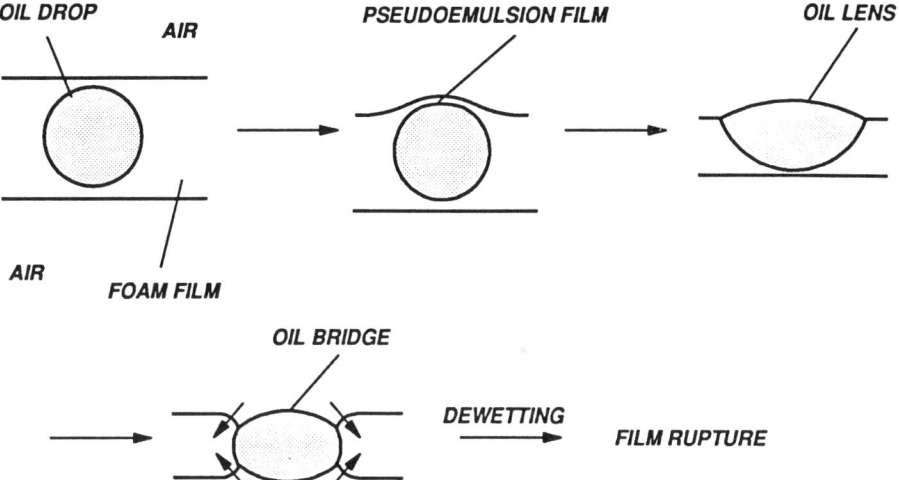

Figure 31. Suggested bridging mechanism of antifoaming in foam film by oil (90). The oil drop first enters one of the film surfaces and forms a lens. On further thinning of the film the lens enters the opposite film surface and an oil bridge is formed. The bridge is unstable because the capillary forces dewet the film from the bridge and the film ruptures. The arrows in the film indicate the direction of capillary forces.

Another possible mechanism of foam-breaking by hydrophobic particles is that the particles can spread on the foam film surface and destabilize the film by decreasing the surface tension of the solution (96). The experiments of Aveyard et al. (86) show that this effect is probably not significant.

A different antifoaming mechanism was suggested by Kulkarni et al. (96). They found that surfactants adsorb on the surface of hydrophobic particles during antifoaming, and this adsorption results in deactivation of the particles. On the basis of this observation, they postulated that the adsorption of surfactants onto the hydrophobic particles is so fast that it results in surfactant depletion around the particle in a foam film, and this effect breaks the film. However, no direct proof was presented on this theory. Moreover, depletion of surfactant would cause the film liquid to flow toward the particle because of the increased surface tension (Gibbs-Marangoni effect), and thus cause a stabilizing effect.

According to the adsorption theory (97), the solid particles are the real foam breakers in the mixed antifoams, and the role of the oil is that it shields the particles from adsorption inside the solution. When the particle, which is shielded by the oil, enters the foam surface, the oil spreads and releases the particles. The problems with this mechanism are similar to those with the adsorption with solid particles alone (88). Moreover,

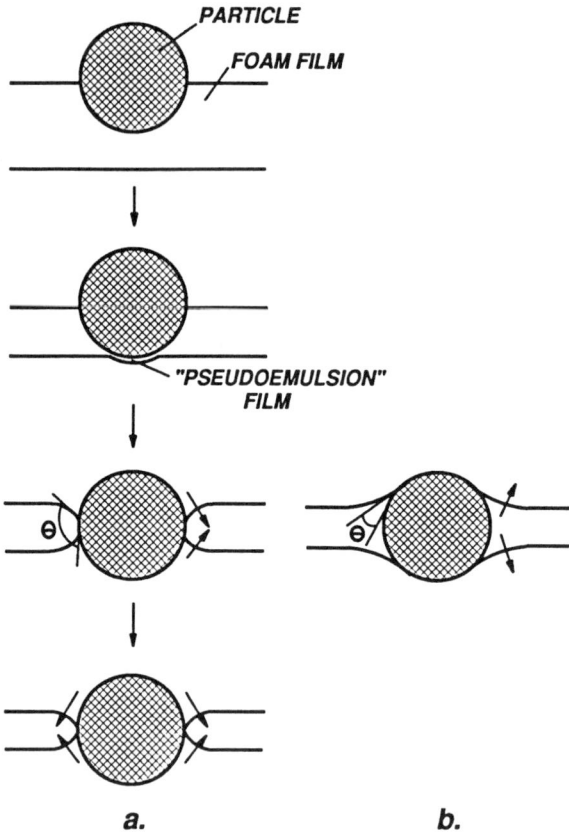

Figure 32. Suggested mechanism of foam film bridging by solid particle (92–94). In part a, the hydrophobic particle bridges the film, the capillary pressure dewets the particle, and the film ruptures. In b, hydrophilic particle's capillary pressure acts in opposite direction and tends to increase the film thickness. The arrows in the film indicate the direction of capillary forces.

the efficiency of an antifoam could be expected to decrease if it is kept in contact with the surfactant solution for a prolonged time. However, experiments show that mixed antifoams retain their efficiency after rinsing in the solution for several hours (86, 98).

Role of the Pseudoemulsion Film in Antifoaming Action. The antifoaming efficiency strongly depends on the stability of the pseudoemulsion film. It must rupture in order for the drop, lens, or particle to enter the air–water foam surface in the bridging mechanisms (Figures 31 and 32a). When oil has spread on the air–water interface, the antifoaming efficiency will also depend on the state of the spread oil: a film or lens with particles or without particles. In the mixed antifoams, the oil is initially (before foam formation) dispersed in the water phase as droplets, and each droplet may contain a few hydrophobic silica particles. The antifoaming efficiency can be explained by the stability of pseudoemulsion and emulsion films. In the previous section, three-phase foams were shown in which the oil stabilized the foam because of the stable pseudoemulsion films in the systems (Figure 29). When a small amount of hydrophobic silica powder with an average size of about 0.2 μm was added to these emulsions, the effect was dramatic. It was impossible to regenerate foam that would remain stable for more than a few seconds, an indication that the particles destabilized the pseudoemulsion film. In general, good antifoam mixtures can be obtained from oil, which alone has no foam destabilizing effect, if they are mixed with hydrophobic particles (*98*).

The stability of pseudoemulsion film was studied directly in our laboratory, by forming such a film from a surfactant solution (0.06 M sodium dodecyl sulfate) on the tip of a capillary (Figure 33). The tip of the capillary, which was filled with antifoam oil, was covered with a small pseudoemulsion film from the surfactant solution. Then, the oil was slowly pushed out of the capillary. In this way, the area of the pseudoemulsion film and the capillary pressure was increased. The pseudoemulsion film between silicone oil containing no particles and air was relatively stable, and it did not rupture before the capillary pressure maximum, (i.e., hemispherical shape). However, when the oil contained hydrophobic silica particles, the film was much less stable. At high particle concentrations (1–6 wt%), the film ruptured shortly after it was formed, (i.e., at very low capillary pressures).

By decreasing the particle concentration in the oil, a critical concentration could be reached at which the film ruptured at intermediate pressures and sizes. Table II shows this critical solid concentration for some systems. A very small amount of particles (0.1–0.01 wt%) is enough to break the pseudoemulsion film. In Figure 34, the foam half-life is plotted as a function of the solid content in mixed antifoams. Comparing Figure 34 and Table II, the foam life sharply increases, (i.e., the foam-breaking efficiency of the mixture indeed diminishes) if the solid content is lower than the critical concentration. The film-destabilizing effect of the particles can be explained by their location. The particles at the oil–aqueous interface partially emerge into the solution (Figure 33) (i.e., the oil–solid–aqueous contact angle is less than 180°), and the edges of the particles penetrating into the aqueous phase can pierce and break the pseudoemulsion film, similar to the manner suggested in the foam-breaking mechanism by hydrophobic particles (*94, 95*). The particle con-

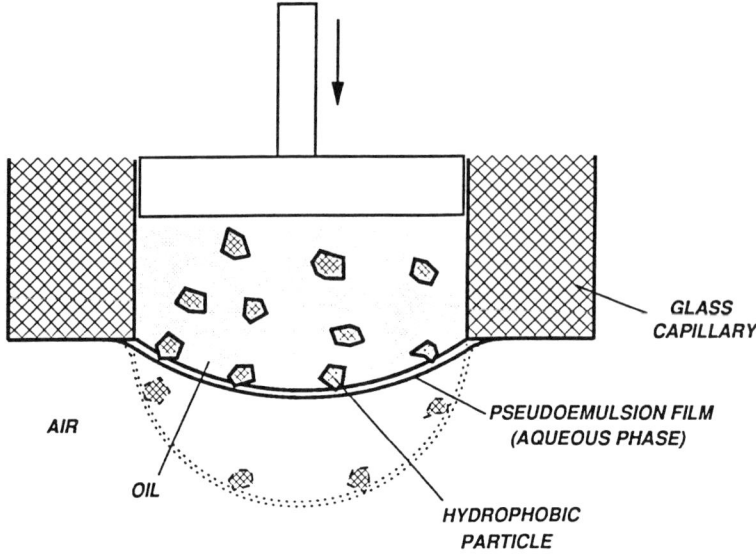

Figure 33. Pseudoemulsion film formed between mixed antifoam (oil with hydrophobic particles) and air, on the tip of a glass capillary (schematic). The film area is increased by pushing the antifoam out of the capillary. The particles destabilize the pseudoemulsion film by partially submerging into the aqueous phase.

Table II. Critical Hydrophobic Solid Concentrations in Silicone Oils To Break the Pseudoemulsion Film at the Tip of Capillary

Oil	Solid	Capillary ID (mm)	Crit. Solid Conc. (wt%)
DMPS-V	T-500	0.22	0.1
DMPS-V	T-250	0.32	0.03
DMPS-2C	T-250	0.22	0.01

NOTE: DMPS-V and -2C are poly(dimethylsiloxane)s with 5- and 200-cSt viscosities, respectively. T-500 and T-250 are Tullanox-500 and -250 (Tulco Co.) hydrophobic silicas with 0.2- and 2-μm particle size, respectively.

centration required to break the foam is very low, because it is enough that the film be pierced at one point to rupture. The orientation of the surfactant molecules on the partially hydrophobic solid surface can also destabilize the pseudoemulsion film.

Garrett (*88*) also suggested that the role of the particles in the synergistic effect of the mixed antifoams is that the particles enhance the rup-

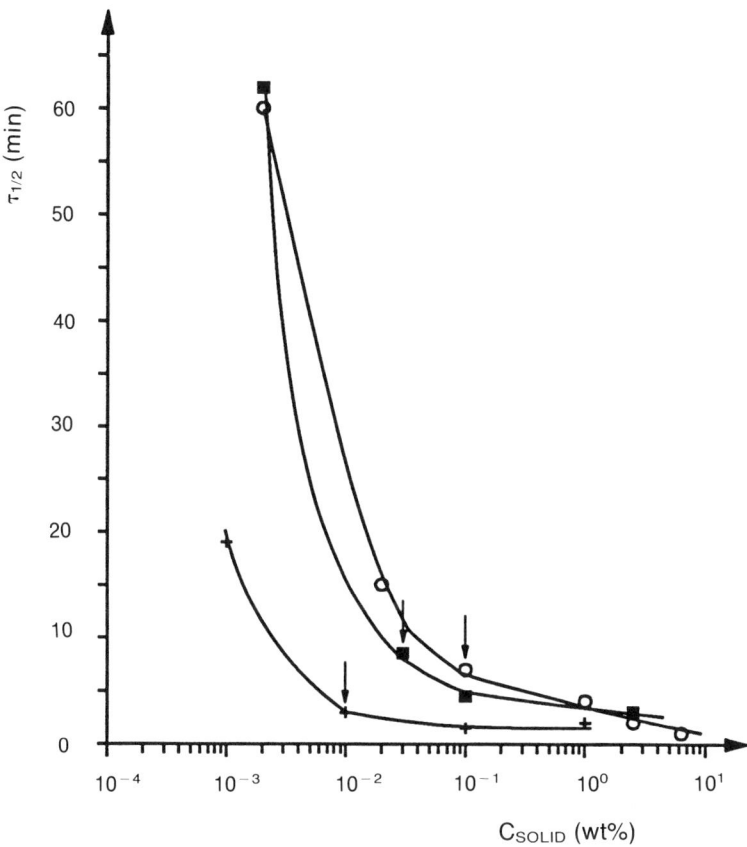

Figure 34. Effect of the solid concentration in mixed antifoams on the foam half-life ($\tau_{1/2}$). Key: + oil was DMPS-2C (200 cP) and solid was T-250 (2 µm); ■ oil was DMPS-V (5 cP) and solid was T-250 (2 µm); ○ oil was DMPS-V (5 cP) and solid was T-500 (0.2 µm). Antifoam concentration was 200 ppm. The arrows show the respective critical solid concentrations to break the pseudoemulsion film at the tip of capillary (Table II).

ture of the oil–water–air film. He explained this on the basis of the contact angles, claiming that an oil drop with particles can rupture the pseudoemulsion film if, for spherical particles:

$$\Theta_{AW} > 180° - \Theta_{OW}$$

where Θ_{AW} is the air–solid–water contact angle, and Θ_{OW} is the oil–solid–water contact angle (Figure 35). Thus, if $\Theta_{OW} > 90°$, $\Theta_{AW} < 90°$

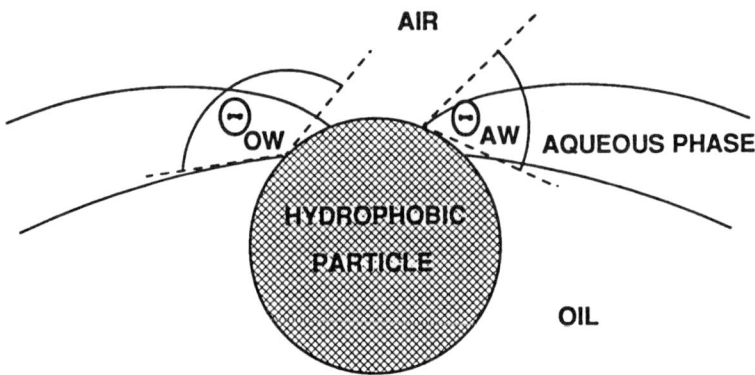

Figure 35. Contact angles at the hydrophobic particles of mixed antifoam.

contact angles satisfy this condition. [For rough particles, similar relations were suggested (*88*)]. Frye and Berg (*90*) and Aveyard et al. (*86*) found that Θ_{OW} is indeed much higher than 90° in mixed antifoam–surfactant solution systems.

Immersion Depth and Position of Antifoams. In addition to the stability of the pseudoemulsion film, several factors influence the efficiency of the various antifoams.

The air–water–solid film is unstable for hydrophobic particles alone, therefore the particles enter the foam surface and only partially immerse into the aqueous phase (Figure 32a). The particles can bridge the film only when they reach the opposite air–water surface. Thus, the time of bridging depends on the immersion depth of the particle and the rate of foam film drainage. As was shown, the contact angle Θ_{AW} should be high in order for high antifoaming efficiency by the hydrophobic particles. The requirement of high immersion depth, however, claims a low contact angle. A particle with $\Theta_{AW} = 180°$ could never bridge the film. In surfactant solutions, however, Θ_{AW} is always well below 180° because of surfactant adsorption, and about half of the particle is immersed into the aqueous phase.

The penetration depth of oil lenses (with or without particles) is generally deeper than that of the particles alone because the shape of the lens is controlled by the air–oil (σ_{AO}) and the oil–water (σ_{OW}) interfacial tensions. In practical systems, σ_{AO} is much higher than σ_{OW}, and thus, most of the lens is immersed into the aqueous phase (Figure 31). This penetration depth difference is a reason that the mixed-type antifoams are much more effective than the hydrophobic particles alone.

Most theories suppose that the antifoams act in the foam films. The position of antifoam lenses and particles was monitored in our laboratory in foams and vertical and horizontal foam lamellae that were formed from micellar surfactant solutions (97). The hydrophobic particles or oil lenses stay in the foam-film region for only a very short time (about a second) because they move out of the thinning foam films into the adjoining menisci (Plateau borders) and move along these borders under the turbulency accompanying foam-film drainage. Film rupture happened when the lenses of the mixed antifoam or the hydrophobic particles stopped, because they got trapped in the thinning Plateau border.

Suggested Antifoaming Mechanism. On the basis of the experimental observations for the particle–oil interactions and the way in which oil droplets carrying fine hydrophobic particles interact with the foam structure during foam drainage, the following mechanism is suggested for antifoaming by nonpolar oil with and without hydrophobic particles (98). During foam formation and in the presence of oil, oil is dispersed into the aqueous phase of the foam, and a polydisperse, oil-in-water emulsion is formed. The average oil-drop size in the emulsion depends on the stress on the drop due to flow, dynamic interfacial tension at the oil–water interface and the viscosity of the oil or the oil–particle mixture. If the oil has a large spreading coefficient at the water–air surface, it can be expected that very fine drops will form during foam generation (shaking, etc.). As was observed, during the foam drainage, the film flow removes the oil droplets from the film area and, depending on the foam bubble size, the oil drops can get trapped and accumulate in the foam canals. During foam formation, the stress due to drainage flow stretches the oil droplets and, in the presence of surfactant, the oil–water interfacial tension (i.e., drop capillary pressure) is too low to resist the drop breakdown process, and fine drops form. The small oil drops flow out from the foam structure (82) with no effect on the foam stability. A simple way to avoid the formation of small drops is to increase the viscosity of the oil phase. Figure 36 shows the effect of oil viscosity on the foam lifetime using silicone oil–hydrophobic silica mixture antifoams. The foam lifetime decreases; that is, the foam-breaking efficiency indeed increases with the oil viscosity.

The presence of fine hydrophobic particles dispersed into the oil phase helps to prevent the formation of very small oil drops and the spreading of the oil at the air–water interface. The particle–particle interactions inside the oil phase were estimated by measuring the settling volume of particles in oil (98). Particles settling under gravity formed a three-dimensional gel-type structure. With small particles, the sediment contained a very low amount of particles, and the final settling volume depended on the type of oil. For example, with 0.2-μm sized, hydrophobic silica particles, the sediment contained 1.2 vol% particles in decane and 1.45 vol% particles in DMPS-V silicone oil. These results show that

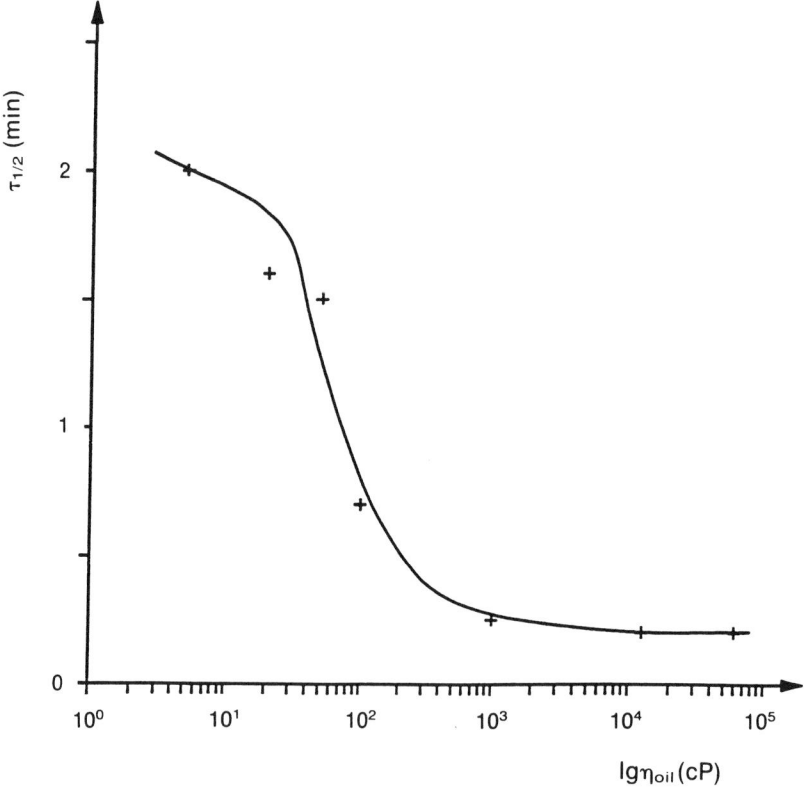

Figure 36. Effect of oil viscosity on the half-lifetime ($\tau_{1/2}$) of foams formed from 0.06 M sodium dodecyl sulfate in the presence of 200 ppm mixed antifoam containing silicone oils [poly(dimethylsiloxane), DMPS] of various kinematic viscosities and 2.5 wt% hydrophobized silica (T-500).

particle–particle interactions lead to the formation of a random structure that can resist the stress due to flow so that the oil can remain dispersed as a droplet. It also shows that, depending on the particle concentration, the drop size can be controlled and the antifoaming effect optimized.

At a given point in the foam drainage, the drops get trapped in the shrinking Plateau borders (Figures 37a and 37b). Consequently, the capillary pressure increases in the Plateau borders and destabilizes the pseudoemulsion film formed between the trapped drop and the bubble, thus allowing the antifoam oil with the particles to enter the foam surface and form a lens (Figure 37c). Without hydrophobic particles, the oil lens spreads at the water–air surface. The spread lens has a low penetration

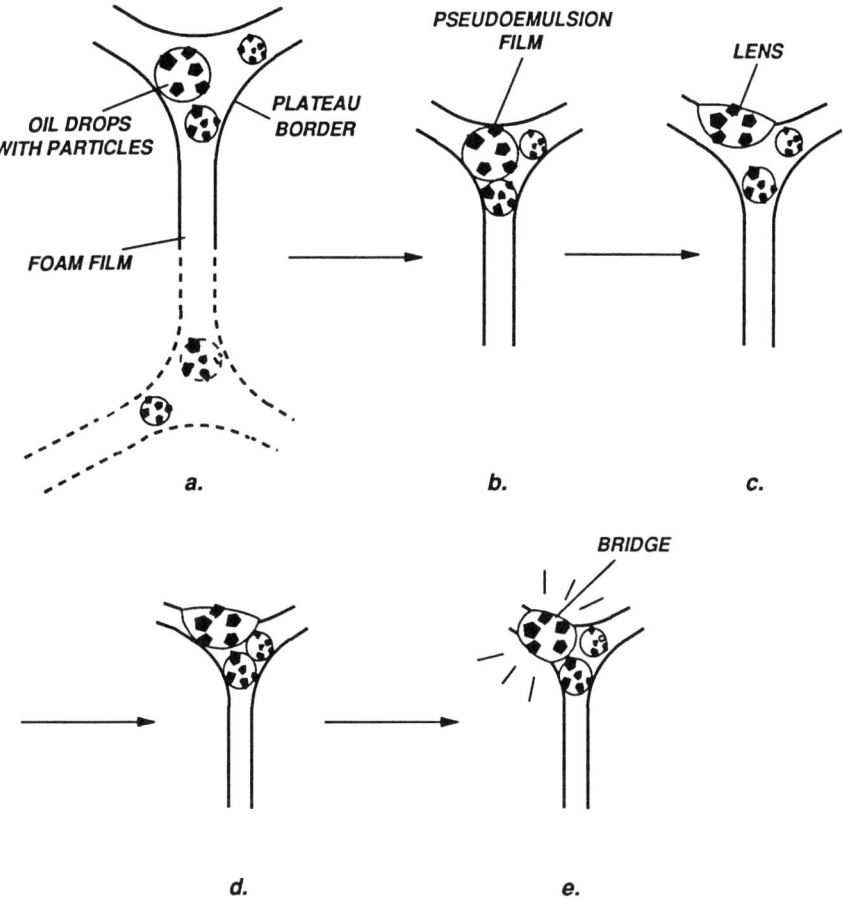

Figure 37. Suggested antifoaming mechanisms for mixed-type antifoams. Key: a, the oil drops containing solid particles collect in the Plateau borders; b, they get trapped in the thinning Plateau border; c, the pseudoemulsion film breaks, a drop enters and forms a solid plus oil lens; d, the lens gets trapped in a later stage of thinning; and e, the lens bridges the film at the Plateau border and the bridge ruptures.

depth into the solution and during foam generation (shaking etc.), and it easily breaks into very small drops. The small drops cannot get trapped in the Plateau borders and flow out of the draining foam. In the presence of oil-wetting particles, the particle structure prevents the oil spreading, and the oil remains at the interface as a lens. On further foam drainage, the lens oil–water interface reaches the opposite foam surface (Figure 37d) and, at a certain capillary pressure, the pseudoemulsion film ruptures, and the oil phase bridges the two air sides of the Plateau border (Figure 37e).

The oil bridge then thins very fast and breaks at a high thickness. At higher antifoam concentrations, the foam breakdown process is faster because, as shown in Figure 37, several drops can get trapped at the same time. Larger drops also get trapped faster than smaller drops; therefore, higher antifoaming efficiency can be expected with a larger antifoam drop size. Moreover, as was shown with the micellar ordering phenomena, the stability of larger pseudoemulsion film is also lower in micellar surfactant solutions, and at the same time the number of drops decreases with increasing drop size. Thus, there is an optimum antifoam dispersity.

Figure 38 summarizes our current knowledge on antifoaming mechanisms.

The effect of oil on aqueous foam stability is controlled by the behavior of the pseudoemulsion films. In the previous sections, two extreme cases of the foam–oil interactions were shown: foam stabilizing when the pseudoemulsion films are stable and antifoaming; that is, fast foam rupture, when these films are very unstable.

In the Chaser SD1000 and SD1020 foams (steam-foam flooding additives manufactured by Chevron Oil Co.) with crude oil, the effect of oil represents intermediate cases because the emulsified oil has both destabilizing (at low oil contents) and stabilizing effects as discussed in following sections.

Several attempts have been made to use aqueous foams for mobility control of steam-flooding in enhanced oil recovery (*99–101*). The application of foam is, however, seriously hindered by the extreme conditions of the process. The temperature can be as high as 80–300 °C, and the pressure can be 60–4000 psi, which is not common in other foam applications. The oil saturation in the porous media can be higher than 15% (*44*), and several authors (*100–102*) found that crude oil is generally detrimental to the foam. Thus, the foam stability in these conditions is a limiting factor in the success of foam flooding.

Suffridge et al. (*103*) found that the effect of crude oil on bulk foams can vary from relatively mild (in a fluorinated surfactant) to essentially catastrophic (in C_{10}-α-olefin sulfonate).

The mechanism of foam stability in bulk foams formed from foam-flooding surfactant solutions in the presence of crude oil was studied in our laboratory at elevated temperatures and pressures. Foam was generated in a glass cylinder by bubbling gas through crude oil emulsions in the surfactant solutions, and the foam decay was observed. In Figure 39, the foam lifetime is plotted as a function of the emulsified crude oil content for foams containing Chaser SD1000, steam-foam flooding additive at three different temperatures (90, 120, and 175 °C). The increase of both the temperature and the oil concentration destabilized the foam. Moreover, the emulsions were unstable, which resulted in the coalescence of drops in the draining foam. The formed, larger drops broke the foam

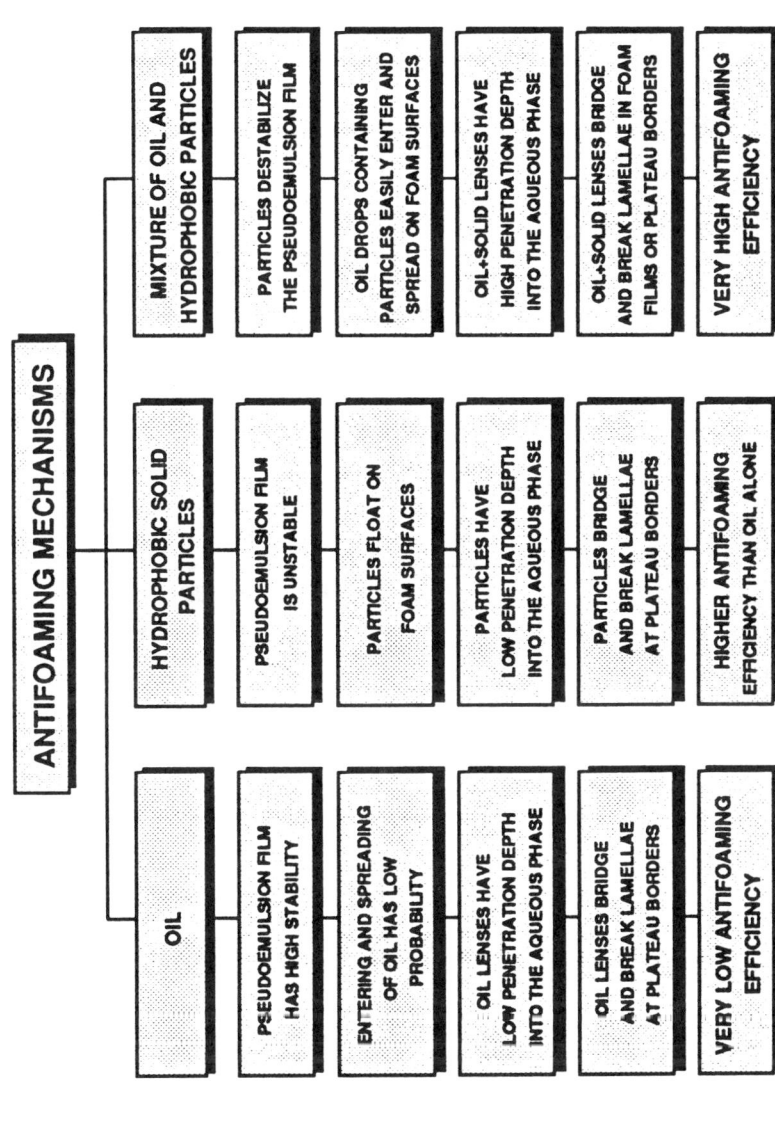

Figure 38. Summary of antifoaming mechanisms for aqueous foams.

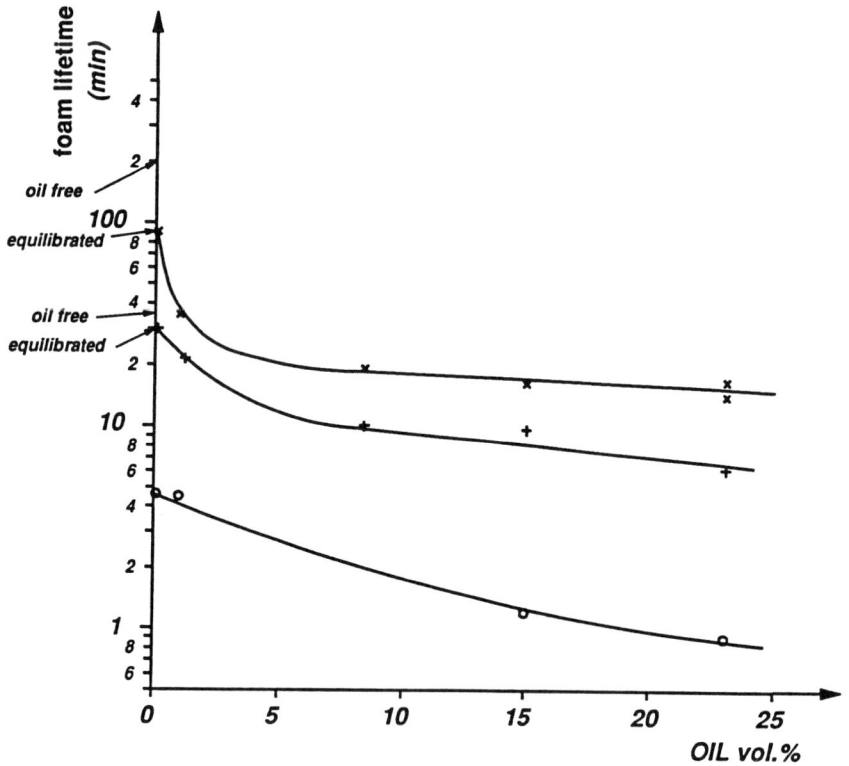

Figure 39. Lifetime of Chaser SD1000 foams in the presence of a California heavy crude oil ("A") at various temperatures as a function of the average, initial, emulsified oil volume fraction, at 200 psi pressure. Temperature: ×, *90 °C;* +, *120 °C; and* ○, *175 °C. Concentration: 0.55 wt% SD1000 and 0.1 wt% NaCl.*

along their path very effectively. At higher oil content (>10 vol%), the foam life decreases much less upon oil addition than at low oil content (Figure 39).

Figures 40 and 41 demonstrate a different foam behavior, for the Chaser SD1020 foam-flooding additive at 90, 120, and 175 °C, at high- (200 psi) and at low- (slightly above the vapor pressures at the given temperatures) pressures. The increase in temperature strongly decreases the stability, similar to the previous system. With increasing oil content, the stability goes through a minimum at 90 and 120 °C and slightly increases at 175 °C. The minimum curves indicate that the stability is controlled by at least two processes.

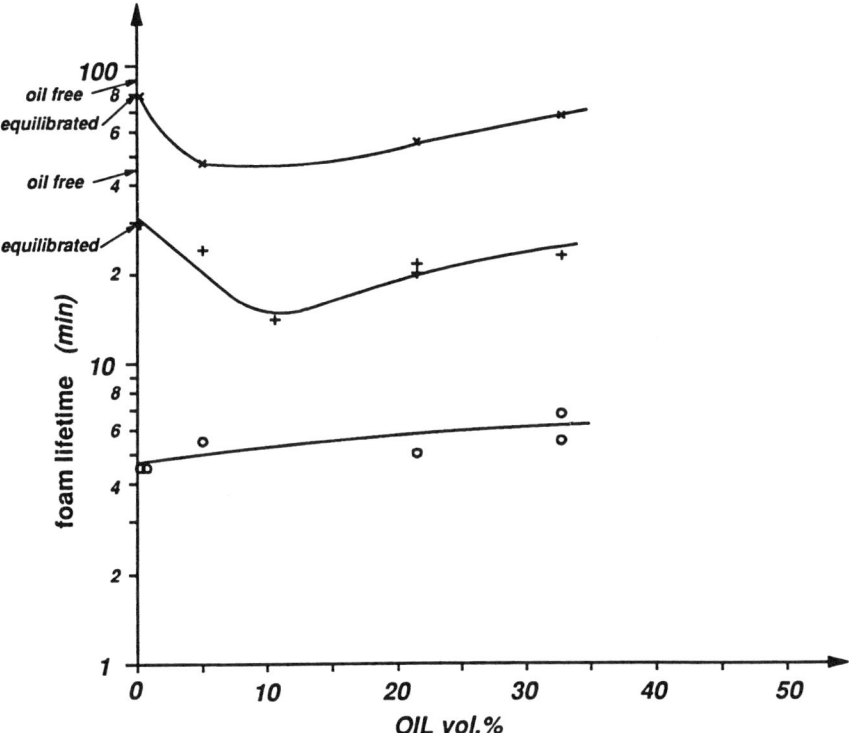

Figure 40. Lifetime of Chaser SD1020 foams in the presence of a California light crude oil ("B") at various temperatures as a function of the average, initial, emulsified oil volume fraction, at 200 psi pressure. Temperature: ×, 90 °C; +, 120 °C; and ○, 175 °C. Concentration: 0.3 wt% SD1020 and 0.3 wt% NaCl.

The mechanism of foam stability in these foams can be described by the stabilities of the foam, pseudoemulsion, and emulsion films (Figure 2).

The increase of temperature enhanced the drainage of the liquid films. As was discussed in the first part of this chapter, the rate of film drainage controls the film stability. At low surfactant concentrations, the film drains by a surface tension gradient mechanism, but at much higher concentrations than the CMC, the micellar structuring controls the process (Figure 4). Table III shows that the CMC of the studied Chaser SD1000 and SD1020 solutions increased with the temperature, but, even at the highest temperature, the applied concentrations were at least an order of magnitude higher than the CMC. Thus, micellar ordering inside

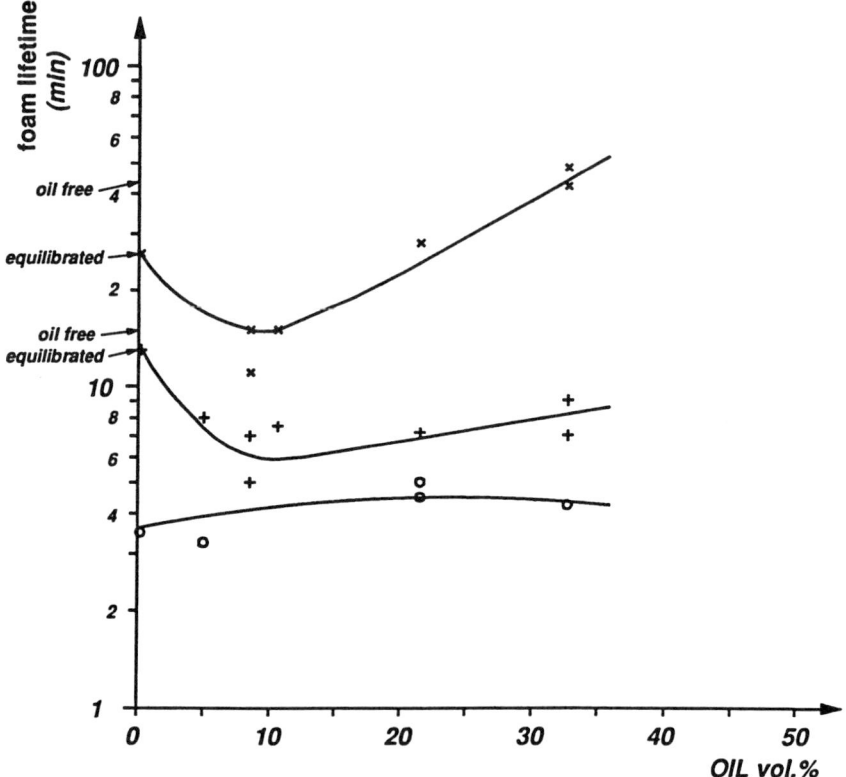

Figure 41. Lifetime of Chaser SD1020 foams in the presence of a California light crude oil ("B") at various temperatures as a function of the average, initial, emulsified oil volume fraction. Temperature and pressure: ×, 90 °C, 25 psi; +, 120 °C, 45 psi; and ○, 175 °C, 135 psi. Concentration: 0.3 wt% SD1020 and 0.3 wt% NaCl.

the liquid films of the studied foam affected their stability. As discussed, the increase of temperature accelerates the film stratification process (Figure 20) and in this way destabilizes the liquid films. The lower foam stability at higher temperatures can be explained by this effect, along with the decreased viscosity of the liquids. Further discussion of the effect of oil viscosity can be found in Chapter 4.

Increasing the pressure enhanced the stability at every temperature (Figures 40 and 41), and this effect can be explained by the mechanism of bubble disproportionation. During foam aging, the small bubbles shrink,

Table III. Critical Micelle Concentration (wt%)
of the Studied Solutions at Different Temperatures (°C)
with Constant NaCl Concentration

Solution[a]	20	90	120	175
SD1000	5×10^{-4}	—	1×10^{-2}	5×10^{-2}
SD1020	1.5×10^{-4}	2×10^{-3}	4×10^{-3}	1×10^{-2}

[a]Salt concentrations: In SD1000 it is 0.1 wt% NaCl; and in SD1020 it is 0.3 wt% NaCl.

NOTE: The CMC of SD1000 was not measured at 90 °C.

and the larger ones expand because the pressure is higher inside the smaller bubbles. The process results in decreasing foam surface area, thus faster foam decay (82). Near to the vapor pressure of the water at a given temperature, the disproportionation proceeds primarily by water evaporation and condensation. At higher pressure, the permanent gas to steam ratio in the gas phase is higher, and the gas diffusion through the foam films becomes rate-controlling, which is slower process than the evaporation–condensation (103). It results in slower bubble disproportionation and slower foam decay.

In the Chaser SD1000 foams (Figure 39), the pseudoemulsion films had a lower stability; therefore, the oil drops could enter the foam surface and break the foam, probably by bridging. In the other system (Figures 40 and 41), the foam breaking effect of the oil drops (as seen at lower concentrations) is overcompensated by the stabilizing effect due to the hydrodynamic resistance of the oil drops which get trapped in the Plateau borders. As was shown, this stabilizing effect of the trapped oil is pronounced, if the pseudoemulsion films are stable and if the oil volume fraction is high (Figure 29). This hydrodynamic effect results in the foam stability decreasing only moderately at large oil contents in the SD1000 foams (Figure 39). However, in the SD1020 system, the pseudoemulsion films were more stable, on a relative scale, therefore, at larger oil content, the stabilizing effect was more pronounced than the foam-breaking process.

Correlation between Pseudoemulsion Film Stability and Improved Oil Recovery. As the pseudoemulsion film controls the stability of bulk foams, it plays an important role also in the stability of foams in porous media containing crude oil.

Nikolov et al. (1) were the first to find a direct correlation between pseudoemulsion film stability and crude oil displacement efficiency in porous media. As discussed in the previous section, the stability of the pseudoemulsion film in α-olefin sulfonate–crude oil systems increased with the chain length of the surfactant as $C_{12}AOS < C_{14}AOS < C_{16}AOS$. Foam-enhanced oil recovery experiments were conducted in Berea sandstone cores (1). For $C_{12}AOS$, Figure 42 shows that after 1 pore volume of surfactant solution injected, the oil saturation dropped below 20% at the first point of the core, and saturation for positions 2–4 dropped to 28–36%. For the rest of the core, the oil saturation remained virtually unchanged during the foam-flooding process. With $C_{14}AOS$, the saturations were lower; hence the frontal spccd of oil was higher. For $C_{16}AOS$, as shown in Figure 43, the oil saturation at the first point dropped below 10% and to about 30% at position 4, and this system produced the fastest oil propagation rate. Similarly, the pressure drop in the core and the degree of oil recovery increased in the $C_{12}AOS < C_{14}AOS < C_{16}AOS$ sequence, the same relationship as in the stability of the pseudoemulsion films.

Manlowe and Radke (77) studied foam–crude oil interactions in a microvisual glass cell, and they also found that the lifetime of the pseudoemulsion film controls the foam stability in foam-flooding systems.

Schramm and Novosad (76) conducted similar studies and found that the stability of the pseudoemulsion film is an important factor in the movement of the crude oil in the porous medium. They also found that the degree of emulsification of the crude oil into the moving foam greatly affects the oil transport. Both the degree of emulsification and the stability of the pseudoemulsion film decreased with the oil–surfactant solution interfacial tension in their systems (mostly zwitterionic surfactants); that

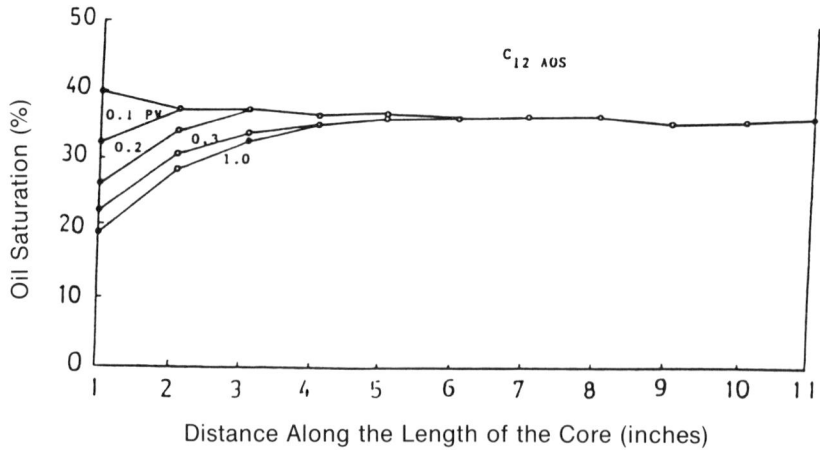

Figure 42. Oil saturation during foam-enhanced oil recovery in Berea sandstone with $C_{12}AOS$.

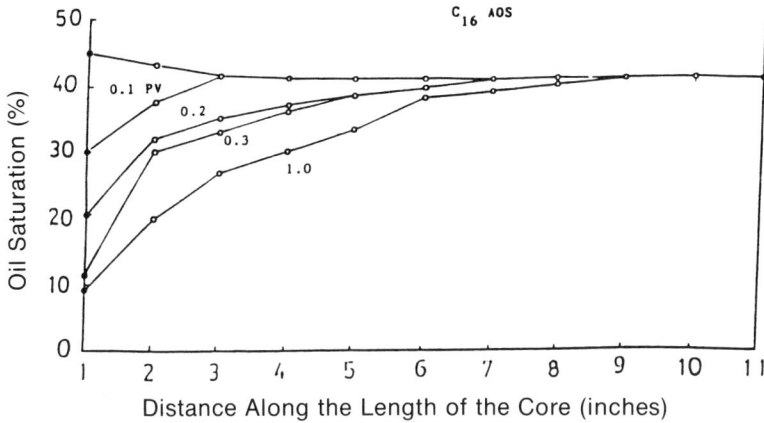

Figure 43. Oil saturation during foam-enhanced oil recovery in Berea sandstone with $C_{16}AOS$.

is, the pseudoemulsion film was less stable with finer emulsions. Thus, they concluded that the oil transport is best in the systems where the degree of emulsification and the stability of the pseudoemulsion film is intermediate.

These observations are in contrast to the findings of Nikolov et al. (*1*), who observed that the stability of the pseudoemulsion film is higher with smaller crude oil drops in the α-olefin sulfonates. We observed that in the foams of the Chaser SD1000 and SD1020 steam-foam additives (also anionic surfactants) with crude oil, higher stability of pseudoemulsion film in the SD1020 system was associated with lower interfacial tension and finer drops. Moreover, in stratifying films (at high surfactant concentrations and low electrolyte concentrations), smaller-size films (i.e., small drops) are more stable because of the inhibited film transition, as was discussed for the vacancy mechanism of micellar film structuring phenomenon. A possible reason for the differing observations is that different types of crude oil might have been used by the different investigators. The effect of crude oil properties, such as viscosity, on foam–oil interactions will be further discussed in Chapter 4.

Influence of Emulsion Film Stability. The stability of the emulsion drops inside three-phase foam systems has an influence on the foam stability. When the oil drops collect (flocculate) in the Plateau borders, the drops are compressed and give rise to the formation of emulsion films (Figures 1 and 2). If the emulsion films are stable, drops do not coalesce, and the foam stability is not affected. However, if these emulsion films are unstable, the oil drops coalesce in the Plateau borders and then, on further foam drainage, the large drops relocate, under the action of the capillary pressure, into the wedges of the Plateau borders, the meeting points of four borders, and form a curved, tetrahedron shape. Figure 44 shows a foam containing crude oil after draining for a long time when the emulsion films were unstable.

If the pseudoemulsion film is stable, the larger drops formed from the unstable emulsion get trapped faster in the Plateau borders and stabilize the foam. On the other hand, if the pseudoemulsion film is unstable, the larger drops can bridge easier and break the foam lamellae faster (Figure 36). Thus, the unstable emulsion film also plays an important role in foam stability in the presence of emulsified oil.

Concluding Remarks

Figure 45 summarizes our current understanding of the possible foam–oil interactions and highlights the role of single-foam lamellae, pseudoemulsion, and emulsion films in foam stability in aqueous foam systems containing solubilized or emulsified oils. Further research is warranted to

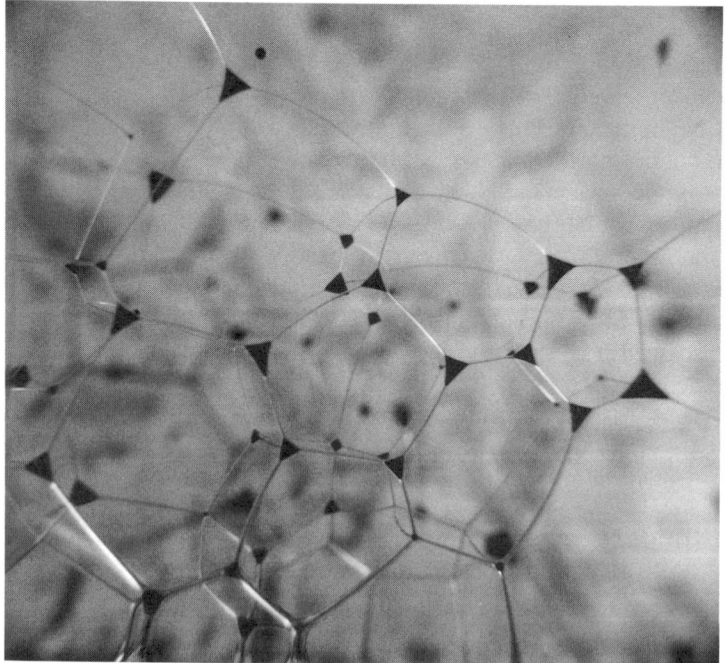

Figure 44. Accumulation of coalesced crude oil emulsion drops in the wedges of the Plateau borders of foams formed from fluorinated surfactant (0.5 wt.% FC751, 2% NaCl) solution.

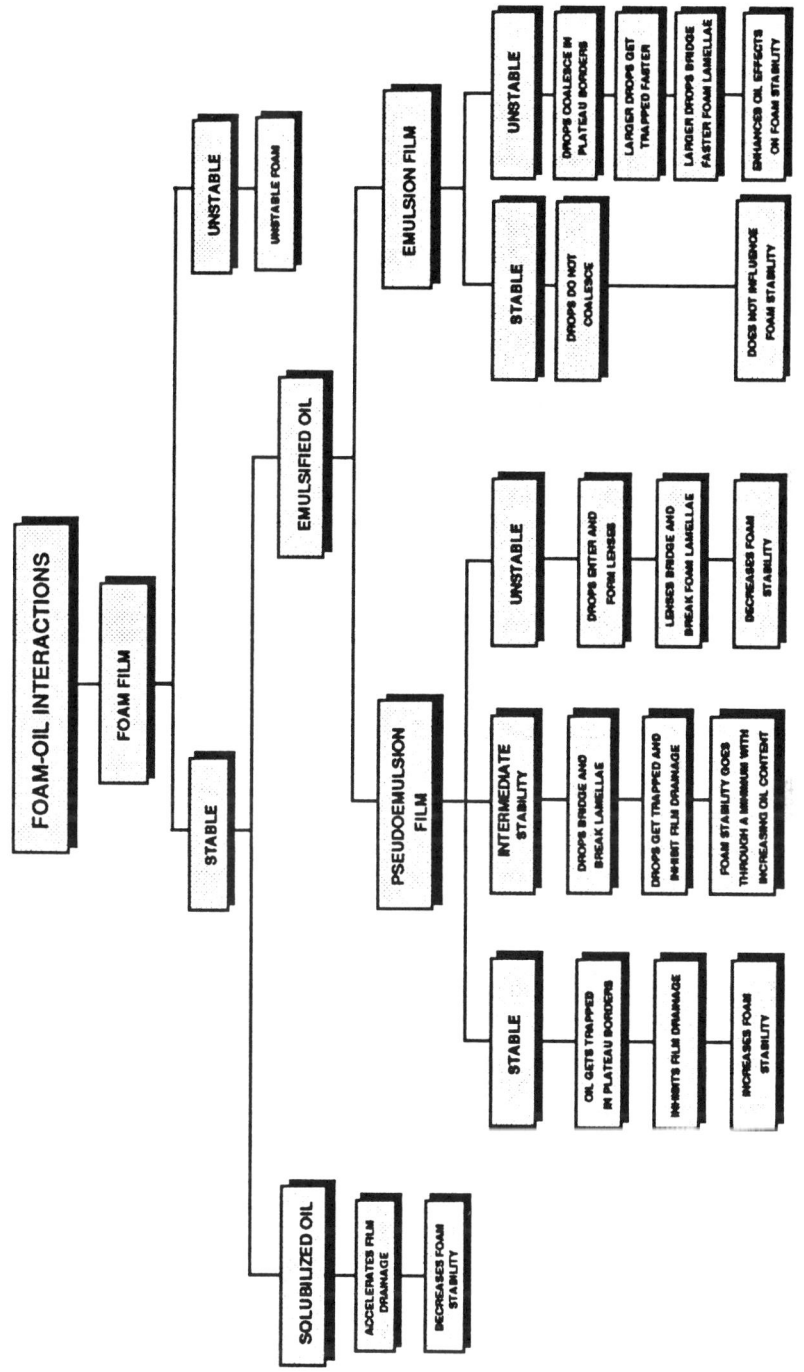

Figure 45. Summary of foam–oil interactions.

better understand the mechanisms of stability in curved foam, pseudoemulsion, and emulsion films and particularly the role of short- and long-range forces in their stability.

List of Symbols

Bo	Boussinesq number
c	concentration
D	diffusivity
D^s	surface diffusivity
E	entering coefficient
E_o	Gibbs elasticity
E_s	dimensionless elasticity number
f	characteristic foam cell dimension
h	liquid film thickness
$\overline{h} = h/f$	dimensionless film thickness
h_{cr}	critical film thickness
h_i	initial film thickness
$\overline{h}_i$	dimensionless initial film thickness
h_f	final film thickness
$\overline{h}_f$	dimensionless final film thickness
h_f^o	initial foam height
K	leading-order dilatational foam viscosity
ΔP_c	capillary pressure
$Pe_s = Uf/\epsilon D^s$	surface Peclet number
R_f	film radius
S	spreading coefficient
t	time
U	rate of film drainage
U_{Re}	rate of film drainage between two, rigid plane-parallel surfaces

Greek

Δ	rate of foam dilatation
ϵ	relative foam film size
θ_{AW}	air–solid–water contact angle
θ_{OW}	oil–solid–water contact angle
κ^s	interfacial dilatational viscosity
μ	bulk viscosity
μ^s	interfacial shear viscosity

$\Pi(h)$	disjoining pressure in a liquid film with thickness h
ρ^s	surface density
σ	interfacial or surface tension
$\sigma_{O/A}$	oil–air surface tension
$\sigma_{W/A}$	water–air surface tension
$\sigma_{W/O}$	oil–water interfacial tension
$\tau_{1/2}$	foam half-life

Acknowledgments

This study was supported in part by the National Science Foundation, the U.S. Department of Energy, and the United States–Hungarian Joint Fund (J.F. No. 11/91).

References

1. Nikolov, A. D.; Wasan, D. T.; Huang D.; Edwards D. A. Presented at the 61st Annual Technical Conference and Exhibition of SPE, New Orleans, LA, October 5–8, 1986; SPE Preprint 15443.
2. Hooke R. "On Holes in Soap Bubbles;" Communication to the Royal Society; London, 1672, see T. Birch, History of the Royal Society; Millard: London, 1757.
3. Newton, I. *Optiks;* Smith and Watford: London, 1704; Book II, Part I, Observation 17.
4. Gibbs, J. W. *The Collected Works of J. Willard Gibbs;* Yale University: New Haven, CT, 1957; Vol. I.
5. Boys, C. V. *Soap Bubbles and the Forces Which Mould Them;* Society for the Promotion of Christian Knowledge, E. and J. B. Young: New York, 1890.
6. Mysels, K. J.; Shinoda, S.; Frankel, S. *Soap Films—A Study of Their Thinning and a Bibliography;* Pergamon: New York, 1959.
7. Bikerman, J. J. *Foams;* Springer-Verlag: New York, 1978.
8. Isenberg, C. *The Science of Soap Films and Soap Bubbles;* Dover Publications: New York, 1992.
9. Scheludko, A. *Adv. Colloid Interface Sci.* **1967**, *1*, 391.
10. McKendrick, C. B.; Smith, S. J.; Stevenson, P. A. *Colloids Surf.* **1991**, *52*, 47.
11. Vrij, A. *Disc. Faraday Soc.* **1966**, *42*, 23.
12. Derjaguin, B. V.; Gutop, V. V. *Kolloidn. Zh.* **1962**, *24*, 431.
13. Derjaguin, B. V.; Prokhorov, A. V. *J. Colloid Interface Sci.* **1981**, *81*, 108.
14. Derjaguin, B. V.; Kusakov, M. M. *Izv. Acad. Sci. U.S.S.R.* **1936**, *1*, 256.
15. Nikolov, A. D.; Wasan, D. T.; Kralchevsky, P. A.; Ivanov, I. B. In *Ordering and Organisation in Ionic Solutions;* Ike, N.; Sogami, I., Eds.; World Scientific: Singapore, 1988.
16. Nikolov, A. D.; Wasan, D. T. *J. Colloid Interface Sci.* **1989**, *133*, 1.
17. Reynolds, O. *Philos. Trans. R. Soc. London A* **1886**, *177*, 157.

18. Ivanov, I. B.; Dimitrov, D. J. *Colloid Polym. Sci.* **1974**, *252*, 982.
19. Zapryanov, Z.; Malhotra, A. K.; Aderangi, N.; Wasan, D. T. *Int. J. Multiphase Flow* **1983**, *9*, 105.
20. Wasan, D. T.; Malhotra, A. K. *AIChE Symp. Ser.* **1986**, *82*, 5.
21. Edwards, D. A.; Brenner, H.; Wasan, D. T *Interfacial Transport Processes and Rheology;* Butterworth–Heinemann: Stoneham, MA, 1991.
22. Huang, D. D.; Nikolov, A. D.; Wasan, D. T. *Langmuir* **1986**, *2*, 672.
23. Radoev. B. P.; Dimitrov. D. S.; Ivanov, I. B. *Colloid Polym. Sci.* **1974**, *252*, 50.
24. Barber, A. D.; Hartland, S. *Can. J. Chem. Eng.* **1976**, *54*, 279.
25. Ivanov, I. B.; Dimitrov, D. S.; Somasundaran, P.; Jain, R. K. *Chem. Eng. Sci.* **1985**, *40*, 137.
26. Malhotra, A. K.; Wasan, D. T. *Chem. Eng. Commun.* **1987**, *55*, 95.
27. Jain, R. K.; Ruckenstein, E. R. *J. Colloid Interface Sci.* **1976**, *54*, 1.
28. Gumerman, R.; Homsy, G. *Chem. Eng. Commun.* **1975**, *2*, 27.
29. Malhotra, A. K.; Wasan, D. T. *AIChE J.* **1987**, *33*, 1533.
30. Brown, A. G.; Thuman, W. C.; McBain, J. W. *J. Colloid Interface Sci.* **1953**, *8*, 491.
31. Davies, J. T.; Rideal, R. K. *Interfacial Phenomena;* Academic: London, 1963.
32. Malhotra, A. K.; Wasan, D. T. In *Thin Liquid Films;* Ivanov, I. B., Ed.; Marcel Dekker: New York, 1990; p 95.
33. Edwards, D. A.; Wasan, D. T. In *Surfactants in Chemical/Process Engineering;* Wasan, D. T.; Ginn, M. E.; Shah, D. O., Eds.; Surfactants Science Series 28; Marcel Dekker: New York, 1988; p 1.
34. Kraynik, A. M. *Proc. Can. Congr. Appl. Mech. 11th* **1987**, *2*, B2–3.
35. Princen, H. M. *Colloid Interface Sci.* **1985**, *105*, 150.
36. Edwards, D. A.; Brenner, H.; Wasan, D. T. *J. Colloid Interface Sci.* **1989**, *130*, 266.
37. Edwards, D. A.; Wasan, D. T. In *Foams;* Prud'homme, R.; Khan, S., Eds.; Marcel Dekker: New York, in press.
38. Princen, H. M.; Aronson, M. P.; Moser, J. C. *J. Colloid Interface Sci.* **1980**, *75*, 247.
39. Princen, H. M.; Levinson, P. *J. Colloid Interface Sci.* **1987**, *120*, 172.
40. Edwards, D. A.; Wasan, D. T. *J. Colloid Interface Sci.* **1990**, *139*, 479.
41. Owete, S. O.; Brigham, W. E. *Flow of Foams through Porous Media;* Department of Energy: Washington, DC, July 1984; SUPRI TR-37 DOE/SF/115646 (Contract No. DE 84012410).
42. Hirasaki, G. J.; Lawson, J. B. *Mechanisms of Foam Flow in Porous Media: Apparent Viscosity in Smooth Capillaries;* April 1985.
43. Marsden, S. S. *Foams in Porous Media;* Department of Energy: Washington, DC, May 1986; SUPRI TR 49.
44. Jensen, J. A.; Friedmann, F. Presented at the SPE California Regional Meeting, Ventura, CA, April 8–10, 1987; paper SPE 16375.
45. Rossen, W. R. Presented at the SPE/DOE Symposium on Enhanced Oil Recovery, Tulsa, OK, April 17–20, 1988; paper SPE/DOE 17358.
46. de Vries, A. S.; Wit, K. *Soc. Pet. Eng. Reservoir Eng.* **1990**, *5*, 5.
47. Friberg, S.; Linden, S. E.; Saito, H. *Nature (London)* **1974**, *251*, 495.

48. Johonnott, E. S. *Philos. Mag.* **1906**, *11*, 746.
49. Perrin, R. E. *Ann. Phys. (Paris)* **1915**, *10*, 160.
50. Bruil, H. G.; Lyklema, J. *Nature (London)* **1971**, *232*, 19.
51. Kenskemp, J. W.; Lyklema, J. *Adsorption at Interfaces;* Mittal, K. L., Ed.; ACS Symposium Series 8; American Chemical Society: Washington, DC, 1975; pp 191–198.
52. Friberg, S.; Linden, Ste.; Saito, H. *Nature (London)* **1974**, *25*, 1494.
53. Manev, E.; Sazdanova, S. V.; Wasan, D. T. *Dispersion Sci. Tech.* **1982**, *3*, 435.
54. Nikolov, A. D.; Wasan, D. T.; Denkov, N. D.; Kralchevsky, P. A.; Ivanov, I. B. *Prog. Colloid Polym. Sci.* **1990**, *82*, 87.
55. Nikolov, A. D.; Wasan, D. T.; Kralchevsky, P. A.; Ivanov, I. B. In *Ordering and Organization in Ionic Solutions;* Ike, N.; Sogami, I. Eds.; World Scientific: Singapore, 1988.
56. Wasan, D. T. *Chem. Eng. Educ.* **1992**, *26*, 104–112.
57. Nikolov, A. D.; Kralchevsky, P. A.; Ivanov, I. B.; Wasan, D. T. *J. Colloid Interface Sci.* **1989**, *133*, 13–22.
58. Derjaguin, B.; Landau, L. *Acta Physicochim.* **1941**, *14*, 633.
59. Verwey, E.; Overbeek, J. Th. G. *Theory of The Stability of Lyophobic Colloids;* Elsevier: Amsterdam, The Netherlands, 1948.
60. Basheva, E. S.; Nikolov, A. D.; Kralchevsky, P. A.; Ivanov, I. B.; Wasan, D. T. In *Surfactants in Solution;* Mittal, K., Ed.; PLenum: New York, 1990; Vol. 11, p 467.
61. Nikolov, A. D.; Wasan, D. T. *Langmuir* **1992**, *8*, 2985, Special J. B. Perrin Issue.
62. Kirkwood, J. G. *J. Chem. Phys.* **1939**, *7*, 919.
63. Alder, B. J.; Wainright, T. E. *Phys. Rev.* **1962**, *127*, 359.
64. Kralchevsky, P. A.; Nikolov, A. D.; Wasan, D. T.; Ivanov, I. B. *Langmuir* **1990**, *6*, 1180.
65. Nikolov, A. D.; Wasan, D. T. Presentation at the American Institute of Chemical Engineers Meeting, Chicago, IL, November 1990.
66. Lobo, L. A.; Nikolov, A. D.; Wasan, D. T. *J. Dispersion Sci. Tech.* **1989**, *10*, 143.
67. Nakagawa, T.; Shinoda, K. In *Colloidal Surfactants;* Shinoda, K.; Nakagawa, T.; Tamamushi, B.; Isemura, T., Eds.; Academic, Orlando, FL, 1963, p 139.
68. Nilsson, G. *J. Phys. Chem.,* **1957**, *64*, 1135.
69. Ross S.; McBain, J. W. *Ind. Eng. Chem.* **1944**, *36*, 570.
70. Prins, A. In *Food Emulsions and Foams;* Dickinson, E., Ed.; Special Publication of the Royal Society of Chemistry No. 58; Royal Society of Chemistry: Letchworth, England, 1987; p 30.
71. Ross, S. *J. Phys. Colloid Chem.* **1950**, *54*, 429.
72. Harkins, W. D. *J. Chem. Phys.* **1941**, *9*, 552.
73. Robinson, S.; Woods W. W. *J. Soc. Chem. Ind.* **1948**, *67*, 361.
74. Tsuge, H.; Ushida, J.; Hibino, S. *J. Colloid Interface Sci.* **1984**, *100*, 175.
75. Kuhlman, M. I. Presented at the SPE/DOE Symposium on Enhanced Oil Recovery, Tulsa, OK, April 17–20, 1988; paper SPE/DOE 17356.
76. Schramm, L. L.; Novosad, J. J. *Colloids Surf.* **1990**, *46*, 21.
77. Manlowe, D. T.; Radke, C. J. *SPE Reservoir Eng.* **1990**, *5*, 495.
78. Wasan, D. T.; Nikolov, A. D.; Huang, D. D.; Edwards, D. A. In *Surfactant-*

Based Mobility Control: Progress in Miscible-Flood Enhanced Oil Recovery; Smith, D. H., Ed., ACS Symposium Series 373; American Chemical Society: Washington, DC, 1988; pp 136–162.
79. Krugljakov, P. M. In *Thin Liquid Films;* Ivanov, I. B., Ed.; Surfactant Science Series 29; Marcel Dekker: New York, 1988; Chapter 11.
80. Hanssen, J. E.; Dalland, M. Presented at the SPE 7th International Symposium on Enhanced Oil Recovery, Tulsa, OK, April 1990; paper SPE/DOE 20193.
81. Lobo. L. A.; Wasan, D. T.*Langmuir* **1993,** *9,* 1668.
82. Koczo, K.; Lobo, L.; Wasan, D. T. *J. Colloid Interface Sci.* **1992,** *150,* 492.
83. *Ullmann's Encyclopedia of Industrial Chemistry*, 5th ed.; VCH Verlagsgesellschaft: Weinheim, Germany, 1988; Vol. A11, pp 466–490.
84. *Kirk–Othmer Encyclopedia of Chemical Technology, Vol. 7;* Wiley-Interscience: New York, 1984; pp 430–447.
85. Pape, P. G. *J. Pet. Technol.* **1983,** *June,* 1197–1204.
86. Aveyard, R.; Cooper, P.; Fletcher, P. D. I.; Rutherford, C. E. *Langmuir* **1993,** *9,* 604–613.
87. Sawicki, G. C. *JAOCS J. Am. Oil Chem. Soc.* **1988,** *65(6),* 1013–1016.
88. Garrett, P. R. In *Defoaming: Theory and Industrial Applications;* Garrett, P. R., Ed.; Marcel Dekker: New York, 1993; Chapter 1.
89. McGee, J. B. *Chem. Eng.* **1989,** *4,* 131–136.
90. Frye, G. C.; Berg, J. C. *J. Colloid Interface Sci.* **1989,** *130(1),* 54–59.
91. Garrett, P. R. *J. Colloid Interface Sci.* **1980,** *76(2),* 587–590.
92. Garrett, P. R. *J. Colloid Interface Sci.* **1979,** *69(1),* 107–121.
93. Dippenaar, A. *Int. J. Min. Proc.* **1982,** *9,* 1.
94. Aronson, M. P. *Langmuir* **1986,** *2,* 653–659.
95. Frye, G. C.; Berg, J. C. *J. Colloid Interface Sci.* **1989,** *127(1),* 222–238.
96. Lucassen, J. *Colloids Surf.* **1992,** *65,* 139–149.
97. Kulkarni, R. D.; Goddard, E. D.; Kanner, B. *Ind. Eng. Chem. Fundam.* **1977,** *16(4),* 472–474.
98. Koczo, K.; Koczone, J. K.; Wasan, D. T. *J. Colloid Interface Sci.,* in press.
99. Mohammadi, S. S.; Van Slyke, D. C.; Ganong, B. L. *SPE Reservoir Eng.* **1989,** *2,* 7–15.
100. Ploeg, J. F.; Duerksen, J. H. Presented at the SPE California Regional Meeting, Bakersfield, CA, March 27–29, 1985; paper SPE 13609.
101. Djabbarah, H. F.; Weber, S. L.; Freeman, D. C.; Muscatello, J. A.; Ashbaugh, J. P.; Covington, T. E. Presented at the 60th California Regional Meeting, Ventura, CA, April 4–6, 1990; paper SPE 20067.
102. Bernard, G. G.; Holm, L. W. *SPE J.* **1964,** *9,* 267–274.
103. Suffridge, F. E.; Raterman, K. T.; Russell, G. C. Presented at the 64th SPE Annual Meeting, San Antonio, TX, October 1989; paper SPE 19691.
104. Falls, A. H.; Lawson, J. B.; Hirasaki, G. J. Presented at the SPE California Regional Meeting, Oakland, CA, April 1986; paper SPE 15053.

RECEIVED for review October 23, 1992. ACCEPTED revised manuscript June 11, 1993.

3

Fundamentals of Foam Transport in Porous Media

A. R. Kovscek and C. J. Radke[*]

Earth Sciences Division of Lawrence Berkeley Laboratory and Department of Chemical Engineering, University of California, Berkeley, CA 94720

Foam in porous media is a fascinating fluid both because of its unique microstructure and because of its dramatic influence on the flow of gas and liquid. A wealth of information is now compiled in the literature that describes. The literature contains conflicting views of the mechanisms of foam generation, destruction, and transport, and on the macroscopic results they produce. By critically reviewing how surfactant formulation and porous media topology conspire to control foam texture and flow resistance, we attempt to unify these disparate viewpoints. Evolution of texture during foam displacement is quantified by a population-balance on bubble concentration that is designed specifically for convenient incorporation into a standard reservoir simulator. Theories for the dominant bubble generation and coalescence mechanisms provide physically based rate expressions for the proposed population-balance. Stone-type relative permeability functions, along with the texture-sensitive and shear-thinning nature of confined foam, complete the model. Quite good agreement is found between theory and new experiments for transient foam displacement in linear cores.

THE MOBILITIES OF CONTINUOUS, NEWTONIAN FLUIDS in reservoir media are inversely proportional to their viscosities. Thus, gas drive fluids for enhanced oil recovery, such as dense carbon dioxide, enriched hydrocarbons, nitrogen, and steam, are highly mobile; this feature causes them

[*]Corresponding author

to channel selectively through zones of high permeability rather than efficiently displace oil. Further, gas drive fluids are also less dense than both brine and crude oil. They rise to the top of the reservoir and override the oil-rich zones. Traditional gas-displacement processes lack mobility control and result in poor volumetric displacement efficiency due to both channeling and gravity override.

In 1961, Fried (*1*) demonstrated that aqueous surfactant-stabilized foam could drastically reduce the mobility of gases in porous media. At that time, foam was studied mainly from a phenomenological perspective. In the intervening 30 years, foam has been recognized as a fluid with unique rheological properties within porous media, and the scope of research has expanded to include local pore-scale phenomena and local microstructure. Because of its dispersed nature, foam profoundly affects the flow patterns of nonwetting fluids within porous media.

In this chapter, we discuss much of the work accomplished since Fried, but without attempting a complete review. Useful synopses are available in the articles and reports of Hirasaki (*2, 3*), Marsden (*4*), Heller and Kuntamukkula (*5*), Baghidikian and Handy (*6*), and Rossen (*7*). Our goals are to present a unified perspective of foam flow in porous media; to delineate important pore-level foam generation, coalescence, and transport mechanisms; and to propose a readily applicable one-dimensional mechanistic model for transient foam displacement based upon gas-bubble size evolution [i.e., bubble or lamella population-balance (*8, 9*)]. Because foam microstructure or texture (i.e., the size of individual foam bubbles) has important effects on flow phenomena in porous media, it is mandatory that foam texture be accounted for in understanding foam transport.

This discussion follows the goals listed previously. First, we describe how foam is configured within porous media, and how this configuration controls foam transport. Next, we review briefly pertinent foam generation and coalescence mechanisms. Finally, we incorporate pore-level microstructure and texture-controlling mechanisms into a population-balance to model foam flow in porous media consistent with current reservoir-simulation practice (*10*). Attention is focused on completely water-wet media that are oil free. Interaction of foam with oil is deferred to Chapter 4.

Foam Microstructure

As demonstrated in Chapter 1, foam is a gas phase dispersed within a liquid phase and stabilized by surfactant adsorbed at the gas–liquid interfaces. In reservoir applications, foams are usually formed by nonwetting gases, such as steam or nitrogen, dispersed within a continuous, wetting aqueous phase containing surface-active agents. Foams formed with dense

CO_2 as the internal phase are strictly emulsions, which led Wellington (5) to coin the phrase "foamulsion". The term foam is retained here. Some nonaqueous foams in porous media have been studied primarily for use as barriers against gas coning through thin oil zones (11, 12) and well stimulation (13, 14). However, because the basic principles appear similar, the discussion is limited to aqueous foams.

The behavior of foam in porous media is intimately related to the connectivity and geometry of the medium in which it resides. Porous media have several attributes that are important to foam flow. First, they are characterized by a size distribution of pore-bodies (sometimes called pores) interconnected through pore-throats of another size distribution. Body and throat size distributions are important as is their possible correlation to determine the distribution of body to throat size ratios. Foam generation and destruction mechanisms in porous media depend strongly on the body to throat size aspect ratio.

Second, pores are not cylindrical but exhibit corners, as illustrated in Figure 1. Under two-phase occupancy, one fluid preferentially wets the pore walls. The wetting fluid completely fills the smallest pores, and resides in the corners of gas-occupied pores and in thin wetting films coating pore walls. Because of the thin films, liquid in the corners of adjacent pores is contiguous with wetting liquid in the smallest pore space. Hence, the wetting phase remains continuous even down to very low saturations. The nonwetting phase resides in the central portion of the largest pores. Accordingly, Figure 1 portrays multiple phases jointly occupying the largest pores.

The third important aspect of porous media is that at the pore level when flow rates are low and tension forces dominate (i.e., for bond and capillary numbers much less than unity), the capillary pressure is constant, set by the local saturation of the wetting phase and the value of the interfacial tension. Local imbalances in capillary pressure tend to equalize pri-

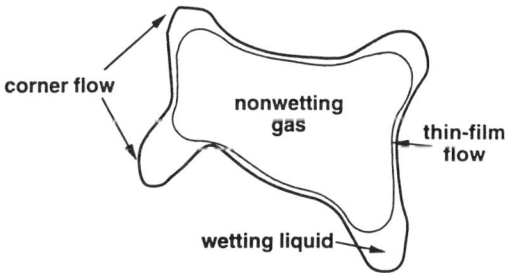

Figure 1. Cross-sectional view of cornered pores. Shaded wetting fluid is held in pore corners. (Reproduced with permission from reference 26. Copyright 1991 Marcel Dekker.)

marily through the interlinked, continuous wetting phase. Local capillary-pressure information is transmitted to foam lamellae or lenses in the same manner. During biphasic flow of continuous fluids, the nonwetting fluid flows in interconnected large pore channels. Wetting fluid flows in interconnected small pore channels and in the corners, and thin films of the nonwetting-phase occupy pores because of pressure gradients in the aqueous phase and viscous traction at the contacting interface.

The mechanisms of foam flow in porous media are not nearly so clear, although they have been scrutinized since the earliest studies (1, 15). Even at small Reynolds numbers, at least four microstructure regimes may be envisioned. When the characteristic length scale of the pore space is much greater than the size of individual foam bubbles, the foam is properly designated as bulk. Bulk foam is further divided into kugelschaum ("ball foam") and polyederschaum ("polyhedral foam") (16). Kugelschaum consists of well-separated, spherical bubbles, whereas polyederschaum consists of polyhedral bubbles separated by surfactant-stabilized, thin liquid films called lamellae. The distinguishing feature is the fraction of foam volume that is gas or the foam quality. High gas fractions correspond to polyederschaum and vice versa. Bulk foam is discussed further in Chapter 2; the flow of bulk foam outside porous media is reviewed by Kraynik (17).

Conversely, when the characteristic pore size of the medium is comparable or less than the characteristic size of the dispersed gas bubbles, the bubbles and lamellae span completely across pores (18). We designate this regime as confined foam. Again, it is possible to distinguish two types of morphology based on the gas content of the foam. However, in situ quality for confined foam is not well defined. It is preferable to classify the two regimes according to gas fractional flow. Almost all researchers who study foam in porous media use quality to represent fractional flow. We also follow this convention. At low gas fractional flow, the pore-spanning bubbles are widely spaced, separated by thick wetting liquid lenses or bridges. Conversely, at high gas fractional flow, the pore-spanning bubbles are in direct contact, separated by lamellae. Hirasaki and Lawson denote this direct-contact morphology as the individual-lamellae regime (18). Because most foam-displacement applications in reservoir media use reasonably high gas fractional flows where the gas–aqueous-phase capillary pressure is high, bulk polyhedral foam and confined, individual-lamellae foam are the two candidate textures.

Marsden et al. (19) were apparently the first to determine the relevant size scale characteristic of foam in porous media. They measured mean bubble sizes for foam exiting sand packs, and concluded that foam bubbles were roughly the same size as pore bodies. Despite the equivalence of bubble and pore sizes, they treated foam as a continuous, non-Newtonian fluid (i.e., a bulk foam). Holm (15), at roughly the same time, injected pregenerated foam into sand packs and measured effluent bubble sizes. He noted effluent steady-state bubble diameters larger (0.6 mm) than the average sand grain size (roughly 0.1 mm) and also larger than the injected bubble diameter (0.4 mm) for a foam of 90% quality and unspecified flow rates. Additionally, he concluded that under steady-state conditions no free gas was present in the medium.

More recently, Ettinger and Radke (20, 21) measured steady-state effluent bubble size distributions exiting 0.5- to 1.0-μm^2 Berea sandstones, and found average bubble sizes to be roughly twice as large as pore dimensions. Bubble diameters ranged from 0.05 to 0.7 mm depending upon the permeability of the sandstone and the gas and liquid injection rates. Average pore-body size in similar Berea sandstones is roughly 0.18 mm (22). Ettinger and Radke confirmed the finding of Holm (15) that pregenerated foam is reshaped by the porous medium. Further, Trienen et al. (23) found bubbles exiting sand packs to be roughly 10 times larger than typical pore size at flow rates close to 1 m/day. Because effluent bubble sizes equal to or larger than pore dimensions are universally reported, it is now generally accepted that single bubbles and lamellae span the pore space of most porous media undergoing foam flow.

Direct observation of foam in transparent etched-glass micromodels confirms that foam bubbles and lamellae generally span entire pores (24–26). A photomicrograph of a foam-filled micromodel (27) is shown in Figure 2. The lightest portions represent the rock matrix. Light gray shading corresponds to the wetting aqueous surfactant solution. The most darkly shaded portions indicate gas. Most important, the dark pore-spanning lamellae terminate in so-called Plateau borders adjacent to the pore walls. A white marker in the lower left of Figure 2 corresponds to a scale of 0.1 mm. Clearly, the foam in Figure 2 is not bulk. The foam does not enter the large water-filled pore indicated by the black arrow because gas can not enter the barely visible, small pore-throat just below it. In spite of the dispersed state of the gas, it remains the nonwetting phase. Finally, cushions of wetting aqueous liquid reside next to the walls in the gas-occupied pores. These correspond to the wetting corner liquid pictured in Figure 1.

Figure 2 illustrates what is coined a discontinuous-gas foam (2, 9), in that the entire gas phase is made discontinuous by lamellae, and no gas channels are continuous over sample-spanning dimensions. Gas is encapsulated in small packets or bubbles by surfactant-stabilized aqueous films. These packets transport in a time-averaged sense through the porous medium (20).

Not all confined foams are discontinuous (9). A continuous-gas foam is illustrated schematically in Figure 3. In continuous-gas foam the medium contains one, or several, interconnected gas channels that are uninterrupted by lamellae over macroscopic distances. As in discontinuous-gas

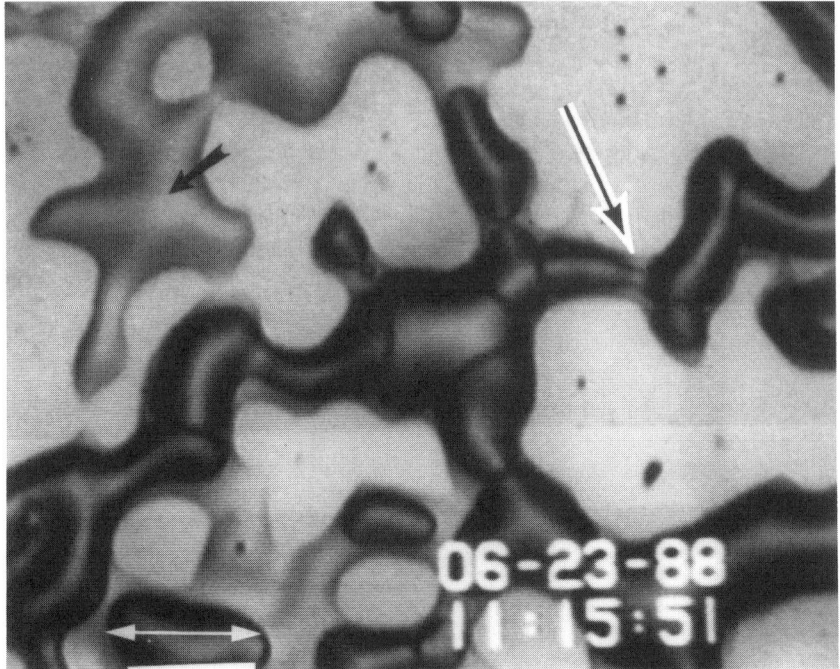

Figure 2. Photomicrograph of foam in a transparent-glass micromodel. A white scale marker at lower left corresponds to 100 μm. The leftmost black arrow marks a large liquid-filled pore connected to the foam-flow channel by a small pore-throat. The black arrow to the right of center indicates the growth of a wetting collar (27).

confined foam, the wetting aqueous phase again fills the smallest pores (not shown in Figure 3). Continuous-gas foams (9) have been observed in bead packs constructed from 0.6- to 0.8-mm sintered glass beads. Further evidence of continuous-gas foam is provided by Hanssen (11, 28, 29).

Typically, a discontinuous-gas foam forms under conditions of coinjection of wetting surfactant solution and gas, where ample wetting phase is present and available for foam-generation events. Continuous-gas foams can form when the wetting-phase saturation falls sufficiently low, so that significant lamella generation ceases, such as in the later stages of gas injection. Continuous-gas channels result. Nevertheless, stationary lamellae and bubbles remain and block gas transport (Figure 3) through the remaining portions of the pore network (9, 11). Gas mobility is still greatly reduced.

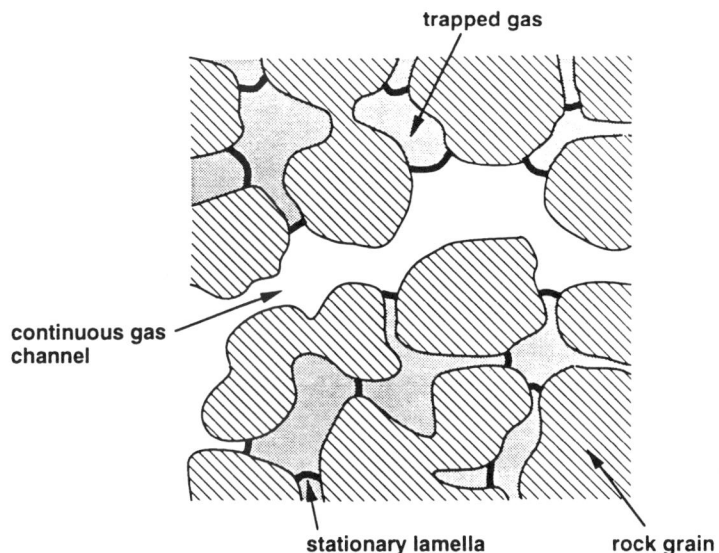

Figure 3. Schematic of a continuous-gas foam in porous media. Rock grains are hatched; gray shading represents gas trapped by stationary lamellae, shown as thick dark lines. A continuous-gas channel is pictured unshaded.

Foam Transport

The basic morphology of confined foam in porous media, shown in Figures 2 and 3, suggests that gas mobility appearing in Darcy's law may be conceptually divided into effects on the gas permeability and viscosity. Because foam in porous media is strictly a nonlinear fluid, this separation may not be formally rigorous (5, 30). Nevertheless, this idea seems necessary to make progress toward modeling foam flow. Thus, stationary gas is equivalent to lowering the gas effective permeability and allowing gas flow in the open channels. Conversely, bubbles that flow in interconnected channels contribute extra resistance to transport that is best described by an effective gas viscosity.

Effective Permeability. Bernard and co-workers (31, 32) pursued pioneering studies to quantify gas and liquid permeabilities in the presence of foam. They either coinjected surfactant solution and nitrogen, or used alternating slugs of each. For consolidated porous media with absolute permeabilities that range from 0.1 to 0.25 μm^2, they found several hundred-fold permeability reductions to gas. For a sand pack with an absolute permeability of 3.89 μm^2, gas permeability was reduced to less than 10^{-3} μm^2 in the presence of foam.

Two important experimental observations were made at this early stage. Bernard et al. (*32*) varied aqueous-phase flow rates, and measured aqueous saturations and overall pressure drops. Water relative permeabilities calculated with Darcy's law were surprisingly unaffected by foam. When compared under identical flow-rate conditions, steady-state water saturations were different between foam and surfactant-free, two-phase flow cases, but the aqueous-phase relative permeability, k_{rw}, versus aqueous saturation, S_w, relation was unchanged. Because the wetting aqueous phase is primarily concentrated in small pores, it is unaffected by the dispersed nature of the nonwetting phase at any given saturation (Figure 2). Subsequently, the independence of the wetting-phase relative permeability on the presence of foam has been reconfirmed numerous times (*15, 33–36*).

Second, after foam flooding cores, Bernard et al. (*32*) flushed with water or brine to estimate trapped-gas saturation. They assumed that water or brine filled the pore space through which gas flowed but did not substantially alter the fraction of gas trapped. Their trapped saturations ranged from 10 to 70% depending upon the surfactant type and the presence of oil in the porous medium during the foam flood. Such measured saturations apply only to trapped gas following a waterflood, and not to dynamic or steady-state foam flooding.

Knowledge of foam trapping remains incomplete. Gas-phase tracer experiments, however, can measure trapped gas saturation of foams at steady state. Nahid (*37*) apparently employed the technique first. More recently, gas-phase tracer experiments have revealed trapped gas saturations for nitrogen foams at steady state in Berea sandstones that range from roughly 80% to nearly 100% over a variety of flow rates (*38, 39*). Friedmann et al. (*39*) used a krypton tracer for a range of frontal advance rates between 25 and 130 m/day. The fraction of gas trapped as a function of gas-phase velocity at constant fractional flow was measured. Interestingly, little change was found in the fraction trapped (approximately 85%) even though the gas velocity was varied by over 2 orders of magnitude. To correct for partitioning of the tracer into the trapped fraction, Gillis and Radke (*38*) used sulfur hexafluoride and methane tracers simultaneously. Total flow rates ranged from 0.5 to 4 m/day. No consistent trend of trapped fraction with liquid or gas velocity was reported. Remarkably, even when only gas was injected into a Berea sandstone at connate saturation of aqueous surfactant solution, an appreciable fraction of the gas (in excess of 70%) remained trapped. A continuous-gas foam was likely to be operative under these conditions.

Flumerfelt and Prieditis (*40*) performed a similar gas-only injection into a 7-μm^2 bead pack. A foam was first generated under conditions of simultaneous injection of gas and surfactant solution at a variety of gas rates but at fixed liquid rates. After steady state was reached, liquid flow was discontinued, and the foam was allowed to decay until continuous gas was produced. It was demonstrated that the permeability of the bead pack to gas at the first appearance of effluent continuous gas was 2 orders of magnitude less than the foam-free case, and that this permeability was independent of gas and initial liquid flow rates. It was concluded that the number of channels available to carry gas was 100 times less in the presence of foam than in the foam-free case.

These macroscopic measurements of gas trapping are confirmed by visual observations in transparent etched-glass micromodels and bead packs (*24–26, 41*). Trapped foam severely reduces the effective permeability of gas moving through a porous medium by blocking all but the least resistive flow paths. Hence, trapped gas reduces the void volume of the porous medium available for flow. Thus, higher flow resistances are measured, and lower permeabilities to gas are computed. This trapped gas accounts for some, but not all, increased resistance to flow.

Visual observation in etched-glass micromodels (*26*) additionally shows that foam trapping is an intermittent process. At steady state, only a portion of the foam flows during any moment. Primary channels, known as backbone channels, carry the major portion of flowing gas. These are relatively few in number. Leading off of the backbone channels are secondary or dendritic channels. None of these channels are always open to flow or blocked at all times. Sporadically, a series, or train, of foam bubbles mobilizes and flow begins. The identity of the individual bubbles in the series changes constantly. Bubbles may join or leave the series, and individual lamellae may be broken or generated. Later, this series of bubbles may cease flowing and block a channel because of a switch in the flow path. The primary characteristic of porous media that permits this switching of flow paths is a high degree of interconnectedness. Clearly, in both continuous and discontinuous foam, trapped gas constitutes the majority of the gas volume in the medium, and must be accounted for in any modeling effort.

On a theoretical level, foam mobilization and trapping is perhaps best tackled with percolation models. Such models (*42–47*), coupled with micromodel visualization, support the preceding mechanistic view of foam trapping. Results indicate that the pressure gradients required to maintain foam flow are quite high, on the order of 2 to >200 kPa/m (1–10 psi/ft) (*42–44*). Rossen (*45*) suggested that the ease of initiating flow increases with the degree of interconnectedness and lower interfacial tension, but decreases with gas compressibility. The most important factors that affect bubble trapping are pressure gradient, gas velocity, pore geometry, foam texture, and bubble-train length (*42*).

Effective Viscosity. Considerable evidence indicates that in some gas-occupied channels, confined foam bubbles transport as bubble-trains. Effluent bubble sizes from 0.8-μm^2 Berea sandstone reflect expected sizes and their predicted shift with flow velocity (*20*). Likewise, pregenerated foam is reshaped to the same average exiting bubble size quite independent of the average inlet size (*20*). As with trapped foam, there is ample direct visual documentation of flowing foam bubble-trains in both micromodels (*26*) and in bead packs (*9, 48*). The flow resistance of transporting bubble-trains is best addressed in terms of an effective gas viscosity.

Initially several researchers measured the effective viscosity of bulk foam using rotational or capillary viscometers (*1, 49*) with the hope of applying their results to porous media. On the basis of the earlier discussion of foam morphology in porous media, such data are inappropriate (*50*). Interaction of elongated bubbles and pore-spanning lamellae with pore walls determines the effective viscosity of the flowing portion of foam. Such interactions are simply not mirrored in bulk foam viscometry.

Bretherton (*51*) provided the cornerstone study for understanding the effective viscosity of confined foam. Long surfactant-free gas bubbles were generated and were flowed through small, liquid-filled, cylindrical capillary tubes such that the bubbles completely spanned the capillary diameter. The experiments demonstrated that for strong liquid-wet capillaries, a thin liquid film deposits on the capillary wall. Film thickness increases with increasing bubble velocity, U. From a theoretical hydrodynamic analysis, Bretherton established that such bubbles indeed slide over a constant thickness film. The film thickness divided by tube radius scales as $Ca^{2/3}$, where $Ca = \mu U/\sigma$ is the capillary number for a liquid of viscosity, μ, and equilibrium surface tension, σ. Over the range $10^{-6} < Ca < 10^{-2}$, where inertia is unimportant, Bretherton's theory and experiments are in satisfactory agreement (*51*). Additionally, the pressure drop to drive a single bubble nondimensionalized by the ratio of surface tension to tube radius varies as $Ca^{2/3}$. As a result, the effective viscosity of an elongated, inviscid gas bubble, defined from Poiseuille flow, has an inverse 1/3 dependence on bubble velocity, and at low capillary numbers is actually larger than that of an equivalent volume of liquid. The reason is that the shear rate of the liquid in the thin films near the front and back of the bubble is larger than that for simple parabolic flow of the liquid phase. Wong (*52*) extended the Bretherton analysis to square tubes to consider the role of pore corners noted in Figure 1. Aside from detailed differences in the nature of the thin wetting films deposited on the flat portions of the tube walls, the pressure drop to drive the bubble again scales as $Ca^{2/3}$. Straight, cornered pores still obey Bretherton's basic theory but with somewhat different scaling constants.

Hirasaki and Lawson (18) studied elongated bubble- and lamella-trains in aqueous surfactant solutions flowing through cylindrical capillaries. This work provides important insight into the effective or apparent viscosity of foam in porous media. Surfactants play a role in bubble transport whenever they are limited by mass transfer or sorption kinetics from maintaining a constant equilibrium tension around the bubble. During flow, the front bubble interface stretches toward the capillary wall, and the rear interface contracts toward the capillary centerline. Accordingly, surfactant depletes at the bubble front leading to a surface tension above the bulk equilibrium value, whereas surfactant accumulates at the bubble rear leading to a surface tension below the equilibrium value. A surface tension gradient arises that is directed towards the bubble front and retards bubble motion. Hirasaki and Lawson (18) demonstrated theoretically and experimentally that such surfactant effects are important. Neglecting the surface tension gradient underestimated effective viscosity by a factor of 8 (where $0.05 < U < 7$ cm/s).

Consideration of flowing bubble-trains separated by lamellae leads to the important finding that, except for very short bubbles and very slow sorption kinetics, the effective viscosity of confined foam scales quite linearly with bubble density and, in concert with Bretherton, is inversely proportional to the capillary number raised to the 1/3 power. The proportionality constant is a strong function of surfactant properties. The result is that flowing, confined foam is shear-thinning, an observation first noted by Fried (1), and that finer textured foam (i.e., larger bubble or lamellae densities) causes larger flow resistance. These ideas are utilized in our later modeling effort.

Two notable approaches have been used to include the role of pore constrictions in the pressure gradient required to drive lamellae through porous media. Falls et al. (48) added a viscous resistance that accounted for pore constrictions and that acted in series with the straight-tube flow resistance of Hirasaki and Lawson (18). Prieditis (41) and Rossen (42–44, 46) computed the static curvature resistance to the movement of a single bubble and also trains of bubbles through a variety of constricted geometries. Rossen considered the role of bubble compressibility (43), asymmetric lamella shapes (44), and stationary lamellae (46) on foam mobilization.

Interestingly, the two analyses are at odds. Lamella stretching is a dynamic effect and is not included in static arguments. Prieditis (41) argued that, as a lamella moves through a pore space that serially diverges and converges, the energy that is used to stretch the lamella and allow it to flow through the diverging section is completely recovered as the lamella squeezes to move through the converging section. Falls et al. (48) believe the opposite. Energy is consumed both by viscous resistance to lamellae flow and by nearby stationary lamellae that oscillate in response to pressure fluctuations of the flowing portion of the foam. Notwithstanding viscous dissipation in nearby oscillating lamellae, the static energy to squeeze a lamella through a constricted pore must mostly be dissipated through an increased local velocity upon entering the pore throat. Thus, inclusion of pore constrictions and even pore corners does not alter the basic shear-thinning and texture-dependent behavior of flowing, confined foam.

Synopsis. On the basis of the preceding information, Radke and Gillis (38) proposed Figure 4 as a summary of the pore-level microstructure of foam during flow through porous media. Figure 4 applies specifically to a confined, discontinuous-gas foam in the individual-lamella regime. In this highly schematic picture, hatched circles reflect water-wet sand grains. Wetting fluid is shown as the dotted phase. Foam bubbles are either unshaded or shaded gray to indicate whether they are flowing or trapped, respectively. Purely for illustrative purposes, the largest pore channels lie near the top of the picture, and intermediate and smaller sized pores are located sequentially nearer to the bottom.

Because of the dominance of capillary forces, wetting surfactant solution flows as a separate phase in the smallest pore spaces. Minimal wetting liquid transports as lamellae. Accordingly, the wetting-phase relative permeability function is unchanged in the presence of foam. Flowing foam transports in the largest pores where it encounters the smallest flow resistance relative to other possible flow paths. Because the smallest pore channels are occupied solely by wetting liquid, and the largest pore channels carry flowing foam, bubble trapping occurs in the intermediate-sized pores. If we refer to the foam as a phase, then the flowing portion of the foam phase is the most nonwetting, and the trapped portion of the foam phase is of intermediate wettability. Thus, following the reasoning behind Stone-type models (53), the relative permeability function of the nonwetting flowing foam and that of the flowing wetting phase are unaffected by the presence of the intermediate-wettability phase. Flowing-foam relative permeability becomes solely a function of the gas saturation of flowing bubbles and is much reduced by the trapped foam saturation. Of course, the relative permeability of the intermediate-wettability trapped foam is zero.

Foam bubbles that move in the largest backbone channels are coupled together through lamellae and lenses. They parade in series as trains. Individual bubbles comprising these trains are relentlessly destroyed and recreated, so that the train is in a constant state of rearrangement. Regardless of whether bubbles are generated externally or in situ, they are

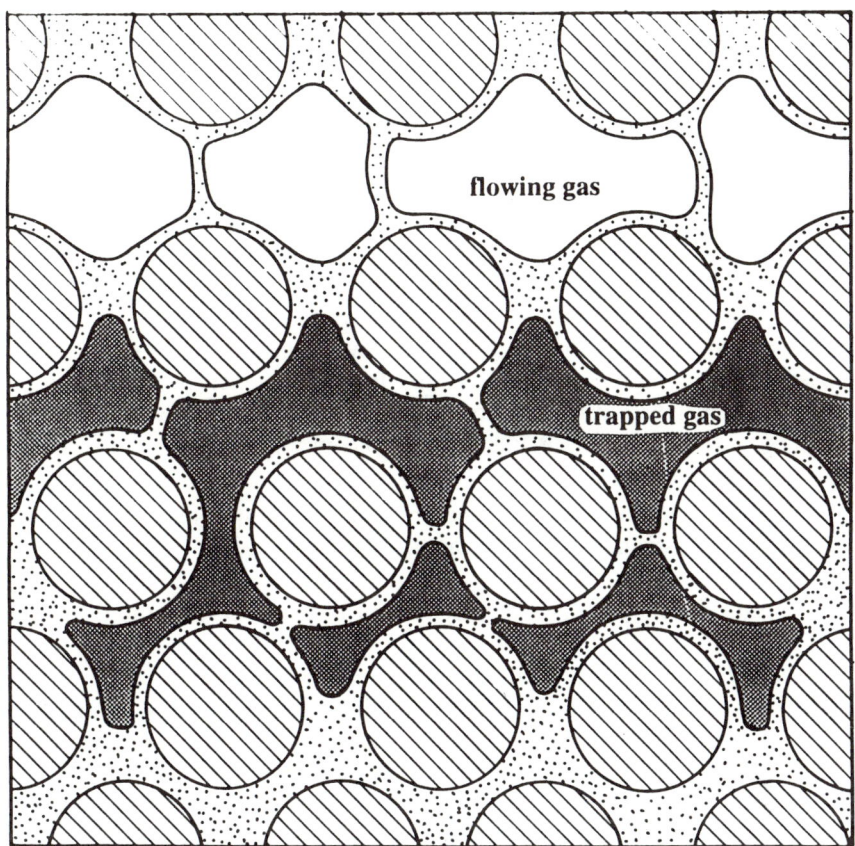

Figure 4. Pore-level schematic of fluid distribution for a discontinuous-gas flowing foam. Flowing bubbles are unshaded, and trapped gas is darkly shaded. (Reproduced with permission from reference 38. Copyright 1990 Society of Petroleum Engineers.)

molded and reshaped by the porous medium (*20, 26*). Bubbles and lamellae transport some distance, perhaps through several pore-bodies and pore-throats, are destroyed, and then reformed. Further, trains halt when the local pressure gradient is insufficient to keep them mobilized, and other trains lurch into motion. The identity of a single bubble or train is not conserved over any large distance. Bubble-trains exist only in a time-averaged sense.

Continuous-gas foams are readily accommodated within the previously mentioned model. During the later stages of gas-only injection, flowing lamellae may collapse and not be regenerated. Thus, some, or all, of the flowing bubble-trains are replaced by continuous gas. Trapped lamellae then become mainly responsible for increased flow resistance. Confined, continuous-gas foam may be viewed as a subcase of discontinuous-gas foam.

Mechanisms of Foam Formation and Decay

Foam texture in Figures 2 through 4 arises because of strong lamella generation and coalescence forces. The interplay of bubble generation and coalescence determines foam microstructure and, hence, gas mobility in porous media. Knowledge of these pore-level events is necessary to derive physically meaningful rate expressions for foam generation and coalescence. The exposition of Chambers and Radke (26) provides visual documentation and explanations for these various mechanisms.

Foam Formation. Three fundamental pore-level generation mechanisms exist: snap-off, division, and leave-behind.

Snap-off. Snap-off is a very significant mechanism for bubble generation in porous media. This phenomenon was first identified and explained by Roof (54) to understand the origin of residual oil. Snap-off is not restricted to the creation of trapped oil globules. It repeatedly occurs during multiphase flow in porous media regardless of the presence or absence of surfactant. Hence, snap-off is recognized as a mechanical process.

Figure 5a illustrates a gas finger that enters a pore constriction initially filled with wetting liquid. The pore is considered cornered in cross-section with local transverse inscribed radius R_c. Upon reaching the

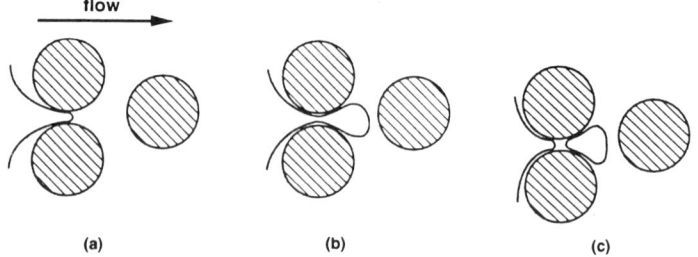

Figure 5. Schematic of snap-off mechanism (gas is unshaded) showing (a) gas entry into liquid filled pore-throat, (b) gas finger and wetting collar formation prior to breakup, and (c) liquid lens after snap-off. (Reproduced with permission from reference 60. Copyright 1989 Society of Petroleum Engineers.)

throat, the interface curvature and corresponding capillary pressure rise to the equilibrium entry value. As the bubble front enters the downstream body, wetting liquid remains in the corners (Figure 1), and the curvature and corresponding local capillary pressure at the bubble front fall with expansion of the interface. The resulting gradient in capillary pressure initiates a gradient in liquid pressure directed from the pore-body toward the pore-throat. Liquid is driven along the corners into the pore-throat where it accumulates as a collar (Figure 5b). Close examination of Figure 2 also reveals such a collar demarked with the black arrow just right of center. Collar growth to snap-off ensues in Figure 5c, provided that the downstream bubble front has a mean radius of curvature that is larger than about twice R_c, and provided that the constriction is not extremely sharp. If the constriction is too sharp, a stable collar can form. Thus, snap-off or germination sites (26) in porous media must exhibit a body–throat size aspect ratio larger than about 2 and be gently sloped. They also require sufficient wetting liquid. Snap-off creates gas bubbles that are approximately the size of pore-bodies.

The growth wavelength of the unstable collar in Figure 5 is close to $2\pi R_c$, which means that snap-off always creates a lens (Figure 5c). A lamella cannot be generated directly at a pore-throat by snap-off. Chambers and Radke pointed out that lamellae form downstream of the constriction, as a newly formed bubble collides with other previously created bubbles (Figures 19 and 20 of reference 26). If stabilizing surfactant is not present in the wetting phase, snapped-off bubbles quickly coalesce (26, 45, 55) so that, on the average, continuous-gas channels exist. Snap-off, followed by coalescence events is a standard ingredient of multiphase flow.

Three varieties of snap-off exist depending upon whether local liquid saturation is increasing or decreasing, and upon the pore-body to pore-throat aspect ratio (26). Figure 5 portrays neck, or Roof, snap-off, which is most prevalent at high wetting-liquid saturations. Rectilinear and preneck constriction snap-off contribute when both phases are flowing near steady state at lower wetting-liquid saturations. Preneck snap-off occurs when a gas bubble lodges upstream of a pore-throat and blocks liquid flow, and thus causes a deformation in the gas–liquid interface. Depending on the detailed geometry of the pore, snap-off ensues when an adequate amount of liquid accumulates (26). Rectilinear snap-off occurs in long (length $>2\pi R_1$), straight pores in a fashion akin to snap-off in straight, cornered capillary tubes. In all three types of snap-off, a wetting liquid lens forms first. Also, all three types occur regardless of the presence of surfactant. Snap-off depends on liquid saturation or, equivalently, on the medium capillary pressure, in addition to pore geometry and wettability (56–58), but, except for alteration of solution properties such as surface tension, is sensibly independent of surfactant formulation.

Lamella Division. Lamella or bubble division proceeds by subdividing foam bubbles or lamellae. Thus, mobile foam bubbles must preexist. Division is illustrated in Figure 6. A translating foam bubble encounters a point where flow branches in two directions (Figure 6a). The interface stretches around the branch point and enters both flow paths. The initial bubble divides into two separate bubbles (Figure 6b) that continue to move downstream.

Whether or not a lamella divides is governed by several factors. First, Chambers (27) observed that foam bubbles smaller than the pore-body size do not divide when they encounter a branch point. The bubble flows down one or the other of the channels unaltered. If, however, the bubble size is larger than that of the pore body, so that the foam lamella spans the pore space, division generally occurs. Moreover, because a lamella that encounters a division site must form two new Plateau borders (Figure 6b), the lamella may be drained of liquid and coalesce in the process (59). Second, Prieditis (41) showed that the likelihood of bubble division depends upon the occupancy of surrounding pores. In the absence of foam bubbles or lamella that surround the branch point, division proceeds. However, nearby trapped foam bubbles greatly reduce the number of branch points. Stationary bubbles or lamellae divert the once branching flow down one path by acting as flexible pore-walls. Division is thus prohibited.

Leave-behind. Figure 7 illustrates the third foam generation mechanism (60). Leave-behind begins as two gas menisci invade adjacent

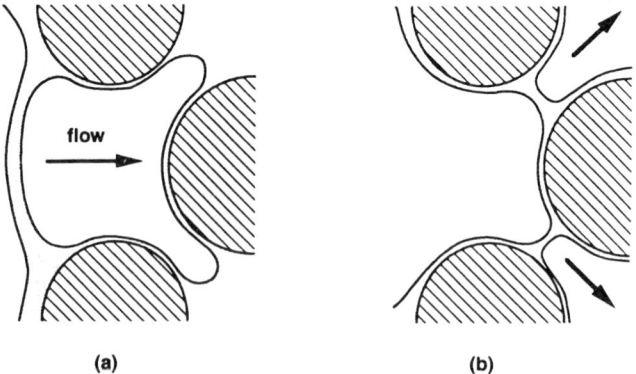

Figure 6. Schematic of division mechanism. A lamella is flowing from the left to the right. (a) Gas bubble approaching branch point. (b) Divided gas bubbles. (Reproduced with permission from reference 60. Copyright 1989 Society of Petroleum Engineers.)

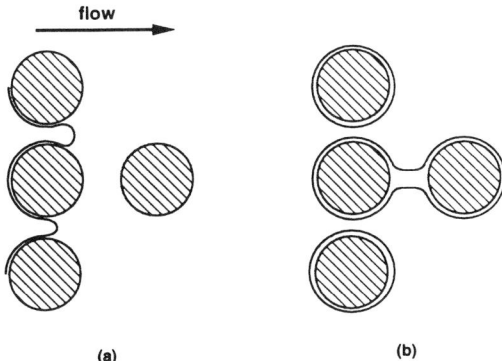

Figure 7. Schematic of leave-behind mechanism showing (a) gas invasion and (b) stable lens. (Reproduced with permission from reference 60. Copyright 1989 Society of Petroleum Engineers.)

liquid-filled pore-bodies (Figure 7a). A lens is left behind as the two menisci converge downstream. As long as the capillary pressure of the medium is not too high, and the pressure gradient is not too large, a stationary stable lens emerges (Figure 7b). Later, the lens may drain to a thin film.

Lenses created by leave-behind are generally oriented parallel to the local direction of flow (i.e., the pore-level flow that created them), and do not make the gas phase discontinuous. If leave-behind is the only form of lens or lamella generation, a continuous-gas foam results. Ransohoff and Radke (*60*) found that foam generated solely by leave-behind gave approximately a five-fold reduction in steady-state gas permeability, whereas discontinuous-gas foams created by snap-off resulted in a several hundred-fold reduction in gas mobility (*20, 61*).

Dominant Mechanism. We assert that snap-off is the dominant foam generation mechanism, especially under conditions of coinjection of surfactant solution and gas. The effluent bubble-size measurements from Berea sandstone of Ettinger and Radke (*20*), referred to earlier, show that foam coarsens with increasing gas rate. The rate of division events is greater at higher gas flow rate. Thus, if bubble generation by division were significant, foam should not be shear-thinning, as is routinely observed (*20, 21, 61, 62*). In light of the significant fraction of foam that is trapped (*38, 39*), of foam shear-thinning behavior, and of Prieditis's observations (*41*) noting the reduced availability of bubble division sites in the presence of stationary foam bubbles, division does not appear to be the major source of flowing lamellae in our studies with sandstones. Leave-behind is a nonrepetitive process that alone cannot account for the high flow resistances measured with foam. Finally, direct observations of foam generation in transparent, etched-glass micromodels (*25–27*) suggest the dominance of the snap-off mechanism.

Foam Destruction. Net foam generation cannot continue unchecked. It is balanced by foam destruction processes. Chambers and Radke (*26*) enunciated two basic mechanisms of foam coalescence: capillary-suction and gas diffusion. Because capillary-suction coalescence is the primary mechanism for lamellae breakage, we focus on it, and only briefly touch upon foam coarsening by gas diffusion.

Capillary-Suction Coalescence. In stark contrast to snap-off, capillary-suction coalescence is strongly affected by surfactant formulation. Thin lamellae are not thermodynamically stable. Their existence is due to excess normal forces within the films originating from long-range concerted intermolecular interactions. Derjaguin and co-workers (*63, 64*) first introduced this idea in terms of a film disjoining pressure, Π, which is a function of film thickness, *h*. Positive values of Π reflect net repulsive film forces, and negative values of Π indicate net attractive forces. Adsorption of ionic surfactant at each gas–liquid surface of the film causes the excess repulsive forces. The identically charged surfaces repel each other through overlap of their double-layer ionic clouds. At small film thicknesses, protrusion, and/or hydration forces give rise to a very steep repulsion. These two stabilizing forces are sensitive to surfactant concentration and structure, and to ionic content of the aqueous solution. Here surfactant formulation comes into play for designing effective foamers. Additionally, attractive van der Waals forces tend to destabilize the film. Combination of these three forces leads to an S-shaped disjoining pressure isotherm that is similar, in form, to a pressure–volume isotherm for real gases and liquids. Many sources elaborate on the origin and behavior of disjoining forces in surfactant-stabilized foam films (*26, 55, 63–71*).

A recently measured disjoining-pressure isotherm of an isolated lamella is shown in Figure 8 for the surfactant sodium dodecyl sulfate (SDS) at 10^{-3} kmol/m^3 in aqueous 0.01 kmol/m^3 sodium chloride brine (*65*). A solid line connects the data points for three independent experimental runs, shown by various symbols. The negative, attractive portion of the isotherm between thicknesses of about 4 and 5 nm is not sketched because equilibrium measurements are not possible there. The measured isotherm indeed obeys the classic S-shape. Film meta-stability demands that the slope of the isotherm be negative (*26, 72*). For positive slopes, even the slightest, infinitesimal disturbance ruptures the film. Thus, the lamella in Figure 8 can exist only along the two repulsive branches near 4 nm and above 7 nm. The thicker branch or common black film arises from electrostatic overlap forces, and the inner branch or Newton black

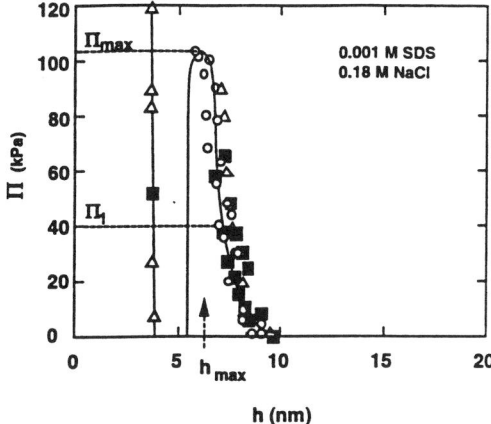

Figure 8. Experimental disjoining pressure isotherm at ambient temperature for sodium dodecyl sulfate (0.001 M) in brine (0.18 M NaCl). (Reproduced from reference 65. Copyright 1992 American Chemical Society.)

film arises from strong protrusion–hydration forces. van der Waals dispersion forces explain the middle unstable branch.

The line drawn through the disjoining-pressure isotherm at Π_1, reveals that three equilibrium film thicknesses are possible, roughly 4, 5, and 9 nm. Which film thickness is operative for lamellae in porous media is determined primarily by the augmented Young–Laplace equation (67–69):

$$P_c = 2\sigma C_m + \Pi(h) \tag{1}$$

where P_c is the local capillary-pressure, C_m is the mean interfacial curvature of the thin film, and σ is the bulk surface tension. In the limit of thick films, the disjoining pressure approaches zero, and the classic Young–Laplace equation is recovered.

For the static, trapped lamellae, the film reaches an equilibrium thickness set by the local capillary pressure and the film curvature in obedience to equation 1. The capillary pressure, in turn, depends on the wetting-liquid saturation; the film curvature depends on the particular location within the pore structure dictated by an approximately 90° contact angle with the pore wall. As the capillary pressure in the porous medium rises during drainage, the film thickness decreases along the common black branch until Π_{max} is reached, and a Newton black film of 4-nm thickness emerges.

At still higher imposed capillary pressures, a disjoining pressure is attained, Π_{rup}, not labeled on Figure 8, where the film eventually ruptures.

In a foam-laden porous medium, the capillary pressure of an equivalent undispersed two-phase system that corresponds to Π_{rup} is termed here the critical capillary pressure for rupture. For bulk systems, Π_{rup} is a well-documented parameter that controls the stability of the foam (71). As shown in Figure 8, even a very dilute SDS surfactant solution exhibits a critical capillary pressure for rupture greater than 100 kPa (i.e., greater than 1 atm) and creates highly robust foam films. However, not all surfactant-stabilized foam films display an inner branch. In this case, the critical capillary pressure for rupture equals Π_{max}.

The message from Figure 8 is that static lamellae are stable to small disturbances until a critical capillary pressure is attained; then coalescence is catastrophic. In porous media, the liquid saturation, absolute permeability, and surface tension control this critical capillary pressure through the Leverett J-function (73). Of course, static lamellae may coalesce at lower capillary pressures, if they are subjected to large disturbances. Figure 8 also reveals that static lamellae in equilibrium with the imposed capillary pressure are amazingly thin.

Coalescence behavior of flowing foam bubbles is more complicated than that of static lamellae. Khatib et al. (62) directly measured capillary pressures in 70- to 9000-μm^2 glass-bead packs during steady foam flow over a wide range of gas fractional flows from 0.1 to 0.99. For a given surfactant system, they observed a drastic foam coarsening at a specific capillary pressure (typically near 3 kPa), called the limiting capillary pressure, P_c^*. Above P_c^*, coalescence of flowing lamellae is significant, and below P_c^*, it is minimal. Limiting capillary pressures varied with gas flow rate and absolute permeability in addition to surfactant formulation (62).

The experimentally determined limiting capillary pressures of Khatib et al. are likely connected to the critical disjoining pressures for rupture from Figure 8. To make this connection, Jiménez and Radke (55) proposed a simple hydrodynamic stability theory that describes the thickness evolution of a lamella translating through a periodically constricted tube. Figure 9 presents one such lamella at successive times t_1, t_2, and t_3. As the lamella translates from left to right, it is squeezed upon entering the constriction at time t_2. Film thickness increases to conserve liquid mass, and the disjoining pressure is correspondingly low. Jiménez and Radke

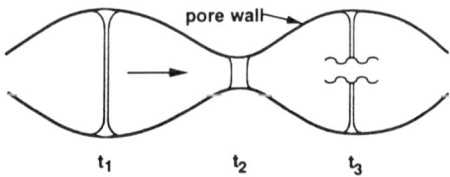

Figure 9. Foam lamella translating from left to right in a periodically constricted tube. Coalescence occurs at t_3.

assumed the ensemble-averaged curvature of the flowing thin films to be negligible compared to the capillary-suction pressure exerted by the porous medium on the Plateau border, so that the film drains to equalize P_c and Π. Drainage is driven by the pressure difference $(P_c - \Pi)$. Fluid resistance in the film is inversely proportional to the local film thickness to the third power, so drainage is not instantaneous.

When the lamella moves out of the pore constriction, it is stretched upon expansion into the downstream body. The film thins to conserve mass, and the disjoining pressure is now high. Here the film fills in an attempt to equalize P_c and Π, again driven by the pressure difference $(P_c - \Pi)$. Thus, the thickness of the transporting lamella oscillates about the equilibrium thickness established in the stationary lamellae in a sequence of squeezing–stretching and draining–filling events. Thickness oscillations are wider the higher the gas flow rate, and the larger the pore-body to pore-throat aspect ratio, R_b/R_c. Jiménez and Radke (55) argued that translating lamellae break instantaneously whenever the film thickness diminishes to h_{max}, corresponding to Π_{max} in Figure 8. Thus, if the film is stretched too rapidly for healing surfactant solution to flow into the film and stabilize it, rupture ensues. The capillary pressure at which the translating lamella ruptures is defined as P_c^*, the limiting capillary pressure.

The theory of Jiménez and Radke predicts that P_c^* lies below Π_{max} to an extent that increases strongly with larger pore-body to pore-throat aspect ratio, and increases weakly with larger gas flow rate. For a given gas flow rate and surfactant composition, specific pore-throat–pore-body combinations in the medium, called termination sites (26, 55), lead to capillary-suction coalescence. At low wetting-phase saturation, corresponding to high capillary pressure, large numbers of termination sites are unveiled because even small aspect ratios force rapid enough stretching of the lamellae to cause rupture. Simply not enough wetting phase is available to heal and stabilize the films.

The proposed model of Jiménez and Radke (55) is oversimplified in that it does not account for the details of the actual curved-film breakup (74), or for any surface tension gradients and elastic effects as the film stretches and squeezes. Nevertheless, it does correctly explain the gas-velocity dependence of P_c^* measured by Khatib et al. (62). Moreover, Huh et al. (75) studied the behavior of CO_2–aqueous surfactant foams in glass micromodels and reported visual observations of the stretching–squeezing mechanism. Lamellae coalesced in the pore bodies of the micromodel. Likewise, Chambers and Radke present photomicrographs of the identical mechanism for coalescence of N_2–aqueous surfactant foams in etched-glass micromodels (Figure 26 of reference 26).

The basic premise underlying capillary-suction coalescence is surfactant solutions that exhibit large rupture disjoining pressures lead to strong foam in porous media with large flow resistance. Aronson et al. (76) recently investigated this premise. Disjoining pressures of single foam films for SDS at several concentrations and brine levels similar to that in Figure 8 were measured; with the identical surfactant solutions, the steady pressure drops of N_2 foams at gas fractional flows of 90% in 2.3-μm^2 glass-bead packs were measured. Figure 10 summarizes the findings.

Open symbols in the figure correspond to the measured rupture disjoining pressures of the single aqueous SDS foam films as a function of NaCl concentration at two surfactant concentrations. Upward directed arrows on some of the experimental points indicate that the rupture pressure is larger than the value shown. Dashed lines simply sketch in the observed trends that increased surfactant and salt concentrations yield increased values of Π_{rup}. Conversely, closed symbols correspond to the measured steady pressure gradients for the same solutions. Solid lines again indicate the trends.

The postulate that large rupture disjoining pressures, above 30 kPa in Figure 10, give rise to strong foam in porous media is clearly confirmed. Aronson et al. (76) estimated that the capillary pressures in their bead packs during steady foam flow are near 1 to 5 kPa. Figure 10 then reveals that once the rupture pressure of the foam films exceeds the capillary pressure of the medium, low mobility foam emerges. It is also fascinating that the low SDS concentrations of 10^{-3} M can produce very large pressure gradients. Finally, Khatib et al. (62), in their direct capillary-pressure measurements, also found that increased brine content increased P_c^*, con-

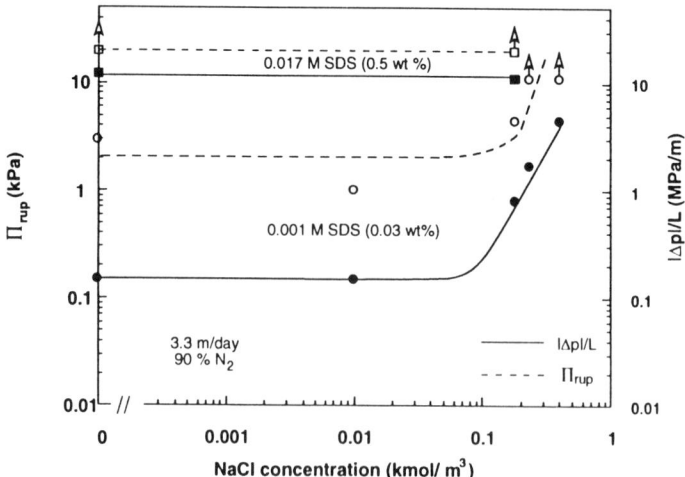

Figure 10. Comparison of foam lamella rupture pressure and bead pack pressure gradient during steady-state foam flow. Upward directed arrows indicate that the actual rupture pressure is greater than the value indicated.

sonant with the disjoining-pressure measurements in Figure 10. The study of Aronson et al. (76) provides striking confirmation of the origin of limiting capillary pressures and their role in foam coalescence in porous media.

Gas Diffusion. The second mechanism for foam coalescence in porous media, gas diffusion, pertains primarily to the stagnant, trapped bubbles. According to the Young–Laplace equation, gas on the concave side of a curved foam film is at a higher pressure and, hence, higher chemical potential than that on the convex side. Driven by this difference in chemical potential, gas dissolves in the liquid film and escapes by diffusion from the concave to the convex side of the film. The rate of escape is proportional to film curvature squared and, therefore, is rapid for small bubbles (16, 26).

Bulk foams coarsen with larger bubbles growing at the expense of smaller ones that eventually disappear. However, confined foam in porous media does not coarsen in a similar fashion because bubble volume is not directly related to film curvature. Rather, in porous media, lamella curvature depends on pore dimensions and on location within the pore space. Gas diffusion still proceeds from the most highly concave bubbles forcing the lamellae to diminish their curvatures by translation toward pore-throats. In the absence of an imposed pressure gradient, it is possible for gas diffusion to drive all lamellae to pore-throats to achieve an equilibrium state of zero curvature. Coalescence occurs only when two lamellae happen to reach the same pore-throat. With steam foam, coarsening is more rapid because water can condense on one side of a lamella and evaporate from the other. A noncondensible gas may be added to retard coarsening (3). Nevertheless, coarsening of the trapped foam by gas diffusion is expected. As the resulting texture diminishes, a portion of the trapped bubbles may remobilize (26).

Foam Flow at Limiting Capillary Pressure

Most foams in porous sand packs and sandstones achieve steady flow at or near P_c^*, provided that the gas fractional flow is high but below unity, and provided that flow rates are fixed (20, 36, 61, 62, 77–80). In the limiting capillary-pressure regime, the wetting-liquid saturation is sensibly constant, independent of gas and liquid velocities over a rather large range. This limiting saturation is thought to reflect a constant P_c^*, which from the preceding discussion, is set primarily by the disjoining-pressure isotherm of the stabilizing surfactant.

Foam-flow behavior in the limiting capillary-pressure regime is rather remarkable (20, 36, 61, 62, 78–81). When liquid velocity is held constant and gas velocity is adjusted, pressure drop is independent of gas flow rate. Varying liquid velocity and holding gas velocity constant usually yields a linearly increasing pressure drop response. Increasing both liquid and gas velocity and holding the fractional flow constant yields a linear response when pressure drop is plotted versus total flow rate. Finally, steady-state saturations are independent of gas fractional flow. These observations cannot be explained by Darcy's law for multiphase flow in porous media. In particular, foam flow is not solely a function of fractional flow, but depends on individual gas and liquid velocities.

The consequences of foam flow in the limiting capillary-pressure regime are important. Khatib et al. (62) pointed out that whenever P_c^* is achieved in a porous medium, relative gas mobility is easily calculated. Because P_c is constant, aqueous-phase saturation consequently remains constant and so does the relative permeability to the wetting liquid. Because pressure gradients in both the wetting and nonwetting phases (foam) are identical at steady state, Darcy's law for the aqueous phase determines foam-flow behavior. Persoff et al. (61) expanded this point and demonstrated experimentally that within the limiting capillary-pressure regime, only one or two measurements of foam pressure drop are needed to predict the entire spectrum of steady-state results. Rossen and co-workers (30, 82), by incorporating these observations into fractional flow theory, demonstrate how powerful the limiting capillary-pressure regime is in understanding foam flow.

Not all foam-flow behavior falls in the limiting capillary-pressure regime, even for constant-rate injection at steady state. Deviations are evident at low gas fractional flow (i.e., low quality foam). In a 1.3-μm^2 Boise sandstone, nitrogen gas fractional flows spanning from 0.7 to 0.995 were found to be within the limiting capillary-pressure regime (61). In fact, Persoff et al. (61) could not find a lower limit to the regime. Likewise, Ettinger and Radke (20) probed fractional flows between 0.7 and 0.9, and found no lower limit to limiting capillary-pressure behavior. However, the data of Khatib et al. (62) for 70- to 9000-μm^2 bead packs and those of De Vries and Wit (36) for sand packs (4.2 μm^2) and a Bentheim sandstone (1.2 μm^2) show that at lower gas fractional flows, pressure drop increases as gas flow rate is increased at a fixed liquid flow rate. This behavior continues up until a break point or maximum gas fractional flow where the limiting capillary pressure is reached. Above fractional flows of roughly 0.92, pressure drop for fixed liquid flow rates is essentially independent of gas flow rate, indicative of the limiting capillary-pressure regime.

More recently, Osterloh and Jante (77) probed a wide range of flow rates and fractional flows for foam in a 6.2-μm^2 sand pack. They distinguished two regimes. At gas fractional flows above 0.94, the pressure gradient was reasonably independent of gas velocity (at fixed liquid velocity) but varied with liquid velocity (at constant gas velocity) to roughly the

1/3 power. Also, liquid saturation was nearly constant at 6%. At lesser gas fractional flows, the converse was found. Pressure-gradient response was negligible with increased liquid velocity but increased with gas velocity to the 0.31 power. It was surmised that the transition between the two regimes occurs at the point where the limiting capillary pressure is attained.

The variation in the range of fractional flows yielding limiting capillary-pressure behavior is likely due to the different surfactant systems employed. Different surfactant structures and conditions such as concentration and temperature lead to different disjoining-pressure isotherms for single foam films and thus different limiting capillary-pressure characteristics. Also, the various porous media have differing capillary-pressure versus aqueous-phase saturation relationships. Although the limiting capillary-pressure regime is by no means general for all conditions of foam flow in porous media, it is an important one. Our modeling effort (to follow) is directed toward predicting how this behavior emerges in a transient, one-dimensional foam displacement.

Population-Balance Modeling of Foam Flow in Porous Media

A variety of methods have been proposed for modeling of foam flow and displacement in porous media. These range from population-balance methods (*8, 9, 20, 39, 78–80, 83*) to percolation models (*42–45, 47, 84*), and from semiempirical alteration of gas-phase mobilities (*85–92*) to applying so-called fractional flow theories (*30, 82*). Semiempirical models are computationally simple, but lack generality. The last method may be unsuitable for modeling of foam flooding because fractional flow theory is approximate when applied to compressible phases (*30*), severe extrapolations from available data are needed to fit model parameters (*30*), and strong foam behavior is not, in general, a unique function of fractional flow (*20, 36, 61, 62, 75, 81*). Constructing fractional flow curves for foam flow in porous media, as such, may be inappropriate as absolute flow rates determine foam-flow behavior.

Of these four methods, only the population-balance method and network, or percolation, models arise from first principles. Network models that allow replication of pore-level mechanisms have the decided disadvantage of requiring large amounts of computation time and providing results on a prohibitively small grid. It seems unlikely that network or percolation models can be useful in transient displacements that demand tracking of saturation, surfactant concentration, and foam on laboratory scales, let alone field scales.

The population-balance method for modeling foam flow (8, 9) was originally proposed because it incorporates foam into reservoir simulators in a manner that is identical to calculating the transport of mass and energy in porous media. Further, the method is mechanistic in that it can account for the actual pore-level events described in the previous sections. In its minimal form, the population-balance simply adds another component to a standard multicomponent simulator. By analogy to balances on surfactant or other chemical species, a separate conservation equation is written for the concentration of foam bubbles. Our goal, in this section, is to map out a population-balance that is easy to implement, fits simply into the framework of current reservoir simulators, and employs a minimum of physically meaningful parameters descriptive of the dominant pore-level events. We focus on transient displacement by strong foam that at steady state achieves the limiting capillary-pressure regime.

Conservation Equations. Mass balance equations for the gaseous and aqueous phases are written in standard reservoir simulator form (10, 93). For the nonwetting foam or gas phase in a one-dimensional medium,

$$\frac{\partial [\phi \rho_g S_g]}{\partial t} + \frac{\partial (\rho_g u_g)}{\partial x} = Q_g \qquad (2)$$

where t denotes time, x gives the axial location, ϕ is the porosity of the porous medium, ρ_g is the gas mass density, S_g is the saturation of the gas phase, u_g is the superficial or Darcy velocity, and Q_g is a source–sink term for gas used here to apply boundary conditions (10). The companion mass balance for the aqueous phase is written by interchanging the subscript g, which denotes gas phase for w, which denotes the liquid phase.

A mass balance on surfactant is also required, which written in standard form becomes

$$\frac{\partial [\phi (C_s S_w + \Gamma_s)]}{\partial t} + \frac{\partial (u_w C_s)}{\partial x} = Q_s \qquad (3)$$

where C_s is the number or molar concentration of surfactant in the aqueous phase, Γ_s is the amount of surfactant adsorption on the rock surfaces in units of moles per void volume, and Q_s is the source–sink term for surfactant in units of moles per unit volume per unit time.

Because the mobility of the foam phase is a strong function of texture (9, 18, 20, 26, 33, 48), mechanistic prediction of foam flow in porous media is impossible without a conservation statement that accounts for the evolution of foam bubble size (8). Following Patzek (8) and others (9, 39), we write a transient population-balance on the mean bubble size,

$$\frac{\partial [\phi(S_f n_f + S_t n_t)]}{\partial t} + \frac{\partial (u_f n_f)}{\partial x} = \phi S_g (r_g - r_c) + Q_b \quad (4)$$

In equation 4, the subscripts f and t refer to flowing and trapped foam, respectively, and n_i is the foam texture or bubble number density. Thus, n_f and n_t are, respectively, the number of foam bubbles per unit volume of flowing and stationary gas. The total gas saturation is given by $S_g = 1 - S_w = S_f + S_t$, and Q_b is a source–sink term for foam bubbles in units of number per unit volume per unit time. The first term of the time derivative is the rate at which flowing foam texture becomes finer or coarser per unit rock volume, and the second is the net rate at which foam bubbles trap. The spatial term tracks the convection of foam bubbles. The usefulness of a foam bubble population-balance, in large part, revolves around the convection of gas and aqueous phases.

On the right of equation 4, the generation and coalescence rates, r_g and r_c, are expressed on a per volume of gas basis. These two terms are fundamental, for they control bubble texture. At steady state, far from any sources or sinks, and where rock properties are constant (e.g., absolute permeability, relative permeability, and capillary-pressure functions), bubble size is set by $r_g = r_c$. That is, the rate of bubble generation by snap-off balances the rate of bubble coalescence by capillary-pressure suction (20). To proceed, kinetic expressions are needed for r_g and r_c.

Generation. Snap-off in germination sites determines the rate expression for bubble generation following the process pictured in Figure 5. Bubble leave-behind is neglected. The division mechanism for producing new lamellae yields a rate that is indistinguishable in form from that of coalescence and is included there. Earlier studies (9, 39, 94) argue that the frequency of snap-off events is inversely proportional to the sum of the time to displace a newly formed lens out of a constriction, and the time for wetting liquid to drain back along the pore corners to initiate pinch-off of another lens. The proportionality constant counts the number of active germination sites. By extending the hydrodynamic analysis of Ransohoff et al. (94) for constricted, cornered pores to include imposed wetting liquid flow, Kovscek (80) found that snap-off frequency may be expressed as linearly proportional to liquid velocity and to gas velocity raised to a power less than unity. The liquid-velocity dependence originates from the net imposed liquid flow, and the gas-velocity dependence arises from the time for the lens to exit the pore. Accordingly,

$$r_g = k_1 v_f^a v_w^b \tag{5}$$

where $v_w = u_w/\phi S_w$ is the local interstitial liquid velocity, and $v_f = u_f/\phi S_f$ is the local interstitial velocity of the flowing foam. These velocities depend upon the local saturation of flowing liquid or gas and the local pressure gradient, which can include capillary-pressure and gravitational effects; a and b are power indices, with the index b close to unity. Equation 5 suggests that bubbles are produced only in the portion of the foam that transports. The generation rate constant, k_1, reflects the number of foam germination sites. Intuitively, as liquid saturation falls, the number of germination sites falls. We take k_1 as a constant here. Of course, the bubble generation rate does vary implicitly with liquid saturation through the dependencies on liquid and gas velocities. No surfactant properties appear in equation 5 consistent with the mechanical origin of snap-off.

Falls et al. (9) and Friedmann et al. (39) pointed out that if a lamella arrives at a germination site prior to the total elapsed time for snap-off, then snap-off is precluded. An upper limit is then placed on the evolution of foam texture. We find in our systems that strong coalescence forces come into play before this upper limit is attained.

Some researchers have found a so-called critical velocity for the onset of foam generation (39, 45, 60). Friedmann et al. (39) generated foam in sandstone cores at different initial surfactant-laden water saturations after steady gas and surfactant-free liquid flow is established. Critical onset velocities increase with decreasing saturation of the water phase, S_w. Velocities up to several hundred meters per day are reported when the initial water saturation is low. Once steady two-phase flow is established, high gas velocities are apparently required for the gas to build a sufficient pressure gradient and enter into wetting liquid-filled pores (e.g., as in Figure 5).

The existence of a critical velocity for the appearance of strong foam is linked, apparently, to the initial condition of the porous medium. Recent experiments in glass-bead packs and in Boise and Berea sandstones have not confirmed that a critical gas velocity or pressure drop must be exceeded for successful foam generation (21, 61, 78, 95, 96). In all of these cases, the porous media were completely saturated with surfactant solution prior to any gas injection. With initial high water saturations, foam readily generates. In the experiments and modeling calculations to follow, the initial water saturation is 100%. For this reason, a critical onset velocity or pressure gradient is not included in equation 5.

Coalescence. A pore-level-based rate expression for capillary-suction coalescence is readily obtained. Figure 9 illustrates that foam lamellae are destroyed in proportion to their flux (i.e., $v_f n_f$) into termination sites. Hence, (20)

$$r_c = k_{-1}(S_w) v_f n_f \qquad (6)$$

where $k_{-1}(S_w)$ is a coalescence rate constant that varies strongly with local aqueous-phase saturation, S_w, and n_f is again the number of bubbles per unit volume of the flowing foam. Additionally, the coalescence rate constant also varies with surfactant concentration and formulation. Equation 6 teaches that higher interstitial gas velocities lead to increased foam coalescence, because rapidly stretched lamellae are more vulnerable to breakage. As discussed previously, sufficient time does not exist for surfactant solution to flow into a rapidly stretched lamella and heal it.

As coalescence depends upon P_c^*, the coalescence rate constant depends strongly upon surfactant formulation, concentration, and S_w. Weak surfactants or low concentrations make $k_{-1}(S_w)$ quite large. Additionally, the saturation dependence of $k_{-1}(S_w)$ is quite dramatic. Khatib et al. (62) showed that for strongly foaming solutions, $k_{-1}(S_w)$ is small for high aqueous phase saturations, but rises steeply as S_w falls to a value corresponding to the limiting capillary pressure. If the aqueous-phase saturation falls near or below that corresponding to the limiting capillary pressure, $k_{-1}(S_w)$ approaches infinity. Cognizant of these observations,

$$k_{-1}(S_w) = k_{-1}^o \frac{(1 - S_w)}{(S_w - S_w^*)} \qquad (7)$$

where S_w^* is the saturation corresponding to the limiting capillary pressure, and k_{-1}^x is a scaling constant. Equation 7 allows the coalescence rate to increase monotonically as the porous medium desaturates. As desired, the coalescence rate becomes infinite at S_w^*. It is important to emphasize that the sensitivity of foam to surfactant formulation is embodied in S_w^*. Foamers with high critical-rupture disjoining pressures result in high limiting capillary pressures, P_c^*. The capillary-pressure curve for the particular medium then sets a low value of S_w^* and, accordingly, a low foam mobility.

Figure 11 displays qualitatively the dependence of both coalescence and generation rates on the wetting-liquid saturation. S_w^*, the aqueous saturation corresponding to the limiting capillary pressure, is not the same

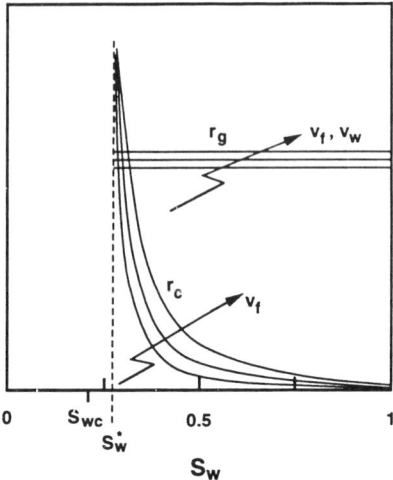

Figure 11. Schematic of generation and coalescence rate versus aqueous-phase saturation. Arrows indicate how generation and coalescence rates change with increasing interstitial velocities.

as the traditional connate water saturation and may be either smaller or larger than S_{wc} depending on the surfactant formulation and the nature of the porous medium. Near S_w^*, the rate of foam coalescence rises steeply as P_c in the medium approaches Π_{rup}. The intersection of the two rate curves determines the relationship between the steady state liquid saturation and foam texture from equations 5 through 7. Because of the steepness of the coalescence rate with S_w, changes in gas or liquid velocity have little effect on the steady saturation. This condition explains the lack of sensitivity of the wetting-liquid saturation to flow rates seen in the limiting capillary pressure foam-flow regime (20, 36, 61, 62, 77, 78).

It is readily argued that the intersection of coalescence and generation rates in Figure 11 leads to a stable steady state. If the system is perturbed away from this point, it naturally returns. Consider a small, positive perturbation in the local liquid saturation. The coalescence rate then declines, and the foam texture becomes finer. This change causes an increased flow resistance that returns the liquid saturation back to the stable operating point. The converse negative saturation perturbation is similarly argued to be stable.

The rate of bubble division, the second mechanism for creating foam, is proportional to the flux of lamellae into division sites (20). Thus, the rate of foam generation by division is formally identical to equation 6. Further, both rate constants share the property of being small when S_w is high because more division sites become available as S_w drops. It is difficult to distinguish between division and coalescence when writing mechanistic rate expressions, and thus, we do not do so here. Additionally, equating capillary-suction coalescence and bubble-division generation rates teaches that, if division is the primary foam generation mechanism, foam texture at steady state is independent of gas rate. From the discussion on foam microstructure, this case is not the experimental result (*20, 21, 62*). Finally, as gas diffusion coarsening of the trapped phase is not currently well understood, it is not included it in the simulations.

Gas Mobility. In addition to bubble kinetic expressions, the mass balance statements in equations 2 through 4 demand flow rate relationships for the foam and wetting liquid phases. From the discussion of Figures 2 through 4, the confined foam is divided into intermediate-wetting trapped and nonwetting flowing portions. For the flowing foam, the structure of Darcy's law is retained:

$$u_f = \frac{K k_{rf}}{\mu_f} \left(\frac{-\partial p_g}{\partial x} \right) \tag{8}$$

where K is the absolute permeability, k_{rf} is the relative permeability to the flowing foam, and μ_f is the foam effective viscosity. Equation 8 does not imply Darcy flow because μ_f is not a constant. Based on the theoretical studies of Bretherton (*51*) and Hirasaki and Lawson (*18*), the following expression for the non-Newtonian foam effective viscosity is adopted:

$$\mu_f = \mu_g + \frac{\alpha n_f}{v_f^c} \tag{9}$$

where α is a constant of proportionality dependent primarily on the surfactant system. Others have written similar expressions (*20, 39, 48*). According to equation 9, foam viscosity increases with finer textured foams but decreases with increasing interstitial velocity. In the absence of flowing foam bubbles (i.e., $n_f = 0$), the gas viscosity of a continuous foam is recovered. Friedmann et al. (*39*) reported an empirical value of 0.29 for the exponent c. The Bretherton-based theoretical value is 1/3 (*18, 48, 51, 97*).

Again, because the portion of foam that actually flows partitions into the largest and, hence, least resistive channels, the trapped fraction partitions into the intermediate-sized pores, and wetting liquid flows in the smallest most resistive channels (Figure 4), a Stone-type model (*53*) for relative permeability is appropriate. That is, the relative permeability of the most nonwetting phase (i.e., flowing foam) is a function of the saturation of the most nonwetting phase. In accordance with reference 53,

$$k_{rf} = k_{rg}^o S_{fd}^g \tag{10a}$$

where

$$S_{fd} = X_f(1 - S_{wd}) \tag{10b}$$

and

$$S_{wd} = \frac{(S_w - S_{wc})}{(1 - S_{wc})} \tag{10c}$$

$X_f = S_f/S_g$ is the fraction of the foam phase that is flowing. Flowing foam relative permeability is a function of the saturation of flowing gas and, consequently, is greatly reduced compared to the case of a free gas propagating through the porous medium at the total gas saturation. Standard Corey exponent models are adopted for the relative permeability functions with g representing the exponent for gas flow (98). The subscript d indicates that the aqueous-phase saturation is normalized over the saturation range where two-phase flow occurs. S_{wc} is the connate aqueous-phase saturation, and k_{rg}^o is the gas relative permeability at S_{wc}. k_{rf} ($= k_{rg}$) is obtained from relative-permeability measurements for continuous gas–liquid flow in the porous medium.

Consonant with Figure 4, Stone's model for relative permeability also implies that the relative permeability for the aqueous wetting phase is unaffected by the presence of foam. Hence, Darcy's law in equation 8 is written for the wetting liquid with

$$k_{rw} = k_{rw}^o S_{wd}^f \tag{11}$$

where k_{rw}^o and f are the Corey scaling constant and exponent for liquid flow, respectively. This framework confirms the experimental result that during foam flow, the aqueous-phase distributes into its own separate wetting channels (15, 33–36). Again, $k_{rw}(S_w)$ is known from studies on continuous two-phase flow in the medium.

Clearly, the relative permeability of the trapped foam is zero. However, knowledge of the fraction of foam trapped in the porous medium is needed to complete the flow model. In general, the fraction of foam trapped, $X_t = S_t/S_g$, is a function of pressure gradient, capillary pressure, aqueous-phase saturation, and pore geometry. So far, the trapped gas fraction has been measured only for experimental systems at steady state (38, 39). Percolation models, on the other hand, hold promise for determining the functional dependence of X_t (42–47). We write the trapped fraction as a function of the trapped texture, n_t:

$$X_t = X_{t,\max}\left(\frac{\beta n_t}{1 + \beta n_t}\right) \quad (12)$$

where $X_{t,\max}$ is the maximum fraction of trapped foam, and β is a trapping parameter. Equation 12 demands no trapping when the trapped texture is zero, and a smooth rise to a maximum trapping for finer textured bubbles. The trapped fraction, $X_t = 1 - X_f$, strongly influences the foam-flow resistance by reducing gas-phase relative permeability through equation 10b.

To relate the flowing and trapped textures, the work of Friedmann et al. (39) is followed and local equilibrium is assumed. We argue that during coinjection of gas and liquid, the trapped fraction is dynamic. Some portion of the trapped bubbles coarsen and remobilize to be replaced by subsequent trapping of flowing bubbles. Flowing and stationary texture are thus approximately the same. In the simulations to follow, n_f is set equal to n_t.

To complete the model, phase-equilibria information is required. The aqueous surfactant phase is assumed incompressible and nonvolatile; the gas (i.e., N_2) in the foam phase is insoluble in the liquid phase and obeys the ideal gas law. In our experiments, surfactant is present in equal concentration throughout the aqueous phase, and rock adsorption is satisfied. Thus, the surfactant balance in equation 3 is automatically satisfied.

In principle, the framework of separating foam mobility into effective viscosity and relative permeability components also permits description of continuous foam. If only free gas flows, then n_f is zero, and the bulk gas viscosity emerges naturally from equation 9. It is, however, no longer possible to couple the flowing and trapped textures. An independent theory for X_t is required.

"Making and Breaking". Individual bubbles in the foam phase do not retain their identities over macroscopic distances. Rather, they coalesce and reform by the pore-level "making and breaking" processes outlined earlier. Some authors (75, 84, 99), however, use the phrase "making and breaking" in a strict sense attributed originally to Holm (15). After a lamella is produced at a pore constriction, it translates only a short distance not exceeding the exit pore-body before it ruptures. The process then repeats so that on the average the pore is blocked for a fraction of time depending on the stability of the lamellae. Careful reading of Holm's discussion, however, reveals that no such specific meaning is attached to the term "making and breaking". To avoid possible misinterpretation, this phrase is not used here.

Nevertheless, it is important to point out that a lamella cannot be created directly at a pore-throat. Rather, a lens forms first with lamella creation occurring upon expansion into the adjacent pore-body, provided surfactant is available (*see* the discussion of foam-generation mechanisms). During two-phase flow without stabilizing surfactant present, lenses are still created by snap-off in Roof sites (*54, 60*) followed by expansion and rapid coalescence in the downstream pore-body, once the lens thins to a film. If stabilized lamellae are pictured to rupture before exiting the immediate downstream pore-body, they are not much longer lived than unstable lenses. Such processes are accounted for in measurements of continuum relative permeabilities.

Experimental Details

The centerpiece of the apparatus is a vertically mounted, 60-cm long, 5.1-cm diameter, 1.3-μm^2 Boise sandstone core with a porosity of 0.25 mounted in a stainless steel sleeve. Nitrogen gas and foamer solution are injected at the top of the apparatus. Experiments are conducted at back pressures in excess of 5 MPa (700 psia) and at ambient temperature. In situ water saturation measurements are provided by gamma-ray densitometry. A translating carriage holding the radioactive source and detector allows sampling of saturations along the entire length of the core. Pressure taps are also located at 10-cm intervals along the core. Considerable experimental details are available elsewhere (*61, 78, 80*).

Two modifications were made to the apparatus to allow exploration of lower flow rates: Gas injection is controlled by a Brooks 5850C 100 SCCM (standard cubic centimeters per minute) mass flow controller (Emerson Electric, Hatfield, PA), and liquid injection is controlled by an ISCO 500 D syringe pump (Instrumentation Specialties Company, Lincoln, NE). Gas superficial velocities now span 0.30 to 2.13 m/day (1 to 7 ft/day) at 5 MPa (700 psia) back-pressure, and liquid velocities as low as 9 mm/day (0.03 ft/day) are possible.

The foamer solution is a saline solution containing 0.83 wt% NaCl (J. T. Baker, reagent grade) with 0.83 wt% active C_{14-16} α-olefin sulfonate surfactant (Bioterg AS-40, Stepan). Water is provided by a Barnstead Fi Streem II glass still (Barnstead Thermolyne Corp. Dubuque, IA). The solution surface tension is 33 mN/m measured by the Wilhelmy plate method, and the solution viscosity is 1 mPa·s. Bottled nitrogen is the gas source.

The core is initially completely saturated with aqueous foamer solution with rock adsorption satisfied. Nitrogen and aqueous surfactant solution are then injected at fixed flow rates until steady state is achieved. Transient pressure and aqueous saturation profiles are monitored for a wide range of gas and liquid flows. Only one transient foam displacement is reported here. Additional results are available elsewhere (*78–80*).

Comparison of Theory and Experiment

To model the measured transient foam displacements, equations 2 through 12 are rewritten in standard implicit-pressure, explicit-saturation (IMPES) finite difference form, with upstream weighting of the phase mobilities following standard reservoir simulation practice (*10*). Iteration of the nonlinear algebraic equations is by Newton's method. The three primitive unknowns are pressure, gas-phase saturation, and bubble density. Four boundary conditions are necessary because the differential mass balances are second order in pressure and first order in saturation and bubble concentration. The outlet pressure and the inlet superficial velocities of gas and liquid are fixed. No foam is injected, so Q_b is set to zero in equation 4. Initial conditions include $S_w = 1$, $n_f = 0$, $C_s = 0.83$ wt%, and a fixed back-pressure. Calculations require less than 1 cpu minute on a VAX 6420 computer. The thesis of Kovscek provides additional numerical details (*80*).

Model Parameters. Table I lists the model parameters, 18 in all. Those applicable to standard, two-phase flow are shown to the left. They include the absolute rock permeability and porosity, phase viscosities, and Corey exponents and scaling constants for the continuous relative permeabilities. Information on the Boise core, including the relative permeabilities of nitrogen and water, is available from the experiments of Persoff et al. (*61*). Equations 10 and 11 are fit to those independently measured relative permeabilities.

Nine additional parameters are demanded to predict foam displace-

Table I. Parameter Values

Two-Phase Flow Parameters		Population-Balance Parameters	
Parameter	Value	Parameter	Value
K	1.3 μm_2	k_1	1.4×10^5 $s^{1/3}cm^{-13/3}$
ϕ	0.25	k_{-1}^o	9.0×10^{-1} cm^{-1}
f	3.0	$S_w^{'*}$	0.26
k_{rw}^o	0.70	a	0.33
g	3.0	b	1.0
k_{rg}^o	1.0	α	4.0×10^{-6} mPa s$^{2/3}$ cm$^{10/3}$
S_{wc}	0.25	c	0.33
μ_w	1.0 mPa·s	$X_{t,max}$	0.90
μ_g	0.018 mPa·s	β	1.0×10^{-3} cm^{-3}

ment, as listed to the right in Table I. They include the generation and coalescence rate constants, the exponents a and b for the generation rate expression, the saturation that corresponds to the limiting capillary pressure, S_w^*, the proportionality constant and velocity exponent for the foam effective viscosity, and the parameters for the trapped foam fraction. All have clear physical meaning. Thus, the exponent for the effective viscosity, c in equation 9 is set to 1/3 following extensions of the Bretherton analysis (18, 97). Also the theoretical calculations of Roof snap-off behavior in constricted, cornered pores by Kovscek (80) teach that $b = 1$ and that a is less than unity.

All but one of the remaining population-balance parameters are determined from steady-state behavior of foam flow. Fortunately, this exercise drastically limits the choice of parameter values. Thus, for our strong foamer solution, we chose a value of $S_w^* = 0.26$, which is slightly above connate saturation (20, 61, 78), and $X_{t,max} = 0.9$, based on the experimental tracer studies of trapped gas saturations (38, 39).

Next, the exponent a is needed to specify the gas-velocity dependence of foam generation in equation 5. As pointed out earlier, in the limiting capillary-pressure regime with strong foamers, the steady foam-flow pressure drop is sensibly independent of gas flow and varies linearly with liquid velocity (20, 36, 61, 80). When $r_g - r_c = 0$, equations 5 and 6 along with the foam-flow rheology predicted by equations 8 and 9 reveal that $a = 1/3$, and confirm the restriction of $a < 1$. The theoretical value of $b = 1$ demands a linear dependence of steady foam pressure drop on liquid velocity. The choices of $a = c = 1/3$ and $b = 1$ predict that foam texture must coarsen at higher gas velocity. This result, though not immediately obvious, is confirmed by experiment (20). We discover here the origin of the unique flow behavior of foam in porous media as texture alteration with changing gas and liquid velocities, in addition to shear-thinning rheology. It is because of changing textures that classical fractional flow theory does not apply to foam.

The important ratio, k_1/k_{-1}^o, sets the general magnitude of the bubble density. We choose steady-state textures on the order of 100 mm^{-3} or an equivalent undistorted bubble radius of about 130 μm, in agreement with the measurements of Ettinger and Radke (20). Equations 8 and 9, combined with the steady-state texture, now set the magnitude of the steady pressure drop and, consequently, α. It remains to specify the individual magnitudes of k_1 and k_{-1}^o. These magnitudes are adjusted to confine the region of net texture refinement, close to the inlet face of the porous medium (20). Thus, of the nine population-balance parameters, eight are preset by results of steady-state measurements. Finally, the simulations prove insensitive to the trapping parameter β, which is chosen such that $X_t = (1 - X_f)$ is 85% when n_f is 20 mm^{-3}.

Transient Displacement. Experimental displacement results for the simultaneous injection of aqueous surfactant solution and nitrogen into a core initially saturated with a surfactant solution are shown in Figures 12 and 13. Darcy velocities relative to the exit pressure of 4.8 MPa are 0.43 m/day (1.4 ft/day) for gas and 0.046 m/day (0.15 ft/day) for liquid yielding a gas fractional flow or foam quality of 90%. Figure 12 provides the transient liquid saturation profiles. Experimental data points are connected by dashed lines. Time is expressed nondimensionally in pore volumes, PV, which is the ratio of total volumetric flow rate (at exit pressure) multiplied by elapsed time and divided by the void volume of the core.

Experimentally, steep fronts are seen in Figure 12, whereby the aqueous saturation upstream of the front is approximately 30%, about 5 saturation units above connate saturation, and downstream it is 100%. From the saturation profiles, it appears that foam moves through the rock in a piston-like displacement when the porous medium is initially saturated with surfactant solution. After the front has passed a particular location, saturation changes very little. Foam clearly provides a very efficient displacement of the aqueous phase. Figure 12 indicates that even though nitrogen and surfactant solution are injected separately, rapid foam generation and liquid desaturation occur near the inlet. Gas breakthrough is at roughly 0.80 PV, and by 1.5 PV the saturation profile ceases to change. In general, the aqueous desaturation is complete in about 1 to 2 PV for all cases. The experimental transient data always follow the general forms shown in Figures 12 and 13 when the core is presaturated with surfactant solution (*78–80*).

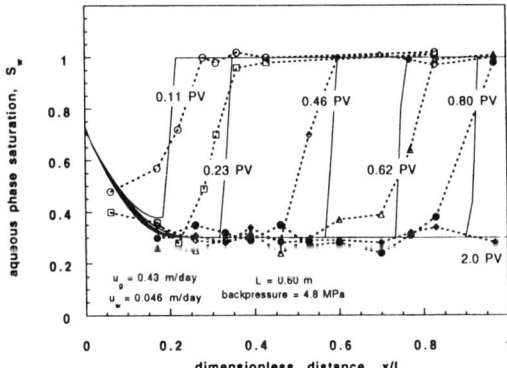

Figure 12. Experimental and model transient aqueous-phase saturation profiles. Model results are shown with solid lines. Experimental data points (symbols) are connected by dashed lines.

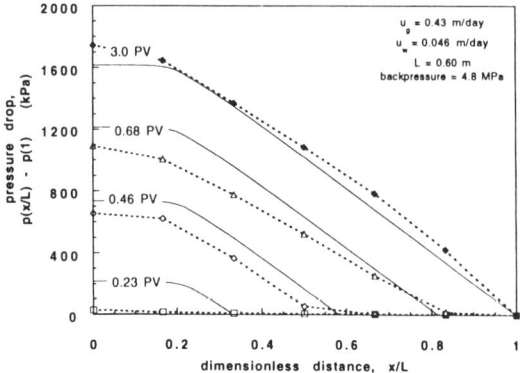

Figure 13. Experimental and model transient pressure profiles. Model results are shown with solid lines. Experimental data points (symbols) are connected by dashed lines.

The theoretical saturation profiles, shown as solid lines in Figure 12, track the experimental results well. Because the dissipative action of capillary-pressure gradients is not included in the model formulation, calculated fronts remain steep and sharp. The population-balance model predicts that S_w is high at the core inlet. Aqueous saturation is around 76% at x/L equal to zero, but drops rapidly to approximately 30% by x/L equal to 0.2, where L is the length of the core. Because no foam is injected, n_f is zero at the inlet, and the foam effective viscosity is equal to the gas viscosity. Consequently, S_w is high. Foam texture, however, rapidly increases and produces a low foam mobility. Unfortunately, due to saturation scanning limitations, few experimental data are available directly at the inlet to verify the model predictions. However, Minssieux (*100*) did detect such a region of high S_w near the inlet of a sand pack during foam displacement.

A region of net foam generation near the core inlet is also witnessed in the transient pressure profiles of Figure 13. Both the experimental data and model calculations show that pressure gradients near the inlet are definitely shallow, indicating that flow resistance there is small. Steep gradients are found closely downstream of the inlet region. Again, this observation demonstrates that foam texture is coarse near the core inlet.

Figure 14 reports the calculated transient foam bubble density, n_f, as a function of dimensionless distance. At all time levels, foam bubbles are coarsely textured near the inlet, but within the first fifth of the core, texture becomes much finer. Beyond the first fifth of the core, the limiting capillary-pressure regime develops; foam texture in this region is nearly constant as is the liquid saturation in Figure 12. Foam texture also increases rapidly with respect to time. At 0.23 PV, foam bubble density im-

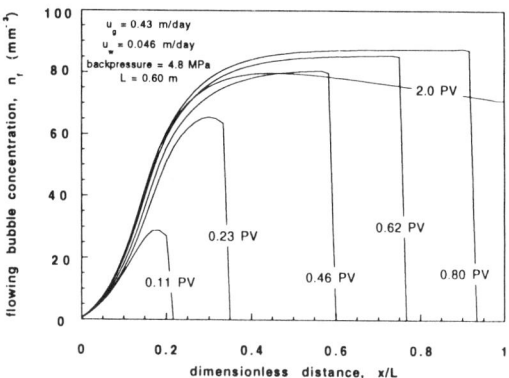

Figure 14. Calculated flowing foam texture.

mediately upstream of the front (i.e., $x/L = 0.35$) is within 80% of its steady-state value. Figure 14 confirms that foam moves through the column in a piston-like fashion consistent with the experimental data in Figures 12 and 13. Unfortunately, no experimental method currently exists to measure bubble density in situ.

Figures 12 through 14 further demonstrate that both model and experimental transient saturation profiles of Figure 12 are exactly tracked by the pressure and foam texture profiles of Figures 13 and 14. High pressure gradients and fine foam textures are seen where liquid saturation is low and vice versa. Careful examination of Figures 12 through 14 points out a deficiency of the model calculations. Predicted saturation and pressure profiles build too quickly. Theoretical steady-state pressure drop and liquid saturation agree well with experiment. Transient pressures, however, are overpredicted in some regions by roughly 100 kPa (15 psi), and the foam front slightly leads the experimental one. These discrepancies are likely a result of the imposition of instantaneous equilibrium between flowing and trapped bubbles, that at an early time may overpredict the amount of bubble trapping. Additionally, the proportionality constant for foam effective viscosity, α, increases with Plateau border curvature *(18)*, and hence, with decreasing water saturation. Accordingly, a constant value of α may increase the viscosity of the foam too much at high S_w. It is also possible that the generation rate constant is reduced somewhat at high S_w *(26)*. Nevertheless, good agreement is found between experiments for transient foam-displacement and the proposed population-balance model *(78–80)*.

Steady Behavior. New steady-state experimental and model results are compared in Figures 15 and 16 *(see* reference 61). Here, over-

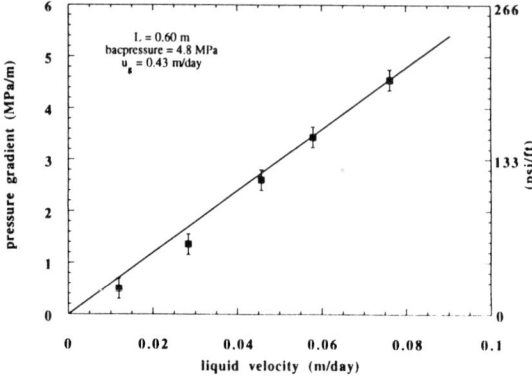

Figure 15. Experimental and model steady-state pressure drop versus liquid velocity. Gas-phase velocity is held constant. Symbols are experimental data; solid lines are model predictions. Error bars are shown.

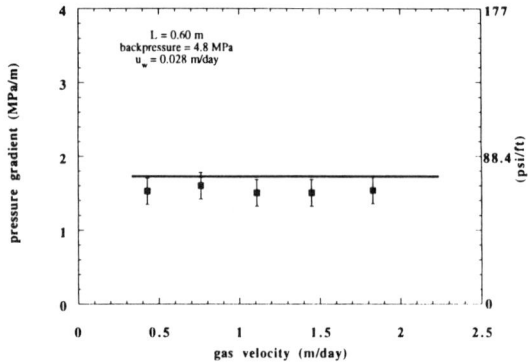

Figure 16. Experimental and model steady-state pressure drop versus gas velocity. Aqueous-phase velocity is held constant. Symbols are experimental data while solid lines are model predictions. Error bars are shown.

all pressure gradients (i.e., the ratio of pressure drop to length) rather than core pressure drops are given. Figure 15 portrays pressure-gradient behavior when gas injection velocity is held constant, and the liquid flow rate is varied. Figure 16 shows the results from holding liquid injection rate constant but varying gas flow rates. In Figure 15, the pressure gradient increases linearly from roughly 0.4 to 4.5 MPa/m as liquid velocity is varied between 0.012 and 0.076 m/day (0.04 and 0.25 ft/day). Except for the slight depression of the single experimental datum at roughly 0.028 m/day, the experimental points all fall on the theoretical curve within experimental error. For comparison, the pressure drop of water flowing at

0.43 m/day in 1.3-μm^2 Boise sandstone is 4.1 kPa/m (0.2 psi/ft). When compared to continuum, gas flow foam reduces gas mobility by factors approaching 5000!

Figure 16 illustrates the independence of steady foam pressure gradient to gas velocity. Model prediction of pressure drop, 1.7 MPa/m (77 psi/ft), is slightly greater than the experimental result, 1.5 MPa/m (68 psi/ft). This discrepancy is understood by comparing the constant liquid velocity (0.028 m/day) used in Figure 16 to the results in Figure 15. The only experimental data point that did not fall on the model-predicted line lies at 0.028 m/day. The data, taken during that particular experiment, appear to have slightly depressed pressure drops.

The steady-state pressure drop versus phase velocity trends corresponding to Figures 15 and 16 strikingly reconfirm the earlier findings of Ettinger and Radke (20) and Persoff et al. (61), obtained with differing surfactants and differing sandstone cores. The data are characteristic of the limiting capillary-pressure regime because measured steady-state liquid saturations corresponding to Figures 15 and 16 are constant near 0.3 (80). Adjustment of foam texture with changing flow rates explains the unusual behavior seen. When gas velocity increases under constant liquid flow rate conditions, foam texture coarsens, effective viscosity decreases, and constant pressure drop is maintained. Conversely, when liquid velocity increases, and gas rates are held constant, foam texture and effective viscosity increase to yield a linearly rising pressure drop. Foam texture adjusts in obedience to generation and coalescence rate laws, compatible with a limiting capillary pressure and a sensibly constant and low aqueous-phase saturation at steady state. Without accounting for bubble texture, such results are difficult to rationalize.

Summary

Foam flow in porous media is a complex, multifaceted process. Macroscopic results are the ensemble average of many pore-scale events that lead to bubble evolution and pore-wall interaction during multiphase flow. Foam in porous media is best understood when the undergirding pore-level phenomena are elucidated and quantified.

A porous medium shapes foam to its own liking as confined, pore-filling bubbles and lamellae. Foam in porous media is not a continuous fluid. The three mechanisms of foam generation (snap-off, division, and leave-behind) are all pore geometry specific. Snap-off is a mechanical process that occurs in multiphase flow without surfactant. For successful gas-bubble snap-off, the pore-body to pore-throat constriction ratio must be sufficiently large (roughly 2) and gently sloped. Otherwise stable wetting collars form in pore-throats and obviate foam generation (26, 54). Sufficient liquid supply must also be available for fluid accumulation in pore-throats prior to pinch-off. Lamellae are never generated directly. Rather, snap-off creates a lens of aqueous fluid that drains later under the action of capillary suction. Division proceeds by subdividing previously generated moving foam bubbles or lamellae at points where flow branches. Thus, division requires the porous medium to be relatively free of stationary lamellae or foam bubbles that greatly reduce the number of branch points. Leave-behind generates stationary aqueous lenses when aqueous-phase saturation is high. It is a nonrepetitive process that alone cannot account for the large reduction in gas mobility seen with foam.

Foam generation does not continue unchecked. Surfactant-stabilized lamellae are only metastable. Coalescence ensues when a translating lamella moves out of a sharply constricted pore-throat into a pore-body, and the lamella is stretched too rapidly for healing flow of foamer solution. Whereas foam generation by capillary snap-off is independent of surfactant formulation, coalescence of foam lamellae strongly depends on surfactant formulation, concentration, and salinity.

Coalescence of flowing foam is complicated. Yet coalescence in porous media does correlate directly with rupture of single, static lamellae. If an isolated foam film can withstand large capillary-suction pressures, that particular foamer solution produces a strong foam in porous media. Thus, films that exhibit large rupture disjoining pressures lead to foams with large flow resistance in porous media. A large disjoining pressure for rupture also explains the limiting capillary-pressure flow regime and its sensitivity to surfactant formulation.

The mobility of gas dispersed as strong foam in porous media is low, many orders of magnitude smaller than that of the parent gas phase. Only a small percentage of the overall foam gas saturation actually flows, typically 1–15% at steady state. The stationary portion blocks intermediate-sized flow paths and lowers the effective permeability of the rock to gas. Of the remaining portion of foam that actually flows in the largest pore channels, interactions of foam bubbles with pore walls determines an effective viscosity that is larger than that for water filling the same channels. Detailed analyses of bubble flow in constricted capillaries with noncircular cross section show quite universally that foam is shear-thinning and texture dependent.

To predict foam behavior mechanistically and quantitatively, it is necessary to account for bubble-size evolution. A foam bubble population-balance provides the necessary framework by including foam as a simple component in a standard reservoir simulator. Reduced gas mobility is modeled by a lowered gas relative permeability and a raised effective viscosity. A Stone-type model for relative permeability provides the requisite rules for modeling the relative permeability of both wetting liquid and nonwetting foam. Theory applied to the snap-off generation and capillary-suction coalescence pore-level events, garners the specific forms of the generation and coalescence rate expressions.

Only the case of steady coinjection of surfactant solution and gas into a one-dimensional core initially filled with surfactant solution is addressed. Calculated transient foam displacement well represents both the measured wetting liquid saturations and pressure profiles with physically meaningful parameter values. It is predicted and experimentally verified that foam moves in a piston-like fashion through a linear porous medium presaturated with surfactant solution. Moreover, the proposed population-balance predicts the entire spectrum of unique steady foam-flow behavior in the capillary-pressure regime.

The population-balance is a powerful tool for modeling foam displacement and flow in porous media because it correctly predicts the evolution of foam microstructure from well documented pore-level events, and because it merges with current reservoir simulation practice. Perhaps the main power of the population-balance approach is its general framework. As understanding of mechanistic detail improves, this information may be incorporated in the modeling effort.

Acknowledgment

This work was supported by the Assistant Secretary for Fossil Energy, Office of Oil, Gas, and Shale Technologies of the U. S. Department of Energy, under Contract No. DE–AC03–76FS00098 to the Lawrence Berkeley Laboratory of the University of California. P. Persoff provided invaluable assistance for the experimental program.

Nomenclature

a, b, c	velocity exponents
Ca	capillary number, ratio of viscous to surface tension forces
C_m	mean interfacial radius of curvature
f	wetting-phase relative permeability exponent
g	nonwetting-phase relative permeability exponent
h	film thickness
k	rate constant
k_r	relative permeability
K	permeability
L	length of porous medium

n_f	number density of flowing foam (number of bubbles per unit volume of flowing foam)
p	phase pressure
PV	pore volume
P_c	capillary pressure, $p_{nw} - p_w$
Q	source–sink term in conservation equations
r	foam generation–coalescence rate (number of bubbles per unit time per unit volume of gas)
R	pore radius
S	phase saturation
t	time
u	superficial velocity
U	bubble velocity
v	interstitial velocity
x	spatial variable
X	foam fraction

Greek

α	proportionality constant for the effective viscosity of flowing foam
β	trapping parameter in equation 12
ϕ	porosity
Π	disjoining pressure
ρ	mass density
σ	equilibrium surface tension
μ	viscosity

Subscripts

1	generation rate constant
−1	coalescence rate constant
b	body or bubble
c	coalescence or constriction throat
f	flowing foam
fd	normalized flowing foam saturation
g	gas phase or generation
max	maximum or local maximum
nw	nonwetting phase
rup	rupture
s	surfactant

t	trapped foam
T	straight tube
w	wetting phase
wc	connate saturation
wd	normalized wetting-phase saturation

Superscripts

| o | scaling or reference value |
| * | value corresponds to the limiting capillary pressure |

References

1. Fried, A. N. *The Foam Drive Process for Increasing the Recovery of Oil;* Report 5866; U.S. Dept of Interior, Bureau of Mines: Washington, DC, 1961.
2. Hirasaki, G. J. *The Steam Foam Process;* Society of Petroleum Engineers: Richardson, TX, 1989; Supplement to SPE 19505,
3. Hirasaki, G. J. *J. Pet. Tech.* **1989**, *41(5)*, 449–456.
4. Marsden, S. S. *Foams in Porous Media;* Report SUPRI TR–49; U.S. Department of Energy: Washington, DC, July 1986.
5. Heller, J. P.; Kuntamukkula, M. S. *Ind. Eng. Chem. Res.* **1987**, *26(2)*, 318–325.
6. Baghidikian, S. Y.; Handy, L. L. *Transient Behavior of Simultaneous Flow of Gas and Surfactant Solution in Consolidated Porous Media;* Topical Report; U.S. Department of Energy: Washington, DC, July 1991.
7. Rossen, W. R. In *Foams: Theory, Measurements and Applications;* Prud'Homme, R. K.; Khan. S., Eds.; Marcel Dekker: New York, Chapter 12, in press.
8. Patzek, T. W. In *Surfactant-Based Mobility Control: Progress in Miscible-Flood Enhanced Oil Recovery;* Smith, D. H., Ed.; ACS Symposium Series 373; American Chemical Society: Washington, DC, 1988; Chapter 16, pp 326–341.
9. Falls, A. H.; Hirasaki, G. J.; Patzek, T. W.; Gauglitz, P. A.; Miller, D. D.; Ratulowski, T. *Soc. Pet. Eng. Res. Eng.* **1988**, *3(3)*, 884–892.
10. Aziz, K.; Settari, A. *Petroleum Reservoir Simulation;* Applied Science Publishers: London, 1979; pp 125–199.
11. Hanssen, J. E.; Dalland, M. Presented at SPE 7th Symposium on Enhanced Oil Recovery, Tulsa, OK, April 1990; paper SPE/DOE 20193.
12. Hanssen, J. E.; Haugum, P. Presented at SPE International Symposium on Oil Field Chemistry, Anaheim, CA, February 1991; paper SPE 21002.
13. Clark, H. B.; Pike, M. T.; Rengel, G. L. Presented at SPE International Symposium on Oil field and Geochemistry, Houston, TX, January 1979; paper SPE 7894.

14. Holcomb, D. L.; Callaway, R. E.; Curry, L. L. Presented at SPE Rocky Mountain Regional Meeting, Casper, WY, May 1980; paper SPE 9033.
15. Holm, L. W. *Soc. Pet. Eng. J.* **1968**, *8(4)*, 359–369.
16. Bikerman, J. J. *Foams;* Springer–Verlag: Berlin, Germany, 1973; pp 33–64, 184–213.
17. Kraynik, A. M. *Ann. Rev. Fluid Mech.* **1988**, *20*, 325–357.
18. Hirasaki, G. J.; Lawson, J. B. *Soc. Pet. Eng. J.* **1985**, *25(2)*, 176–190.
19. Marsden, S. S.; Earligh, J. J. P.; Albrecht, R. A.; David, A. *The Proceedings of the 7th World Petroleum Congress;* Mexico City; April 1967; pp 235–242.
20. Ettinger, R. A.; Radke, C. J. *Soc. Pet. Eng. Res. Eng.* **1992**, *7(1)*, 83–90.
21. Ettinger, R. A. M.S. Thesis, University of California, Berkeley, CA, 1989.
22. Wardlaw, N. C.; Li, Y.; Forbes, D. *Transport in Porous Media* **1987**, *2*, 597–614.
23. Trienen, R. J.; Brigham, W. E.; Castanier, L. M. *Apparent Viscosity Measurements of Surfactant Foam in Porous Media;* Topical Report SUPRI TR–48; U.S. Department of Energy: Washington, DC, October 1985.
24. Mast, R. F. Presented at 47th SPE Annual Meeting, San Antonio, TX, October 1972; paper SPE 3997.
25. Owete, O. S.; Brigham, W. E. *Soc. Pet. Eng. Res. Eng.* **1987**, *2(3)*, 315–323.
26. Chambers, K. T.; Radke, C. J. In *Interfacial Phenomena in Petroleum Recovery;* Morrow, N. R., Ed.; Marcel Dekker: New York, 1991; Chapter 6, pp 191–255.
27. Chambers, K. T. M.S. Thesis, University of California, Berkeley, CA, 1990.
28. Hanssen, J. E. *J. Pet. Sci. Eng.* **1993**, *10(2)*, 117–134
29. Hanssen, J. E. *J. Pet. Sci. Eng.* **1993**, *10(2)*, 135–156.
30. Rossen, W. R.; Zhou, Z. H.; Mamun, C. K. Presented at 66th SPE Annual Technical Conference, Dallas, TX, October 1991; paper SPE 22627.
31. Bernard, G. G.; Holm, L. W. *Soc. Pet. Eng. J.* **1964**, *4(3)*, 267–274.
32. Bernard, G. G.; Holm, L. W.; Jacobs, W. L. *Soc. Pet. Eng. J.* **1965**, *5(4)*, 295–300.
33. Friedmann, F.; Jensen, J. A. Presented at the SPE California Regional Meeting, Oakland, CA, April 1986; paper SPE 15087.
34. Sanchez, J. M.; Schechter, R. S.; Monsalve, A. Presented at the 61st SPE Annual Technical Conference, New Orleans, LA, October 1986; paper SPE 15446.
35. Huh, D. G.; Handy, L. L. *Soc. Pet. Eng. Res. Eng.* **1989**, *4(1)*, 77–84.
36. De Vries, A. S.; Wit, K. *Soc. Pet. Eng. Res. Eng.* **1990**, *5(2)*, 185–192.
37. Nahid, B. H. Ph.D. Thesis, University of Southern California, Los Angeles, CA, 1971.
38. Gillis, J. V.; Radke, C. J. Presented at the 65th SPE Annual Technical Conference, New Orleans, LA, September 1990; paper SPE 20519.
39. Friedmann, F.; Chen, W. H.; Gauglitz, P. A. *Soc. Pet. Eng. Res. Eng.* **1991**, *6(1)*, 37–45.
40. Flumerfelt, R. W.; Prieditis, J. In *Surfactant-Based Mobility Control: Progress in Miscible-Flood Enhanced Oil Recovery;* Smith, D. H., Ed.; ACS Symposium Series 373; American Chemical Society: Washington, DC, 1988; Chapter 15, pp 295–325.
41. Prieditis, J. Ph.D. Thesis, University of Houston, Houston, TX, 1988.
42. Rossen, W. R. *J. Colloid Interface Sci.* **1990**, *136(1)*, 1–16.

43. Rossen, W. R. *J. Colloid Interface Sci.* **1990**, *136(1)*, 17-37.
44. Rossen, W. R. *J. Colloid Interface Sci.* **1990**, *136(1)*, 38–53.
45. Rossen, W. R.; Gauglitz, P. A. *Am. Inst. Chem. Eng. J.* **1990**, *36(8)*, 1176–1188.
46. Rossen, W. R. *J. Colloid Interface Sci.* **1990**, *139(2)*, 457–468.
47. de Gennes, P. G. *Rev. Inst. Fr. Pet.* **1992**, *47(2)*, 249–254.
48. Falls, A. H.; Musters, J. J.; Ratulowski, J. *Soc. Pet. Eng. Res. Eng.* **1989**, *4(2)*, 155–164.
49. Marsden, S. S.; Khan, S. A. *Soc. Pet. Eng. J.* **1966**, *6(1)*, 17–25.
50. Heller, J. P.; Lien, C. L.; Kuntamukkula, M. S. Presented at the 57th SPE Annual Technical Conference, New Orleans, LA, September 1982; paper SPE 11233.
51. Bretherton, F. P. *J. Fluid Mech.* **1961**, *10*, 166–188.
52. Wong, H. Ph.D. Thesis, University of California at Berkeley, Berkeley, CA, 1992.
53. Stone, H. L. *J. Pet. Tech.* **1970**, *22(2)*, 214–218.
54. Roof, J. G. *Soc. Pet. Eng. J.* **1970**, *10(1)*, 85–90.
55. Jiménez, A. I.; Radke, C. J. In *Oil-Field Chemistry: Enhanced Recovery and Production Stimulation;* Borchardt, J. K.; Yen, T. F., Eds.; ACS Symposium Series 396; American Chemical Society: Washington, DC, 1989; Chapter 25, pp 460–479.
56. Kanda, M.; Schechter, R. S. Presented at the 51st SPE Annual Technical Conference, New Orleans, LA, October 1976; paper SPE 6200.
57. Li, Y.; Wardlaw, N. C. *J. Colloid Interface Sci.* **1986**, *109(2)*, 461–472.
58. Sanchez, J. M.; Hazlett, R. D. *Soc. Pet. Eng. Res. Eng.* **1992**, *7(1)*, 91–97.
59. Gauglitz, P. A. Presented at the AIChE Annual Meeting, Chicago, IL, November 1990.
60. Ransohoff, T. C.; Radke, C. J. *Soc. Pet. Eng. Res. Eng.* **1988**, *3(2)*, 573–585.
61. Persoff, P.; Radke, C. J.; Pruess, K.; Benson, S. M.; Witherspoon, P. A. *Soc. Pet. Eng. Res. Eng.* **1991**, *6(3)*, 365–371.
62. Khatib, Z. I.; Hirasaki, G. J.; Falls, A. H. *Soc. Pet. Eng. Res. Eng.* **1988**, *3(3)*, 919–926.
63. Derjaguin, B. V.; Obukhov, E. V. *Acta Physicochim. URSS* **1936**, *5(1)*, 1–22.
64. Derjaguin, B. V.; Kussakov, M. M. *Acta Physicochim. URSS* **1939**, *10(1)*, 25–44.
65. Bergeron, V.; Radke, C. J. *Langmuir* **1992**, *8*, 3020–3026.
66. Exerowa, D. R.; Kolarov, T.; Khristov, K. H. R. *Colloids Surf.* **1987**, *22*, 171–185.
67. Derjaguin, B. V.; Kussakov, M. M. *Acta Physicochim. URSS* **1939**, *10(2)*, 153–174.
68. Derjaguin, B. V.; Churaev, N. V.; Muller, V. M. *Surface Forces;* Consultants Bureau: New York, 1987; pp 25–52, 327–367.
69. Hirasaki, G. J. In *Interfacial Phenomena in Petroleum Recovery;* Morrow, N. R., Ed.; Marcel Dekker: New York, 1990; Chapter 2, pp 23–76.
70. Khristov, K. H. R.; Krugljakov, P.; Exerowa, D. R. *Colloid Polym. Sci.* **1979**, *257*, 506–511.
71. Khristov, K. H. R.; Exerowa, D. R.; Krugljakov, P. M. *Colloid J. USSR* **1981**, *43*, 80–84.
72. Vrij, A. *Discuss. Faraday Soc.* **1966**, *42*, 23–33.

73. Leverett, M. C. *Trans. AIME* **1941**, *195*, 159–172.
74. Jiménez-Laguna, A. I. Ph.D. Thesis, University of California at Berkeley, Berkeley, CA, 1991.
75. Huh, D. G.; Cochrane, T. D.; Kovarik, F. S. *J. Pet. Tech.* **1989**, *41*, 872–879.
76. Aronson, A. S.; Bergeron, V.; Fagan, M. E.; Radke, C. J. *J. Colloids Surf. A* **1994**, *83*, 109–120.
77. Osterloh, W. T.; Jante, M. J., Jr. Presented at 8th SPE/DOE Symposium on Enhanced Oil Recovery, Tulsa, OK, April 1992; paper SPE/DOE 24179.
78. Kovscek, A. R.; Radke, C. J. Presented at DOE/NIPER Field Application of Foams for Oil Production Symposium, Bakersfield, CA, February 1993; paper FS-9.
79. Kovscek, A. R.; Patzek, T. W.; Radke, C. J. Presented at 68th SPE Annual Technical Conference, Houston, TX, October 1993; paper SPE 26402.
80. Kovscek, A. R. Ph.D. Thesis, University of California at Berkeley, Berkeley, CA, 1994.
81. Lee, H. O.; Heller, J. P. *Soc. Pet. Eng. Res. Eng.* **1990**, *5(2)*, 193–197.
82. Zhou, Z.; Rossen, W. R. Presented at the SPE/DOE 8th Symposium on Enhanced Oil Recovery, Tulsa, OK, April 1992; paper SPE/DOE 24180.
83. Chang, S. H.; Owusu, L. A.; French, S. B.; Kovarik, F. S. Presented at the 7th SPE/DOE Symposium on Enhanced Oil Recovery, Tulsa, OK, April 1990; paper SPE/DOE 20191.
84. Chou, S. I. Presented at the 7th SPE/DOE Symposium on Enhanced Oil Recovery, Tulsa, OK, April 1990; paper SPE/DOE 20239.
85. Fisher, A. W.; Foulser, R. W. S.; Goodyear, S. G. Presented at SPE/DOE 7th Symposium on Enhanced Oil Recovery, Tulsa, OK, April 1990; paper SPE/DOE 20195.
86. Liu, D.; Brigham, W. E. *Transient Foam Flow in Porous Media with Cat Scanner;* Topical Report; U.S. Department of Energy: Washington, DC, March 1992.
87. Patzek, T. W.; Koinis, M. T. *J. Pet. Tech.* **1990**, *42(4)*, 496–503.
88. Patzek, T. W.; Myhill, N. A. Presented at SPE California Regional Meeting, Bakersfield, CA, April 1989; paper SPE 18786.
89. Mohammadi, S. S.; Coombe, D. A. Presented at SPE California Regional Meeting, Bakersfield, CA, April 1992; paper SPE 24030.
90. Mohammadi, S. S.; Coombe, D. A.; Stevenson, V. M. *J. Can. Pet. Tech.* **1993**, *32(10)*, 49–54.
91. Marfoe, C. H.; Kazemi, H. Presented at the 62nd SPE Annual Meeting, Dallas, TX, September 1987; paper SPE 16709.
92. Mahmood, S. M.; Tariq, S. M.; Brigham, W. E. Presented at SPE California Regional Meeting, Oakland, CA, April 1986; paper SPE 15076.
93. Hillestad, J. G.; Mattax, C. C. In *Reservoir Simulation;* Mattax, C. C.; Dalton, R. L. Eds.; SPE Monograph Series Vol. 13; Society of Petroleum Engineers: Richardson, TX, 1990; Chapter 6, pp 57–73.
94. Ransohoff, T. C.; Gauglitz, P. A.; Radke, C. J. *AIChE J.* **1987**, *33(5)*, 753–765.
95. Chou, S. I. Presented at the 66th SPE Annual Technical Conference, Dallas, TX, October 1991; SPE 22628.
96. Fagan, M. E. M.S. Thesis, University of California at Berkeley, Berkeley, CA, 1992.

97. Ginley, G. M.; Radke, C. J. In "Oil-Field Chemistry: Enhanced Recovery and Production Stimulation; Borchardt, J. K.; Yen, T. F., Eds.; ACS Symposium Series 396; American Chemical Society: Washington, DC, 1989; Chapter 26, pp 480–501.
98. Corey, A. T. *Producer's Monthly* **1954**, *19(1)*, 38–41.
99. Yang, S. H.; Reed, R. L. Presented at the 64th SPE Annual Meeting, San Antonio, TX, October 1989; paper SPE 19689.
100. Minssieux, L. *J. Pet. Tech.* **1974**, *26(1)*, 100–108.

Note: All papers with SPE numbers are available through the Society of Petroleum Engineers, P.O. Box 833836, Richardson, TX 75083–3836.

RECEIVED for review October 14, 1992. ACCEPTED revised manuscript August 1, 1993.

4

Foam Sensitivity to Crude Oil in Porous Media

Laurier L. Schramm

Petroleum Recovery Institute 100, 3512 33rd Street N.W., Calgary, Alberta T2L 2A6, Canada

This chapter addresses the sensitivity of foams flowing in porous media to interaction with crude oil. The interaction effects can range from imperceptible to strong and have a corresponding influence on foam stabilities. An outline of predictive models, microvisualization observations, and core-flood measurements is given and used to identify a framework for physical modeling of foam–oil interactions in porous media. A number of comparisons of model predictions with actual core-flood experiments that were conducted under a variety of conditions are presented. In this context, the influence of oil gravity and rock wettability on foam stability is reviewed. Although some gaps in fundamental understanding exist, it is possible to make a number of useful generalizations about the sensitivity of foams flowing in porous media to crude oil.

Applications of Foam in Enhanced Oil Recovery

The production of oil from a petroleum reservoir involves first the primary and secondary production modes, which may recover less than half of the oil originally in place. To recover additional oil, it is necessary to apply enhanced oil recovery (EOR) techniques such as miscible or immiscible gas displacement (CO_2, hydrocarbon gases, etc.). However, major problems occur in these EOR methods because the displacing agent has high mobility and low density compared with that of reservoir fluids. Fingering (channeling) and gravity override reduce the sweep efficiency, contribute to early breakthrough of injected fluid, and thus reduce the amount of oil recovered.

The use of surfactant-stabilized foams to counteract these kinds of problems was suggested several decades ago (*1, 2*) and has recently become actively pursued in laboratory and field tests (*3–8*). The use of foam is advantageous compared with the use of a simple fluid of the same nominal mobility because the foam, which has an apparent viscosity greater than the displacing medium, lowers the gas mobility in the swept or higher permeability parts of the formation. This lowered gas mobility diverts at least some of the displacing medium into other parts of the formation that were previously unswept or underswept. From these underswept areas, the additional oil is recovered. Because foam mobility is reduced disproportionately more in higher permeability zones, improvement in both vertical and horizontal sweep efficiency can be achieved.

There is an abundant literature on the subject of bulk foams and thin films much of which is reviewed in Chapters 1 and 2 of this book. The extension to foams in porous media is discussed in Chapter 3. In the development of suitable foam-forming surfactant solutions for EOR processes, researchers have focused on finding ways to produce foams that have the necessary stability and viscosity to function as diverting and mobility-reducing agents in the reservoir (*9, 10*). As the increasing number of publications in this area shows, it is crucial to know how to select surfactants or surfactant mixtures to provide effective foam-forming solutions. This selection involves assessing not only foam stability but also such properties as chemical stability, solubility, and adsorption by the rock (*11–13*). These properties are of particular concern where reservoirs considered for foam injection consist of different rock types and can have very high salinity (e.g., 300 g/L), hardness (e.g., 25 g/L), and temperature. These and other specific considerations that are related to designing foam injection processes that use CO_2, steam, and hydrocarbon gases are discussed in Chapters 5 through 8.

This chapter will focus on the stability of foams flowing in porous media when in the presence of crude oil. Many laboratory investigations of foam-flooding have been carried out in the absence of oil, but comparatively few have been carried out in the presence of oil. For a field application, where the residual oil saturation may vary from as low as 0 to as high as 40% depending on the recovery method applied, any effect of the oil on foam stability becomes a crucial matter. The discussion in Chapter 2 showed how important the volume fraction of oil present can be to bulk foam stability. A recent field-scale simulation study of the effect of oil sensitivity on steam-foam flood performance concluded that the magnitude of the residual oil saturation was a very significant factor for the success of a full-scale steam-foam process (*14*).

Foam Stability in the Presence of Oil

Oils, such as castor oil, are among the earliest chemicals to be used for foam inhibition and foam breaking. Accordingly, considerable literature exists on the subject (15–20). Some theories for the mechanisms of antifoaming action, together with an abundance of qualitative observations, have emerged from this work. In general, matters are not simple. Apparently, foams can be destabilized by oils by several mechanisms, and more than one mechanism may be operative in a given case.

Some possibilities for the mechanisms of foam destabilization by a given oil phase include

- Foam-forming surfactant may be absorbed or adsorbed by the oil, especially if there is emulsification, causing depletion in the aqueous phase and hence from the gas–liquid interface.
- Surfactants from the oil may be adsorbed by the foam lamellae, form either a mixed or replaced adsorption layer, and produce a less favorable state for foaming.
- Components from the oil may be adsorbed by the porous medium altering the wettability of the solid phase, and this alteration makes it more difficult for foam to be generated and regenerated.
- The oil may spread spontaneously on foam lamellae and displace the foam stabilizing interface.
- The oil may emulsify spontaneously and allow drops to breach and rupture the stabilizing interface.

Although the physical properties of foams and foam-forming solutions for enhanced oil recovery have been studied intensively, only relatively recently has attention been paid to the effect of the oil on the foams. In the present work, we will not focus attention on the first two mechanisms listed, which concern phase-behavior. The surfactant absorption (partitioning) may be important in some cases but has been shown to be negligible for a number of commercial foam-forming surfactant–crude oil combinations in which quite hydrophilic surfactants were involved (21). Generally, if foam-forming surfactant adsorptions (on the oil and mineral surfaces) have been satisfied, then physical changes such as spreading are responsible for foam destabilization.

Ample evidence suggests that crude oil can have an effect on foams applied to enhanced oil recovery. Rendall et al. (21) investigated the behavior of several commercial surfactant-stabilized foams in the presence of crude oils. On the basis of dynamic bulk foaming tests, gas mobility reduction factors measured in reservoir cores, and observations in a microvisual apparatus, it was found that all but one of the foams studied were destroyed when brought into contact with oil. Similarly, Kuhlman (*22*) and Manlowe and Radke (*23*) observed in micromodel studies that oils were capable of destroying foams. Foam sensitivity to oil is also manifested as an increased difficulty of forming and propagating foams through porous media containing oil (*21, 24–28*). For example, Novosad and Ionescu (*11*) found lower mobility reduction factors in foam-floods conducted in cores containing residual oil compared to the same floods conducted in cores that were completely free of oil. The lower mobility reduction factors were interpreted to be due to some kind of foam destruction by the oil.

The literature clearly shows that some foams are stable in the presence of oil and others are not. The column foaming tests show a gradual decrease in bulk foam stability as increasing amounts of emulsified oil are contacted, but it is uncertain in these tests how the oil is destabilizing the foam, and whether surfactant is being adsorbed by the increased oil surface area. The foam-flooding experiments conducted in core samples show that the presence of residual oil can significantly lower the mobility reduction that would otherwise have been achieved, but the mechanism by which this happens is not readily judged from such experiments. The microvisualization studies probably provide an intermediate degree of realism between these two kinds of tests and also provide qualitative and semiquantitative information about the possible mechanisms of interaction.

Experimental Studies of Foam–Oil Interaction in Porous Media

Bulk Foam–Oil Sensitivity Tests. The experimental techniques used to investigate the influence of oil on foam stability in porous media can be categorized broadly as bulk foam stability tests, microvisual simulations, and core-flood tests. The first category, bulk foam tests, are not directly related to the situation of foam flowing in porous media at all. They do have the advantages of simplicity and common availability of suitable apparatus. In a typical dynamic foam test, foam is continuously generated by flowing gas through a porous orifice into a test solution (*15, 29, 30*). In a typical static foam test, a blender is used to generate a column of foam whose sequential half-lives are determined after shutting off the blender. Again, there are many variations of this kind of test (*31–34*).

One approach to determining the influence of an oil phase involves bulk foam stability tests conducted in the presence and absence of specified volume fractions of oil. An example of this kind of testing, in the context of foam sensitivity to oil in porous media, is the work of Suffridge et al. (*35*). Bulk foam tests may impart so much shear that artificial mechanisms may dominate. Generally, a considerable concern is that the physical situation in bulk foam tests is too far removed from what happens during foam flowing in porous media, especially regarding the foam quality, texture, flow rate, and effective shear rate. For example, Hanssen and Dalland (*36*) found that their core-flood foam test results were not predictable from bulk foam stability tests. Although the experiments are more difficult, microvisual simulations and core-flood tests have been used to more closely represent the practical physical situation.

Microvisual Simulations. Glass microvisual apparatus of several kinds have been used to visualize the interaction of constrained foam lamellae with oil. One example of such an apparatus is shown in Figure 1 where the glass cell consists of two glass plates, one having a flow pattern etched into it and the other having the fluid entry and exit ports. The etched patterns are generally intended to represent a region of porous medium, although some, such as that depicted in Figure 1, are extremely simplified. The foams are either pregenerated by pumping gas together with surfactant solution through a foam generator or are allowed to form in situ. Somewhere in the cell, foam and oil will be observed as they come into contact with each other. In the pattern shown in Figure 1, advancing foam and residual oil are brought into contact in specific reproducible locations in the cell.

Kuhlman (*22*) conducted microvisual experiments for CO_2 foams and crude oil and observed foam destruction due to spreading of oil over foam. Manlowe and Radke (*23*) used microvisual experiments to study foams made from steam-foamers with air in the presence of pure alkane oils and observed foam destruction due to pseudoemulsion film thinning and rupture. Schramm and Novosad (*37*) used a microvisual apparatus of the design shown in Figure 1 to study foams made from hydrocarbon gas foaming agents in the presence of crude oil. In the work by Schramm and Novosad, a range of foam sensitivities to crude oil was observed, and three qualitative types of behavior were distinguished.

1. Type A Foams. These foams showed little interaction with crude oil, although as foam lamellae passed over the oil surface, a certain amount of oil would frequently be drawn up and pinched off into large droplets that would be carried along in Plateau borders along cell walls. This behavior is suggested in Figures 2 and 3.

2. Type B Foams. Other foams showed significant interaction with crude oil. Upon contact with these foams, the oil became emulsified into smaller droplets in the Plateau borders. Some of

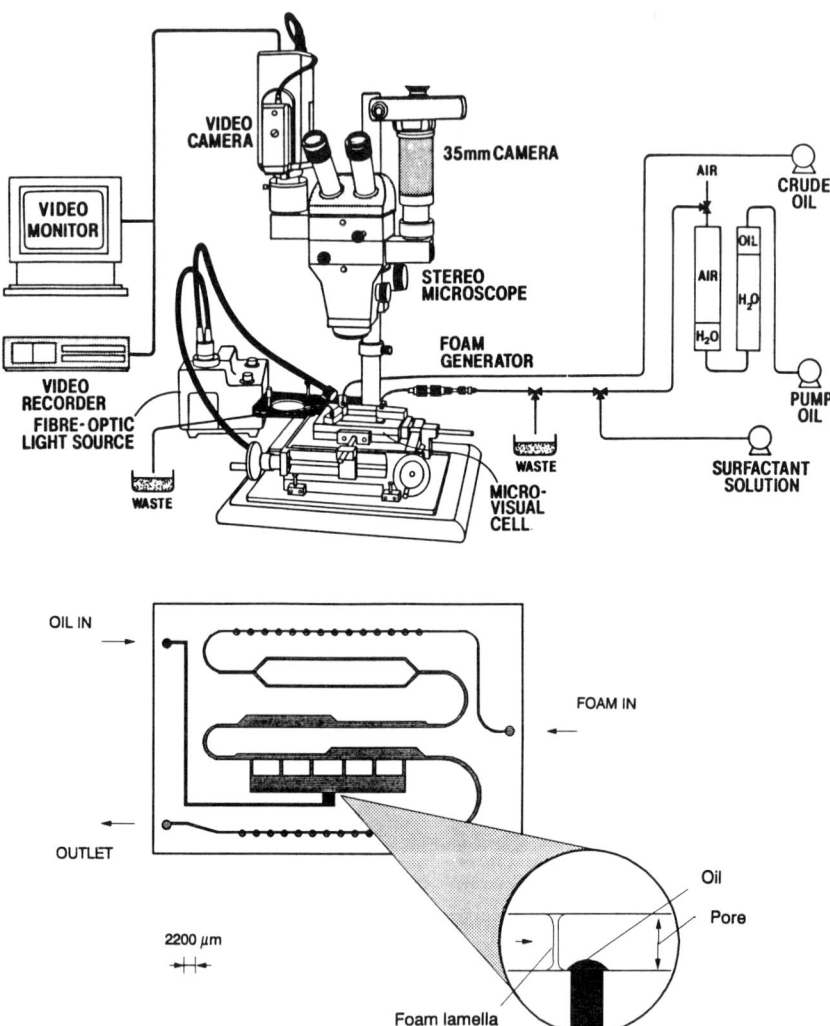

Figure 1. Example of a microvisual apparatus. A microvisual cell plate is shown in the lower portion of the figure and illustrates an etched flow pattern and the arrangement of phases at initial contact. (Reproduced with permission from reference 37. Copyright 1990 Elsevier Science Publishers.)

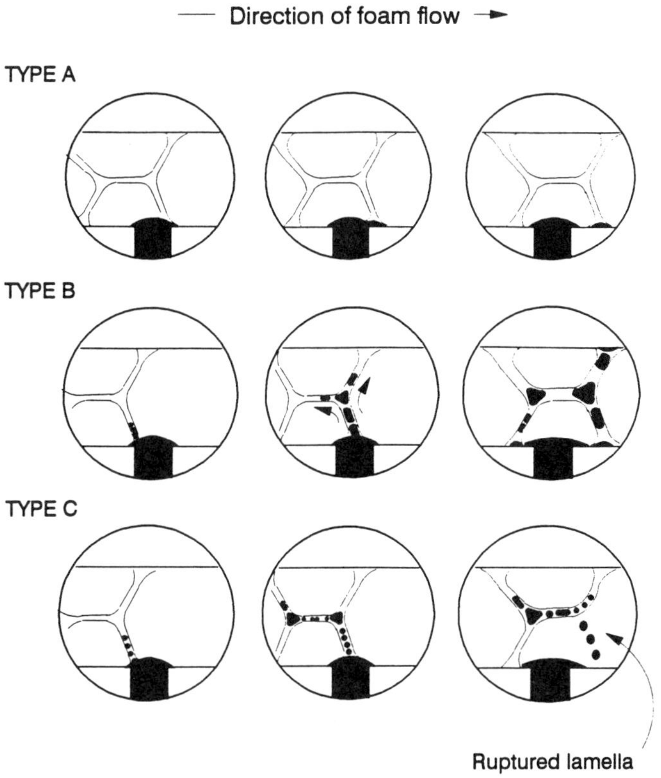

Figure 2. Types of foam behavior after oil contact. (Reproduced with permission from reference 37. Copyright 1990 Elsevier Science Publishers.)

these smaller size droplets were drawn up and transported into the "bulk" foam (Figure 2). These foams had a moderate stability to collapse; oil-containing lamellae would carry oil droplets some distance before rupturing and releasing their oil. Subsequent lamellae would then sweep up this oil and carry the oil some distance again, and so on. Figure 4 shows an advancing foam lamella sweeping up oil left behind from a previous lamella rupture.

3. Type C Foams. These foams showed a severe interaction effect. Upon contact with foam, the oil emulsified into very small droplets that were drawn up into the foam as described. However, in this case, the smaller oil droplets filled even the thinner lamellar regions, and lamellae were observed to rupture frequently as

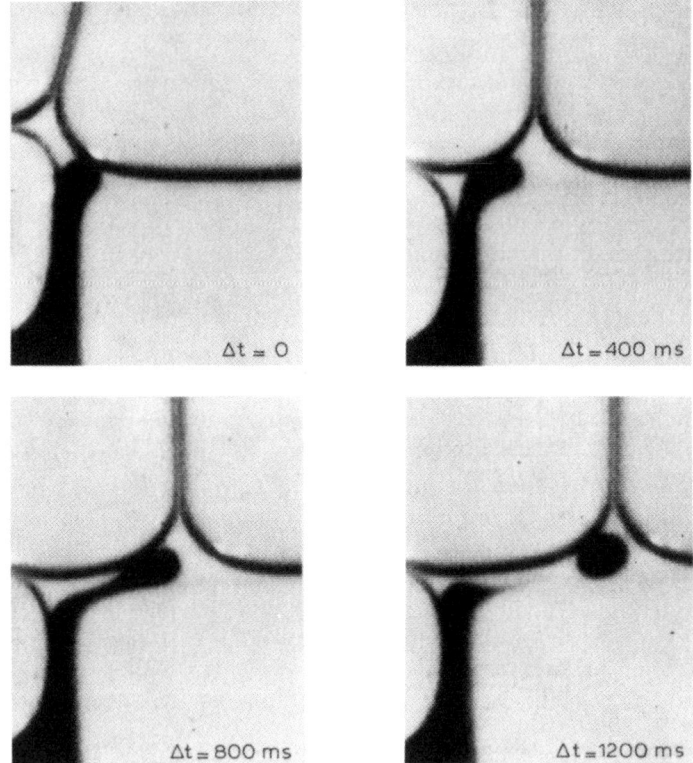

Figure 3. High-speed video photomicrographic sequence that shows microdisplacement of oil by a foam lamella. (Reproduced with permission from reference 37. Copyright 1990 Elsevier Science Publishers.)

the oil-filled foam continued to flow in the apparatus. The sequence is illustrated in Figure 2. These foams were quite unstable in the presence of crude oil.

In the foregoing, all instances of foam lamella rupture (types B and C foams) appeared to result from the imbibition (after emulsification) of oil droplets into the foam lamellae. Together with oil spreading (e.g., Kuhlman's observations) and pseudoemulsion film thinning (e.g., Manlowe and Radke's observations), the emulsification and imbibition brings forward a third possible mechanism of foam sensitivity to oil, each of which has been observed in microvisual experiments. These will be described further.

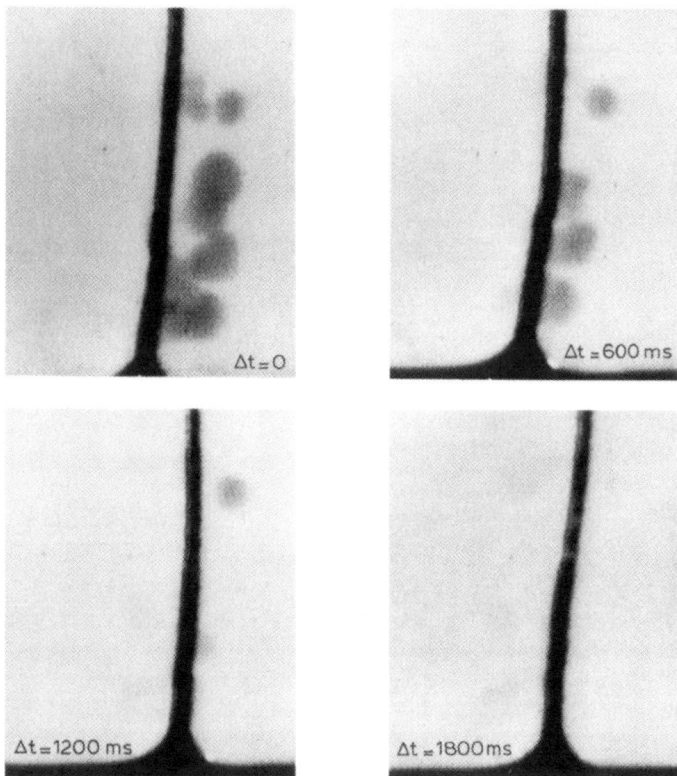

Figure 4. High-speed video photomicrographic sequence that shows microdisplacement (sweep-up) of oil by a foam lamella. In this case the microdisplacement occurs at 90° to that illustrated in Figure 3. (Reproduced with permission from reference 37. Copyright 1990 Elsevier Science Publishers.)

Core-Flood Experiments. One example of an apparatus for core-flood testing of foams is illustrated in Figure 5. In general, fluid flow and pressure transducer fittings are secured to sections of porous rock (cores) that are then encased in a constraining sleeve. Foam is prepared either in situ by coinjection of gas and surfactant solution, or the foam is pregenerated by passing gas and surfactant solution through a foam generator prior to injection into the rock. During foam tests, the pressures generated and the oil and aqueous phase productions are measured. These tests are conducted at either constant imposed rates of gas and liquid injection (as in the apparatus shown) or at constant imposed overall pressure drop. In either case, some time elapses before the pseudosteady

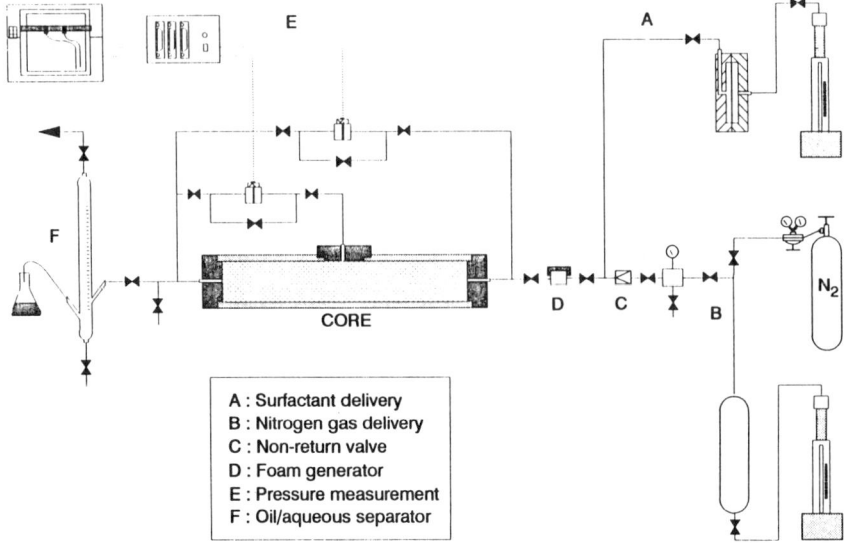

Figure 5. Schematic drawing of a low-pressure core-flood apparatus for studying the performance of foams in porous media in the presence of a residual oil saturation.

state is reached (in which the average pressure drop or average injection flow rate, respectively, become constant).

There have been a few studies of foam performance in oil-free cores and cores containing residual oil, but comparisons between the results are difficult. A diversity of parameters have been chosen to measure the efficiency of different foams with regard to their capacity to improve volumetric sweep efficiency. Mobility reduction factors (MRF), permeability reduction factors (PRF), or relative apparent viscosities (RAV) are measured for foams flowing through porous media. The MRF is a ratio of the pseudosteady-state pressure drops across a core with foam and with only gas and brine flowing at rates equivalent to those in the foam. Similar definitions are used for PRFs and RAVs. On the other hand, the experimental procedures adopted by different investigators have been very different, involving for example:

- injection of preformed foam into either a surfactant preequilibrated or nonequilibrated core
- injection of gas into a surfactant-saturated core
- alternate injection of gas and surfactant solution

- different sequences of foam initiation and flow (increasing or decreasing order of differential pressures)
- oil saturations equal or greater than residual oil saturation

When this diversity is coupled with the range of surfactants, gases, and oils used, comparisons are difficult. Some examples may be cited as follows.

Jensen and Friedmann (25) tested foams in cores at high velocities, a simulation of the phenomena occurring near a wellbore. In cores at residual oil saturation, initial foam injection [10–100 pore volumes (PV)] yielded pressure drops that were about 10 times lower than in the absence of oil. On the other hand, after continued foam injection, the final pressure drops attained were the same as for oil-free cores of the same permeability (but many hundreds of pore volumes had to be injected). The effect of residual oil saturation on foam propagation was strongly surfactant-specific; some foams did not propagate when the oil saturation was above 10–15%. McPhee and Tehrani (38) observed MRFs of 27–30 in cores containing residual oil, and the MRF was about 30 times greater in oil-free cores. Maini (39) conducted foam tests with viscous oils at high temperature and pressure and found a wide range of MRF values for oil-free cores and for cores containing residual oil. Most of the foams studied by Jensen and Friedmann, McPhee and Tehrani, and Maini exhibited sensitivities to residual oil that were about an order of magnitude in MRF (a factor of 10–30).

Holt and Kristiansen (26, 27) obtained similar results for foams flowing in cores under North Sea reservoir conditions in that the presence of any of a number of residual oils (including a crude oil and a variety of pure hydrocarbon oils) reduced the effectiveness of flowing foams. Raterman (28) measured the pressure drops obtainable for several foams flowing in sandstone cores under moderate pressure and in the presence of a residual pure alkane oil phase and found that the foams were destabilized by the oil. Schramm et al. (40) conducted foam-floods in sandstone cores and found a range of sensitivities to residual crude oil from oil-tolerant foams through to oil-sensitive foams.

Another little-studied aspect concerns the effect of the amount of oil present in the porous media. Friedmann and Jensen (11) interpreted their foam-flood results in terms of an apparent maximum residual oil saturation above which their foams were not stable. Isaacs et al. (42) tested a number of steam-foam floods in sand packs saturated with heavy crude oil and also found that foam effectiveness depended on residual oil saturation. Their foams were effective at saturations below about 10% and were not effective at oil saturations above about 15%. This finding has been incorporated into reservoir foam-flood simulations (14) using the concept of "critical foaming oil saturation", residual oil saturations above which a foam will not propagate *(14, 41)*. Yang and Reed *(43)* also found a residual oil saturation level sensitivity in their CO_2 foam-flood studies in sandstone and carbonate rocks. Their foam was effective at residual oil saturations of less than about 5% and virtually nonexistent at oil saturations of about 20%. Similarly, Hudgins and Chung *(44)* could generate foams in core-floods at very low residual crude oil saturations but not at a higher residual saturation of 23%.

The available literature shows that foams can reduce gas mobility in porous media but are sensitive to the presence and saturation level of residual oil. Several of the core-flood studies suggest that incremental oil recovery may be possible because of foam-flooding, and that oil-in-water emulsions may be formed in the process *(24, 25, 39, 40)*. There have also been relatively few attempts to connect core-flood results with micromodel studies and basic physical properties.

Theories of Foam–Oil Interaction

Having reviewed the main experimental approaches, some theories of foam–oil interaction will be examined. Figure 6 illustrates the principal different interaction behaviors that can occur when foam lamellae move along a thin channel and come into contact with an oil phase in the channel. In the first place, there may be no significant interaction, in which case the foam would simply advance unchanged over the oil. As long as the foam spreads over the oil, some oil could become emulsified into the aqueous phase, and films of aqueous solution separating oil from gas (pseudoemulsion films) will be produced. Alternatively, the oil could spread over the foam. The usual starting point for describing foam–oil interactions in porous media has its origins in the basis for destroying bulk foams using defoaming agents. Such agents frequently act by spreading out over the bubble or lamella surfaces or by penetrating and rupturing the surfaces from within the lamellae.

Spreading and Entering Coefficients. From thermodynamics, a defoamer would be predicted to spread as a lens over a foam if its "spreading coefficient" is positive, *(17, 45)*. The spreading coefficient, S, for an oil–foam system is given by

$$S = \gamma^\circ_F - \gamma_{OF} - \gamma^\circ_O \qquad (1)$$

where γ°_F is the foaming solution surface tension, γ_{OF} is the initial foam-

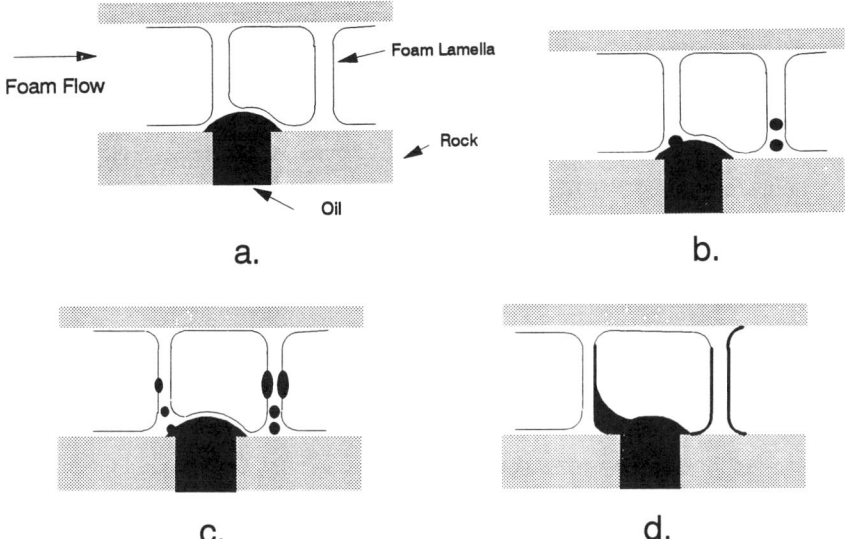

Figure 6. Illustration of interaction behaviors that can occur when foam lamellae move along a thin channel (pore) and come into contact with an oil phase.

ing solution–oil interfacial tension, and $\gamma°_O$ is the surface tension of the oil. The process of spreading is illustrated in Figure 7. When oil spreads out over the gas–aqueous interface, a certain amount of both gas–oil and aqueous–oil interface is created, and some gas–aqueous interface is eliminated. For unit surface area, the Gibbs free energy change (ΔG) is given by $-S$. The spreading is predicted to be favored when ΔG is negative and S is positive. Kuhlman (22) suggested a more complex equation to account for complete spreading of oil drops around foam bubbles of different sizes. His equation approaches equation 1 when the foam bubble radii are greater than the oil drop radii.

Rapid spreading of a drop of oil that has a low surface tension over the lamella can cause rupture by providing a weak spot (46). The spreading oil lowers the surface tension, increases the radius of curvature of the bubbles, alters the original surface elasticity, and also changes the surface viscosity. Thus the interfacial film loses its foam-stabilizing capability. If S is negative, then the oil should not spread at the interface.

Because the interfacial tension can change with time after an initial spreading of oil, S may be time-dependent, and it follows that, in some cases, oils may act as defoamers only for a limited amount of time. The dynamic interfacial tensions were studied (37, 40, 47) for various crude oil and foam-forming surfactant solution combinations. Some of these systems exhibited dynamic interfacial tensions, but typical variations over up

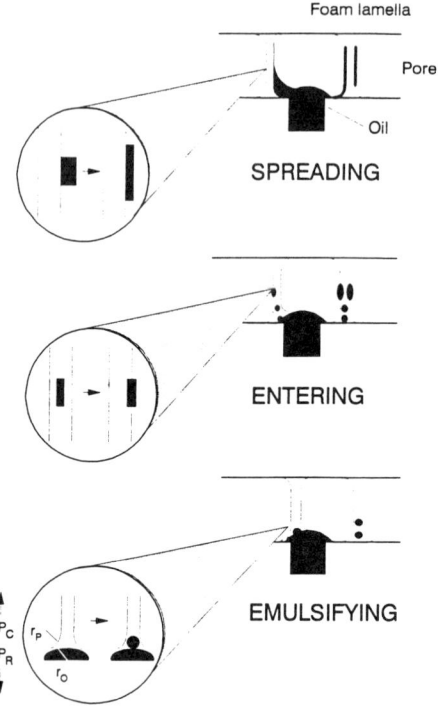

Figure 7. Spreading, entering (pseudoemulsion film rupture), and emulsifying processes related to the interaction behaviors shown in Figure 6. (Reproduced with permission from reference 37. Copyright 1990 Elsevier Science Publishers.)

to 24 h were of the order of 1–2 mN/m. Because the values of S and E (the entering coefficient) are frequently less than 10 mN/m, such changes could significantly change the values of these coefficients, but the choice of which time-dependent value to use would not, in most cases, change the sign of either S or E.

Another mechanism of rupture involves the drawing up of a droplet of defoamer into the lamellar region between two adjacent bubbles so that it can bridge the two bubbles. The breaching of the aqueous–gas interface by the oil from within the lamellar liquid is termed "entering" (Figure 7). This process is thermodynamically favorable if the "entering coefficient" is positive (originally defined by Robinson and Woods (*18*) as a rupture coefficient, R). For a foam–oil system, the entering coefficient, E, is given as

$$E = \gamma^\circ_F + \gamma_{OF} - \gamma^\circ_O \qquad (2)$$

where the symbols are defined as for equation 1. For unit surface area, the Gibbs free energy change is given by $-E$. The entering is predicted to be favored and spontaneous when ΔG is negative and E is positive.

The terms "penetrates" or "breaches" are used to mean that the pseudoemulsion film thins to the rupture point. The new local oil–gas interfacial region has an increased radius of curvature. This increase may cause some net increase in radii of curvature of the adjoining bubbles and thereby cause thinning, and again this process alters the original surface elasticity and surface viscosity of the lamellae. Thus the interfacial film can lose its foam-stabilizing capability and thin to the rupture point, although Ross (19) pointed out that entering does not always destabilize a foam. If E is negative, the oil should be ejected from the lamellar region and not destabilize the foam by this mechanism. Again, because E depends on the same surface properties as S, the value of E may be time-dependent.

In the enhanced oil recovery literature, much use is made of the spreading coefficient, but the entering coefficient (22, 23, 48) is ignored. Many of the correlations with foam behavior attributed to the spreading coefficient would also correlate with the entering coefficient (37, 40, 49). Ross (19) suggested predicting the defoaming action of an insoluble chemical by considering the possible combinations of values of S and E. These are interrelated through equations 1 and 2 as shown in equation 3.

$$E = S + 2\gamma_{OF} \tag{3}$$

This equation shows that E is always greater than or equal to S, but that three combinations of effects can occur:

1. Type A. If E is negative, then S must be negative, and oil would neither be drawn through nor spread at the foaming solution–gas interface. It would not be expected to act as a defoamer by the previously discussed mechanisms.

2. Type B. If E is positive and S is negative, then the oil would be drawn through but would not be expected to spread at the interfaces. This condition should cause destabilization but may not, depending on whether small enough oil drops are drawn up to seriously reduce the coherence of the lamellae, and whether the oil remains as a lens or is ejected outside the lamella.

3. Type C. If both E and S are positive, then the oil will be drawn through and then spread as a film along the lamellae surfaces. This behavior leads to lamellae ruptures.

Schramm and Novosad (37) showed that the three types of behavior can be represented in a foam "phase-behavior" diagram by fixing γ°_O (choosing a single oil) and plotting γ_{OF} versus γ°_F. This representation is shown in Figure 8 where both S and E increase toward the right side of the diagram, and three regions are distinguished by solving equations 1 and 2 for the conditions $E = 0$ and $S = 0$ to produce the two line boundaries. The four phase-behavior diagrams of Figure 8 correspond to one alkane and three crude oils of increasing gravity, or "heaviness", from top to bottom in the figure legend. In each case, the data correspond to solutions of a range of foaming agents. Schramm and Novosad's phase-behavior approach has been used with some success by others as well (27).

Some authors use a coefficient to describe the spreading of aqueous solution over oil, for example, $S_w = \gamma^\circ_O - \gamma^\circ_F - \gamma_{OF}$ (22, 27). The process described is equivalent to the reverse of entering, and $E = -S_w$. Thus the region marked "Type A" in Figure 8 could be considered to represent the conditions under which foaming solution will spread over oil, as illustrated in Figure 6a.

Pseudoemulsion Film Model. Although much attention is traditionally focused on foam lamellae, the thin liquid films bounded by gas on each side and denoted as air–water–air (A/W/A), a distinct but related thin liquid film, may be the most important for foam stability in the presence of oil. That is, the thin liquid films bounded by gas on one side and by oil on the other denoted air–water–oil (A/W/O). These films are illustrated in Figure 6a–6c. They are also referred to as pseudoemulsion films (48). Their importance stems from the fact that it is possible for the pseudoemulsion film to be metastable in a dynamic system, even when the thermodynamic entering coefficient is greater than zero.

Wasan and co-workers (48, 50, 51) have found pseudoemulsion film stability to be influenced by micelle structuring effects, Marangoni surface effects, and the presence of oil droplets. (see the discussion in Chapter 2) They have also found that in some systems, the emulsification and imbibition of oil can actually stabilize foams. Manlowe and Radke's results (23) were interpreted in terms of pseudoemulsion film stabilities depending mainly on electrostatic and dispersion forces. Undoubtedly, interfacial viscosity could also be important.

In one approach, the stability of pseudoemulsion films can be predicted by calculating the total interaction energy for the thinning of a liquid film bounded by an air–water and a water–oil interface along the lines described earlier in Chapters 1 and 2. From this interaction energy, the disjoining pressure, which is the net pressure difference normal to the surface between the gas and oil phases and the bulk liquid from which the lamellae extend, can be calculated. Estimates of the disjoining and capil-

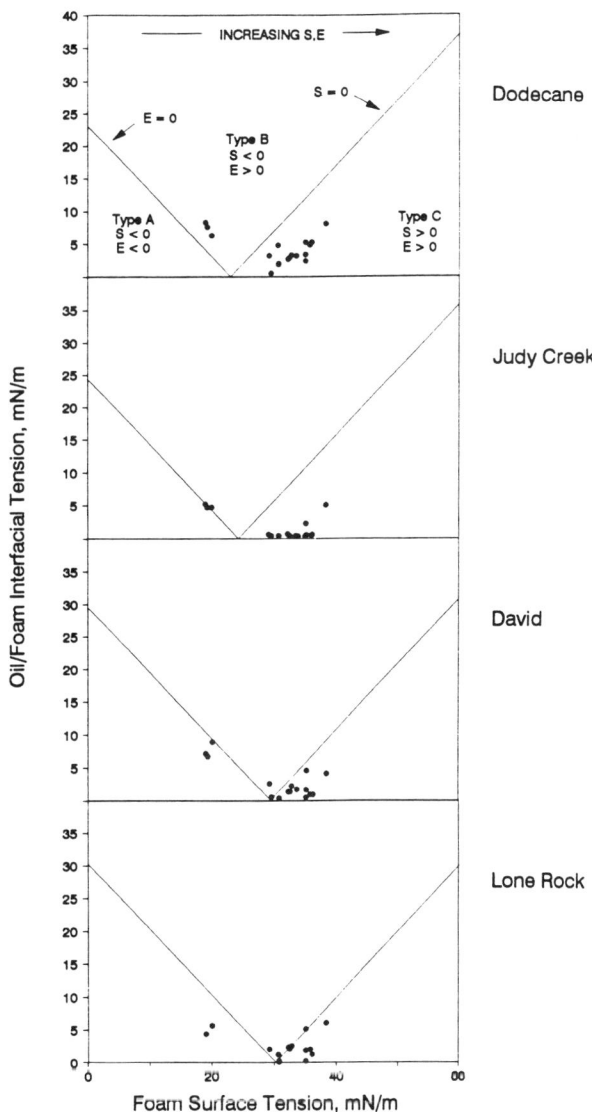

Figure 8. Foam "phase-behavior" diagrams for a series of foam-forming surfactants and four different oils. (Reproduced with permission from reference 47. Copyright 1992 Elsevier Science Publishers.)

lary pressures for foam lamellae in porous media can then be used to make predictions about the probable stabilities of the lamellae (23). Raterman (28) directly measured film disjoining pressures and pseudoemulsion film stabilities and compared these with the results of coreflood tests. His results showed that foam systems containing thermodynamically stable pseudoemulsion films are oil tolerant, but except for thermodynamically stable pseudoemulsion film systems, he found no trend in foam–oil stability and disjoining pressures. On the other hand, Fagan (52) found a good correspondence between high pseudoemulsion film disjoining pressures and good oil tolerance on the part of foams flowing in porous media and vice versa.

The disjoining pressure approach has some important limitations. Disjoining pressures typically will be important only for very thin lamellae such as films thinner than 100 nm. There is evidence to suggest that foam lamellae in porous media extend to at least an order of magnitude larger than this. Raterman (28) found a much better correlation between pseudoemulsion film thinning and foam stability for constant gas injection in situ foam generation tests than for constant preformed foam injection tests. The foam lamellae would have been thinner in the foams that are generated by constant gas injection in situ and much thicker in the foams that are generated by constant preformed foam injection. Also, the disjoining pressure itself is usually most significant in systems that contain appreciably charged interfaces, that is, where ionic surfactants are present at the interfaces. When nonionic or zwitterionic surfactants are involved, the disjoining pressures will be significant only for even thinner lamellae or if, as suggested by Wasan et al. (48, 50), the surfactant type and concentration are such as to produce micellar structuring effects.

Emulsification and Imbibition Model. Nikolov et al. (48) concluded, in part, that "Foam destabilization ... may often involve the migration of emulsified oil droplets from the foam film lamellae into the Plateau borders where critical factors, such as the magnitude of the Marangoni effect in the pseudoemulsion film, the pseudoemulsion film tension, the droplet size, and number of droplets may all contribute to destabilizing or stabilizing the three-phase foam structure." Raterman (28) also concluded that oil droplet size should be an important determinant of the relative importance of pseudoemulsion film stability. Further, in typical antifoaming processes, fine emulsions (small drops) make better antifoaming agents (18, 53), and indeed, the colloid chemical description of foam-breaking outlined in equations 1 and 2 has, as a starting point, the introduction of oil droplets as an emulsion that have drop diameters smaller than the thickness of the foam lamellae. In the foams contacting oil in a porous medium, the oil may not be emulsified or may not have such small drop sizes. If this is the case, then the spreading and entering coefficients may not directly apply, and any pseudoemulsion films will have a much lower interfacial area. This is one of the reasons there can be a different influence on foam stability between oil as a macrophase and oil as emulsified droplets. The observations of a number of researchers (22, 24, 25, 37, 39, 40, 48) indicate that emulsification of oil into the foam, as depicted in Figure 6b, may be a very important factor.

To describe the process of an oil phase emulsifying into foam lamellae, one could adopt an equilibrium thermodynamic approach analogous to that employed to obtain S and E. Such an approach produces the result that emulsification is only favored (negative ΔG) for negative values of the interfacial tension. In an alternative approach, the tendency of an oil phase to become emulsified and imbibed into foam lamellae has been described through a simplified balance of forces by the lamella number, L (37). For foam lamellae flowing in porous media, it is predicted that oil will be drawn in and pinched off to produce emulsified drops inside foam lamellae when $L > 1$, where

$$L = \frac{\Delta P_C}{\Delta P_R} = \frac{(\gamma°_F) r_o}{(\gamma_{OF}) r_p} \quad (4)$$

Here ΔP_C and ΔP_R are the pressure difference between inside the Plateau border and inside the laminar part of the lamella, and the pressure difference across the oil–aqueous interface, respectively, and r_o and r_p are the radii of the oil surface with which the lamella comes into contact and that of the lamella Plateau border, respectively (Figure 7). If L is scaled by substituting one-half the lamella thickness for r_o, then $L > 1$ indicates that small enough oil droplets will be produced to permit their movement within a lamella. Experimental work with a microvisual cell of fixed dimensions has shown that equation 4 can be simplified to a useful approximate form. For a wide variety of foams, oils, flow rates, and foam qualities in the range 85–95%, all studied in the microvisual cell depicted in Figure 1, equation 4 was found to simplify to (37, 40, 47)

$$L = \frac{0.146(\gamma°_F)}{(\gamma_{OF})} \quad (5)$$

The physical basis for the lamella number is that a combination of capillary suction in the Plateau borders and the influence of mechanical shear cause the oil-phase distortion and pinch off into droplets. This result is in accord with the observations of emulsification and imbibition by Lobo et al. (50), French et al. (54), and Schramm and Novosad (37). The mechanical shear may come from several sources, including the flow of the foam lamellae through the pores, the stretching and contracting of lamellae as they pass through pores and throats, and any Marangoni flows within the foam lamellae. None of these sources of shear are accounted for in the simplified description given by equations 4 or 5. Some important limitations of the lamella number are that it will be important only for foam lamellae that are thick enough to accommodate realistic emulsion droplet sizes, and that an explicit factor for the effect of mechanical shear is missing.

Comparisons between Theory and Experiment

Microvisual Experiments and the Proposed Mechanisms.

The spreading and entering coefficients have correlated with foam sensitivity to oil in a number, but not in all, of the cases. As already noted, the degrees of foam sensitivity to oil (Figure 2) observed in microvisual experiments have been compared to the thermodynamic predictions based on S and E in Figure 8 (*37, 40, 47*). In these comparisons, the predictions were not always borne out: Depending on the oil studied, on the order of one-half of the surfactant solutions predicted to produce foams of type C actually produced foams of type B. The predictions of which surfactants would produce foams of type A were much better for the heavier oils than for the lighter oils. Even quantitative measurements of lamella rupture frequency in the microvisual experiments showed that satisfactory correlations with S or E were not obtained (*37, 40, 47*).

Although a very large negative spreading coefficient can be used to distinguish the type A foams from the rest, neither S nor E could be used to distinguish the B and C foams from each other. Thus it appears that even the most stable foam–oil systems cannot reliably be predicted from the thermodynamic relationships for S and E. In Kuhlman's work (*22*), spreading of oil over the foam lamellae was observed and correlated with foam stability. In this work, foam sensitivity to the crude oil was also associated with emulsification of the oil into small droplet sizes. Conversely, Hanssen and co-workers (*36, 55*) could not find a consistent link between oil spreading and foam sensitivity to oil in porous media.

In Wasan et al. (*48, 50, 51*) and Manlowe and Radke's (*23*) studies, spreading and entering were not found to be important causes of foam destruction. Instead, thin film (pseudoemulsion film) drainage between solution and oil were proposed.

The microvisual experiments of Roberts et al. (*53*) and Schramm et al. (*37, 40, 47*) showed that foam destruction was associated with the ability or degree of emulsified oil droplets to travel within the foam lamellae. For example, on the basis of the work of Schramm et al., the frequency of foam lamella ruptures, f_b, a measure of foam stability in microvisual experiments, is shown plotted versus the logarithm of the lamella number from equation 5 for each of four different oils in Figure 9. In each case, there is a fairly good correlation between stability and lamella number, especially considering the uncertainty in the measured values of f_b (±14%). The qualitative and quantitative results can be related as in the regions marked A, B, and C. In region A ($L < 1$), it was predicted and observed that oil would not be emulsified to a size that would allow ready access to the foam lamellar structure, and type A foam behavior (Figure 2) would result. In region B ($1 < L < 7$), it was predicted and observed that the emulsification of oil into smaller droplets would become more favorable, and type B foam behavior would result. In region C ($L > 7$), it was predicted and observed that emulsification into smaller drops would occur that would allow ready access to the foam structure and result in type C

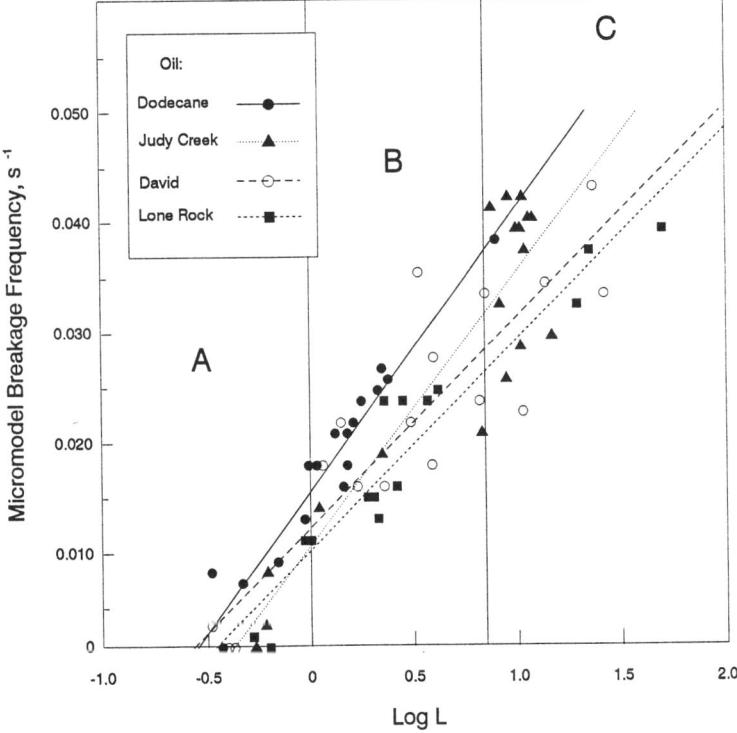

Figure 9. Relation between foam breakage frequencies in a microvisual cell and the lamella number for four different oils. (Reproduced with permission from reference 47. Copyright 1992 Elsevier Science Publishers.)

behavior. The quantitative measurements and the qualitative observations matched in 65 of the 67 foam–oil systems investigated (47).

Regarding the limitation of the lamella number description, that it applies to foam lamellae that are thick enough to accommodate realistic emulsion droplet sizes, the microvisual studies reported are consistent with this result. Schramm et al. (37, 40, 47) observed emulsification into lamellae that have thicknesses on the order of 50–80 μm, as did Kuhlman (22) for foam lamellae on the order of 40-μm thick. Nikolov et al. (48) argued that the presence of emulsified oil droplets themselves has a thickening effect on foam lamellae, and can even have a stabilizing effect on foams if the droplets sizes are small enough. (Foams that are stabilized by micellar structuring can apparently be further stabilized when emulsion droplets are incorporated into the micelles, making them swell.) Another possible stabilizing mechanism is the existence of a strong electric double-layer repulsion between oil drops and aqueous–gas foam interfaces, which causes a large disjoining pressure. Such stabilization by disjoining pressure would be expected only for ionic surfactant systems. On the other hand, some (52) have suggested that the thickness of foam films in porous media is less than about 100 nm. If this is true for some foams, then in these cases, it is unlikely that oil droplets would be emulsified into such thin films and move at typical reservoir injection flow rates unless the interfacial tensions were extremely low.

Core-Flood Experiments and the Proposed Mechanisms.

The mechanisms for foam sensitivity to oils can also be compared to the results from core-flood experiments in which foams were made to flow through porous rock in the presence of residual oil. Holt and Kristiansen (26, 27, 56) studied foams flowing in cores under North Sea reservoir conditions and found that the presence of residual oil could reduce the effectiveness of flowing foams. They compared their results with the spreading and entering coefficients and found foam sensitivity to be correlated with the (oil) spreading coefficient.

Raterman's (28) core-flood studies for foams flowing in Berea sandstone cores at moderate pressure and in the presence of a residual oil phase led to the conclusion that oil spreading is of only secondary importance to foam sensitivity to oil. For gas injection that causes in situ foam generation and flow, he found a good correlation between foam stability and pseudoemulsion film stability. However, for constant quality preformed foam injection, he found pseudoemulsion film stability to be less directly connected with foam sensitivity. This finding is probably due to the fact that the foam lamellae in the constant quality preformed foam injection floods were too thick for disjoining pressures to be the most important factor. Fagan (52) found a good correspondence between high pseudoemulsion film disjoining pressures and good oil tolerance on the part of foams flowing in porous media and vice versa.

The emulsification–imbibition of oil into foam, which the lamella number is intended to describe, has been noticed or predicted by a number of authors and was illustrated in Chapter 2. Lobo et al. (*50*) emphasized the importance of this phenomenon for foam stability. Raterman (*28*) predicted that emulsification–imbibition would be important in constant quality preformed foam injection floods. In the core-flood studies of French et al. (*54*), they observed that the contacting of foam with crude oils produced emulsified droplets of oil within the foam lamellae. Schramm et al. (*40*) determined MRFs for a number of foams flowing in Berea sandstone cores containing residual light crude oil and found a strong correspondence with the micromodel results (Figure 10). This work was the first to show that foams that are quite stable to oil in the micromodel are also quite effective in core-floods and vice versa.

Some of these results can be brought together in Figure 11. Hanssen and Dalland (*36*) measured gas permeabilities for different foams (variable quality) in synthetic seawater flowing in relatively high-permeability ($K_a =$

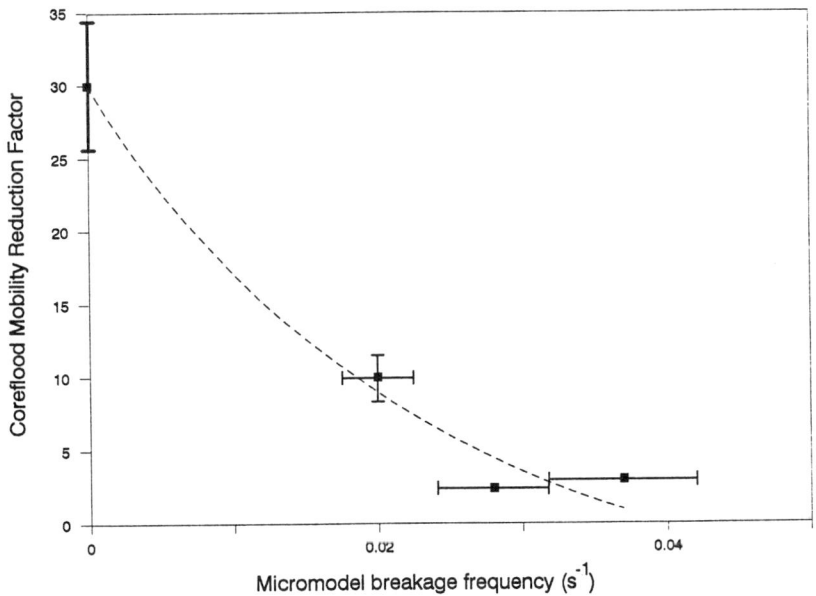

Figure 10. Mobility reduction factors for foams flowing in Berea sandstone at residual oil saturation versus the breakage frequencies of foam lamellae flowing in a microvisual cell when in contact with the same oil. (Reproduced with permission from reference 40. Copyright 1990 Society of Petroleum Engineers.)

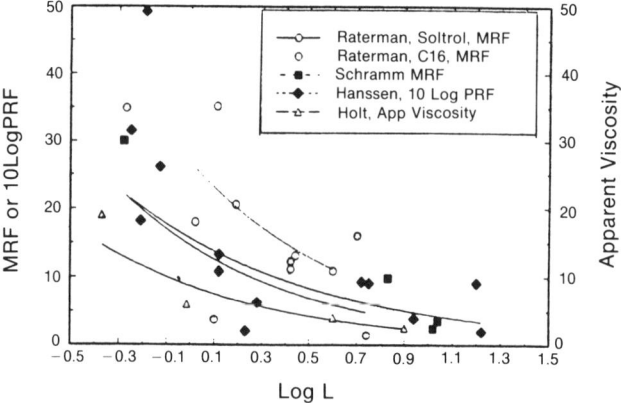

Figure 11. Foam mobility (MRF) or permeability (PRF) reductions measured in various core-floods that involve different oils plotted versus log L.

8.5 μm^2) glass-bead packs in the presence of a crude oil at residual oil saturation at 70 °C and under various high pressures. Permeability reduction factors (PRF) were calculated from Hanssen and Dalland's reported permeability data. The permeability reduction factors were scaled to facilitate comparison with MRF data and are reported as 10 Log(PRF). Raterman (28) measured the pressure drops obtainable for several foams (80% v/v quality) in 15% by mass synthetic brine flowing in Berea sandstone cores held at moderate (400 psig) back-pressure and in the presence of a waterflood residual oil phase of either an odorless mineral oil (Soltrol 130) or hexadecane.

The MRFs shown in Figure 11 were calculated from Raterman's pressure drop data. Schramm et al. (40) determined MRFs at ambient temperature and pressure for a number of foams flowing in Berea sandstone cores that contained residual light crude oil. These are plotted directly in the figure. Holt and Kristiansen (27) studied foams flowing in the presence of residual oil (either a nonane–xylene mixture or crude oil) in cores under North Sea reservoir conditions of temperature and pressure. They measured apparent viscosities for the foams flowing in the cores. These measurements are reported in Figure 11. In all cases, lamella numbers were calculated from data reported in each paper by using equation 5.

Figure 11 shows all of these core-flood foam performance results plotted versus the logarithm of the lamella number. A strong correspondence is obtained with the logarithm of lamella number. That the trend between mobility reduction and lamella number is so consistent among the different systems is remarkable considering that the core-flood results were obtained for widely different conditions of oil nature, surfactant types, brine compositions, foam qualities, flow rates, core permeabilities, temperatures, and pressures. These results are taken as support for the mechanism of emulsification and imbibition of oil drops into the foam structure that was proposed on the basis of the micromodel studies and advanced in the Theory section.

Dalland et al. (57) compared a number of foam–oil combinations in core-flood tests against the spreading and entering coefficients and a pseudolamella number. They found that none of these parameters, individually or in combination, could consistently predict oil tolerance on the part of the foams. Examination of their data for flowing foams in porous media (i.e., not including their data for static foams in porous media) shows that the mobility reduction factors measured correspond to a narrow range of lamella number values that fall within the spread of the curves in Figure 11. Fagan (52) also compared four foam–oil combinations against spreading and entering coefficients and the lamella number. He also concluded that none of these correlated with oil tolerance, although again, three out of four of his systems would fall within the spread of the curves in Figure 11. The fourth system clearly did not follow this trend. Fagan found a more consistent correlation with pseudoemulsion film disjoining pressure.

A final point concerns the possible increase in microscopic displacement efficiency by foams that would be expected from emulsification, imbibition, and transportation of oil droplets by the type B foams described earlier. Figure 12 shows incremental oil recoveries due to foam injection into cores, expressed as a percentage of the residual oil in place after waterflooding. Reading the graph from left to right, the type A foam (for which the micromodel lamella number, $L = 0.53$) produced a negligible incremental oil recovery, but the type B ($L = 6.7$) and type C ($L = 10.5$ and 10.9) foams produced significant oil recoveries. Emulsification and imbibition occurs for both types B and C foams, but the type C foams collapse more quickly and do not propagate as well. Thus the type B foam yielded the best incremental oil recovery.

In these homogeneous cores, any small changes in sweep efficiency due to the differences in MRF among the foams should not have had a significant influence on oil recovery, surfactant solutions alone had no significant influence on oil recovery, so the additional oil recovered by the type B foam is interpreted as being caused by the transport of oil droplets by the emulsification and imbibition mechanism. As pointed out earlier in this chapter, quite a number of authors have speculated on some role for oil-in-water (O/W) emulsions produced by foams. Maini (39) and Farrell and Marsden (58) suggested that improved oil recoveries could be achieved by foam injection if some kind of emulsification–microdisplacement of oil by foam can occur such as that just described. The selection of foams for this purpose is described in a recent patent (59).

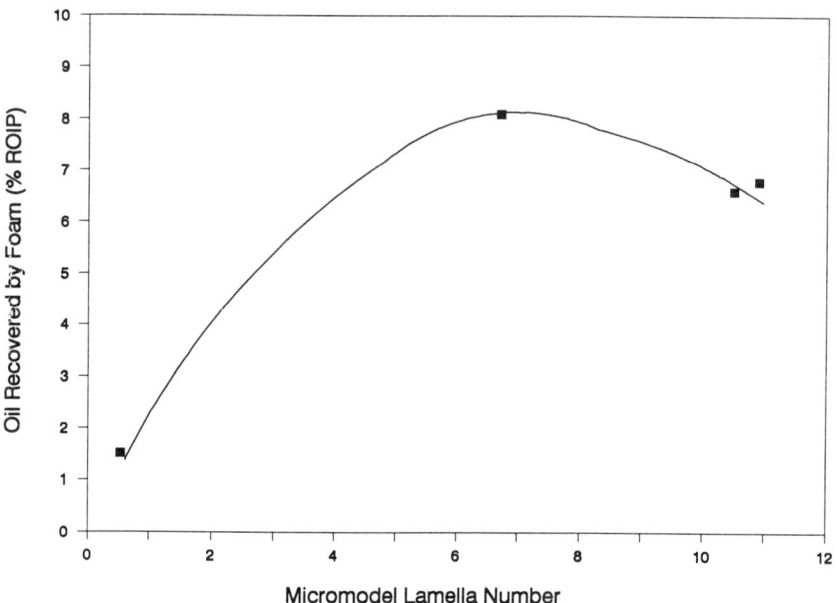

Figure 12. Incremental oil recovered by different types of foams (A, B, and C) flowing in Berea sandstone at residual oil saturation. (From data in reference 40.)

Effect of the Nature of the Oil

One of the impressions that has come from experience with foam-flooding porous media in the presence of different crude oils is that the foams seemed to be destabilized most by the lighter oils. Indeed, Suffridge et al. (35), using a series of pure alkanes, found that higher molecular weight alkanes have less of a destabilizing effect on foams than do lower molecular weight hydrocarbons. On the other hand, Meling and Hanssen (55) also studied a series of pure alkanes but found no general correlation between foam destabilization and oil carbon number (hence molecular weight). In both cases, only pure alkanes were studied, so a comparison among crude oils is of interest.

Figure 13 shows the foam breakage frequencies for the same foam and oil systems as in Figures 8 and 9. The foams have been ranked along the horizontal axis by the average stability of each foam to all four oils. For each oil, the plotted curves represent the foam data fitted to a logarithmic equation. The figure shows that among the three crude oils, the foam stability in the presence of oil decreased as the "heaviness" of the oil decreased for the three crude oils. The four oils had densities ranging

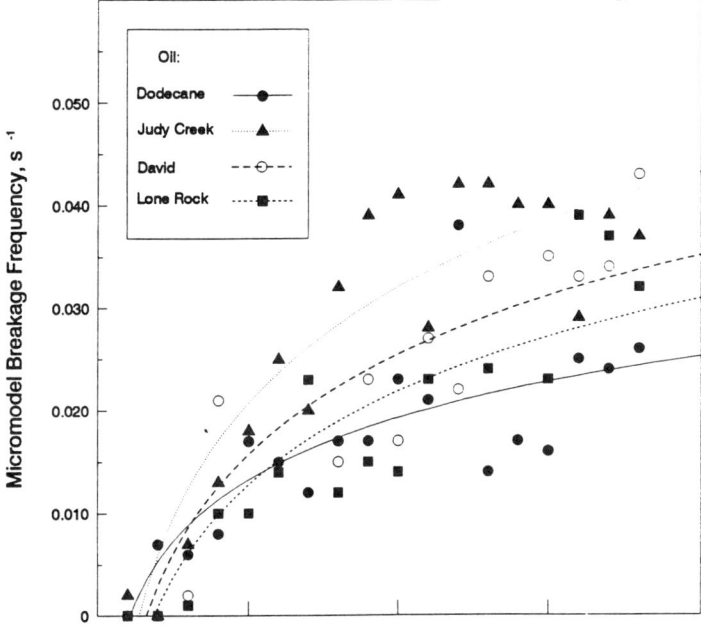

Figure 13. Comparison of foam stabilities to different oils. (Reproduced with permission from reference 47. Copyright 1992 Elsevier Science Publishers.)

from 0.747 to 0.949 g/mL (23.0 °C) and viscosities ranging from 1.4 to 809 mPa·s (23.0 °C) in the same order. The results for the pure alkane oil, dodecane, are not consistent with this trend. This points to a danger in studying foam destabilization by single-component pure hydrocarbon oils if the results are intended to reflect to actions of reservoir crude oils that contain many components. In fact, in the language of Chapter 2, all of the crude oils discussed in this chapter would fall under the category of "mixed type antifoamers". Therefore, there is some limitation on the range of applicability of the trend between foam stability and "heaviness" of oil.

Plotting the same foam stability data versus lamella number (39) shows that for these four oils, the same variation of foam breakage frequency with lamella number is observed. Thus, even though for a given foaming surfactant solution the foam does not have the same stability in the presence of each oil, the stability is still well predicted by the surface properties, as reflected in the lamella number (equations 4 and 5). Also, using the surface properties allows the effects of the dodecane to be described consistently with those of the crude oils. This condition suggests that at least one reason that the dodecane results did not match the trend between destabilization and oil "heaviness" may be ascribed to different effects at the interfaces. Possibly the ordering of oils shown in Figure 13 reflects an influence of the oil viscosity. The more viscous oils would be expected to emulsify more slowly than the others, and this slow emulsification would be expected to retard the rate of foam lamella breakage.

Effect of the Wettability of the Rock

Most of the research on foam sensitivity to oils in porous media, whether in microvisual or core-flood tests, has been concerned with water-wetted pore and throat surfaces. Because petroleum reservoirs are frequently of intermediate, mixed, or oil wettability, it is of considerable interest to understand how rock wettability influences foam stability.

Schramm and Mannhardt (*60*) conducted some oleophilic microvisual cell experiments with several foams and crude oils in the apparatus shown in Figure 1. In this case, the glass microvisual cell surfaces were treated by bonding a fatty acid–trivalent chromium (Quilon C) chrome complex that contains fatty acid chains that orient away from the glass and make it oleophilic. They found that the foams were significantly less stable in the presence of crude oil and oleophilic solid surfaces compared with the same crude oils and hydrophilic surfaces.

The same conclusions have been reached on the basis of core-flood experiments. Suffridge et al. (*35*) studied foam effectiveness in Berea sandstone cores, both untreated (water-wetted) and treated with the Quilon C chrome complex described previously. The treated cores became intermediate to oil-wetted at waterflood residual oil saturation. They found that the foams were more effective (stable) in the water-wet cores than in the oil-wet cores. Holt and Kristiansen (*27, 56*) studied foams flowing in cores under North Sea reservoir conditions that were either partially or completely oil-wetted. They found that foam effectiveness was favored by water-wet conditions; any degree of oil-wet character reduced the effectiveness of the flowing foam.

These results have been expressed in terms of the influence of the wetting condition of the porous medium as foam is flowing through it. A complication is that the foam-forming surfactant may adsorb onto the solid surfaces and may alter the wettability. In the microvisual experiments of Schramm and Mannhardt (*60*), some of the foaming systems investigated appeared to change the wettability back to water-wet, in which case the foam sensitivities to oil reverted back to those appropriate to the water-wet cases. This result is consistent with the core-flood tests of Sanchez and Hazlett (61) in which foaming surfactants apparently caused wettability reversal in oil-wet porous media. This reversal, in turn, was postulated to be the reason that stable foams could be generated and propagated in their oil-wetted cores.

Conclusion

Foams flowing in porous media can be very sensitive to contact with any crude oil that may be encountered. Furthermore, the degree of sensitivity depends upon both the nature of the foam and the nature of the oil. Whereas many foams are quite sensitive to oil contact, it is also true that some foams are very resistant to oil contact.

Some evidence supports the concept that even foams that are sensitive to oil contact in porous media can still be effective if the residual oil saturation is quite low, less than about 10%. The literature suggests that these same foams will be significantly destabilized at higher residual oil saturations, higher than about 20%.

A single model of foam–oil interaction cannot account for all situations. Certain foam–oil sensitivity models can be reconciled with both microvisual studies and core-flood foam effectiveness measurements, all for a wide variety of foams, oils, porous media, and other experimental conditions. However, exceptions are readily found. In an earlier section, the models of emulsification–imbibition, pseudoemulsion film thinning, entering, and spreading were introduced. Cases in favor of, and exceptions to, the applicability of each of these can be found in the literature. Although this situation prompts some inclination to search for additional mechanisms, the truth may be that all the models presented have some validity and that one or another valid mechanism is most significant in a given situation.

An outline of the fundamental steps involved in aqueous foam destabilization by oil, in porous media, is proposed as follows:

1. Oil imbibition (penetration) into the foam lamellae, including prior emulsification if necessary.

2. Perturbation of the lamellae whether due to external shocks, the oil imbibition, foam flow, or lamella expansions and contractions (Marangoni flow).

3. Thinning of the pseudoemulsion films to the point of allowing entering (oil penetration of the gas–aqueous interface). This step may be sufficient for lamella ruptures.

4. Spreading of the oil over the gas–aqueous interface. This step will cause lamella ruptures but is necessary only if the entering process does not cause lamella ruptures.

This mechanism is quite similar to the mechanism for foam destabilization by oil proposed a few years ago by Raterman (*28*), and is consistent with the mechanisms advanced by Wasan et al. (*48, 50, 51*), Radke and Manlowe (*23*), Fagan (*52*), and Schramm et al. (*37, 40, 47*).

Emulsification–imbibition greatly increases the oil–aqueous contact area in the foam and is possibly an important process when constant quality foam injection into waterflooded media at residual oil saturation is practiced, and when the amount of interfacial area constituted by pseudoemulsion films is a rate-determining factor for foam sensitivity. This emulsification–imbibition can act in favor of or against foam stability, depending upon whether the pseudoemulsion films, once formed, are quite stable or quite unstable.

Pseudoemulsion film thinning probably comes into independent prominence when the system is quite dynamic and where the foam lamellae are thin. Pseudoemulsion film thinning, causing entering and subsequent lamella ruptures, is more likely to be a dominant mechanism when continuous gas injection into surfactant flooded media at residual oil saturation is practiced.

List of Symbols

	E	Entering coefficient
	f_b	Foam lamella rupture frequency
	ΔG	Gibbs surface free energy change
	L	Lamella number
	ΔP_C	Pressure difference across Plateau border interface
	ΔP_R	Pressure difference across foaming aqueous solution–oil interface
	r_o	Radius of curvature of an oil surface
	r_p	Radius of curvature at the Plateau border where lamellae wet the pore wall
	S	Spreading coefficient

Greek

	γ_{OF}	Foaming aqueous solution–oil interfacial tension
	γ^o_O	Oil surface tension
	γ^o_F	Foaming aqueous solution surface tension

References

1. Bond, D. C.; Holbrook, O. C. U.S. Patent 2 866 507, December 30, 1958.
2. Fried, A. N. *The Foam-Drive Process for Increasing the Recovery of Oil;* Report of Investigations 5866; U.S. Department of the Interior, Bureau of Mines: Washington, DC, 1961.
3. Lawson, J. B.; Reisberg J. *Proceedings of the SPE/DOE Symposium on Enhanced Oil Recovery;* Society of Petroleum Engineers: Richardson, TX, 1980; paper SPE/DOE 8839.
4. Holm, L. W. *J. Pet. Technol.* **1970**, *22(12)*, 1499–1506.
5. Holm, L. W.; Garrison, W. H. *SPE Res. Eng.* **1988**, *3(1)*, 112–118.
6. Kuehne, D. L.; Ehman, D. I.; Emanuel, A. S. *Proceedings of the SPE/DOE Symposium on Enhanced Oil Recovery;* Society of Petroleum Engineers: Richardson, TX, 1988; paper SPE/DOE 17381.
7. Patzek, T. W.; Koinis, M. T. *Proceedings of the SPE/DOE Symposium on Enhanced Oil Recovery;* Society of Petroleum Engineers: Richardson, TX, 1988; paper SPE/DOE 17380.
8. Liu, P. C.; Besserer, G. J. *Proceedings of the 63rd Annual Technical Conference of SPE;* Society of Petroleum Engineers: Richardson, TX, 1988; paper SPE 18080.
9. Maini, B. B.; Ma, V. *Proceedings of the 59th Annual Technical Conference of SPE;* Society of Petroleum Engineers: Richardson, TX, 1984; paper SPE 13073.
10. Ling, T. F.; Lee, H. K.; Shah, D. O. In *Industrial Applications of Surfactants;* Karsa, D. R., Ed.; Royal Society of Chemistry: London, 1986; pp 126–178.
11. Novosad, J. J.; Ionescu, E. F. *Proceedings of the 38th Annual Technical Meeting of CIM;* Canadian Institute of Mining, Metallurgy, and Petroleum: Calgary, Canada, 1987; paper CIM 87–38–80.
12. Duerksen, J. H. *Proceedings of the California Regional Meeting of SPE;* Society of Petroleum Engineers: Richardson, TX, 1984; paper SPE 12785.
13. Mannhardt, K.; Schramm, L. L.; Novosad, J. J. *Proceedings of the 65th Annual Technical Conference of SPE;* Society of Petroleum Engineers: Richardson, TX, 1990; paper SPE 20463.
14. Law, D. H.; Yang, Z.-M.; Stone, T. W. *SPE Res. Eng.* **1992**, *7(2)*, 228–236.
15. Bikerman, J. J. *Foams;* Springer–Verlag: New York, 1973.
16. Ross, S. In *Kirk–Othmer Encyclopedia of Chemical Technology,* 3rd ed.; Wiley: New York, 1980; Vol. 11, pp 127–145.
17. Harkins, W. D. *J. Chem. Phys.* **1941**, *9*, 552–568.
18. Robinson, J. W.; Woods, W. W. *J. Soc. Chem. Ind.* **1948**, *67*, 361–365.
19. Ross, S. *J. Phys. Colloid Chem.* **1950**, *54(3)*, 429–436.
20. Ross, S.; McBain, J. W. *Ind. Eng. Chem.* **1944**, *36(6)*, 570–573.
21. Novosad, J. J.; Mannhardt, K.; Rendall, A. *Proceedings of the 40th Annual Technical Meeting of CIM;* Canadian Institute of Mining, Metallurgy, and Petroleum: Calgary, Canada, 1989; paper CIM 89–40–29.
22. Kuhlman, M. I. *Proceedings of the SPE/DOE Enhanced Oil Recovery Symposium;* Society of Petroleum Engineers: Richardson, TX, 1988; paper SPE/DOE 17356.
23. Manlowe, D. J.; Radke, C. J. *SPE Res. Eng.* **1990**, *5*, 495–502.

24. Minssieux, L. *Ann. Combust. Liq.* **1971**, *26(5)*, 375–398.
25. Jensen, J. A.; Friedmann, F. *Proceedings of the California Regional Meeting of SPE;* Society of Petroleum Engineers: Richardson, TX, 1987; paper SPE 16375.
26. Holt, T.; Kristiansen, T. S. *Proceedings of the 5th European Symposium on Improved Oil Recovery;* Steering Committee of the European IOR Symposium: Budapest, Hungary, 1989; pp 727–736.
27. Kristiansen, T. S.; Holt, T. *Proceedings of the 12th International Workshop and Symposium of the IEA Collaborative Project on Enhanced Oil Recovery;* AEA Technology: Winfrith, United Kingdom, 1991.
28. Raterman, K. T. *Proceedings of the 64th Annual Technical Conference of SPE;* Society of Petroleum Engineers: Richardson, TX, 1989; paper SPE 19692.
29. Ross, S.; Suzin, Y. *Langmuir* **1985**, *1*, 145–149.
30. *ASTM Designation D 1881–86;* American Society for Testing and Materials: Philadelphia, PA, 1988.
31. *ASTM Designation D 1173–53 (Reapproved 1986);* American Society for Testing and Materials: Philadelphia, PA, 1988.
32. *ASTM Designation D 892–74 (Reapproved 1984);* American Society for Testing and Materials: Philadelphia, PA, 1988.
33. *ASTM Designation D 3601–88;* American Society for Testing and Materials: Philadelphia, PA, 1988.
34. *ASTM Designation D 3519–88;* American Society for Testing and Materials: Philadelphia, PA, 1988.
35. Suffridge, F. E.; Raterman, K. T.; Russell, G. C. *Proceedings of the 64th Annual Technical Conference of SPE;* Society of Petroleum Engineers: Richardson, TX, 1989; paper SPE 19691.
36. Hanssen, J. E.; Dalland, M. *Proceedings of the SPE/DOE 7th Enhanced Oil Recovery Symposium;* Society of Petroleum Engineers: Richardson, TX, 1990; paper SPE/DOE 20193.
37. Schramm, L. L.; Novosad, J. *J. Colloids Surf.* **1990**, *46*, 21–43.
38. McPhee, C. A.; Tehrani, A. D. H.; Joly, R. P. S. *Proceedings of the SPE/DOE Enhanced Oil Recovery Symposium;* Society of Petroleum Engineers: Richardson, TX, 1988; paper SPE/DOE 17360.
39. Maini, B. *Proceedings of the 37th Annual Technical Meeting of CIM;* Canadian Institute of Mining, Metallurgy, and Petroleum: Calgary, Canada, 1986; paper CIM 86-37-01.
40. Schramm, L. L.; Turta, A.; Novosad, J. J. *Proceedings of the SPE/DOE 7th Symposium on Enhanced Oil Recovery;* Society of Petroleum Engineers: Richardson, TX, 1990; paper SPE/DOE 20197.
41. Friedmann, F.; Jensen, J. A. *Proceedings of the 56th California Regional Meeting of SPE;* Society of Petroleum Engineers: Richardson, TX, 1986; paper SPE 15087.
42. Isaacs, E. E.; Jian, L.; Green, K.; McCarthy, C.; Maunder, D. *AOSTRA J. Res.* **1988**, *4*, 267–276.
43. Yang, S. H.; Reed, R. L. *Proceedings of the 64th Annual Technical Conference of SPE;* Society of Petroleum Engineers: Richardson, TX, 1989; paper SPE 19689.
44. Hudgins, D. A.; Chung, T.-H. *Proceedings of the SPE/DOE 7th Symposium on*

Enhanced Oil Recovery; Society of Petroleum Engineers: Richardson, TX, 1990; paper SPE/DOE 20196.
45. Ross, S.; Becher, P. *J. Colloid Interface Sci.* **1992**, *149(2)*, 575–579.
46. Kitchener, J. A. In *Recent Progress in Surface Science;* Danielli, J.F.; Pankhurst, K. G. A.; Riddiford, A. C., Eds.; Academic Press: Orlando, FL, 1964; Vol. 1, pp 51–93.
47. Schramm, L. L.; Novosad, J. J. *J. Pet. Sci. Eng.* **1992**, *7*, 77–90.
48. Nikolov, A. D.; Wasan, D. T.; Huang, D. W.; Edwards, D. A. *Proceedings of the 61st Annual Technical Conference of SPE;* Society of Petroleum Engineers: Richardson, TX, 1986; paper SPE 15443.
49. Lau, H. C.; O'Brien, S. M. *SPE Res. Eng.* **1988**, *3(3)*, 893–896.
50. Lobo, L. A.; Nikolov, A. D.; Wasan, D. T. *J. Dispersion Sci. Technol.* **1989**, *10(2)*, 143–161.
51. Koczo, K.; Lobo, L. A.; Wasan, D. T. *J. Colloid Interface Sci.* **1992**, *150(2)*, 492–506.
52. Fagan, M. E. M.S. Thesis, University of California, Berkeley, August 1992.
53. Roberts, K.; Axberg, C.; Osterlund, R. In *Foams;* Akers, R. J., Ed.; Academic Press: London, 1976; pp 39–49.
54. French, T. R.; Broz, J. S.; Lorenz, P. B.; Bertus, K. M. *Proceedings of the 56th Regional Meeting of SPE;* Society of Petroleum Engineers: Richardson, TX, 1986; paper SPE 15052.
55. Meling, T.; Hanssen, J. E. *Prog. Colloid Polym. Sci.* **1990**, *82*, 140–154.
56. Holt, T.; Kristiansen, T. S. *Proceedings of the SPOR Seminar;* Norwegian Petroleum Directorate: Stavanger, Norway, 1990; pp 59–77.
57. Dalland, M.; Hanssen, J. E.; Kristiansen, T. S. In *Proceedings of the 13th International Workshop and Symposium, IEA Collaborative Project on Enhanced Oil Recovery;* Schramm, L. L., Ed.; Petroleum Recovery Institute: Calgary, Canada, 1992.
58. Farrell, J.; Marsden, S. S. *Foam and Emulsion Effects on Gas Driven Oil Recovery;* U.S. Dept. of Energy: Washington, DC, November 1988; Report DOE/BC/14126-3.
59. Schramm, L. L.; Ayasse, C.; Mannhardt, K.; Novosad, J. U.S. Patent 5 060 727, October 29, 1991.
60. Schramm, L. L.; Mannhardt, K., Petroleum Recovery Institute, Calgary, unpublished results.
61. Sanchez, J. M.; Hazlett, R. D. *SPE Res. Eng.* **1992**, *7(1)*, 91–97.

RECEIVED for review October 30, 1992. ACCEPTED revised manuscript March 31, 1993.

ENHANCING OIL RECOVERY FROM POROUS MEDIA TESTING FOAMS

5

CO_2 Foams in Enhanced Oil Recovery

John P. Heller

New Mexico Petroleum Recovery Research Center, New Mexico Institute of Mining and Technology, Socorro, NM 87801

This chapter reviews the special qualities of CO_2 floods that make foam particularly attractive as a mobility-control agent. These features also impose requirements on the surfactant that forms the foam. Examples are given of high-pressure techniques of measurement that can determine the mobility of CO_2 foam under different surfactant concentration and type, flow velocities, and concentrations of the dense CO_2. A promising result is that some surfactants produce CO_2 foams for which the mobility is less than proportional to the rock permeability, and this result implies that CO_2 foams might be used to "smooth out" the uneven flow otherwise resulting from permeability heterogeneities in the reservoir. Other topics include surfactant adsorption and the durability of high-pressure CO_2 foams. A final topic concerns field operational procedures in application of this promising mobility-control method.

CARBON DIOXIDE is well known as a nontoxic gas that is present in the normal atmosphere at concentrations of about 330 parts per million (ppm). It is also familiar as a chief product of combustion of coal and hydrocarbons. CO_2 is soluble enough in water and aqueous mixtures to be useful in providing effervescence to a multitude of drinkable products. However, at pressures high enough to liquefy it (the critical temperature, pressure, and density constants of CO_2 are respectively, $T_c = 31.04$ °C or 87.87 °F, $P_c = 72.85$ std atm or 1070 psia or 7.381 MPa, $\rho_c = 0.468$ g/cm^3 or 29.22 lb/cu ft), its properties render it useful for another purpose as well.

CO_2 as a Displacing Fluid

Compressed to high density, CO_2 is a unique material with properties that make it valuable as a displacing fluid in oil recovery. Most significant among these helpful qualities is the fact that CO_2 is a dense fluid over much of the range of pressures and temperatures found in many oil reservoirs. This fluid is miscible in all proportions with the lighter hydrocarbon components of crude oil. Depending on the CO_2 density (in the range from 0.5 to 0.9 g/cm^3 as shown in Figure 1 and reference 1), the total miscibility limit extends upward from ethane toward hydrocarbons with 14 or more carbon atoms. At carbon numbers higher than this, the miscibility becomes only partial, but a substantial mutual solubility with CO_2 still persists for hydrocarbons above C_{30}.

The linear displacement of fluid through porous solid material by another fluid that is completely miscible in the first can have an efficiency approaching 100%. This linear displacement is in contrast to immiscible displacement (such as of oil by water) in which a significant fraction of the original fluid remains trapped in the pores. Thus, dense CO_2 has an inherent advantage over immiscible fluids, like water, in the recovery efficiencies that are possible with its use.

Dense CO_2 has another advantageous quality for use in enhanced oil recovery (EOR). It has fairly low solubility in the water that is always present in newly discovered as well as in previously waterflooded reservoirs, so that an excessive amount of it is not lost to the process during

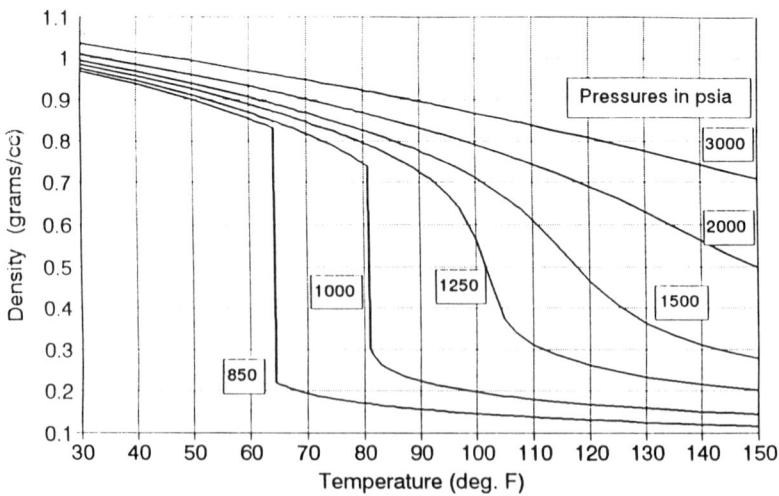

Figure 1. Density of CO_2. Data are from a computer program by Ely and Hanley (1).

the displacement. In addition, the solubility of water in dense CO_2 is also extremely low, so that the miscibility of oil and CO_2 is not greatly reduced.

However, perhaps the greatest advantage of CO_2 as a displacement fluid is its availability at a relatively low cost near many oil fields. In particular, a number of natural reservoirs that contain highly pure CO_2 at high pressure have been discovered at various locations in the Rocky Mountains and other places in the central United States. Three reservoirs in southern Colorado and northern New Mexico are currently supplying CO_2 to operators in the Permian Basin of west Texas and eastern New Mexico.

Despite these advantages, CO_2 is not without its faults as an oil displacement fluid. Chief among these faults is that the viscosity of dense CO_2 is quite low (2). Although it increases with density, as shown in Figure 2, the viscosity is still below that of most crude oils by a factor ranging from 5 to 100 or more. This low-viscosity means that the mobility of the CO_2, which is defined as the ratio of the effective permeability of the fluid to its viscosity, is usually much higher than that of the oil that is displaced. Because of this high mobility ratio, the displacement front is subject to instability when CO_2 displaces crude oil from a reservoir. Because of the instability, protuberances or fingers of the displacing fluid will grow from the displacement front, eventually reach the production well, and cause premature breakthrough. With CO_2 floods, some mitigating influences have been ascribed to various causes. One cause is the lowered relative permeability of the CO_2 in the presence of residual water in the formation. Other causes are the high solubility and high molecular mobility of

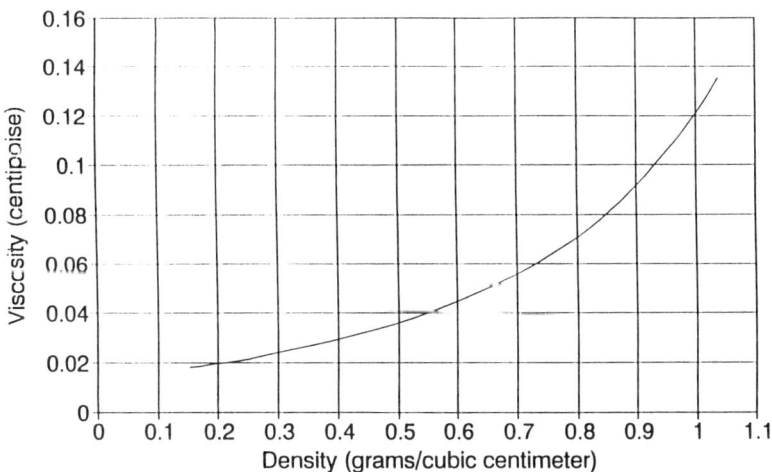

Figure 2. Viscosity of CO_2 at 90 °F. Data are from a program by Chung (2). (Values at other temperatures are practically overlapping.)

CO_2 in hydrocarbons. These features enhance transverse dispersion, which ordinarily suppresses short-wavelength frontal instability patterns, so that even longer wavelength fingers do not grow so rapidly.

A second disadvantage of CO_2 as a displacement fluid is that even high-gravity, light crude oil is a complex mixture, and sometimes it even contains colloidal components. Because of this composition, the displacement is not completely miscible. The situation was described by Hutchinson and Braun (3), who distinguished between "first-contact miscible", in which displaced and displacing fluids are miscible in all proportions, and "multiple-contact miscible", in which complete miscibility can be generated after the flood has progressed some distance at the expense of leaving some heavier components behind the front. The fractions that are left behind may also assist in the instability mitigation by lowering the absolute permeability of those parts of the rock through which the CO_2 has passed.

Despite the natural mitigation of some of the effects of adverse mobility ratio, most of the CO_2 floods that have been performed in the field, and in large core samples in the laboratory, are distinguished by early CO_2 breakthrough and lower displacement efficiency than expected. Thus, CO_2 floods do suffer from both the adverse mobility ratio and from this solvent's incomplete miscibility with all crude oil components.

Although the major difficulties in CO_2 floods may presumably be ascribed to the adverse mobility ratio and the solvent's incomplete miscibility with all of the crude oil components, other oil displacement projects also obtain less than 100% recovery. Even if an ideal miscible displacement fluid were to be used, it could not be perfectly efficient in existing reservoirs. The chief reasons for the low recovery efficiency in even an ideal case are the presence of spatial nonuniformities in permeability and other reservoir properties, which are of largely unknown magnitudes. These heterogeneities increase the contrast among the fluid velocities in different parts of the reservoir and eventually cause the produced fluid to be diluted with the injected fluid at a continually increasing rate that inexorably necessitates abandonment before all the oil is recovered. A second general situation that makes complete displacement infeasible is the fact that both injection and production occur through wells that even from a mathematical viewpoint cannot be expected to completely drain a three-dimensional reservoir in a finite time.

Not all oil reservoirs can be flooded by CO_2 for one reason or another. For a CO_2 flood to be effective, the crude oil must, as noted by Holm and Josendal (4), contain a high enough proportion of light hydrocarbons, and the reservoir must be deep enough so that it can safely contain pressures high enough to keep the CO_2 density high at the existing reservoir temperature. The balance of these factors was discussed by Heller and Taber (5). Additionally, perhaps the most important factor of all is the simple economic requirement. Sufficient quantities of CO_2 to supply the tertiary recovery operation must be available nearby at a cost low enough that the price to be obtained for the extra oil renders its recovery economically attractive.

Although numerous current CO_2 floods are profitable, even the most successful of these projects cannot recover all of the oil remaining in the reservoir. However, mobility control can improve the efficiency by relieving the considerable part of the difficulty that is caused by frontal instability. To explain some of the parameters of the required mobility control, information about the mechanism of instability growth might be useful. Although irregularities on miscible displacement fronts in real reservoirs grow predominantly at locations and along directions determined by variations of permeability, the rate of growth of fingers and channels depends also on the magnitude of the mobility ratio.

The lateral spacing of such protuberances on displacement fronts is important in their growth. It is convenient to distinguish between fingers, which grow as the result of frontal instability, and channels, which are caused by permeability inhomogeneities that are spatially correlated in the direction of flow. In either case, the lateral spacing of the fingers or channels determines the speed with which they can be subdued (and converted into a relatively smooth longitudinal transition) by transverse dispersion. Closely spaced channels and fingers do not endure for long, whereas widely spaced ones can persist throughout the life of the flood. For fingers, in addition to the influence of the transverse spacing on the duration of fingers, one may also consider the causative flow patterns in regions farther away from the front. The depth of the roots of a group of fingers, that is, of the region behind the front where the flow lines start to converge toward them, must be proportional to their transverse dimension. Thus, the more widely spaced and most damaging fingers involve changes in the flow velocity at a considerable distance behind (and ahead of) the displacement front itself.

This fact is important to the design of any mobility-control measure, because it mandates that the zone in which the displacement fluid is thickened must be quite large if the thickening is to suppress the largest of the efficiency-reducing fingers. The concept of "grading" the mobility from the first-injected, higher viscosity solvent back to a less expensive one in a miscible front was patented by Weinaug and Ling (6) and is a useful idea. Unfortunately, it appears that the volumetric extent of the transition zone must be larger than is often anticipated in order to reduce the mobility gradually enough throughout a sufficiently large portion of the reservoir.

The purpose of this discussion is to acquaint the reader with the constraints of CO_2 flooding and the requirements for mobility control. The major requirements are (1) that the thickening that is produced in the displacing fluid should extend back from the displacement front by a distance of the same order as the width of the zone to be swept, (2) that it should enable the production of more dollars worth of oil than it costs, and (3) that the magnitude of the thickening should be no greater than necessary (so as not to reduce inordinately the overall fluid throughput rate).

The use of "foam" with the injected CO_2 causes it to act in the reservoir like a thickened fluid and probably offers the best hope for mobility control in CO_2 floods.

Foam as a Population of Lamellae in the Pores of Reservoir Rock

This section describes foam and the variety of foams that are useful as CO_2 mobility-control agents.

The foam of everyday experience, though different in several ways from the "foam" that occurs and is used in reservoirs, is worthy of some examination. Such everyday foam is a two-phase mixture of gas and liquid, in which the liquid is the continuous fluid and the gas is held in separate cells. (*See also* the discussions in Chapters 1–3 of this book.) To display the distinctive foamlike characteristics, the volume fraction of the discontinuous phase must be greater than about 70%. At this high gas volume fraction (the so-called quality), the bubbles of gas are closely crowded together so that they cannot move independently. They also change in shape, and the walls of the cells become approximately planer, polygonal surfaces that are called lamellae or bubble-films.

The lamellae that separate the cells consist of thin films of the liquid contained between surfactant-stabilized, liquid–gas interfaces. The mass of connected bubbles that forms the foam is confined, if at all, only in a container that is much larger than the size of its cells. If, in addition, the observer is situated far enough away from the mass so that the individual cells cannot be visually resolved, it is common to allude to the foam as a substance or material in its own right. It is certainly an unusual material with rheological properties quite dissimilar from those of Newtonian or even the accepted non-Newtonian fluids. This distinction is especially evident in the manner of its transport through tubes or pipes. In particular, although such measurements are not always reproducible, the ratio of the resistance or pressure drop offered to foam transport through large-diameter tubes to that observed in smaller tubes is different from the corresponding ratio for ordinary fluids. In 1983, Hirasaki and Lawson (7) used Poiseuille's formula to calculate an effective viscosity of the foam and found that this number is not independent of tube size but decreases steadily as the tube size is reduced, as indicated in Figure 3. In Figure 3,

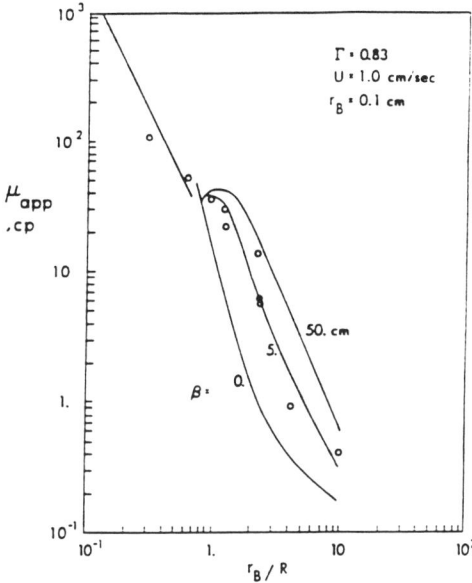

Figure 3. Combined effects of bubble radius, r, and capillary radius R, on apparent viscosity of foam; circles are experimental data, and curves are theoretically calculated. (Reproduced with permission from reference 29. Copyright 1992 Society of Petroleum Engineers.)

the abscissa is the ratio of bubble to tube radii, and β is a parameter that indicates the "surface tension gradient", discussed in reference 7.

The explanation for this behavior, in terms of the known properties of the lamellae surrounding the cells, was given by a number of investigators (8–11). The manner in which the foam undergoes shear consists of a rolling motion that involves the distortion of individual cells from their most stable shapes. Such static distortion depends on the interfacial tension of the films. Geometry dictates that a continual distortion of each bubble's shape would require an ever-increasing force. When this type of strain reaches a critical amount, the bubbles rearrange themselves. Thus, as the shear strain exceeds the stability limits for each cell's distortion, individual readjustments take place that are perceptible as flow. As a result, the total shear strain involved in the flow of the foam through a tube is concentrated in the outermost layer of bubbles, which are in contact with the tube wall. The majority of the foam moves as a plug that travels along the tube supported on a thin film at the wall.

Thus, the transport of foam is markedly influenced by the solid boundaries of the container in which it is confined. The dependence of effective viscosity on tube size shown by Hirasaki and Lawson (7) (who use the term "apparent") and reproduced as our Figure 3 extends down to bubble sizes as small as one-tenth of the tube size. For calculation of effective viscosity in a case where bubble size is one-tenth of the tube size, Hirasaki and Lawson built on Bretherton's (12) analysis of the motion of a bubble in a tube to derive the curve that matched the experimental measurements.

The fact that this relationship applies to small container sizes can be taken to mean that foam in porous media cannot consist of masses of tiny bubbles within the pore space. Such a foam in the pores of the rock would have a greater effective viscosity than would a collection of larger bubbles in the pores, or than simply a relatively few lamellae distributed through the pores. A collection of small bubbles would be extremely difficult to move into, through, or out of the pores. Thus, a foam that is movable by the available pressure gradients in a reservoir can consist only of the small number density of lamellae. Of course, it may seem paradoxical to refer to a dilute collection of only one lamella for each several pores as "foam", but this appears to be the only reasonable view of the type of foam that could be useful in mobility control.

This conclusion suggests the interesting topics of the generation of foam in porous media and the emplacement of the lamellae into the pores of reservoir rock. Two major methods can be used to generate everyday foam. The first is the method by which shaving-cream foam is generated and involves quickly depressurizing a concentrated surfactant solution in which a gas is dissolved. The large concentration of microscopic gas bubbles that are precipitated from the liquid grow continuously as more gas comes out of solution. At the same time, the bubbles of gas are kept from coalescing by the surfactant, which congregates at the surfaces of the bubbles as fast as they are formed. As the bubbles grow larger, their condition becomes more and more crowded, so that the gas fraction of the total volume increases above 70%, and the mass of bubbles acquires the unusual properties of a foam. When one uncaps a bottle of carbonated beverage, a similar process occurs as the bubbles rise to the surface. In carbonated beverages, the concentration of surfactant is thankfully much smaller, and the foam does not persist as long.

Contrasted to the dissolution and expansion process is the old-fashioned way of foam generation in which the mixture of gas and surfactant solution is subjected to violent shear motion. This motion may be produced by the use of a brush, or perhaps by rapid flow of the two-phase mixture through a plug of stainless steel wool. This method is used, for instance, for the generation of fire-fighting foam. These methods work because the shear stretches the gas bubbles into long "tubes" of gas in the fluid. As was described by Boys (13) in 1911, such a configuration of fluids is unstable against periodic, longitudinal variations in diameter. These circumferential ripples rapidly increase in amplitude until they break the stretched bubble into a set of still smaller individual bubbles. The rapid shearing motion produced by the motion of a brush through the mixture is really quite similar, except for a change of the observer's frame of reference, to the flow of the gas–liquid mixture through a set of small apertures.

Although an externally produced foam can be forced into a porous rock, it is evidently not necessary because the foam will be generated there by itself. As the mixture of gas and surfactant solution flows through the porous rock, rapid shear strain occurs and leads quite naturally to the generation and stretching of bubbles within the pores. The texture of the foam, that is, the size of the bubbles, depends mainly on the size of the pores. Similarly, the number of bubbles that exist will be determined by the balance between the rate of generation of lamellae and their rate of decay.

The rate of generation depends on the pore sizes and complexity, and should be roughly proportional to the flow rate. The rate of decay is the result of several simultaneous processes that cause the bubbles to coalesce. First, the coalescence occurs very quickly if the surfactant is ineffective. A stronger (or more effective) surfactant can stabilize the interfaces and postpone the coalescence, whereas with a weak surfactant or none, coalescence occurs immediately when bubbles come into contact (14). Second, the lamellae become thin enough to break only where two bubbles of the gas approach each other closely. The frequency of this event depends on the average gas fraction, on the flow rate, and on the geometry of the pore space. Third, in some locations in the rock, a moving lamella that spans the pore has a low chance of survival. These locations may be distinguished by unfavorable pore size or shape or by residual oil components that interfere with the wetting of the pore around its perimeter.

A further point of difference between CO_2 foam and everyday foam is the fact that in CO_2 foams, the gas, which is a dense fluid, acts more like a liquid than like the gas encountered in everyday foams. This means that the discontinuous phase of the foam is also capable of carrying other materials, such as surfactants, in solution. The capability imposes a new constraint on the surfactant, as we shall see. Perhaps surprisingly, however, the dense state of the gas does not cause any difficulty in the formation of the foam. In fact, it is possible to make analogous foams at low pressure with the same surfactants in water or brine, and with a light hydrocarbon like isooctane substituted for the dense CO_2. Despite this fact, it is common in the foam literature to refer to the discontinuous phase as "gas". Similarly, the brine–surfactant solution that forms the continuous phase is often referred to simply as "water".

The purpose of these qualitative descriptions is to furnish the background for a more quantitative description.

Evaluation of Mobility Reduction

From the time when the use of foam in reservoirs was first proposed in a patent by Bond and Holbrook (15), it was usually implicitly assumed without specific mention, that foam would preferentially impede flow in the higher permeability layers or fractures in the reservoir that had already been swept of their oil. It was assumed without evidence that the unswept portions of the reservoir would remain at least as accessible and available to have their content displaced and forced into the production wells. Because of this assumption, many, if not most, of the early reports of foam investigations included descriptions of core-floods with listings of oil recovered and ranked in the order of those values. This assumption for CO_2-foam floods cannot be examined for validity without more thoroughly considering the processes involved in complete reservoirs. But first, it is necessary to examine the behavior in core samples.

In order to evaluate the usefulness of CO_2 foam as a displacing fluid, more quantitative reference must be made to some of the topics that were introduced more qualitatively and more briefly in the introductory discussions of this chapter. The flow of foam in pipes has been discussed. For foam flowing in pipes, the pipe diameter is much larger than the cell size, and no shear flow occurs throughout most of the cross-section of the mass of foam in the pipe. The shearing strain is concentrated instead in a thin water-film around the inside perimeter of the pipe, and the foam plug rides on this film of water. This circumstance is quite different from that in which foam flows in porous media, in which immobile containing boundaries are close to every part of the flow. The flow of a hypothetical foam in porous media, which consisted of a mass of bubbles in every pore, would be almost impossible. This follows from some of the theoretical work about the motion of two-dimensional arrays of foam cells.

The theoretical works of Prud'homme (8) and Princen (9), and their elaboration by Kraynik (10), were concerned primarily with foam in pipes or other larger containers. They found a limiting shear stress of an assemblage of foam cells that is proportional to the surface tension, σ, and inversely proportional to twice the "radius" of the foam cells, r. This yield stress, τ_{yield}, was predicted and found to vary with the ratio of the interfacial tension to twice the bubble radius. The proportionality constant, α, is geometry-dependent and near unity in magnitude.

$$\tau_{yield} = \frac{\alpha\,\sigma}{2r} \qquad (1)$$

Below this yield stress, no continued shear strain can occur, only a small elastic (dilational) strain that corresponds to the distortion of the polyhedral cells. Above the yield stress, layers of polyhedra "roll over" each other, a motion that constitutes the continuous shear strain. The calculation was made for the two-dimensional "stack of pencils" case because, although that situation is physically impossible, it has the virtue of being exactly calculable. Except for the fact that the three-dimensional polyhedral cells in a pipe-contained foam are not all the same shape, the principle is the same, and the numbers are approximately the same, as shown by Kraynik's 1982 experiments (10).

This limiting value of yield stress cannot be directly applied inside porous media because the cross-sections of pore spaces are generally much too small to contain more than one polyhedral cell within them. But the calculation by Princen (9), as given in equation 1, can show what would happen if there were foam of such fine texture within the rock. The yield stress would be very large because of the small radii of the hypothetical foam cells within the pores. For instance, if these radii averaged 1 μm, and the interfacial tension were 20 dynes/cm (20 mN/m), the yield stress would be in the order of 1×10^5 dynes/cm^2 (15 kPa), or about one-eighth of an atmosphere. The pressure drop required to impose this yield stress on all the bubbles in a core of reasonable length would be quite high. In a straight row, there would be 250 of this size bubbles per centimeter or about 7500 cells per foot. To force such a small-bubbled foam fully into a 1-foot core would require a pressure drop rising to about 750 atm, 76 MPa, or 11,000 psi. Of course, this figure is very approximate, but it is enough to make apparent the difficulty one would face to pump a foam with such fine texture into a formation, or to get one out.

Instead of a number of small bubble-cells within each pore, the foam to be considered in CO_2-displacement applications must consist only of individual lamellae spanning some of the pores in the rock. Whether they all move along at the rate of the overall "foam velocity" is irrelevant, as is the question of the survival of individual lamellae. A relatively small population of such lamellae, created at favorable times and places within the rock, will be sufficient to impede the entire foam in its motion.

Perhaps the most important variable in the description of foam flow through porous rock is the mobility of the foam, the flow achieved for a given pressure drop. This quantity is defined as the simple ratio of the combined superficial flow rate to the imposed pressure gradient. This ratio is indicated in the first part of equation 2, in which the mobility, λ, is given in terms of the combined flow rate, Q, the cross-sectional area of the sample, A, the pressure drop, ΔP, across the sample, and its length, L. As is well known, in the flow of an ordinary fluid through porous media, the mobility can be separated into two factors; one has to do primarily with the properties of the rock, and the other, with the properties of the fluid. By using Darcy's relation, the mobility for the ordinary fluid is computed to be the ratio of effective rock permeability, k, to the fluid viscosity, μ.

The Darcy relation is indicated in the second part of equation 2.

$$\lambda = \frac{QA}{\Delta P/L} = \frac{k}{\mu} \qquad (2)$$

The units of mobility can be given in traditional nomenclature as cm^2/(atm·s), or darcy/cP, or in SI units as m^2/(Pa·s).

Darcy's equation was extended in the early days of petroleum engineering to describe the simultaneous flow through porous media of two immiscible fluids that contain no surfactant. The effective permeabilities and the mobilities of the fluids themselves are distinguished from each other in this extension. This extension is possible for the apparent reason that in the surfactantless case, the fluid–fluid interfaces are transient and play no role in increasing the flow resistance. In the steady state, and within a large range of flow fraction, the two fluids flow in separate paths. Although each of these paths is dedicated to one particular phase, each path contains relatively static accumulations of the other phase as well. The flow resistance for each fluid depends on the geometry of the paths open to it and on the fluid's own viscosity.

The concept of separate mobilities of the fluids can be used to make useful predictions of displacement behavior and has indeed been accepted without question by most of the petroleum-engineering profession. The combined mobility, λ_{total}, in this case is given in equation 3 as the sum of the mobilities of the two individual fluids where each fluid flows through the rock with distinct effective permeabilities, k_{e1} and k_{e2}. The individual fluid viscosities, which are each measurable in bulk, are μ_1 and μ_2.

$$\lambda_{\text{total}} = \left(\frac{k_{e1}}{\mu_1} + \frac{k_{e2}}{\mu_2} \right) \qquad (3)$$

When a surfactant is present, it is not always possible or desirable to separate the two mobilities as is done in equation 3. When surfactant molecules congregate at the water–gas interfaces, these interfaces become stabilized to some extent so that bubbles that come into contact with each other do not immediately coalesce. A sufficient concentration of surfactant causes a distinct change in the flow regime through porous media as well as in pipes. As a result of the surfactant, much of the gas phase present in the rock is unable to flow freely in its own dedicated channels. Instead, flow in these channels is inhibited by lamellae. An individual bubble-film is not strong enough to sustain much pressure difference across it or to prevent flow from occurring, but depending on their number, the bubble-films can markedly limit the flow of gas by dividing up each of the pathways that would otherwise be open to gas, altering the geometries of those paths significantly. Unfortunately, the details of this activity cannot be easily measured.

The surfactant, in changing the flow regime in the porous rock, changes the conditions and requirements for describing the flow resistance. For instance, any measurement of separate effective permeabilities would have to be done by different techniques than those measured for surfactantless phases and could not be considered comparable to them. It is also doubtful that the use of the viscosity of the gas phase in an equation like equation 3 would make sense because much of the gas in a foam does not get the chance to move freely through channels or pathways in the porous medium as discussed previously in Chapter 3.

Instead of attempting such separation, no attempt is made here to partition the mobility on the grounds that no reasonable means for the purpose is available. This approach is justified by the fact that there is no need for these separate quantities. The mobility of CO_2 foam, as used here, is a measure of the flow of its combined components, the surfactant–brine solution and the dense CO_2. It is not implied that in all circumstances the two components of a foam will stay together during flow. For CO_2 foam, however, and considering its manner of use in the reservoir, it is considered that the data from steady-state experiments can supply the needs of reservoir engineering. Except where stated, the mobility discussed has been computed from the first part of equation 2 with Q dense CO_2. The area and length of the core sample and the pressure gradient across it are also used in the calculation.

Another possibility is that an effective viscosity of foam, such as the one measured and calculated by Hirasaki and Lawson (7) for capillary tubes, can be used in equation 3. Although such a choice would require the assumption of some arbitrary value for relative permeability, it could be used to give useful predictions for flow resistance in the reservoir. There is disagreement as to which of the choices steers the best course to follow. It is perhaps wisest to use an empirical approach, to measure only what is possible to measure at this time, and to wait patiently and receptively for future revelations.

The required measurements have been made in different ways by different researchers, and they have sometimes presented their results in different terms. To minimize confusion, several of the different quantities that are also used to describe the flow resistance of CO_2 foam are presented.

Relative mobility, λ_r, is obtained by dividing the foam mobility, λ, by the absolute permeability, k, of the rock in which it is measured.

$$\lambda_r = \frac{\lambda}{k} \qquad (4)$$

The relative mobility is usually given in reciprocal centipoises [or in $(\text{Pa·s})^{-1}$], which suggests another quantity of interest; the effective viscosity of the foam. We could also define the "effective viscosity" of foam, μ_{eff}, as simply the reciprocal of the relative mobility.

$$\mu_{\text{eff}} = \frac{1}{\lambda_r} \quad (5)$$

Thus, the units of this effective viscosity are centipoises (Pa·s). The effective viscosity is not the actual viscosity of any real fluid, it is the viscosity of a virtual fluid simulating the combined flow of CO_2 and surfactant–brine through reservoir rock. The effective viscosity is a number that can be used in Darcy's law, along with the absolute permeability of the rock, to give the ratio of pressure gradient to superficial flow rate. In particular, the effective viscosity defined previously is not to be used with any assumed value of the relative permeability of dense CO_2 in the rock. As will be seen, experiments show that this effective viscosity is not constant, but changes in value as a number of other parameters of the flow are varied.

A major difficulty with the use of effective viscosity described here is that it has been confused with the apparent viscosity that is sometimes used with an estimated value of effective permeability of foam to yield the mobility.

Often the best solution is to present a dimensionless number that indicates the effectiveness of the surfactant. Such an index is the mobility-reduction factor, m_{rf}, which is the ratio of the mobility of the CO_2 foam, $\lambda_{CO_2 \text{ foam}}$, to the total mobility of the mixture of surfactant-free brine, λ_{brine}, and dense CO_2, λ_{CO_2}, at the same gas–liquid volumetric flow ratio.

$$m_{\text{rf}} = \frac{\lambda_{CO_2 \text{ foam}}}{\lambda_{CO_2} + \lambda_{\text{brine}}} \quad (6)$$

The denominator in equation 6 is the sum of the mobilities of dense CO_2 and of brine that does not contain surfactant when the two are flowing simultaneously. In those circumstances when the flow of CO_2 can conveniently be measured separately, a resistance factor, R_f, can also be defined, as in equation 7, as the ratio of the mobility of the CO_2 when it is flowing simultaneously with surfactant-free brine to the mobility of the CO_2 portion of foam.

$$R_f = \frac{\lambda_{CO_2 \text{ flowing with surfactant-free brine mixture}}}{\lambda_{CO_2 \text{ flowing within foam}}} \quad (7)$$

Again, for the definition to make any sense, the fractional flows of CO_2 and the overall velocities must be the same in both cases.

The next question regards the means by which values of any of these quantities to be used in reservoir-engineering calculations may be obtained. There is a continuing history of theoretical attempts to calculate the mobility of foam starting from known quantities and familiar principles of two-phase flow in porous rocks. One of these principally considers the effect of capillary pressure and concludes that this quantity is a principal determinant of the stability and therefore of the population of lamellae. Presuming equilibrium conditions in which the radii of curvature of the Plateau boundaries determines the excess of absolute pressures in the gas over that in the liquid, Khatib et al. (16) computed a limiting value of the capillary pressure. Above this value, the lamellae become too thin for the surfactant to stabilize. Increasing the gas fractional flow decreases the water saturation and raises the capillary pressure.

Using these concepts, Khatib et al. (16) concluded that the gas mobility, when measured at the limiting capillary pressure, will be an increasing function of gas flow rate. Unfortunately, their experiments were not in the range of interest in oil recovery practice, having been performed at high flow rates through sand and bead packs with permeabilities between 70 and 9000 darcies (69 and 8900 μm^2). They observed critical capillary pressures for a pack of 270 darcies (266 μm^2) to be attained at very high gas flow fractions in the neighborhood of 98%. This foam would be uncommonly dry, and again, would be outside the range of practice in CO_2 floods.

Other attempts at constructing theoretical models were made by Friedmann and Jensen (17), Falls et al. (18), Rossen and Gauglitz (19), Zhou and Rossen (20), and others. In addition to considering capillary pressure, they also used the percolation theory. One of their interesting conclusions is that, below a minimum pressure gradient, foam will not flow; in other words, a high enough concentration of lamellae can contain any pressure drop and completely plug off a horizon of a porous formation. The conditions under which this plugging could happen are of special interest to engineers whose production wells have severe water-coning problems or whose injection wells pierce a known "thief zone" or highly permeable stratum of rock. This use is certainly closer to the needs of the CO_2 flood. In such situations, an injected "slug" of high gas content foam has proved itself for the purpose of markedly reducing flow through cores in the laboratory. Chapter 8 gives a thorough discussion of such "gas-blocking" foams. Hanssen and Dalland (21) have experimentally surveyed some 40 surfactants for the purpose. Although none of these surfactants completely stopped the flow of gas through a core, many reduced the flow to very low levels. Hanssen and Dalland (21), who also set the criterion that the foam should be completely immune to degradation by crude oil, identify four of these surfactants as the most effective of the group.

Unfortunately, all of these four were fluorinated surfactants, which would be too expensive for use in CO_2 flooding.

Instead of relying completely on theory for the determination of mobility, most researchers also performed experimental measurements of the quantities of interest. Many of the first experiments on foam were performed with water and gas with the outlet at ambient pressure, and many were simply gas floods of packs or cores saturated with surfactant solution. Although for many such transient experiments, the published data were insufficient for the estimation of the steady-state mobilities required for the estimation of mobility-control effectiveness, this was not true for some of them. Calculated values of mobility and relative mobility were derived by Heller et al. (22), from the data published in six different papers (23–28). The values they found, given in terms of relative mobilities, ranged from 0.001 to 0.6 cP^{-1}, or in terms of effective viscosities from 1000 down to 1.6 cP (1 to 0.0016 Pa·s). Not enough information was available to trace all of the relevant parameters that may have caused these differences.

In that same publication, Heller et al. (22) also designed an experiment to determine the steady-state mobility of a foam consisting of surfactant brine and either isooctane at ambient pressure or liquid CO_2 at about 1000 psia (6.9 MPa). In their reported experiments, the measured relative mobilities of both isooctane and dense CO_2 ranged from somewhat less than 0.1 to 0.7 cP^{-1}, showing an increase in relative mobility with increasing flow rate for combined Darcy velocities from 15 up to 250 ft/day (4.6 to 76 m/day).

These calculations and measurements definitely showed that "foam", brought about by the presence of surfactant dissolved in the discontinuous phase, showed a reduced mobility. Nevertheless, the tests raised more questions than they answered. Among these were questions about the effects on the mobility of the type and concentration of the surfactant, the flow rate, the flowing volumetric fraction of the CO_2, and the rock type and permeability.

Before and during the course of the exploration of these questions, the experimental system used to seek the answers was refined. A recent version of the apparatus was shown by Tsau and Heller (29) and is reproduced here as Figure 4. It depicts a simple experimental flow system in which flow through the core being tested comes from two constant-speed syringe pumps, dense CO_2 from a Ruska pump, and surfactant–brine from a transfer vessel into which is forced oil from an Isco pump. Differential pressures are monitored by transducers connected across the core being tested and across a calibrated capillary tube carrying the liquid CO_2 from its pump. The purpose of the calibrated capillary tube is to enable the instantaneous flow rate of the CO_2 to be measured, because the output of such a compressible fluid from the pump is greatly influenced by the rapid pressure changes experienced at the core. Before entry to the core being

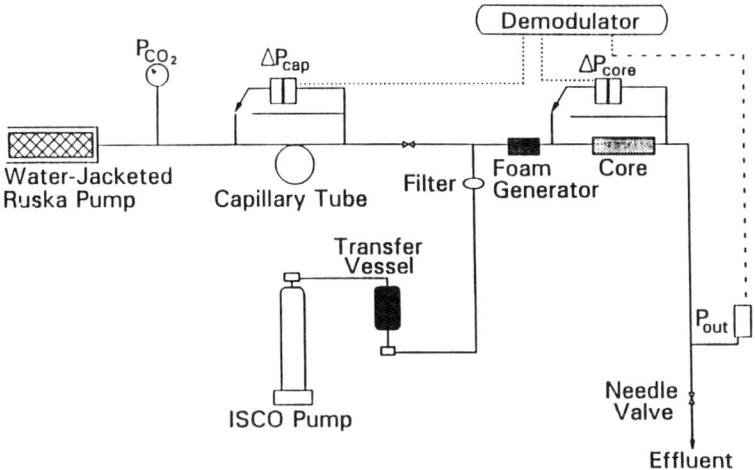

Figure 4. Apparatus for measurement of mobility of CO_2 foam. (Reproduced with permission from reference 29. Copyright 1992 Society of Petroleum Engineers.)

tested, the mixed CO_2 and surfactant–brine are passed through a shorter piece of similar rock. The purpose of this "foam generator" is to support the assumption that the short piece of tubing leading to the input end of the core will contain closely spaced bubbles. That assumption was verified in some of the experiments by using electro-optical detection of the bubbles moving through a glass capillary section of the tubing.

Absolute pressure at the downstream end of the core is also monitored. This output pressure is maintained by a sensitive, manual needle valve that withstands the major pressure drop of the experiment. Because this is a steady-state experiment and not a core-flood, the correct setting of this valve needs to be made very infrequently during the run after steady state has been reached. The three transducer readings are continually recorded on a chart recorder, noted graphically on a computer screen, and periodically written to disk by a dedicated personal computer.

With this apparatus, many measurements have been made in the search for answers to all of the questions listed. These experiments have also uncovered another very significant issue that will require even more observations for its resolution.

Only a limited number of different surfactants have been used in these foam-mobility measurements, and these surfactants do not include any cationic types. The ones chosen to be examined further in the mobility experiments were surfactants that had passed various screening tests, showed promise for other reasons, or had been recommended by other researchers. Altogether, 10 different surfactants have been used in this type of measurement.

The surfactants for which the foam mobility was measured in these experiments were as follows:

- Emulphogene BC-720, a nonionic from GAF (*22*)
- Varion CAS, an amphoteric surfactant from Sherex (*30–32*)
- Enordet X 2001, an anionic from Shell (*30, 31, 33*)
- Alipal CD-128, an anionic surfactant from GAF (*31*)
- Chembetaine BC3, an amphoteric surfactant (*31*)
- AEGS, an anionic surfactant (*32*)
- CD1045, a surfactant mixture from Chevron (*29*)
- CD1050, a nonionic surfactant from Chevron (*29*)
- Avanel S-30, an anionic from PPG–Mazer (*29*)
- NES-25, an anionic from Henkel (*29*)

CO_2-foam mobility was reduced with all of these surfactants, although there were significant differences in the degree of mobility reduction. In particular, both anionic and nonionic surfactants were effective, as were the amphoteric surfactants. The major differences became evident when the amount of their adsorption on rock and their chemical stabilities were tested.

One point of difference among surfactants that demonstrates their effectiveness came when the effect of surfactant concentration on CO_2-foam mobility was measured. Figure 5 is a log–log plot showing the measured

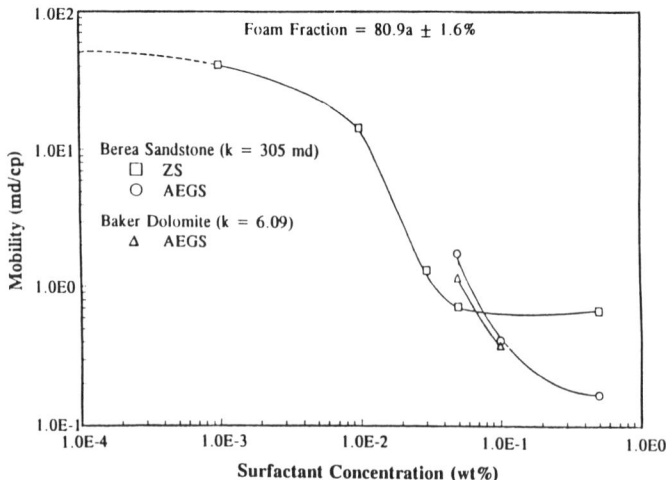

Figure 5. Influence of surfactant concentration on mobility (32).

mobilities versus concentration for three different surfactants. The greatest range of surfactant concentrations is shown for measurements with Varion CAS (designated in the figure as ZS for zwitterionic surfactant) in Berea sandstone. This curve illustrates the general pattern of the variation of foam mobility with surfactant concentration. At very low concentration, the lamellae are nonexistent, and the measured mobility is just the sum of the separate mobilities as is used in the denominator of equation 6 for determining the dimensionless mobility-reduction factor. At slightly higher concentrations, the mobility starts to decrease and then decreases at an increasing rate. Eventually, at much greater concentrations, the mobility reaches its lowest values beyond which it cannot be lowered further by this surfactant. Most of the subsequent work was done in the portion of this graph that encompasses the middle concentrations.

A second variable of great interest in the application of CO_2 foam for oil recovery is the effect of rate of flow on the mobility. Inasmuch as it was convenient to vary the flow rates in these mobility measurements, most of the results of the experiments are plotted in terms of flow velocity. Figure 6, for instance, is a plot showing the mobility of foam made with Shell's Enordet X2001 in a Berea sandstone core. The drier foams, with gas fraction greater than 80%, show a just significant increase of mobility at greater velocities. Presumably, this "shear-thinning" effect is due to a gradually decreasing equilibrium point in the population of lamellae

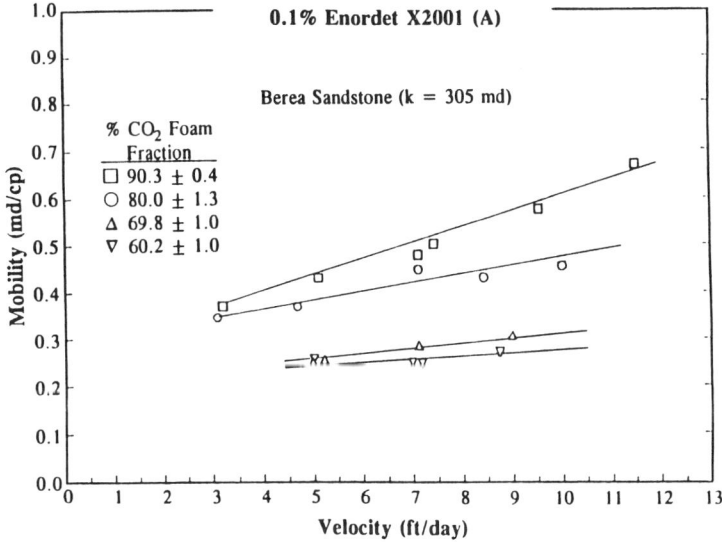

Figure 6. Graph showing the effect on mobility of volumetric CO_2 fraction. (Reproduced with permission from reference 30. Copyright 1988 Society of Petroleum Engineers.)

as the flow rate is increased. Unfortunately, in these experiments, it has not been possible to perform these tests over a wider range of flow rates. A derived value of mobility was obtained by Yang and Reed (34) in coreflood experiments in Berea cores. Their experiments, which extended from a low of 0.15 ft/day, showed a decreasing mobility for velocities up to about 0.5 ft/day followed by an increase in mobility as velocity was increased up to 9 ft/day.

However, the most exciting result of these experiments came when the effect of foam in different rock samples was investigated. The measured mobilities did not vary over nearly as wide a range as the permeabilities of these rocks. However, when the variation of relative mobilities were examined on a log–log plot against the permeabilities, an important trend could be discerned. The reciprocal of relative mobility can be called the effective viscosity of the CO_2 foam. This parameter varies, at least for these conditions, as shown in Figure 7 (33). Apparently, the CO_2 foam under these conditions acts like a more viscous fluid in high-permeability rock than in low-permeability rock. This selective mobility reduction (SMR) is the type of behavior that is needed for CO_2 foam to perform, as had been optimistically assumed in early foam literature.

More recent experiments that were aimed at the selection of a suitable surfactant for use in a southeastern New Mexico oil field have not been so favorable. Apparently, the marked change in effective viscosity with rock permeability, shown in Figure 7, for the obsolete surfactant

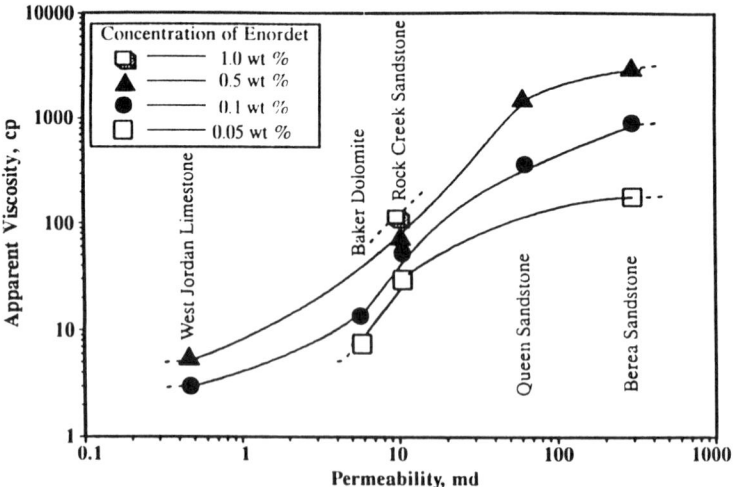

Figure 7. Apparent viscosity of CO_2 foam as a function of rock permeability, with surfactant concentration as parameter. (Reproduced with permission from reference 33. Copyright 1991 Society of Petroleum Engineers.)

Enordet X2001, and which is called SMR, is not as general as might have been supposed.

In very recent work, the above variation was examined for three other surfactants, using pairs of core samples (Type I and Type II) that had been taken from the same well but differed from each other in permeability by factors of about 20. For each of the three surfactants, the mobility is shown on a log–log plot as a function of core permeability. The first of the four lines on each of the three plots are for surfactant-free brine–CO_2 mixture, and the others are for three different low concentrations. Figure 8 is for Chevron Chaser CD-1045, Figure 9 is for Henkel NES-25, and Figure 10 is for Chevron CD-1050.

Had the CO_2 foam in the three tests exhibited SMR like the Enordet X2001 in the previous tests graphed in Figure 7, the mobility would be the same for the different cores, and the lines would be horizontal. Only one of the surfactants showed this behavior in the concentration range depicted.

More experiments along these lines and others are currently being carried out to attempt to resolve the uncertainty. A major question, for instance, is to determine how higher surfactant concentration affects this aspect of the CO_2 foam's behavior. Will new experimental measurements with other surfactants show the SMR property, continuing the behavior shown in Figures 7 and 10, or will the trends shown in Figures 8 and 9 prove to be the more general pattern?

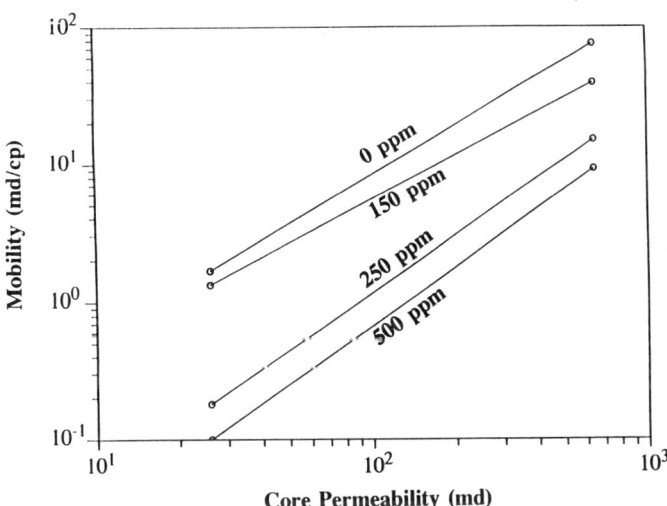

Figure 8. CO_2-foam mobility using CD 1045 in different permeability rocks. (Reproduced with permission from reference 29. Copyright 1992 Society of Petroleum Engineers.)

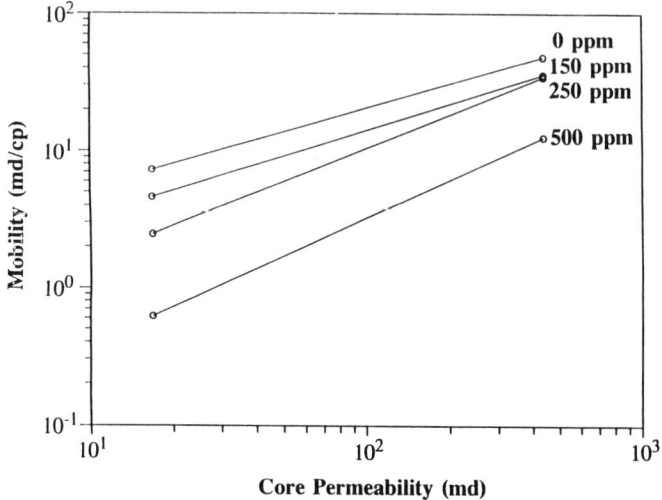

Figure 9. CO_2-foam mobility using NES-25 in different permeability rocks. (Reproduced with permission from reference 29. Copyright 1992 Society of Petroleum Engineers.)

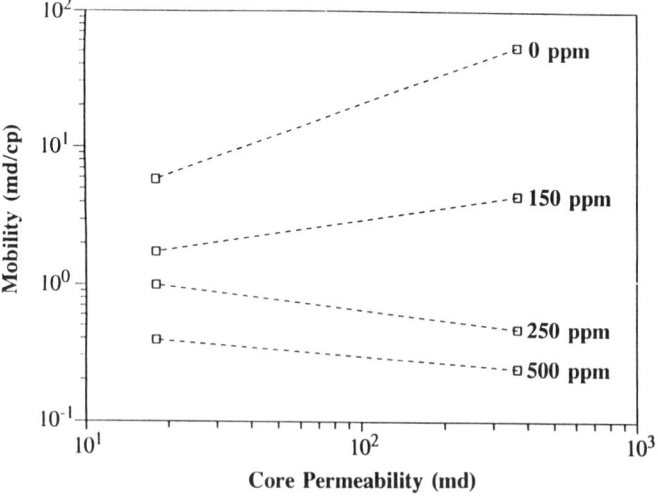

Figure 10. CO_2-foam mobility using CD 1050 in different permeability rocks. (Reproduced with permission from reference 29. Copyright 1992 Society of Petroleum Engineers.)

Surfactants for CO_2 Floods

The previous section recounted experiments that establish the obviously necessary property that for a surfactant to be useful for the purpose being discussed, it should be capable of decreasing the mobility of CO_2 foam. This qualification must be explored in experiments that are performed under conditions duplicating those in the field, especially of temperature and pressure.

A similar situation exists in regard to other properties when there is no accepted theory by which they can be explored in advance. Among these is the simple question of a surfactant's ability to stabilize the lamellae against coalescence. Again, one can only accept experimental determinations of this capability if the work is done in conditions that duplicate the reservoir environment. The ionic and other chemical conditions are also of major importance.

A direct observational procedure has been developed to deal with the paucity of laboratory data. The apparatus constructed for this experiment, which is called the "foam durability test", can be operated at Permian Basin reservoir temperature and pressure. This device is illustrated in Figure 11. Its sapphire visual cell is initially filled with the pressurized sur-

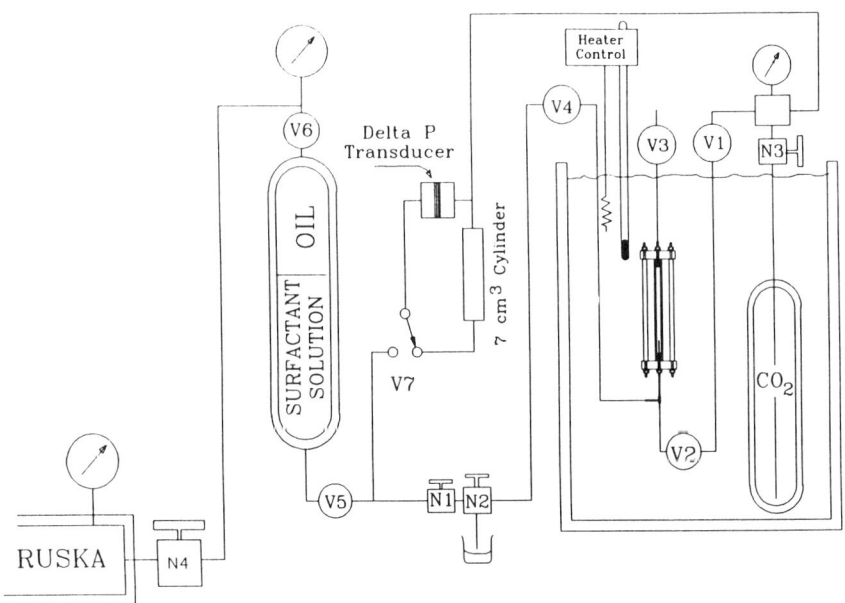

Figure 11. Foam durability apparatus for observing high-pressure CO_2 foam. (Reproduced with permission from reference 33. Copyright 1991 Society of Petroleum Engineers.)

factant being tested. Bubbles of dense CO_2 are then introduced through the central tube of a connection at the lowest end of the tube, and the displaced surfactant solution is withdrawn through the annulus. Because all fluids in the apparatus are at the same high pressure, the CO_2 enters at the same rate that surfactant is withdrawn. Because the CO_2 rate is set by a Ruska pump, and the bubble volume at detachment is determined by the interfacial tension (IFT) between the surfactant solution and the dense CO_2, the IFT can be computed from the interval between the release of bubbles from the needle.

From the plot of the values of IFT measured at different surfactant concentrations, such as shown in Figure 12, the critical micelle concentration (CMC) can be determined as the concentration at which the change of slope occurs. Figure 12 shows such a plot made at reservoir conditions in a particular field. The CMC is a useful reference concentration for a particular surfactant, although its full significance for foam formation and stability within porous media is not yet known.

The usefulness of the CO_2 bubbles for providing further information about the influence of surfactant on the interface has not ended after they are released from the needle tube and counted. The rising bubbles are captured at the top of the sapphire tube where each bubble does one of two things: It either coalesces with the CO_2 bubbles already there, or it stays separate and forms part of a large-cell foam that gradually increases

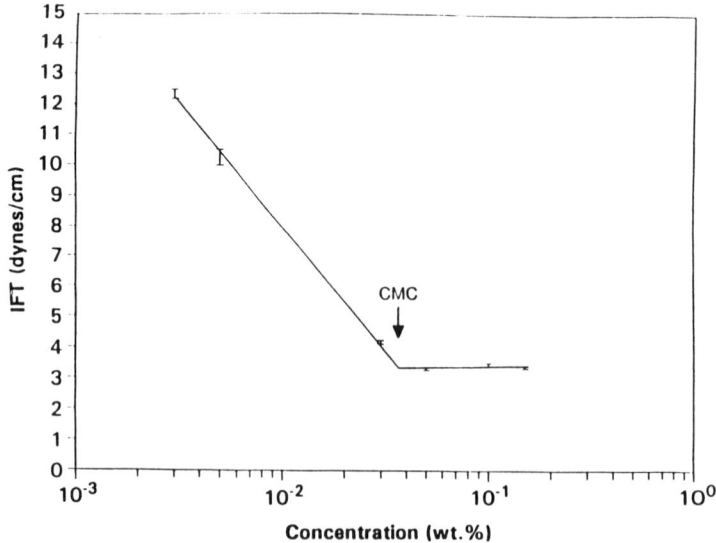

Figure 12. Interfacial tension (IFT) of Witcolate 1276 in brine against dense CO_2, at 104 °F and 2500 psia. (Reproduced with permission from reference 33. Copyright 1991 Society of Petroleum Engineers.)

in volume. Which outcome occurs and how much of this foam is built up depend on the lamella stabilization capability of the surfactant or foamability. The durability of the foam can also be measured simply by observing the time it takes for foam to decay into a simple layer of dense CO_2 at the top of the tube. The record of decay is not simple; for instance, Figure 13 shows the normalized foam height as a function of time for several different surfactant concentrations of the same surfactant described in Figure 12. The durability increases (the foam is more long-lasting) as the concentration increases from 0.03% to 0.05%, but decreases again above 0.05%.

Although surfactants differ widely from each other in both of these properties, the precise significance of surfactant foamability and foam durability, in terms of the behavior of these foams in CO_2 floods, is also not obvious. Clearly, for a surfactant to be used successfully in a CO_2-foam flood, both these measures should be above average. However, these characteristics are not the only ones that are needed. For instance, an unbreakable foam with both high foamability and high foam durability could very well be excessively active, so that the CO_2 at the leading edge of the transition zone would remain tied up in foam and could not get to the oil. Such an excessively active surfactant, for instance, might be quite insensitive to the effect of crude oil, the foam-breaking effect of which varies widely from one surfactant to another, and certainly with different crude

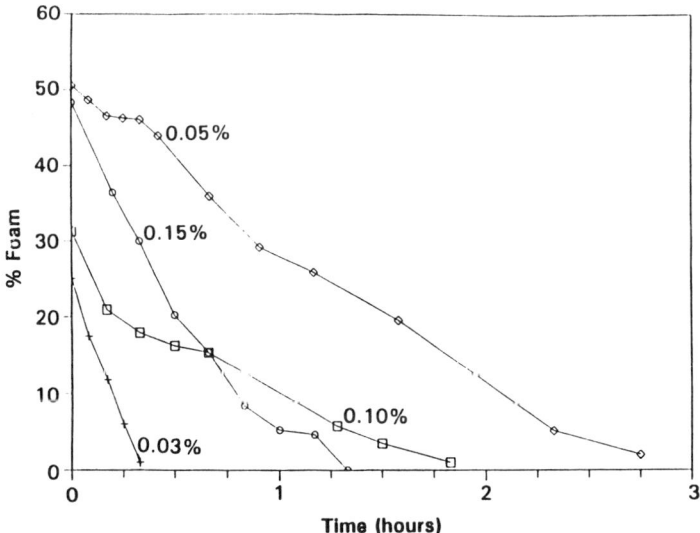

Figure 13. Decay of CO_2 foam in Witcolate 1276 at 104 °F and 2500 psi. (Reproduced with permission from reference 33. Copyright 1991 Society of Petroleum Engineers.)

oils as was shown in Chapter 4. The greatly varied effect on foam durability and foamability for a particular Permian Basin crude was documented by Tsau and Heller (29), and the molecular reasons for the ambivalent effect of crudes on surfactants was discussed by Yang and Reed (34).

Of course, the final test by which to assess the requirement that the surfactant should be capable of maintaining lamellae for mobility control in CO_2 floods is the measurement of foam mobility in cores. Because such tests are time-consuming and expensive, the foam durability test has been used as a useful screening tool for more rapid appraisal of surfactants at reservoir conditions.

A second important surfactant characteristic is its adsorption onto the pore walls of reservoir rock. Perhaps as a consequence of the size of the surfactant molecules, adsorption equilibrium is not immediate but requires appreciable time. The reverse process, desorption into a lower-concentration solution, is even slower. Given a long enough period of contact, surfactant molecules can be expected to adsorb onto, and desorb from, the internal walls of a porous rock according to an equilibrium isotherm resembling the Langmuir curve. However, in many laboratory experiments and at the displacement front in the field such equilibrium may not be attained.

The first consequence of this unfulfilled equilibrium on the process in the field is that a preliminary "pad" of surfactant should be injected into the formation prior to the CO_2 foam to satisfy the adsorption and prevent subsequently injected CO_2 and surfactant solution from running ahead of the surfactant-readied zone. A second consequence is that surfactant will not be leached out from the formation immediately when fresh water or brine is injected to follow the CO_2-foam flood.

The slowness of adsorption and especially of desorption also influence laboratory experiments designed to measure the process. These measurements have been performed in several ways. In the first method, a solution containing a known amount of surfactant is circulated by a pump through a loop that contains the rock sample in a suitable core holder and a means of measuring concentration in the flowing stream. As surfactant is adsorbed, the concentration level in the circulating fluid declines. The amount of adsorption can be calculated from the amount of decline and from the known system volume. This method is a variation of the classical method that can be used when the sample is in the form of loose granules. In that case, the whole experiment can be performed in a beaker with a stirring rod replacing the pump. Unfortunately, grinding up a porous rock sample into granules would expose new surfaces for adsorption and thus give erroneously high readings. An example of the use of this circulation method for core samples was an experiment by Kim et al. (35) with a setup that is illustrated in Figure 14. Typical results from the method are shown in Figure 15.

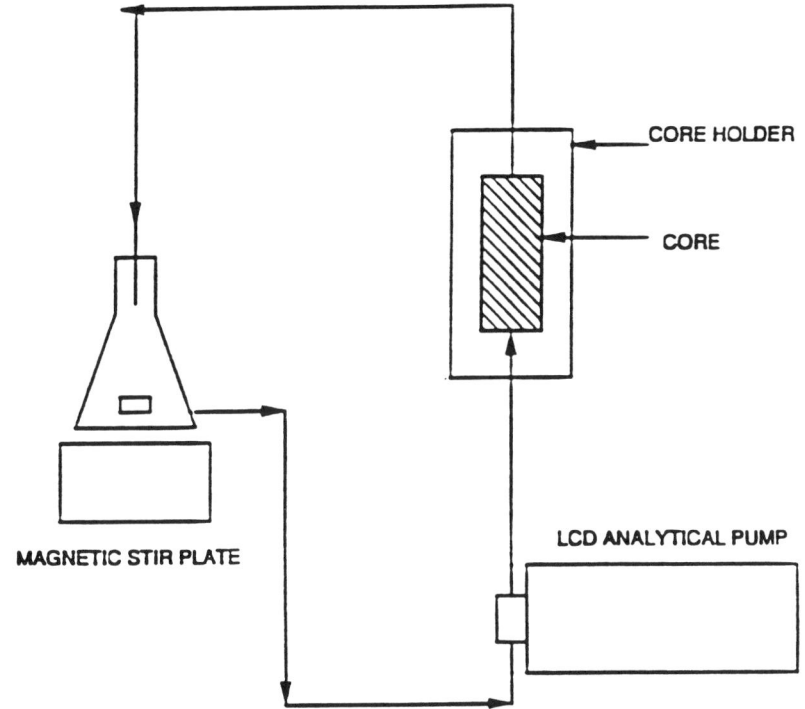

Figure 14. Apparatus for measurement of adsorption of surfactant onto rock core by the recirculation method. (Reproduced with permission from reference 35. Copyright 1992 Society of Petroleum Engineers.)

A second method involves straight-through flow. After brine flow through the core sample is established, a relatively long duration slug of known surfactant concentration is then injected. An on-line detection system continuously measures the surfactant concentration in the outlet stream, and the leading edge of the resulting curve is then integrated to find the amount of surfactant adsorbed. This method is illustrated in Figure 16, and typical results are shown in Figure 17.

In addition to the two main ones, further surfactant selection requirements are concerned mainly with the chemical and thermal stability of the surfactant. It should not decompose, enter into adverse reactions with components of reservoir fluids, or degrade at the high temperature that is characteristic of the environment in which it is expected to maintain its properties for years at a time. The last-named specification is of course impossible to satisfy within reasonable time constraints. Nevertheless, shorter stability tests can be reassuring to some extent. For instance, a solution of the surfactant in the appropriate brine should be maintained at

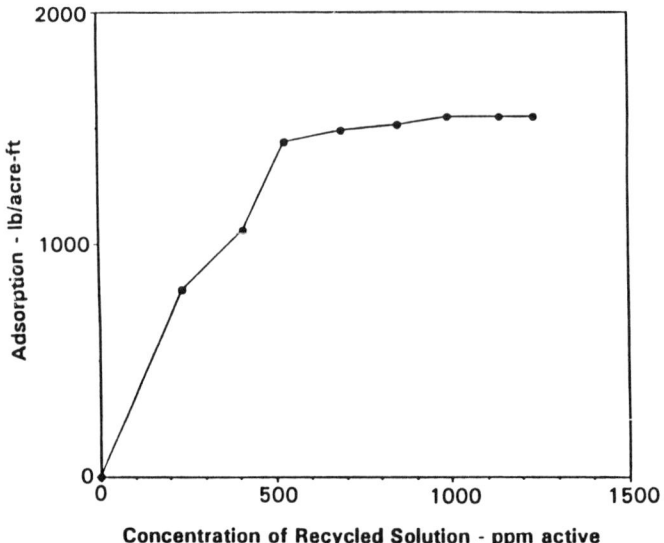

Figure 15. Adsorption isotherm of NES-25 onto dolomitic reservoir rock, obtained from recirculation method. (Reproduced with permission from reference 35. Copyright 1992 Society of Petroleum Engineers.)

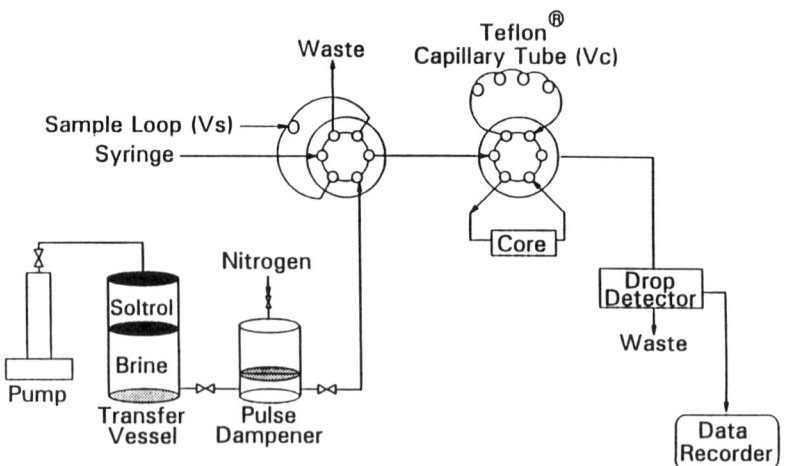

Figure 16. Apparatus for straight-through flow adsorption measurement. (Reproduced with permission from reference 29. Copyright 1992 Society of Petroleum Engineers.)

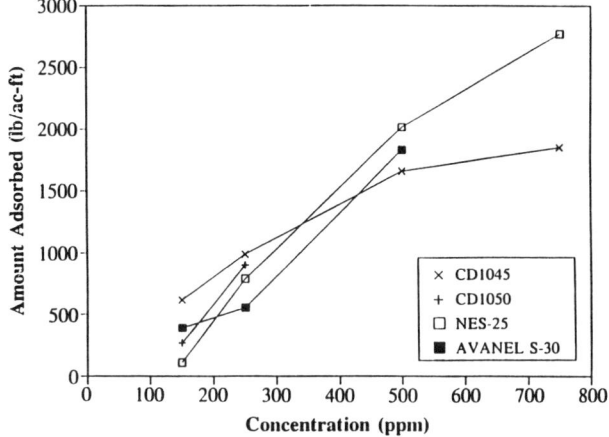

Figure 17. Adsorption isotherms obtained by straight-through apparatus. (Reproduced with permission from reference 29. Copyright 1992 Society of Petroleum Engineers.)

a temperature higher than that in the reservoir for a time as long as convenient and should show no degradation or products of such a process.

Operational Considerations

In addition to all of the problems that determine whether the injected mixture of CO_2 and surfactant solution can perform and survive as a foam in the reservoir rock, the operator of the oil field must design the procedures for injection and other operations in an advantageous way, so that profitability and oil recovery will be maximized. In this pairing of objectives, the oil field's leaseholders and owners are naturally more interested in the best return on their investment than, for instance, in the eventual total recovery. The latter objective, concerned with the overall recovery efficiency, should be the first consideration of the regulating authority. These operational problems entail new factors that transcend those encountered in the laboratory. From the point of view of the operator, the most important considerations are cost and availability of the additional supplies and materials needed, and whether the expected increase in oil production will be more than enough to pay for them.

The manner of injection of the CO_2 is often at question. When no mobility-control additive is used, the choice is usually between continuous CO_2 injection and the so-called WAG (water alternated with gas) method. Continuous CO_2 injection method is generally used only when the reservoir is very tight or water-sensitive, so that water could be injected only at an uneconomically low rate. WAG was originally suggested by Caudle and

Dyes (*36*) and Blackwell et al. (*37*) as a mobility-control method to be used with light hydrocarbon solvent floods. Although the original suggestions were for simultaneous injection of water and solvent, field practicalities changed this to alternation. Because of the low viscosity of CO_2, WAG has become the standard for use in these floods as well, with the only variables being the WAG ratio (volumetric water rate:CO_2 rate) and the WAG cycle time. Most of the CO_2 floods now being operated are WAG floods, often with a cycle time of 1 year and a WAG ratio of 1:1 or 2:1.

If a CO_2 flood is already underway, then the initiation of CO_2-foam mobility control will cause changes in the injection schedule. Because in many cases CO_2 is the major adjustable operating expense, this alone may call for an increase in rate of outlay. On the other hand, industry experience has shown that other things being equal, the rate of oil production is proportional to the rate of injection of CO_2, which would be greater by a factor of 2–3 in a high-CO_2-fraction foam flood over a 2:1 WAG flood. Consequently, this cost increase can be expected to be matched quickly by increasing production.

When CO_2 foam is to be used for mobility control, it would be advantageous to inject (after the surfactant pad) a mixture of about 80% CO_2 and 20% surfactant-containing brine. The operational problem is the same as that faced when the use of WAG was first considered. If the two phases were to be simply pumped simultaneously into the injection well, there would be difficulties of two types. First, because the required injection pressures for the two specified rates would probably be far different from each other, the existing types of pumps would not work well together. Second, because of the appreciable difference in density between CO_2 and brine, the two fluids would probably not enter both upper and lower parts of the formation at the same relative rates. Worse, the extent of such gravity segregation in the wellbore would be unknown.

Without diligent and sophisticated engineering research to enable simultaneous injection to be accomplished simply and with confidence, it seems most reasonable that a cyclic scheme, in which surfactant solution and CO_2 are injected alternately, should be used instead. Such a program can be called SAG, for surfactant alternated with gas. The "SAG ratio" can obviously be set to meet the specifications desired. There remains the question of the cycle time to be used.

The idea behind the compromise that uses alternate instead of simultaneous injection of surfactant is that the volume fraction, and therefore the saturation, of CO_2 should be maintained constant in the region behind the front. Within the pores of a section of the rock in this region, an alternating injection scheme would be expected to cause an oscillating saturation of CO_2 and surfactant solution. The oscillations could be expected to be of higher amplitude for large cycle-time SAG floods than for short cycle-time floods, and this larger swing would cause large cyclic changes of mobility that could have adverse effects. For this reason, as short a cycle time as is practical in a SAG flood should be used. As reported by Martin et al. (35), a 2-week cycle time is used in the CO_2-foam field trial now underway at the East Vacuum Grayburg San Andres Unit by Phillips Petroleum Company and its partners.

Another serious problem that arises in all CO_2 floods, and must be considered here, is corrosion. As is well known, solutions of CO_2 in water dissociate to form carbonic acid. This is a weak acid and does not cause much trouble in fresh water floods. However, in brine instead of pure water, CO_2 does become much more corrosive, and the particular composition of the brine has a great effect. These difficulties can be made worse by the presence of surfactant that can remove coatings emplaced by corrosion inhibitors. This difficulty could occur in both the injection as well as in the production facilities. In many fields, special metallurgy has been needed in critical elements of the hardware. Not much in the way of general principles can be given, except that, in designing a CO_2 flood, the modification of existing problems is to be expected.

When CO_2 foam is used for mobility control, only uneconomic amounts of oil would be expected to remain in the pore space behind a matched-mobility stabilized front. Because of the well-to-well nature of the displacement, however, and because of the presence of reservoir heterogeneities, all segments of the front will not arrive simultaneously at the producers. Consequently, foam will not prevent the breakthrough of the injection fluid before complete recovery. However, the use of CO_2 foam should increase the recovery that will be obtained from the field before its abandonment is mandated by the rising gas and water cut.

Summary

First, some attention was paid to the difference between foam as it exists or flows in containers considerably larger than the cell size, and foam in the pores of reservoir rock. This contrast was drawn to assist in describing the properties of the reservoir foam, which differs from that of more structureless fluids.

The population of bubble-films, or lamellae, in pores was identified as the regulating variable that was responsible for the inability of the high volume fraction fluid (i.e., the gas) to flow at its own viscosity.

An almost irrelevant point of difference in the behavior of foam was whether the discontinuous phase was a gas, a dense gas, or a low-viscosity liquid. The presence of the lamellae across some of the pores gives foam its distinctive properties in porous media.

Numerous examples were given of measured mobilities, and the influence of changing various conditions was explored. The coverage here is not broad and does not include many of the experiments or the theoretical efforts that are described in the literature.

The kinds of different surfactants that are, or have been, of interest for use in CO_2 foams have also been referred to with special emphasis on the problems of measurement of foamability and of the property by which surfactants adsorb onto reservoir rock. A current research question concerns the apparent ability of certain surfactants to generate CO_2 foams that exhibit selective mobility reduction (SMR). The mobility is reduced by a greater fraction in high than in low permeability rocks by such a foam. This property thereby promises to compensate, at least partially, for the natural heterogeneity of reservoirs.

Finally, a survey of the operational problems encountered in the use of CO_2 foam for mobility control was given. This chapter is written to serve merely as an introduction to the field. An engineer responsible for research or for a field project should obviously dig much more deeply into the subject.

List of Symbols

A	area (cm^2 or m^2)
k	permeability (darcy or m^2)
L	length (cm or m)
m_{rf}	mobility reduction factor (dimensionless)
P	pressure (atm or Pa)
Q	flow rate (cm^3 or mL)
r	bubble radius (cm or m)
R_f	resistance factor (dimensionless)
α	proportionality constant (dimensionless)
λ	mobility [darcy/cP or $m^2/(Pa \cdot s)$]
λ_r	relative mobility [cP^{-1} or $(Pa \cdot s)^{-1}$]
μ	viscosity (cP or Pa·s)
μ_{eff}	effective viscosity (cP or Pa·s)
σ	surface or interfacial tension (dynes/cm or mN/m)
τ_{yield}	yield stress (dynes/cm^2 or Pa)

Acknowledgments

I acknowledge the cooperation of Dave Martin and Jyun-Syung Tsau of the Petroleum Recovery Research Center for consultations during preparation of this chapter. I also acknowledge several anonymous reviewers, as well as Laurier Schramm, K. Allbritton, and Jyun-Syung Tsau for their careful reading and checking of this chapter.

References

1. Ely, J. F.; Hanley, J. M. *TRAPP—Transport Properties Prediction [Computer] Program;* Thermophysical Properties Division, National Institute of Standards and Technology: Boulder, CO, 1987.
2. Chung, F.; Nguyen, H. *Carbon Dioxide Thermodynamic Properties [Computer] Program;* National Institute for Petroleum and Energy Research: Bartlesville, OK, 1985.
3. Hutchinson, C. A.; Braun, P. H. *AIChE J.* **1961**, *7(1)*, 64–74.
4. Holm, L. W.; Josendal, V. A. *J. Pet. Technol.* **1974**, *16*, 1427–1438.
5. Heller, J. P.; Taber, J. J. *Proceedings of the Permian Basin Oil and Gas Recovery Conference;* Society of Petroleum Engineer: Richardson, TX, 1986; paper SPE 15001.
6. Weinaug, C. F.; Ling, D. U.S. Patent 2 867 277, 1959.
7. Hirasaki, G. J.; Lawson, J. B. *SPE J.* **1985**, *25*, 176–190.
8. Prud'homme, R. K. *Abstracts of Papers,* 53rd Annual Meeting of the Society of Rheology, Louisville, KY; Society of Rheology: New York, October 1981; paper E-7.
9. Princen, H. M. *J. Colloid Interface Sci.* **1983**, *91*, 160–175.
10. Kraynik, A. M. *Abstracts of Papers,* 54th Annual Meeting of the Society of Rheology, Evanston, IL; Society of Rheology: New York, October 1982; paper B-10.
11. Kuntamukkula, M. S.; Heller, J. P. *Ind. Eng. Chem. Res.* **1987**, *26*, 318–325.
12. Bretherton, F. P. *J. Fluid Mech.* **1961**, *10 (Part 2)*, 166–188.
13. Boys, C. V. *Soap Bubbles—Their Colours and Forces Which Mold Them;* Dover: New York, 1959; (reprint of 1911 work).
14. Bikerman, J. J. *Foams;* Springer Verlag: New York, 1973.
15. Bond, D. C.; Holbrook, O. C. U.S. Patent 2 866 507, 1958.
16. Khatib, Z. I.; Hirasaki, G. J.; Falls, A. H. *SPE Reservoir Eng.* **1988**, *40*, 919–926.
17. Friedmann, F.; Jensen, J. A. *Proceedings of the California Regional Meeting of the Society of Petroleum Engineers;* Society of Petroleum Engineers: Richardson, TX, 1986; paper SPE 15087.
18. Falls, A. H.; Gauglitz, P. A.; Hirasaki, G. J. *Proceedings of the SPE/DOE Fifth Symposium on Enhanced Oil Recovery;* Society of Petroleum Engineers: Richardson, TX, 1986; paper SPE/DOE 14961.
19. Rossen, W. R; Gauglitz, P. A. *AIChE J.* **1990**, *36*, 1176.
20. Zhou, Z.; Rossen, W. R. *Proceedings of the SPE/DOE Eighth Symposium on Enhanced Oil Recovery;* Society of Petroleum Engineers: Richardson, TX, 1992; paper SPE/DOE 24180.
21. Hanssen, J. E.; Dalland, M. *Proceedings of the SPE/DOE Seventh Symposium on Enhanced Oil Recovery;* Society of Petroleum Engineers: Richardson, TX, 1990; paper SPE/DOE 20193.

22. Heller, J. P.; Lien, C. L.; Kuntamukkula, M. S. *SPE J.* **1985**, *25*, 603–613.
23. Bernard, G. G.; Holm, L. W. *SPE J.* **1964**, *4*, 267–274.
24. Marsden, S. S.; Khan, S. A. *SPE J.* **1966**, *6*, 17–25.
25. Holm, L. W. *SPE J.* **1968**, *8*, 359–369.
26. Raza, S. H. *SPE J.* **1970**, *10*, 328–336.
27. Chiang, J. C.; Sanyal, S. K.; Castanier, L. M.; Brigham, W. E.; Sufi, A. *Proceedings of the 50th Annual California Regional Meeting of SPE;* Society of Petroleum Engineers: Richardson, TX, 1980; paper SPE 8912.
28. Bernard, G. G.; Holm, L. W.; Harvey, C. P. *SPE J.* **1980**, *20*, 281–292.
29. Tsau, J-S.; Heller, J. P. *Proceedings of the SPE Permian Basin Oil and Gas Recovery Conference;* Society of Petroleum Engineers: Richardson, TX, 1992; paper SPE 24013.
30. Lee, H. O.; Heller, J. P. *Proceedings of the SPE/DOE Sixth Symposium on Enhanced Oil Recovery;* Society of Petroleum Engineers: Richardson, TX, 1988; paper SPE 17363.
31. Lee, H. O.; Heller, J. P. *Surfactant-Based Mobility Control: Progress in Miscible-Flood Enhanced Oil Recovery;* Smith, D. H., Ed.; ACS Symposium Series 373; American Chemical Society: Washington, DC, 1988; Chapter 19, pp 375–386.
32. Lee, H. O; Heller, J. P. *Oil-Field Chemistry: Enhanced Recovery and Production Stimulation;* Borchardt, J. K.; Yen, T. F., Eds.; ACS Symposium Series 396; American Chemical Society: Washington, DC, 1989; Chapter 27, pp 502–517.
33. Lee, H. O.; Heller, J. P.; Hoefer, A. M. W. *SPE Reservoir Eng.* **1991**, *6(4)* 421–428.
34. Yang, S. H.; Reed, R. L. *Proceedings of the 64th Annual Technical Conference of the SPE;* Society of Petroleum Engineers, Richardson, TX, 1989; paper SPE 19689.
35. Martin, F. D.; Heller, J. P.; Weiss, W. W.; Tsau, J.-S.; Zornes, D. R.; Sugg, L. A.; Stevens, J. E.; Kim, J. E. *Proceedings of the SPE/DOE Eighth Symposium on Enhanced Oil Recovery;* Society of Petroleum Engineers: Richardson, TX, 1992; paper SPE/DOE 24176.
36. Caudle, B. H.; Dyes, A. B. *Trans. AIME* **1958**, *213*, 281–284.
37. Blackwell, R. J.; Terry, W. M.; Rayne, J. R.; Lindley, D. C.; Henderson, J. R. *Trans. AIME* **1960**, *219*, 293.

RECEIVED for review October 14, 1992. ACCEPTED revised manuscript February 25, 1993.

Steam-Foams for Heavy Oil and Bitumen Recovery

E. E. Isaacs, J. Ivory, and M. K. Green

Alberta Research Council, Oil Sands and Hydrocarbon Recovery,
P.O. Box 8330, Station F, Edmonton, Alberta T6H 5X2, Canada

> *Steam-based processes in heavy oil reservoirs that are not stabilized by gravity have poor vertical and areal conformance, because gases are more mobile within the pore space than liquids, and steam tends to override or channel through oil in a formation. The steam-foam process, which consists of adding surfactant with or without noncondensible gas to the injected steam, was developed to improve the sweep efficiency of steam drive and cyclic steam processes. The foam-forming components that are injected with the steam stabilize the liquid lamellae and cause some of the steam to exist as a discontinuous phase. The steam mobility (gas relative permeability) is thereby reduced, and the result is in an increased pressure gradient in the steam-swept region, to divert steam to the unheated interval and displace the heated oil better. This chapter discusses the laboratory and field considerations that affect the efficient application of foam.*

Properties of Surfactants at Elevated Temperature

Both the effectiveness and the economics of steam-foam processes depend critically on surfactant losses, and it is therefore essential to minimize losses due to the chemical and physical phenomena occurring in the reservoir at elevated temperatures.

Thermal Stability. The thermal stability of surfactants has been investigated in a number of studies (*1–5*). Surfactants with sulfate moieties decompose rapidly at temperatures above 100 °C, and surfactants stable above 200 °C have, almost exclusively, sulfonate groups. A large number of anionic sulfonate surfactants are thermally stable in the range 100–300 °C (*3, 6*). Sulfonates decompose via hydrolytic desulfonation of the sulfonate group. The decomposition of the sulfonate produces acid, and the reaction proceeds according to autocatalytic kinetics (*5*) as

$$ArSO_3^- + 2H_2O \xrightarrow{H^+} ArH + SO_4^{2-} + H_3O^+ \qquad (1)$$

where Ar represents an alkylaryl group. At low pH and temperatures near 200 °C, this reaction is relatively fast and leads to substantial decomposition within hours or days, but hydrolysis is almost completely inhibited at basic pH. For example, a petroleum sulfonate was only about 10% decomposed after 28 days at pH 9 (not buffered) and 205 °C (*2*). Thermal stability of sulfonates increases in the order (*7*):

petroleum sulfonates < alpha-olefin sulfonates < alkylarylsulfonates

p-Dialkylarylsulfonates are more stable than the meta-isomers (*8*), and one material showed essentially no degradation after 10 days at 299 °C and pH 4 (*9*). The effect of pH (3.1 to 11.0) on the decomposition of an alkylarylsulfonate (Suntech IV) surfactant at 299 °C is shown in Figure 1. Surfactant thermal stability tests carried out in the presence of reservoir sand show that the reservoir rock had little or no effect on the chemical reactivity of the surfactant (*1*).

Retention in Porous Media. Anionic surfactants can be lost in porous media in a number of ways: adsorption at the solid–liquid interface, adsorption at the gas–liquid interface, precipitation or phase-separation due to incompatibility of the surfactant and the reservoir brine (especially divalent ions), partitioning or solubilization of the surfactant into the oil phase, and emulsification of the aqueous phase (containing surfactant) into the oil. The adsorption of surfactant on reservoir rock has a major effect on foam propagation and is described in detail in Chapter 7 by Mannhardt and Novosad. Fortunately, adsorption in porous media tends to be, in general, less important at elevated temperatures (*10, 11*). The presence of ionic materials, however, lowers the solubility of the surfactant in the aqueous phase and tends to increase adsorption. The ability of cosurfactants to reduce the adsorption on reservoir materials by lowering the critical micelle concentration (CMC), and thus the monomer concentration, has been demonstrated (*12, 13*).

Both the precipitation and partitioning of anionic surfactants increase with increasing temperature. For a C_{16}–C_{18} alkylaryl surfactant, such as Suntech IV, surfactant losses due to partitioning were in the range

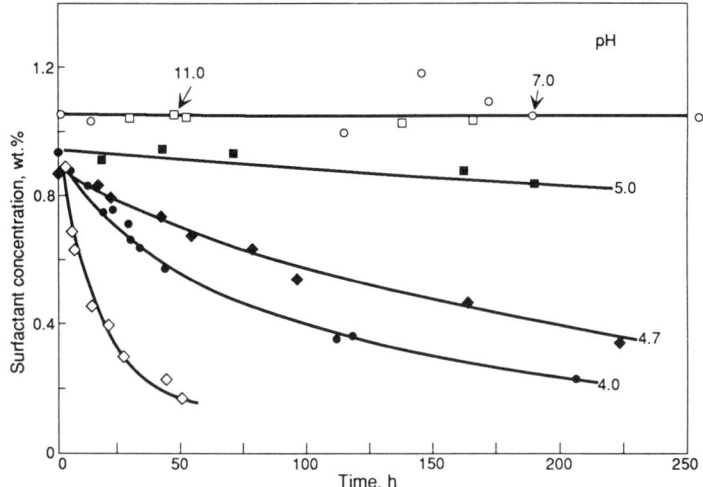

Figure 1. Effect of pH on the rate of reaction at 299 °C. (Data are taken from reference 5.)

of 25–40% at 125 °C (*14*) and 20–30% at 205 °C (*1*) depending on initial surfactant concentration. As a general guide, partitioning and precipitation of surfactant increases with increasing temperature, increasing brine concentration, and increasing surfactant chain length. However, the phase behavior of surfactants is a complex phenomenon, and the interaction of divalent (and trivalent) cations with sulfonate surfactants causes surfactant precipitation followed by dissolution of the precipitate at higher concentrations(*15*). The precipitate redissolution phenomenon is not observed with monovalent ions.

The emulsification of the aqueous phase in oil is likely to have a detrimental effect only in situations where the heavy oil phase is also flowing in porous media. Reservoir applications of foams invariably assume that the oil phase is at residual saturation (S_{or}). Consequently, emulsification is not generally considered a serious problem in most foam applications.

Interfacial Tension Behavior. Reduction in the residual oil saturation over and above that obtained by steam injection is desirable and, in many heavy oil reservoirs, essential to ensure efficient foam formation during application of steam-foam processes (*13*). The extent of heavy oil desaturation is, however, dependent on the reduction in interfacial tension between oil and water. Thus, foam-forming surfactants can improve their own cause by reducing interfacial tensions at steam temperature.

Low interfacial tensions (<0.1 mN/m) are, in general, difficult to maintain at elevated temperatures. At a given NaCl concentration, Handy et al. (*16*) observed little or no temperature dependence (25–180 °C) for interfacial tensions (IFT) between California heavy oil and aqueous sulfonate surfactants. In contrast, for a pure hydrocarbon or mineral oil and the same surfactant systems, an abrupt decrease in interfacial tension was observed at temperatures in excess of 120 °C (*8*). In a similar manner, Babu et al. (*17*) obtained little effect of temperature on interfacial tensions; however, values of about 0.02 mN/m were obtained for a light crude (39 °API), and were about an order of magnitude lower than those observed for a heavy crude (14 °API) with the same aqueous formulations.

For Clearwater (Canada) and Karamay (China) heavy oils, Isaacs et al. (*18*) observed that at a given NaCl concentration with Suntech IV surfactant, an increase in temperature (25–200 °C) resulted in an increase in interfacial tension. In contrast, in the presence of relatively high amounts of divalent ion, interfacial tensions decreased with increasing temperatures (Figure 2). Thus, in designing steam-foam processes, the range of temperature and water chemistry to be encountered in the reservoir is important. The use of surfactants that reduce the oil–water interfacial tension in the design of a steam-foam process is discussed in a following section (Prefoam Slug Technology).

Foam Stability and Propagation at Elevated Temperatures

Bubble Coalescence. The stability of a foam at any temperature depends on the balance between bubble-forming and bubble-destroying

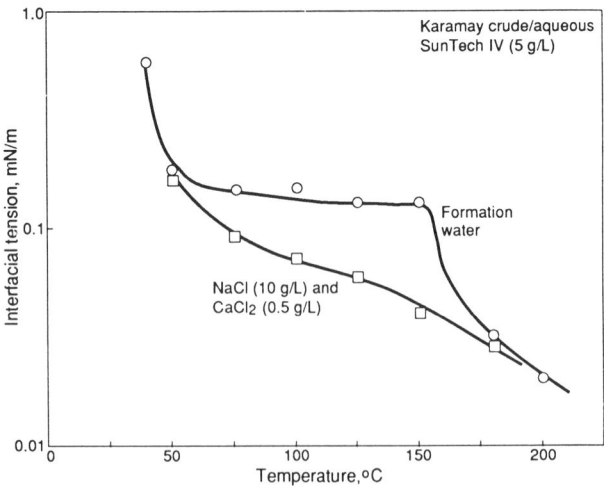

Figure 2. Effect of temperature on interfacial tension of Karamay crude oil in synthetic brine and formation water (0.25 g/L divalent and 2 g/L monovalent cations).

processes. Bubble coalescence occurs as the film between bubbles drains and thins to the point where the film becomes unstable and ruptures (see Chapter 3). A number of complex interactions are involved, but any process that impedes the flow of fluid from the film to the borders or increases the repulsion between the two gas–fluid interfaces will enhance the stability of the foam toward coalescence.

Any foaming surfactant has an optimum performance range, above which its ability to reduce gas-phase mobility drops off. In some cases, this decrease in blocking ability correlates very well with decreasing water viscosity (19). Robin (20) reported that adding polymer to increase the liquid–phase viscosity improved high-temperature foaming performance. Foaming ability at high temperature varies for different classes of sulfonate surfactants; alpha-olefin sulfonates generally produce more effective blocking than alkylarylsulfonates. Within a class, foaming performance at high temperature appears to increase with the length of the alkyl chain. This increase in the high-temperature limit to performance is accompanied by an increase in the low-temperature limit and a decrease in solubility at low temperature.

The ability of a foam to resist coalescence may be assessed by measuring the coalescence pressure, that is, the pressure required to force two bubbles to coalesce. Figure 3 illustrates the effect of NaCl brine on the coalescence of two bubbles at 125 °C measured in a spinning drop apparatus according to methods developed by Flumerfelt (21).

Effect of Noncondensible Gas. Theoretical considerations (22), core studies (23, 24), and field data (25, 26), have demonstrated that foams are more effective when noncondensible gas is present even in small concentrations. The major attraction of injecting surfactants without a noncondensible gas is economic. Steam-foams without noncondensible gases are viable in reservoirs where channeling is relatively close to the wellbore and heat losses are small, as well as in reservoirs having substantial gas production. The higher pressure inside small bubbles creates a tendency for the smaller bubbles to shrink and larger ones to grow. In a noncondensible gas, the gas phase must diffuse through the liquid lamellae to alter bubble size. However, in steam-foams, bubble size may be altered by a process of condensation into small bubbles, accompanied by evaporation into larger bubbles. This process occurs much more quickly, because it relies on heat, rather than mass, transfer through the liquid films. In foams at high temperature, the stabilizing effect of noncondensible gas is thus attributed to its ability to increase elastic stability of lamellae and slow condensation within the bubble. The critical amount of noncondensible gas required to stabilize steam-foams has been predicted (22) and is dependent on temperature and pore radius, as shown in Figure 4.

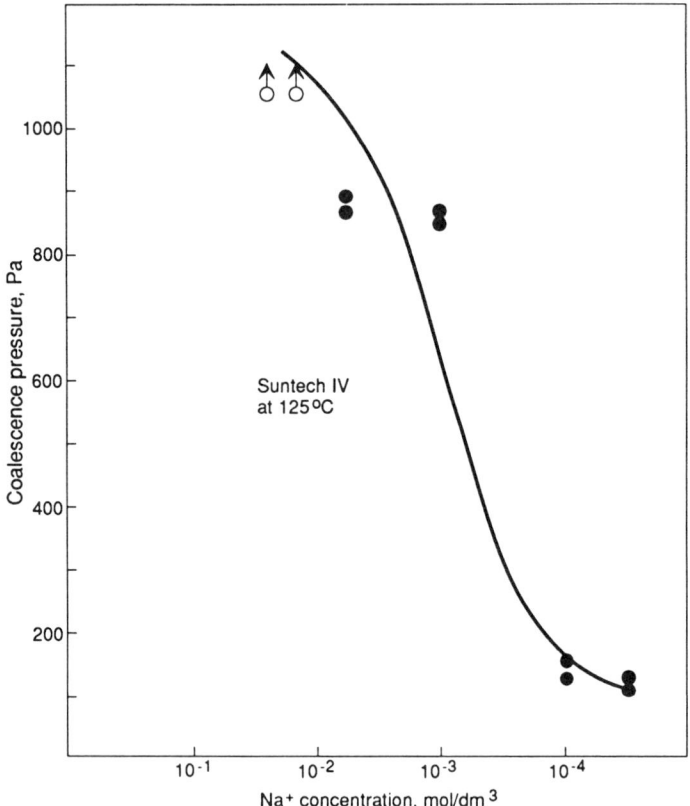

Figure 3. Effect of Na^+ concentration on bubble–bubble coalescence at 125 °C (open circles represent conditions under which no coalescence took place at the maximum pressure imposed).

Flow Resistance of Foams in Porous Medium. Foam flows in the reservoir by "making" and "breaking" processes. Because of its dispersed nature, foam exhibits low flow mobilities depending on texture (bubble size and distribution) in relation to the imposed injection and reservoir conditions (27). To perform successfully, the foam must withstand a vast range of conditions and still maintain its integrity and flow resistance properties. For a given surfactant and reservoir type, therefore, it is important to consider a range of gas and liquid velocities, and liquid volume fraction (LVF).

The effect of gas velocity at a constant liquid velocity on the relative permeability to gas (k_{rg}) is shown in Figure 5, for a sand pack of 2-darcy absolute permeability containing residual oil after steam-flooding (15 pore volume) at 180 °C (28). The k_{rg} is a measure of foam flow resistance; the

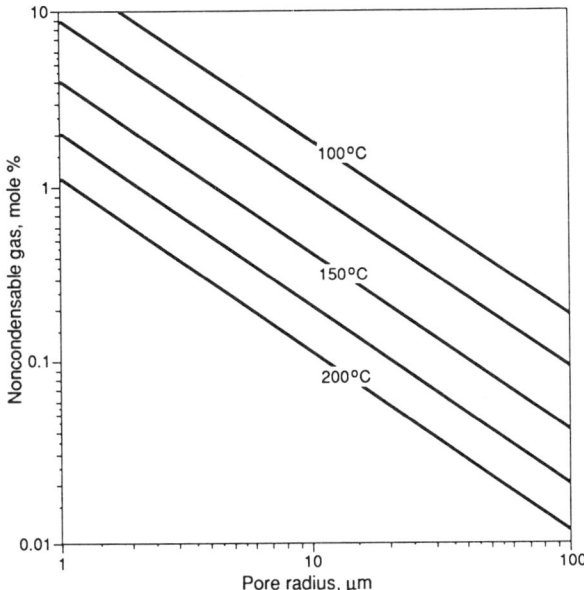

Figure 4. Noncondensible gas required to steam-foam bubbles as a function of average pore size and temperature. (Reproduced with permission from reference 22. Copyright 1988 Society of Petroleum Engineers.)

higher the k_{rg}, the lower the foam resistance. Clearly the foam flow resistance decreases with increasing gas velocity. Because the liquid velocity has been held constant, the LVF of the injected foam (given in parenthesis) is changing. As the LVF decreases, the foam becomes drier and its resistance to flow declines. This result is consistent with other work (27, 29) carried out at ambient temperatures.

Figure 6 shows the reverse case where the gas is held constant and the liquid velocity is changed (28). The resistance of the foam (decreasing k_{rg}) increases with increasing liquid velocity. As before, the wetter the foam (higher LVF), the more resistant it is to flow. When k_{rg} is plotted against LVF (not shown), it is apparent that the foam flow resistance is not a unique function of LVF. This finding is in agreement with the work of Persoff et al. (27).

Figure 7 shows the effect of a simultaneous change in gas and liquid velocities keeping the LVF constant (0.15), where k_{rg} is plotted against fluid velocity at two surfactant concentrations (28). At high velocities, a shear-thinning effect occurs and is associated with "flowing foam"; whereas, at low velocities the foam is much more resistant to flow with a tendency toward being a "trapped foam". As depicted in the photographs taken of bulk-foam exiting the sand pack and included in Figure 7, the

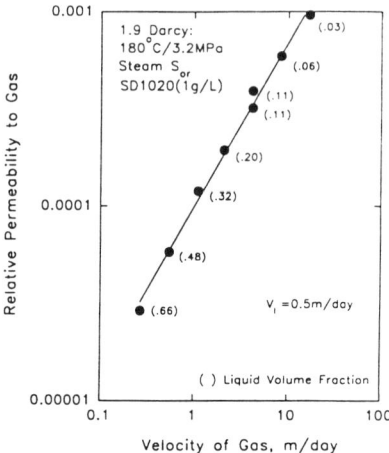

Figure 5. Effect of gas velocity on the foam flow resistance.

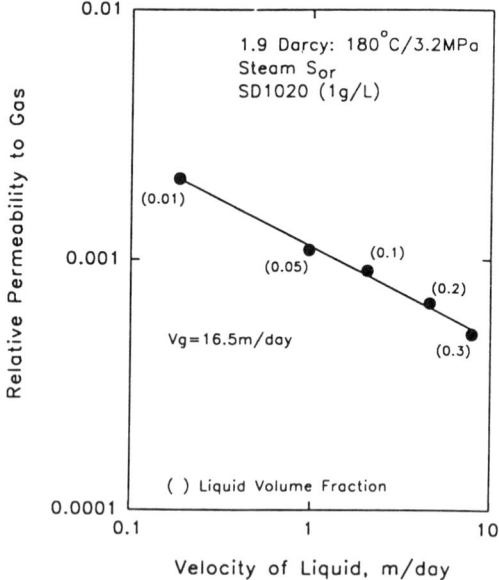

Figure 6. Effect of liquid velocity on the foam flow resistance.

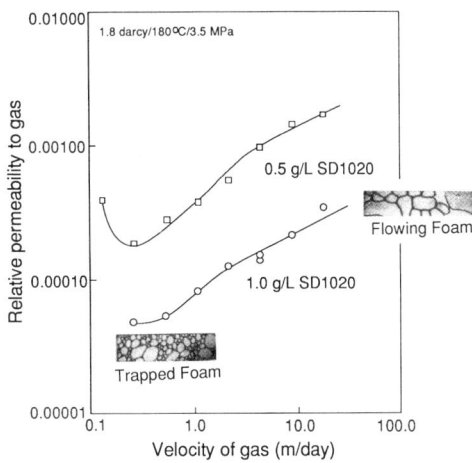

Figure 7. Effect of surfactant concentration and fluid velocity at constant LVF on the foam flow resistance.

difference between the "flowing" and "trapped" foams is the texture. The higher the surfactant concentration, the greater the foam resistance at all flow rates. Thus, if desired, the surfactant concentration can be used to compensate for the effect of fluid velocity. Furthermore, foams can withstand drastic changes in fluid velocity, and their velocity dependence is exactly what is required in the reservoir, that is, relatively low resistance near the injection well, where velocities are high, and higher resistance further into the reservoir, where velocities are relatively low.

Steam-Foam Performance in Laboratory Cores

Constant Flow Rate. Classic experimental evaluation of mobility reduction due to steam or gas foams involves injecting gas and liquid at constant flow rate and monitoring the pressure drop behavior across several subsections (to avoid capillary end effects). Although the results of core-flood studies have been positive, there is concern regarding their applicability to field situations. Under constant flow-rate conditions, the diversion of steam that follows surfactant injection is accompanied by a large increase in pressure drop across the core. It is unreasonable to expect that such a large increase in pressure gradient would occur in a reservoir. In fact, if the decrease in mobility caused by surfactant resulted in a drop in flow rate rather than an increased change in pressure (ΔP), the foam could collapse, because conditions may fall below a minimum flow velocity for foam formation (*30*). An additional concern in constant flow experiments is that when saturated steam is injected and a constant back-pressure is used, an increase in ΔP leads to an increase in injection pressure, which results in a higher injection temperature. Thus the effects of higher temperature are mixed in with the effects of diversion.

Constant ΔP. Injection of steam under constant pressure-drop conditions provides a better representation of steam injection into a well than constant flow rate, because field injection pressures will be limited by the temperature–pressure capabilities of the injection system in relation to the reservoir volume. Implementation is somewhat more complicated than for a constant flow experiment because rates of addition of surfactant solution, steam, and noncondensible gas must be varied in concert to maintain the set point ΔP, but computer control simplifies the task greatly. In this case, mobility reduction is indicated by a drop in flow rate, rather than an increase in ΔP. Generally, similar trends in mobility reduction with parameters such as surfactant concentration are observed (*31*). However, as illustrated in Figure 8, the danger of an excessively strong foam is highlighted; blocking was strong enough to limit flow into the system as a whole to the point where oil production from the core was reduced.

Cyclic Steam. The idea of using foams in cyclic-steam operations to control injection profiles is well understood and has been demonstrated in the field (next section). However, the role of a foaming surfactant dur-

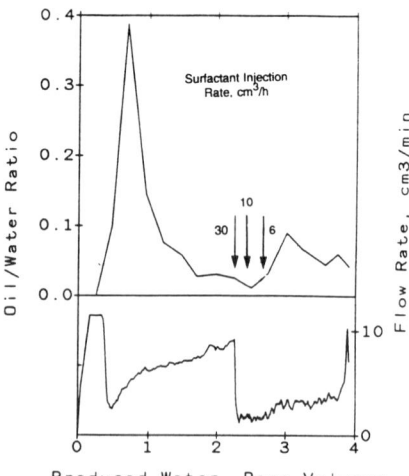

Figure 8. Produced oil–water ratio and flow rate during a steam followed by steam-surfactant injection into an Athabasca core.

ing the drawdown phase of a cyclic process is less clear. Steam-foam could conceivably play a beneficial role by restricting the flow of flashing steam into the well, with a consequent improvement in water–oil ratio. Strong foam formations, however, could be detrimental, by impeding flow of all fluids to the well.

When surfactant is injected with steam, in the absence of noncondensible gas, into an oil sand core while the production pressure is cycled, some mobility reduction due to foam is observed during the drawdown phase, but the greatest effect is seen during periods of steady flow (Figure 9) *(32)*. This result suggests that at least a portion of the recovery enhancement seen during cyclic operations in the field may be attributed to foaming effects during drawdown.

Prefoam Slug Technology. All foaming surfactants are sensitive to oil (Chapter 4). There is considerable interest in developing surfactants with improved oil resistance and in developing strategies to overcome this difficulty. One strategy that appears to have potential is the injection of a prefoam slug to mobilize residual oil ahead of the foaming surfactant and to allow for rapid foam formation *(27)*. As an example, Figure 10 compares three steam-foam experiments in 2-darcy-packs that contain Cold Lake bitumen and have been steam-flooded to residual oil saturation (S_{or}). In the absence of a prefoam slug, the foam has taken about 10 pore volumes (PV) to generate, even using an oil-tolerant surfactant (Chaser SD1020). Injection of a 0.25PV slug of either diesel solvent or a surfactant that reduces interfacial tension (IFT) has considerably hastened the evolution of the foam. The diesel solvent is miscible with the residual oil, reducing its viscosity and hence making it more amenable to mobilization

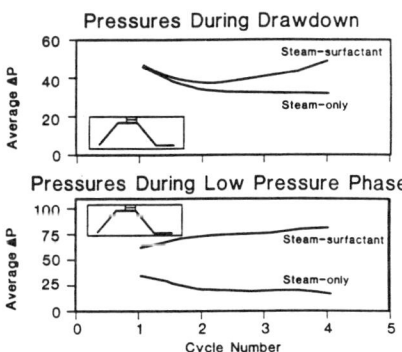

Figure 9. Comparison of pressure drop (ΔP) behavior during the drawdown and low-pressure phases, between steam-only and steam–surfactant runs using Athabasca oil sand cores. The inset indicates the period of surfactant injection.

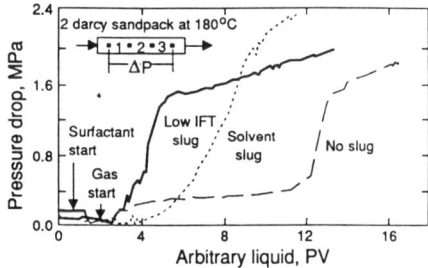

Figure 10. The effect of prefoam slug injected prior to foam formation on the rate of foam propagation.

by steam. The low-IFT slug was designed to mobilize the residual oil by reducing oil–water interfacial tensions below 10^{-2} mN/m at run temperatures.

An alternative approach to minimizing the effect of residual oil is the injection of air with steam ahead of the foam front. Air converts the residual oil to coke; the result is improved mobility reduction behavior of the ensuing steam-foam (*33*).

Field Experience with Steam-Foams

Foam has been used in field applications involving both cyclic and steam-drive processes. Many of the steam-foam tests have been performed in Kern County, California, where most of the U.S. heavy oil is produced. In many situations, foam has successfully increased both volumetric sweep efficiency and oil recovery rates (*34*). Generally, the application of foam has been considered to be a technical success but economically suspect.

The surfactant has been injected as high-concentration slugs [10 wt% active (*35*)] or continuously at a lower concentration (0.1–1.0 wt% active). Noncondensible gas (at a concentration of 0.5–1.0 mol% in the gas phase) or NaCl (1–4 wt% in the aqueous phase) may be coinjected with the surfactant and steam in order to stabilize the foam. Figure 4 can be used to estimate the minimum concentration of noncondensible gas required. Extra water may be added to the steam to maintain the liquid volume fraction at the desired value (typically above 0.01).

Application of Foams. Foam injection is a simple addition to a steam process. It basically requires a surfactant storage tank or drums, a surfactant pump, a gas compressor, and methods of measuring surfactant-injection rates(e.g., changes in liquid level) and gas -injection rates (e.g., orifice meter). A schematic of a foam-injection system used in one field test (*36*) is shown in Figure 11.

A number of tests are used to evaluate the effectiveness of foam in field applications. It is crucial that stable baseline measurements for steam-only injection be obtained so that the effect of foam can be determined. In addition, influences from wells outside the test pattern should be minimized and estimated.

Oil production generally increases as a result of the improved volumetric sweep caused by foam injection. Incremental oil production from down-dip wells is likely to be higher than that from up-dip wells because of a reduction in steam override. Changes in the steam–oil ratio (SOR) can provide a reliable indication of foam blocking if steam-drive is the primary displacement method.

As a result of an increased flow resistance, foam injection is often accompanied by an increase in the injection-well bottomhole pressure (BHP). Temperature surveys at injection, production, and observation wells indicate whether foam is successfully diverting steam. By minimizing gravity override, a successful foam application increases the temperature

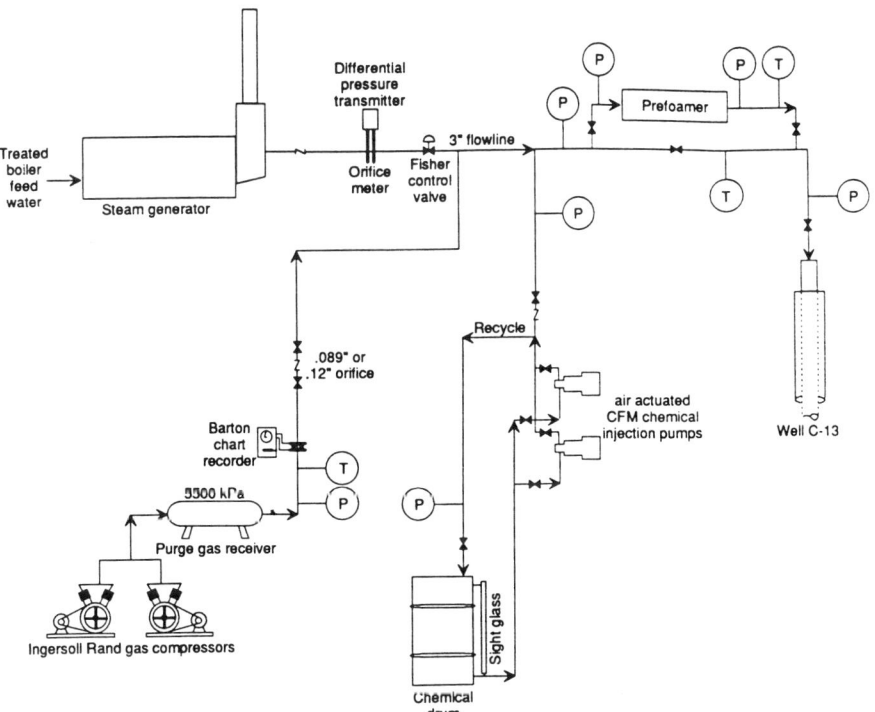

Figure 11. Schematic of foam-injection system.

near the bottom of the formation. In addition, a decrease in steam channeling is indicated by a reduction in the temperature at the well to which steam previously channeled and by an increase in the temperature at other production wells. The temperature difference (postfoam minus prefoam) at various depths may be integrated over the depth to indicate the effect of foam at a particular observation well (37). Temperature surveys at observation wells at the Gregoire Lake In Situ Steam Pilot (GLISP) site (38) showed that foam successfully reduced the effects of gravity override and diverted steam downward. As a result, the temperature at the center of the oil-rich zone [200 meters from Kelley Bushing (mKB)] increased from 65 °C to 145 °C over the first 74 days of foam injection (Figures 12 and 13).

A velocity shot survey is used to determine if the steam flow pattern is being altered near the well-bore. This survey involves the injection of cold water at the surface and the injection of a water-based tracer (e.g., iodine-131) down-hole. From the time it takes the tracer to travel between two detectors, it is possible to estimate the perforation depths at which the water preferentially enters the formation.

An injection profile survey (39) involves the injection of a radioactive tracer (e.g., iodine-131) at the surface. The well-bore is subsequently logged using a gamma-ray tool. The retained radioactivity adjacent to the well at a given depth is assumed to be proportional to the volume of the liquid or gas (depending on the tracer) entering the formation at that depth. Retained radioactivity lasts only for a few minutes, so logging must be performed quickly after injection of the tracer. These profiles may be in error if some of the retained reactivity is due to adsorption or chemical reactions.

Neutron logs (37) have been used to determine changes in the liquid

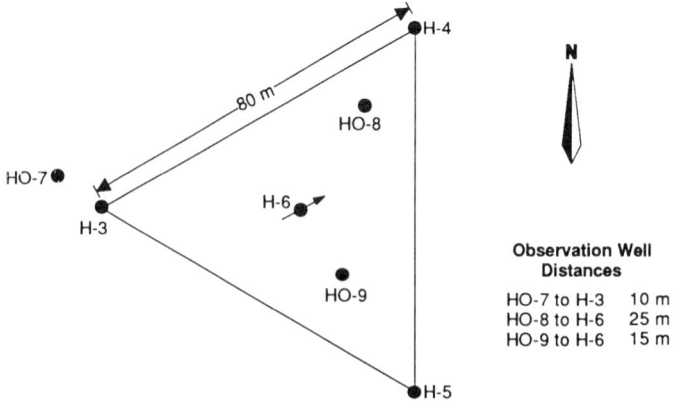

Figure 12. GLISP well configuration (reference 38).

6. ISAACS ET AL. *Steam-Foams for Heavy Oil and Bitumen Recovery*   249

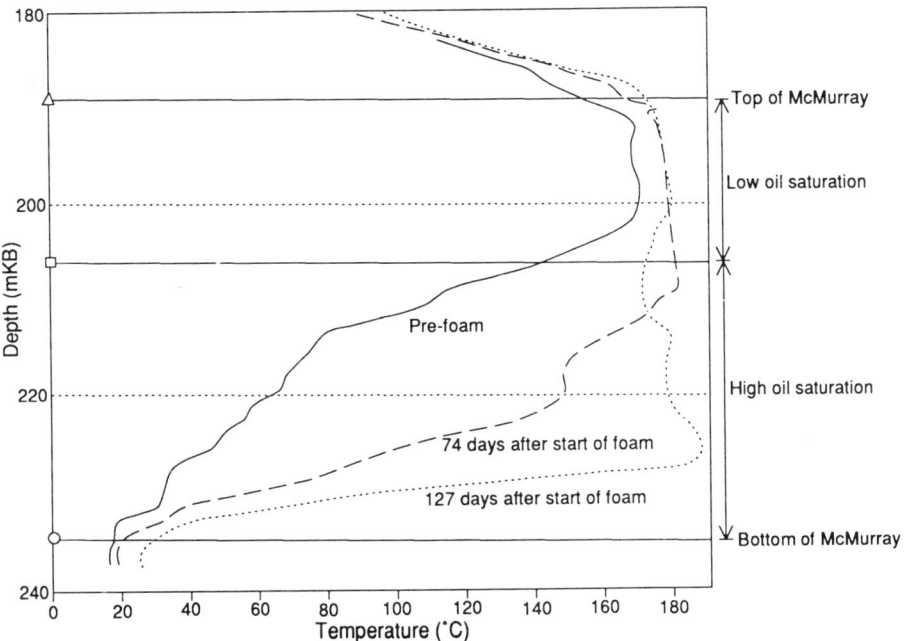

Figure 13. Well HO9 temperature profile (reference 38).

saturation near observation wells. A reduction in liquid saturation is an indication of foam propagation. Carbon–oxygen logs provide an estimate of oil saturation (*39*).

Steam diversion may also be indicated by changes in the produced water composition. For example, if the Cl⁻ ion concentration in the injected aqueous phase is less than that in the formation, then an increasing Cl⁻ ion concentration indicates less channeling and greater contact with "fresh" reservoir. Changes in the produced gas composition (e.g., CO_2 or CH_4 concentrations) may also indicate steam diversion.

Tracers (*40, 41*) are used primarily to estimate volumetric sweep and to locate steam channels. The tracer should not react with the reservoir. The four main categories of tracers are radioisotopes (e.g., tritium and krypton), which are used in steam floods; salts (e.g., sodium bromide and sodium nitrate); fluorescent dyes (e.g., uranian); and water-soluble alcohols (e.g., methanol, ethanol, and 2-propanol). The surfactant itself may be used as a tracer. Its concentration at production wells may be determined by using methods such as liquid chromatography, colorimetry, and titration.

In addition to these tests, pressure falloff and buildup tests may be performed to determine the effect of foam on steam-zone volume and transmissibility. The pressure falloff test involves monitoring the decay of the BHP after the end of the injection period. In a buildup test, the increase in BHP is recorded after the end of the production period. The rate of pressure change in both tests depends on the fluid transmissibility in the reservoir.

Cyclic Processes. During the first cycle, steam follows the path of least resistance (e.g., high permeability and low oil saturation) and preferentially depletes portions of the reservoir. In subsequent cycles, the flow resistance is even lower in the depleted region. Thus, in every cycle, the steam will preferentially flow to the depleted zone. By significantly increasing the flow resistance in this zone, foam diverts the steam to regions of higher oil saturation. This action results in higher oil recovery rates. As a result of steam diversion, the volume of the depleted zone increases with each cycle. Therefore, the amount of foam injected in each cycle must be increased to maintain satisfactory oil production rates.

For cyclic steam-foam injection, it is important that the foam breaks down in the presence of oil or after prolonged exposure to high temperature. In this way, the resistance to the flow of production fluid will not be substantially increased. A concern with a cyclic foam injection process is that the low mobility foam will displace oil further (as compared to steam-only injection) from the well during the injection portion of the cycle. The oil will then have a greater distance to flow to the well during production. Thus, initial water cuts may actually increase in cyclic steam-foam tests (26). The oil recovery may also be low initially but then increase to a level higher than that obtained from a steam-only cycle.

Some cyclic field foam test results have been reported as follows:

1. On the basis of more than 4000 field tests (mostly in Kern County, California) performed by Chemical Oil Recovery Company (CORCO) (35, 42):

    - 1980–1984: 1.1 m^3 of oil/L of surfactant solution
    - 1985–1986: 0.3 m^3 of oil/L of surfactant solution
    - 1990: 0.2 m^3 of oil/L of surfactant solution

2. The results of two foam tests (using SD1000) performed by Chevron (43) were

    - at Midway-Sunset field, the increase in injection pressure was 0.6– 1.1 MPa, and the incremental oil production was 0.4–1.2 m^3/kg of active SD1000
    - at Bradley-Canyon field, the increase in injection pressure was 3.6 MPa, and the incremental oil production was 0.12 m^3/kg of active SD1000

Drive Processes. A number of foam-drive tests are summarized in Table I. A standard application involves an initial active surfactant concentration in the aqueous phase of about 1 wt% before it is reduced to about 0.2–0.5 wt% after a number of days. The noncondensible gas mole fraction in the vapor phase is in the range 0.5–1.0%. NaCl is sometimes added to reduce ion exchange. The steam quality may be reduced to increase the LVF. If steam diversion results, then the injection pressure increases by about 1 MPa. In addition, about 0.3 m³ of incremental oil is produced per kilogram of active surfactant. However, the behavior of the foam is highly dependent on the condition of the reservoir (permeability streaks, depletion, etc.). Assuming a cost of $10/kg of active surfactant (including noncondensible gas), the cost of incremental oil is $33/m³ ($5/barrel). However, in many tests the incremental oil production is not high enough to be economical.

Numerical Modeling of Field Foam Tests

Surfactant propagation in the reservoirs has been modeled (*44, 45*) by allowing for surfactant adsorption, oil partitioning, and first-order surfactant decomposition; all of these variables are functions of temperature. The foam mobility reduction is taken into account by reducing the gas relative permeability as follows:

$$k_{rf}S_g = k_{rg}S_g / (1 + \text{MRF}_{\text{eff}}) \quad (2)$$

$$\text{MRF}_{\text{eff}} = \text{MRF}\ (W_s/W_s^m)^{es}((S_o^m - S_o)/S_o^m)\ e^o(N_c/N_c^r)^{ev} \quad (3)$$

where MRF is maximum mobility reduction factor, which is determined from history matching (*50*); W_s is surfactant concentration (0.0001875 mol fraction); S_o is oil saturation to foam (0.25); N_c is the capillary number (10^{-6}); S_g is gas saturation; the superscript m and the subscript eff denote maximum and effective, respectively; and the superscript r denotes the term reference.

A value of 1.0 has been used for the exponents es, eo, and ev, which are the effect of surfactant, effect of oil concentration, and effect of velocity, respectively.

The following surfactant properties are also specified:

- oil–water k value (0.0)
- maximum adsorption at initial temperature (2.56 mol/m³)
- maximum adsorption at injection temperature (0.38 mol/m³)
- surfactant half-life at initial temperature (∞)

Table I. Summary of Field Foam

Company	Field	Oil Gravity (°API)	Foaming Agents
CLD (50)	Midway-Sunset, California	11	Thermophoam BW-D and air
SUPRI, CORCO (51, 52)	Kern River, California	12	Suntech IV and N_2
Mobil (53)	Kern County, California	13	Linear toluene sulfonate and N_2
Shell (49, 54–57)	Kern River	13	AOS-1618 and N_2
Unocal (58)	Guadulupe, California	9	Alkyl toluene sulfonate and N_2
Unocal (36, 59)	Midway-Sunset	11	AOS-1618, NaCl and N_2
Chevron (60, 61)	Midway-Sunset	14	SD1000 and N_2
Chevron (41)	Midway-Sunset	—	SD1020 and N_2
Amoco (48)	Fremont County, Wyoming	14	Suntech IV and CH_4
GLISP (38, 62–64)	Athabasca, Alberta	10	SD1020 and noncondensible gas

[a]For 2 days/week for tests 1 and 2.

- surfactant half-life at injection temperature (∞)
- injected surfactant concentration (eg., 0.0001875 mole fraction)
- surfactant molecular weight (480)

Equations are based on experimental observations of the effect of surfactant concentration, oil saturation, and velocity. The values in parentheses were used in an actual field simulation (44).

The following conclusions have been drawn from the model results:

- If foam is applied late in a steam drive, for an aerially isotropic reservoir, its effect will be minimal because of the large distances it must move to cause an effect.
- Even in thin reservoirs, foam can be effective by reducing gravity override. In these reservoirs the growth of the foam zone is spherical.

Applications in Steam-Drive Processes

Surfactant Conc. (wt % Active in Aqueous Phase)	Noncondensible Gas Conc. (mol %)	NaCl Conc. (wt %)	Increase in Injection Pressure (MPa)	Incremental Oil Production Active (m^3/kg Surfactant)
Slugs	—	0.0	—	0.3
Slugs	2	0.0	0.5	>0.1
1 for 1–2 weeks 0.5 for 12–13 months	0.5	0.0	0.5	0.4
0.5	0.06	4.0	—	0.02
0.4–0.45 for 3 years 0.24 for 6 months	0.5	0.0	1.4	0.3
0.31 for 3 years 0.24 for 6 months	0.8	2.3	—	0.07
0.5 for 2 days/week for tests 1 and 2	1.0^a	0.0	1.0 during slugs	Test 1: 1.2 Test 2: 0.9
0.27 after initial 3-day slug	1	0.0	1.4	0.1
0.4–2.5	2	0.0	<0.4	Test 1: 0.2
2.5–40	2	0.0	<0.4	Test 2: 0.0
0.5 for 2 weeks, 0.2 for 8.5 months followed by 0.1	1 for 2 weeks, 0.1–1	0.0	1.8 1.8 1.3	— 0.05 0.14

- In multizone reservoirs, foam affects steam distribution between zones and within a zone.
- Injection of surfactant into the bottom zone limits foam blocking to the upper zones because the liquid surfactant stays in the bottom zone.
- Separation of gas and liquid due to gravity results in less surfactant available to form foam in the gas regions.

The injection pressure rise and decay in a mini-foam test over about 1–2 weeks may be used to determine MRF_{eff}. The numerical model can then be used to predict the effect of long-term foam injection.

Another steam-foam model was developed at the Alberta Research Council (46, 47). It considers surfactant transport and flow resistance to foam. Thermal degradation is assumed to be first-order. The rate constant is dependent on both temperature and pH. Surfactant adsorption is described by Langmuir adsorption. The partitioning of surfactant between oil and water phases is based on the assumption that the surfactant concentration immediately reached equilibrium in both phases. A Langmuir model yields

$$X_s = Z_s / (C_1 + C_2 Z_s) \qquad (4)$$

where X_s is surfactant concentration in the oil phase, Z_s is surfactant concentration in the liquid phase, and C_1 and C_2 are constants.

The gas-phase relative permeability is an empirically determined function of superficial gas-phase velocity, superficial water-phase velocity, and surfactant concentration in the water phase.

This model was used (47) to evaluate the foam-injection strategy and concentration that resulted in the highest profit. The study was based on the "discounted net profit" (DNP) due to surfactant injection which was defined as

DNP = value of incremental oil minus cost of surfactant

$$\text{value of incremental oil} = \sum_{1}^{N} (\text{IOP}_t - \text{IOP}_{t-1}) / (1 + d)^{t-0.5} \times \text{oil price}$$

$$\text{cost of surfactant} = \sum_{1}^{N} (S_t - S_{t-1}) / (1 + d)^{t-0.5} \times \text{cost of surfactant}$$

where d is the semi-annual discount rate, N is the total number of half-year periods, IOP_t is the incremental oil produced at the end of t periods (m^3), and S_t is the cumulative surfactant injected at the end of t periods (kg).

Shell (48) used a simple foam model (49) for their Bishop Fee pilot. The foam generation rate was matched by using an effective surfactant partition coefficient that took into account surfactant losses and foam generation inefficiencies. The value of this coefficient was selected so that the numerical surfactant propagation rate was equal to the actual growth rate. Foam was considered to exist in grid blocks where steam was present and the surfactant concentration was at least 0.1 wt%. The foam mobility was assumed to be the gas-phase relative permeability divided by the steam viscosity and the MRF. The MRF increased with increasing surfactant concentration. The predicted incremental oil production [5.5% of the original-oil-in-place (OOIP) due to foam and 3% due to infill wells] was in good agreement with the actual value.

References

1. Al-Khafaji, A.; Wang, P. T.; Castanier, L. M.; Brigham, W. E. *Proceedings of the 52nd California Regional Meeting;* Society of Petroleum Engineers: Richardson, TX, 1982; paper SPE 10777.
2. Isaacs, E. E.; McCarthy, C.; Smolek, K. F. *Proceedings of the 2nd European Symposium on Enhanced Oil Recovery;* Society of Petroleum Engineers, Richardson, TX, 1982; p 549.
3. Keijzer, P. P. M.; Muijs, H. M.; Janssen-van Rosmalen, R.; Teeuw, D. *Proceedings of the 5th SPE/DOE Symposium on Enhanced Oil Recovery;* Society of Petroleum Engineers, Richardson, TX, 1986; paper SPE 14905.
4. McPhee, C. A.; Tehrani, A. D. H.; Jolly, R. P. S. *Proceedings of the 6th SPE/DOE Enhanced Oil Recovery Symposium;* Society of Petroleum Engineers: Richardson, TX, 1988; SPE/DOE paper 17360.
5. Angstadt, H. P.; Tsao, H. *SPE Reservoir Eng.* **1987**, *2,* 613.
6. Castanier, L. M. *J. Pet. Sci. Eng.* **1989**, *2,* 193.
7. Ziegler, V. M. *Proceedings of the 59th SPE Annual Technical Conference and Exhibition;* Society of Petroleum Engineers, Richardson, TX, 1974; paper SPE 13071.
8. Handy, L. L.; El-Gassier, M.; Ershaghi, I. *Proceedings of the 5th Symposium on Oil Field and Geothermal Chemistry;* American Institute of Mining, Metallurgical and Petroleum Engineers: New York, 1980; paper SPE 9003.
9. Angstadt, H. P. U.S. Patent 4 682 653, 1987.
10. Ziegler, V. M.; Handy, L. L. *Soc. Pet. Eng. J.* **1981**, *21,* 218–228.
11. Novosad, J.; Maini, B. B.; Huang, A. *J. Can. Pet. Tech.* **1986**, *25,* 42.
12. Dawe, B.; Oswald, T. *J. Can. Pet. Tech.* **1991**, *30,* 133–137.
13. Mannhardt, K.; Novosad, J. *Chem. Eng. Sci.* **1991**, *46,* 75–83
14. Isaacs, E. E.; Jian, L.; Green, M. K.; McCarthy, C.; Maunder, D. *AOSTRA J. Res.* **1988**, *4,* 267–276.
15. Celik, M. S.; Somasundaran, P. *J. Colloid Interface Sci.* **1988**, *122,* 163.
16. Handy, L. L.; Amaefule, J. O.; Zeigler, V. M.; Ershaghi, I. *Proceedings of the 4th Symposium on Oil field and Geothermal Chemistry;* American Institute of Mining, Metallurgical and Petroleum Engineers: New York, 1979; paper SPE 7867.
17. Babu, D. R.; Hornof, V.; Neale, G. *Can. J. Chem. Eng.* **1984**, *62,* 156.
18. Isaacs, E. E.; Maunder, J. D.; Jian, L. *Oil-Field Chemistry: Enhanced Recovery and Production Stimulation;* Borchardt, J. K.; Yen, T. F., Eds.; ACS Symposium Series 396; American Chemical Society: Washington, DC, 1987; pp 325–344.
19. Green, M. K.; Isaacs, E. E. Presented at the 37th Petroleum Society of Canadian Institute of Mining (CIM) Meeting, Calgary, Canada, 1986; paper 86–37–02.
20. Robin, M. *Proceedings of the 62nd SPE Annual Technical Conference and Ex-*

hibition; Society of Petroleum Engineers: Richardson, TX, 1974; paper SPE 16729.
21. Flumerfelt, R. W. Report under Sandia Contract No. 62–7230; U.S. Department of Energy: Sandia, NM, 1982.
22. Falls, A. H.; Lawson, J. B.; Hirasaki, G. J. *J. Pet. Technol.* **1988**, *40*, 95–104.
23. Green, M. K.; Isaacs, E. E. *AOSTRA J. Res.* **1988**, *4*, 133–142.
24. Muijs, H. M.; Keijzer, P. P. M.; Wiersma, R. J. *Proceedings of the SPE/DOE Enhanced Oil Recovery Symposium;* Society of Petroleum Engineers: Richardson, TX, 1988; paper SPE 17361.
25. Janssen-van Rosmalen, R.; de Boer, R. B.; Keijzer, P. P. M. *Proceedings of the 3rd European Meeting on Improved Oil Recovery;* Rome, Italy, 1985; Vol. 1, p 329.
26. Keijzer, P. P. M.; Muijs, H. M; Janssen-van Rosmalen, R.; Teeuw, D.; Pino, H.; Avila, J. *Proceedings of the 1986 SPE/DOE Enhanced Oil Recovery Symposium;* Society of Petroleum Engineers: Richardson, TX, 1986; paper SPE 14905.
27. Persoff, P.; Radke, C. J.; Pruess, K.; Benson S. M.; Witherspoon, P. A. *SPE Reservoir Eng.* **1991**, *6*, 365–372.
28. Isaacs, E. E.; Green, M. K.; Jossy, W. E.; Maunder, J. D. *Proceedings of the 2nd Latin American SPE Conference;* Society of Petroleum Engineers: Richardson, TX, 1992; paper SPE 23754.
29. Khatib, Z. I.; Hirasaki, G. J.; Falls, A. H. *SPE Reservoir Eng.* **1988**, *3*, 919–926.
30. Friedmann, F.; Chen, W. H.; Gauglitz, P. A. *Proceedings of the SPE/DOE Symposium on Enhanced Oil Recovery;* Society of Petroleum Engineers: Richardson, TX, 1988; paper SPE 17357.
31. Green, M. K.; Isaacs, E. E.; Smid, J. J. *Fuel* **1991**, *70*, 1303.
32. Green, M. K.; Isaacs, E. E.; Non Chom, K. *Can. J. Tech.* **1991**, *30*, 28.
33. Ivory, J.; Isaacs, E. E.; Maunder, J. D. Alberta Research Council, unpublished results.
34. Hirasaki, G. J. *J. Pet. Technol.* **1989**, May, 449.
35. Cooke, R. W.; Eson, R. L. *Proceedings of the International Thermal Operations Symposium;* Society of Petroleum Engineers: Richardson, TX, 1991; paper SPE 21531.
36. Ivory, J.; van der Voet, G.; Best, D. S.; Isaacs, E. E.; Gunter, W. D. *Proceedings of the 1993 NIPER/DOE Symposium;* National Institute for Petroleum Engineering Research: Bartlesville, OK, 1993.
37. Mohammadi, S. S.; Van Slyke, D. C.; Ganong, B. L. *SPE Reservoir Eng.* **1989**, *4*, 7–16.
38. Sander, P. R.; Clark, G. J.; Lau, E. C. *Proceedings of the 66th Annual Technical Conference and Exhibition of SPE;* Society of Petroleum Engineers: Richardson, TX, 1991; paper SPE 22630.
39. Ferrell, H. H.; Plumb, L. B.; Lowery, P. H. U.S. Department of Energy Report DOE/BC/10830-2 Bartlesville, OK, 1986.
40. Wagner, O. R. *J. Pet. Technol.* **1977**, 1410–1416.
41. Friedmann, F.; Smith, M. E.; Guice, W. R.; Gump, J. M.; Nelson, D. G. *Proceedings of the SPE Western Regional Meeting;* Society of Petroleum Engineers: Richardson, TX, 1991; paper SPE 21780.

42. Eson, R. L. United Nations Conference on Heavy Oil and Tar Sands, Long Beach, CA, 1985; preprint.
43. Chevron Chemical Company, Chevron Bulletin, California, 1986.
44. Mohammadi, S. S.; Coombe, D. A. Presentation at the UNITAR/UNDP 5th International Conference on Heavy Crude and Tar Sands, Caracas, Venezuela, 1991; paper 113.
45. Mohammadi, S. S.; Coombe, D. A.; Stevenson, V. M. Presentation at the CIM/AOSTRA Technical Conference, Banff, Alberta, 1991; paper CIM/AOSTRA 91-77.
46. Law, D. H.-S.; Yang, Z.-M.; Stone, T. W. Presentation at the UNITAR/UNDP 4th International Conference on Heavy Crude and Tar Sands, Edmonton, Alberta, August 1988.
47. Law, D. H.-S. Presentation at the 3rd Technical Meeting of South Saskatchewan Section, Petroleum Society of CIM, Regina, Canada, 1989; paper 17.
48. Yannimaras, D. V.; Kobbe, R. K. *Proceedings of the SPE/DOE Enhanced Oil Recovery Symposium;* Society of Petroleum Engineers: Richardson, TX, 1988; paper SPE/DOE 17382.
49. Patzek, T. W.; Myhill, N. A. *Proceedings of the California Regional Meeting;* Society of Petroleum Engineers: Richardson, TX, 1989; paper SPE 18786.
50. Doscher, T. M.; Kuuskaraa, V. A.; Hammershaimb, E. C. *Proceedings of the 58th Annual Technical Conference and Exhibition;* Society of Petroleum Engineers: Richardson, TX, 1983; paper SPE 12057.
51. Brigham, W. E.; Castanier, L. M.; Markov, J.; Malito, O. P.; Sanyal, S. K. Presentation at the 5th Annual Advances in Petroleum Recovery and Upgrading Technology Conference, Calgary, Alberta, 1984; Session 3, paper 3.
52. Brigham, W. E. Final Report of U.S. Department of Energy, San Francisco, CA, DOE/SF/11445-4, 1986.
53. Djabborah, N. F.; Webber, S. L.; Freeman, D. C.; Muscatello, J. A.; Asbaugh, J. P.; Covington, T. E. *Proceedings of the 60th California Regional Meeting;* Society of Petroleum Engineers: Richardson, TX, 1990; paper SPE 20067.
54. Patzek, T. W.; Koinis, M. T. *Proceedings of the SPE/DOE Enhanced Oil Recovery Symposium;* Society of Petroleum Engineers: Richardson, TX, 1988; paper SPE/DOE 17380.
55. Falls, A. H.; Lawson, J. B.; Hirasaki, G. J. *Proceedings of the 56th California Regional Meeting of SPE;* Society of Petroleum Engineers: Richardson, TX, 1986; paper SPE 15053.
56. Dilgren, R. E.; Owens, K. B. Canadian Patent 1 172 160 (Issued 840807).
57. Dilgren, R. E.; Doemer, A. R.; Owens, K. B. *Proceedings of the 1982 California Regional Meeting of SPE;* Society of Petroleum Engineers: Richardson, TX, 1982; paper SPE 10774.
58. Mohammadi, S. S.; McCollum, T. J. *Proceedings of the 56th California Regional Meeting of SPE;* Society of Petroleum Engineers: Richardson, TX, 1986; paper SPE 15054.
59. Mohammadi, S. S.; Tenzer, J. R. *Proceedings of the SPE/DOE 7th Symposium on Enhanced Oil Recovery;* Society of Petroleum Engineers: Richardson, TX, 1990; paper SPE/DOE 20201.
60. Lee, W.; Kamilos, G. N. *Pet. Eng. Intl.* **1985,** *2,* 36, 40, 44, 48, 57.

61. Ploeg, J. F.; Duerkson, J. H. Proceedings of the SPE 1985 California Regional Meeting; Society of Petroleum Engineers: Richardson, TX, 1985; paper SPE 13609.
62. Kular, G. S.; Low, K.; Coombe, D. S. *Proceedings of the 64th Annual Technical Conference and Exhibition of SPE;* Society of Petroleum Engineers: Richardson, TX, 1989; paper SPE 19690.
63. Reimond, S. A.; Stasiuk, N. F. *Alberta Heavy Oil, Oil Sands and Enhanced Recovery Experimental Pilot Projects;* Alberta Oil Sands Technology and Research Authority: Edmonton, Canada, 1992.
64. Lau, E. C.; Coombe, D. A. Presented at the Canadian Institute of Mining (CIM) 1992 Annual Technical Conference, Calgary, Alberta, 1992; paper CIM 92-58.

RECEIVED for review December 4, 1992. ACCEPTED revised manuscript April 20, 1993.

7

Adsorption of Foam-Forming Surfactants for Hydrocarbon-Miscible Flooding at High Salinities

Karin Mannhardt and Jerry J. Novosad

Petroleum Recovery Institute 100, 3512 33rd Street N.W., Calgary, Alberta T2L 2A6, Canada

> *This chapter reports adsorption data for a number of surfactants suitable for mobility control foams in gas-flooding enhanced oil recovery. Surfactants suitable for foam-flooding in reservoirs containing high salinity and hardness brines are identified. The results of adsorption measurements performed with these surfactants are presented; surfactant adsorption mechanisms are reviewed; and the dependence of surfactant adsorption on temperature, brine salinity and hardness, surfactant type, rock type, wettability, and the presence of an oil phase is discussed. The importance of surfactant adsorption to foam propagation in porous media is pointed out, and methods of minimizing surfactant adsorption are discussed.*

The Significance of Hydrocarbon-Miscible Flooding

Hydrocarbon-miscible flooding refers to an oil recovery process in which a "solvent", usually a mixture of low and intermediate molecular-weight hydrocarbons (methane through hexane), is injected into a petroleum reservoir. Several mechanisms contribute to oil recovery in this process: displacement of oil by solvent through the generation of miscibility between solvent and oil, oil swelling with a resulting increase in oil saturation and therefore in oil relative permeability, and reduction of oil viscosity. When solvent and oil remain immiscible, a reduction of gas–oil interfacial tension leads to improved oil recovery.

Depending on temperature, pressure, and oil and solvent composition, the injected solvent may be completely miscible (first-contact miscible) with the oil or may develop multiple-contact miscibility by continuous mass transfer of components between the solvent and the oil during flow through the porous medium. Two processes for the development of multiple-contact miscibility are distinguished. If the oil is poor in intermediate (C_2 to C_6) components, natural gas that is enriched in these components may be used as the solvent. Continuous transfer of the intermediate components from the solvent to the oil during flow through the porous medium leads to the development of a zone of solvent–oil miscibility. This process is referred to as condensing gas drive. For oils that are rich in intermediate components, a lean gas may be used as the solvent. In this process, referred to as vaporizing gas drive, multiple-contact miscibility is developed by continuous transfer of intermediate components from the oil to the gas.

Solvent injection is usually followed by the injection of a cheaper drive fluid. The solvent slug may be displaced miscibly with another gas, for example, natural or flue gas, or immiscibly with water.

Most of the world's enhanced oil production comes from the injection of a gas or vapor phase into a petroleum reservoir (Table I): 83% in Canada, 99% in the United States, and 99% in all other countries (1). Injected gases are usually steam, light and intermediate hydrocarbons, or CO_2. Large density and viscosity contrasts between displacing and displaced fluids in gas-injection processes result in poor sweep efficiency because of gravity override and viscous fingering. As a result of the widespread application of gas-flooding and the large potential for increasing the oil recovery from gas-injection schemes by improving the sweep efficiency, interest in mobility control foams has risen sharply over the past decade.

The proportion of enhanced oil production that is caused by hydrocarbon-miscible and, to a lesser extent, immiscible flooding is evident from Table I. Hydrocarbon-flooding dominates in Canada, where it accounts for 81% of Canada's enhanced oil recovery (EOR) production. Although steam-flooding is the most commonly applied EOR process in the United States, oil production from hydrocarbon-flooding is significant and has increased by more than the production from any other EOR process over the past 2 years. In countries other than the United States and Canada, most of the oil recovered by hydrocarbon-flooding comes from Libya, but hydrocarbon injection is also applied in the Soviet countries and in the United Arab Emirates (1).

Table I shows some statistics on the types of reservoir subjected to enhanced recovery operations. Most EOR projects in Canada, particularly hydrocarbon-miscible projects, are being conducted in carbonate reservoirs, and hydrocarbon injection in the United States is mostly in sandstones. As will be evident from this chapter, the type of reservoir rock is of significance to surfactant propagation during foam-flooding.

Table I. Enhanced Oil Recovery Statistics

Enhanced Oil Recovery Process	EOR Production		Number of Projects				
	barrels per day	% Increase since 1990[a]	Sandstone[b]	Carbonate	Other	Total	% Increase since 1990[a]
Canada							
Hydrocarbon miscible–immisc.	122,812	−3	5	44	0	49	−4
Steam[c]	2,745	+84	3	0	0	3	−57
CO_2 and other gas	270	0	5	1	0	6	+20
Total EOR	150,930	−1	24	45	0	69	−9
United States							
Hydrocarbon miscible–immisc.	113,072	+104	20	4	1	25	+9
Steam	454,009	+2	117	0	0	117	−15
CO_2 and other gas	184,948	+37	28	29	7	64	−6
Total EOR	760,907	+16	205	35	8	248	−16
Others							
Hydrocarbon miscible–immisc.	40,300	+1	0	2	0	2	+100
Steam	440,887	+81	54	0	0	54	0
CO_2 and other gas	17,895	−20	5	3	0	8	−38
Total EOR	502,503	+63	72	5	0	77	−8

[a] Percentages were calculated using 1990 values as base.
[b] Includes unconsolidated sand.
[c] Excludes heavy oil.
SOURCE: Reference 1.

Foams for Hydrocarbon-Miscible Flooding in Reservoirs Containing High-Salinity Brines

The reason for the large number of hydrocarbon-miscible flooding projects in Canada is the preponderance of reasonably priced gas throughout the province of Alberta. Also, a large number of gas plants separates intermediate components, such as ethane, propane, and butane, and allows custom design of individual solvents tailored to specific reservoir conditions. Because of the large number of hydrocarbon-miscible projects in Canada, the application of mobility-control foams seems an attractive means to significantly increase oil production.

The Gilwood formation and the Beaverhill Lake group provide excellent examples of pools that are candidates for hydrocarbon-miscible floods. There are 19 producing pools in the Gilwood formation and 27 in the Beaverhill Lake group. The range of properties encountered in these pools is listed in Table II.

All the pools in these reservoirs contain light oil and brine of extreme salinity and hardness. The total dissolved solids (TDS) concentrations of the formation brines range from 100,000 to almost 300,000 ppm (10 to 30 mass %) with divalent ion contents from 5,000 to 25,000 ppm (0.5 to 2.5 mass %). The electrolyte concentrations in these brines are not far from the saturation limits. However, many of the pools may no longer contain the original formation brine because they have been waterflooded with fresh water. Typically, fresh water was injected to initial water breakthrough, after which the produced water was mixed with additional fresh water as required to maintain desirable injection rates. Thus, testing the effectiveness of foam-forming surfactants requires performance measurements over the full range of brine salinities from injection water to the original formation brine.

No universally accepted practices exist for selecting foam-forming surfactants for specific rock–fluid systems. In addition, information dealing with foam performance at brine salinities as high as those found in the pools mentioned previously is not readily available in the literature. The selection process is based on the following "common sense" criteria (2):

1. Solubility: The surfactant must be fully soluble in formation as well as in injection brines and in all brines of intermediate salinity. Solubility must be maintained both at room temperature (simulating surface injection) and at reservoir temperature.

2. Chemical stability: The surfactant must maintain its interfacial activity at reservoir temperature for several months.

3. Mobility reduction factor (MRF) in porous media: Foam generated by

Table II. Range of Characteristics of Reservoirs in the Gilwood and Beaverhill Lake Formations

Characteristics	Gilwood (Sandstone)		Beaverhill Lake (Carbonate)	
	Lowest	Highest	Lowest	Highest
Area per pool (ha)	64	29,248	64	23,367
Oil in place per pool (m^3/m^3)	0.040	0.165	0.033	0.068
Porosity (%)	9	34	5	9
Permeability (md)	16	217	10	81
Oil density (kg/m^3)	806	842	797	839
Oil viscosity (mPa.s)	0.60	1.2	0.19	0.91
Formation water TDS (ppm)	101,600	296,400	179,000	245,400
Formation water divalent cation content (ppm)	5,000	25,000	5,000	25,000
Reservoir temperature (°C)	41	127	67	113
Reservoir pressure (MPa)	6.5	34.9	21.6	37.9

NOTE: In the Gilwood pools, the number of producing pools was 19, 3 of which were using waterflood EOR. In the Beaverhill Lake pools, there were 27 producing pools, 16 of which used EOR. In the Beaverhill Lake, 14 of the EORs used waterflooding, and 2 used solvent flooding.

SOURCE: Reproduced with permission from reference 2. Copyright 1987 Canadian Institute of Mining, Metallurgy, and Petroleum.

the surfactant must be effective in reducing gas mobility under as wide a range of conditions as those at which the solubility was measured.

4. Retention: The surfactant must be able to propagate in porous media with minimal retention losses.

Surfactant Solubility. Surfactant solubility was determined by simple visual observation of 1% surfactant solutions at temperatures from 23 to 130 °C and in brines ranging in salinity from injection to formation brine. The requirement of complete solubility in near-saturated salt solutions is rather stringent, and therefore, the solubility tests eliminated the large majority of the 157 surfactants initially selected for screening. Only nine surfactants representing three chemical classes passed the solubility tests. List 1 (2) and Table III list the 14 chemical classes tested and the nine surfactants that passed the tests, respectively. Most of the surfactants

List 1. Chemical Classes of Surfactants Tested for Solubility

- Alkanolamides
- Amine oxides
- Betaine derivatives
- Ethoxylated and propoxylated alcohols and alkylphenols
- Ethoxylated and propoxylated fatty acids
- Ethoxylated fatty amines
- Fatty acid esters
- Fluorocarbon-based surfactants
- Phosphate derivatives
- Polymers and copolymers
- Quaternary ammonium chlorides
- Sulfate derivatives
- Sulfonate derivatives

    — alkylarylethoxysulfonates

    — alkylarylsulfonates

    — alkylethoxysulfonates

    — alkylsulfonates

    — diphenyl ether disulfonates

    — lignin derivatives

    — olefin sulfonates

    — petroleum sulfonates

- Unknown chemical classes

SOURCE: Reproduced with permission from reference 2. Copyright 1987 Canadian Institute of Mining, Metallurgy, and Petroleum.

Table III. Surfactants Passing Solubility Criteria

Chemical Class	Trade Name	Manufacturer
Amine oxides	Empigen OS	Albright & Wilson
Betaine derivatives (amphoteric)		
Alkylbetaines	Marchon DC 1803	Albright & Wilson
	Empigen BB	Albright & Wilson
Amide-modified alkylbetaines	Empigen BS	Albright & Wilson
	Empigen BT	Albright & Wilson
	Varion CADG-HS	Sherex
Alkylsulfobetaine	Varion HC	Sherex
Amide-modified alkylsulfobetaine	Varion CAS	Sherex
Sulfonate derivatives		
Diphenyl ether dilsulfonate–α-olefin sulfonate mixture	Dow XS84321.05	Dow Chemical

that passed the solubility tests are amphoteric betaine derivatives. Only one anionic surfactant, a diphenyl ether disulfonate–α-olefin sulfonate mixture, was fully soluble.

Surfactant Chemical Stability. Two approaches were used in assessing surfactant degradation over time. The first consisted of monitoring the pH of surfactant solutions that were in contact with pieces of reservoir rock over several months. Because only commercially available surfactants were tested and almost all of them contained secondary components, the pH data were rather inconclusive. The fact that reservoir solids have some buffering capacity made the interpretation of pH trends even more difficult.

The second approach was based on monitoring the interfacial characteristics of surfactant solutions kept at reservoir temperature. Foaming capacity was chosen as a measure of surfactant interfacial activity, and all nine surfactants followed the pattern shown in Figure 1 and indicated no deterioration over time.

Foam Effectiveness in Porous Media. No generally accepted correlations exist between foam characteristics measured outside the porous medium and foam effectiveness as a gas mobility-reducing agent in porous media. The performance of the nine surfactants that passed the solubility criteria was therefore evaluated in porous media under typical reservoir conditions. The results of such an evaluation can be expressed in several ways. One of the simplest measures of foam effectiveness, and arguably the most straightforward one, is the mobility-reduction factor (MRF). The MRF is defined as the ratio of pressure gradients across a

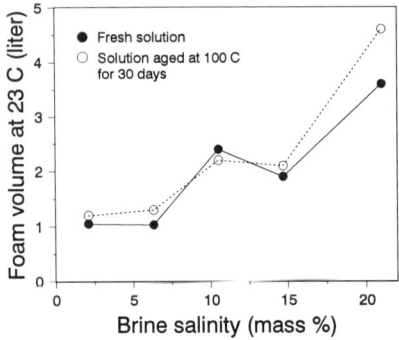

Figure 1. Comparison of foaming capacity of fresh and aged surfactant solutions. (Reproduced with permission from reference 2. Copyright 1987 Canadian Institute of Mining, Metallurgy, and Petroleum.)

core that results from simultaneous flow of gas and liquid in the presence and in the absence of the foam-forming surfactant in the liquid phase:

$$\text{MRF} = \frac{\Delta P_{\text{surfactant-gas}}}{\Delta P_{\text{brine-gas}}}$$

where ΔP is the pressure drop measured across the core. An MRF equal to unity indicates an ineffective surfactant. The higher the MRF, the more effective the foam is in reducing gas mobility. Because the fluid saturations in a core are most likely different during brine–gas and foam injection, MRF values represent a combination of two factors: foam rheology and two- or three-phase relative permeability effects.

The accuracy of MRF measurements is dependent on several factors that may be difficult to assess: constancy of liquid injection rate at extremely low flow rates (less than 0.25 mL/h in some cases), maintenance of constant pressure at the core outlet during two- or three-phase flow, stability of the pressure transducers over wide ranges of pressure and over long periods of time, and homogeneity of the cores. Thus, individual values of MRF can be subject to significant errors. On the basis of information from repeated experiments, 10–20% variations in MRF seem to be within realistic limits for these types of measurements.

Figures 2 to 5 show examples of mobility reduction factors measured in oil-free Berea cores containing high-salinity, high-hardness brines under reservoir conditions. (An explanation of surfactant names used in these figures appears in the Appendix.) Nitrogen was used as the gas phase. MRF values presented in these figures were obtained from pressure gradients measured after pseudosteady-state flow through the linear cores had been achieved. That is, the pressure gradients were measured only after the core was fully saturated with the surfactant, the foam had propagated throughout the core (as indicated by intermediate pressure taps along the core), and the flow of foam had stabilized. Therefore, neither surfactant adsorption nor the kinetics of foam generation are reflected in the MRF measurements presented here.

The MRF increases with increasing surfactant concentration (Figure 2), even when the surfactant concentration is above the critical micelle concentration (CMC). (All surfactant solutions tested were at concentrations well above the CMC.) The MRF increases and then levels off when the foam quality is decreased (Figure 3), and this leveling-off indicates that down to a foam quality of about 80%, nitrogen foams with a higher liquid content are more effective in reducing gas mobility. Presumably, the MRF will go through a maximum when the foam quality is reduced

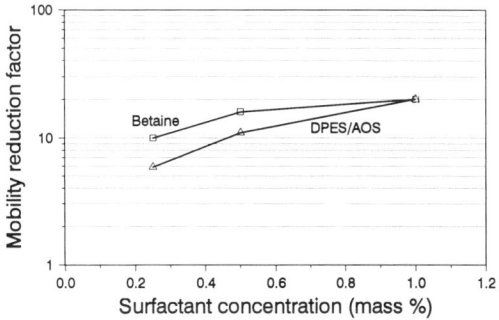

Figure 2. The dependence of mobility-reduction factor on surfactant concentration in Berea sandstone at 80 °C and 98% foam quality in 210,000 ppm (21 mass %) reservoir brine.

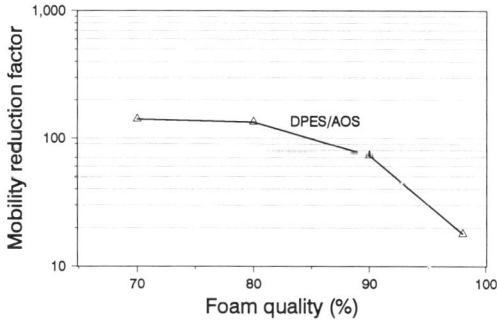

Figure 3. The dependence of mobility-reduction factor on foam quality. Anionic surfactant (diphenyl ether disulfonate–α-olefin sulfonate) in Berea sandstone at 85 °C in 21,000 ppm (2.1 mass %) reservoir brine.

further, because the porous medium will then contain a relatively high bulk liquid saturation rather than a large population of the surfactant-stabilized liquid films that give a foam its high effective viscosity and render it an effective gas mobility control agent. Also, one of the attractive features of foams is that only a small amount of liquid is required in order to achieve a substantial increase in effective gas viscosity. Therefore, optimization of a foam-flood requires a balance between the degree of mobility reduction desired and the cost of the injected surfactant solution.

In general, trends in MRF with surfactant concentration and foam quality are consistent for all surfactants, but the effects of brine salinity (Figure 4) and temperature (3) vary from surfactant to surfactant. Different surfactants are also affected to different degrees by the presence of an oil phase, as discussed in Chapter 4 of this book. The MRF increases with increasing permeability (Figure 5), as also noted by Lee and Heller (4) and described earlier in Chapter 5. This effect could be very beneficial to foam performance, because it leads to better mobility control in high-permeability zones.

The most important conclusion from the core-flood experiments is that the selected surfactants produced effective nitrogen-based foams at extreme conditions of salinity and hardness in oil-free porous media under reservoir conditions (2).

Because the surfactants described previously were selected for hydrocarbon-miscible flooding, the effect of hydrocarbon solvent on foam performance should be included in the surfactant screening process. Data on the characteristics of foams generated with light hydrocarbons as the gas phase are not readily available in the literature. Limited data comparing nitrogen foams with hydrocarbon solvent foams are shown in Table

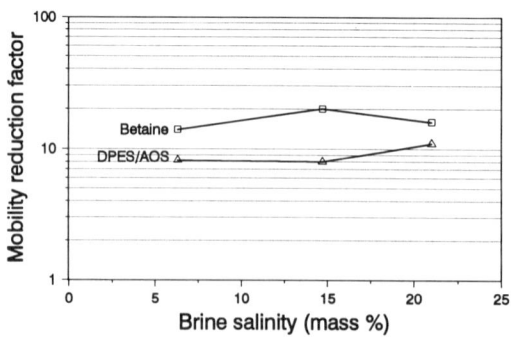

Figure 4. The dependence of mobility-reduction factor on brine salinity in Berea core at 80 °C and 98% foam quality.

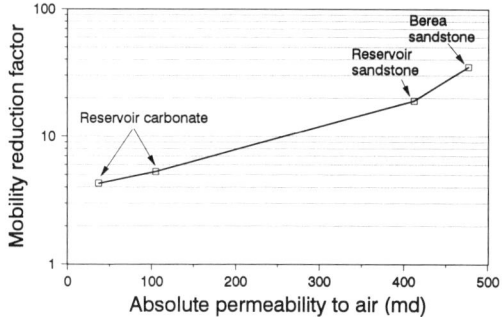

Figure 5. The dependence of mobility-reduction factor on permeability for an amphoteric surfactant at 80 °C.

IV. The listed MRFs were measured with hydrocarbon gas mixtures in reservoir sandstones with a variety of surfactants under a variety of reservoir conditions. Both surfactants are among those that passed the solubility screening. Although the data in Table IV indicate higher MRFs with hydrocarbons when compared to nitrogen, additional data collected under private contract and some data reported in the literature (5) have shown the opposite trend. A conclusive comparison of hydrocarbon solvent foams with nitrogen foams is not possible from the limited data available. However, the data do show that effective foams can be generated with light hydrocarbons.

A similar conclusion can be drawn from a limited number of field tests with hydrocarbon solvent foams (5, 6). Injection pressures in both field tests indicated the generation of foam in the formation. Both tests can be considered successful because they report either increased oil production (6) or reduced gas–oil ratio at the producing wells (5).

Table IV. Comparison of Mobility-Reduction Factors Measured with Nitrogen and with Light Hydrocarbon Solvent Mixtures

Surfactant	Gas	MRF
0.5% Sulfobetaine in 128,000 ppm brine	Nitrogen	49
	Solvent 1	61
0.5% Diphenyl ether disulfonate–α-olefin sulfonate mixture in 24,700 ppm brine	Nitrogen	16
	Solvent 2	29

NOTE: Solvent 1 composition (mol%): nitrogen, 5; methane, 44; ethane, 24; and propane, 27. Solvent 2 composition (mol%): nitrogen, 4; carbon dioxide, 0.6; methane, 47.4; ethane, 23.5; propane, 13.8; butane, 7.4; pentane, 2.0; and hexanes+, 1.3.

Surfactant Adsorption. Surfactant propagation is crucial to foam propagation. The data compiled in later sections of this chapter show that surfactants that are similarly effective as gas mobility reducing agents may have significant differences in adsorption levels. The level of surfactant adsorption and its dependence on parameters such as brine salinity and hardness may then be the deciding factors in surfactant selection for a specific application.

This section has demonstrated that some commercially available surfactants are soluble in brines of extreme salinity and hardness and also form effective mobility control foams under these conditions. The remainder of this chapter is devoted to the development of a better understanding of the adsorption properties of foam-forming surfactants, mainly those for high-salinity conditions. It is hoped that this discussion will contribute to the development of a systematic approach for selecting or formulating surfactants with minimal adsorption levels.

Limits of Foam Propagation

The effectiveness of a foam as a mobility-controlling agent is affected equally by the foam's flow characteristics and by the distance that the foam propagates into the reservoir. Although foam formation, stability, and flow properties are governed by the mechanisms of lamella generation and collapse, the distance that a foam can travel is ultimately dictated by surfactant retention during flow through the reservoir. If a foam is to propagate, propagation of the surfactant itself is necessary. Ultimately, economics will make or break any EOR process, and an extremely important part of the design of a foam-flood is the minimization of surfactant requirements and surfactant losses.

A number of mechanisms can contribute to surfactant retention in a reservoir. The most important of these is, arguably, adsorption at the solid–liquid interface, because it cannot be eliminated completely. However, measures can be taken to minimize adsorption.

The evaluation of surfactant adsorption is particularly important when foams for high salinity reservoirs, such as many Canadian reservoirs subjected to hydrocarbon-miscible flooding, are considered. This puts stringent requirements on the solubility, foaming, and adsorption properties of surfactants that may be considered for foam applications, and severely limits the types of surfactant that may be used.

Some anionic foam-forming surfactants do not adsorb appreciably on sandstone from low-salinity (0.5 mass %) brine (7). However, adsorption increases steeply when the salinity is increased to several mass percent.

Moreover, these surfactants precipitate at moderate salinities, making them unsuitable for foam applications in many Canadian reservoirs currently subjected to hydrocarbon-miscible flooding. Although surfactants that remain soluble and form effective mobility-control foams in near-saturated salt solutions have been identified (2), these surfactants adsorb at moderate to high levels under some conditions. Therefore, consideration of adsorption is extremely important in the selection of foam-forming surfactants.

Figure 6 shows adsorption levels measured for 96 different surfactant–brine–rock combinations. The data were either taken directly from references 7–12 and from some unpublished results, or were calculated from the data given in these references. All adsorption levels were determined from core-flood experiments, most of the surfactants being suitable for high-salinity conditions. A number of different solids, including outcrop sandstones and carbonates as well as reservoir cores, some at waterflood residual oil saturation, were used. Temperatures ranged from 23 to 150 °C, and the aqueous phase included brines with total dissolved solids contents up to 250,000 ppm (25 mass %), some with significant levels of divalent ions. Each bar in Figure 6 shows an adsorption level at a surfactant concentration of 1 mass %, each measured in a single core-flood. Surfactant concentrations of the order of 1 mass % are typically applied in foam-flooding, and surfactant adsorption is usually at or near its plateau level at this concentration. The bars in Figure 6 have been arranged in order of increasing adsorption.

Figure 6 shows that adsorption levels of foam-forming surfactants can vary widely from near-zero (measurements numbered 1 and 2) to almost

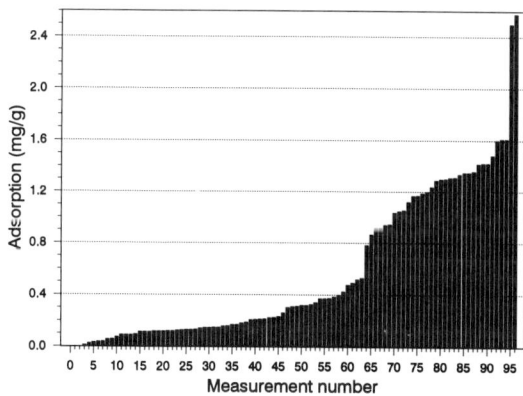

Figure 6. Adsorption levels measured in 96 core-flood experiments with foam-forming surfactants.

2.6 mg/g (measurements numbered 95 and 96). These adsorption levels can be roughly converted to reservoir units using the following conversion factors:

$$1 \text{ mg of surfactant/g of rock} = 2.1 \text{ kg of surfactant/m}^3 \text{ of reservoir}$$

$$= 5600 \text{ lb of surfactant/acre ft of reservoir}$$

Figure 7 shows how the measured adsorption levels translate into distance of surfactant propagation in a homogeneous reservoir with radial flow from the injector. A 40-acre five-spot pattern was assumed in calculating the pore volume (PV), and other assumptions are listed in the figure. The calculation of radial distance is based on a simple material balance using the following equation:

$$r = \left[\frac{m_s}{\pi q h}\right]^{1/2} = \left[\frac{\phi R^2 V_f (1 - Q) C}{q}\right]^{1/2}$$

where r (m) is radial distance of propagation, m_s (kg) is mass of surfactant injected, q (kg/m^3) is adsorption normalized to unit volume of reservoir, h (m) is the thickness of the target formation, ϕ is porosity, R (m) is the radius of the five-spot, V_f (PV) is foam volume injected, Q is foam quality or the volume fraction of gas in the injected foam, and C (kg/m^3) is surfactant concentration. A foam quality of 90% is assumed, that is, only 10% of the total fluid volume injected is surfactant solution.

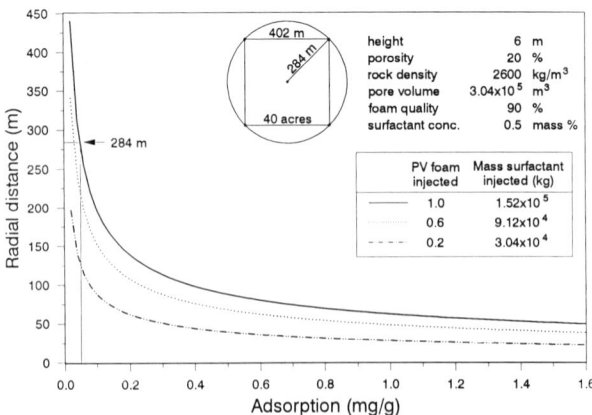

Figure 7. Example of radial surfactant propagation in a homogeneous reservoir.

Figure 7 shows that, even at the highest injection volume, a very low level of adsorption (0.05 mg/g) is required if the surfactant is to propagate the full 284 m from the injector to the producers. Very few of the surfactants in Figure 6 fulfill this requirement. However, the assumption of cylindrical, homogeneous flow throughout the thickness of the pay zone is not likely to hold in a reservoir that is considered for foam applications. In fact, if radial flow occurred, there would be no need for sweep improvement by foam. Figure 8 shows the results of similar calculations for a layered reservoir with two layers of different permeability. A foam might be used in such a system to block the high-permeability layer or to reduce gas mobility in this layer in order to divert the injected fluids to the lower permeability layer. Fluid intake into each layer is assumed to be proportional to each layer's permeability–height product; no cross-flow between layers and radial flow in each layer are assumed. The following equation was used to calculate the distance of propagation in each layer:

$$r_i = \left[\frac{m_s \dfrac{k_i h_i}{k_1 h_1 + k_2 h_2}}{\pi q h_i} \right]^{1/2}$$

where the subscript i refers to layer 1 or 2, and k_i and h_i are the permeability and thickness of layer i, respectively.

Figure 8 gives the radial distance that the surfactant can propagate in each layer when one pore volume of foam (0.1 PV of surfactant solution) is injected. An adsorption level of 0.2 mg/g or less is required for the surfactant to propagate all the way to the production well in the upper layer.

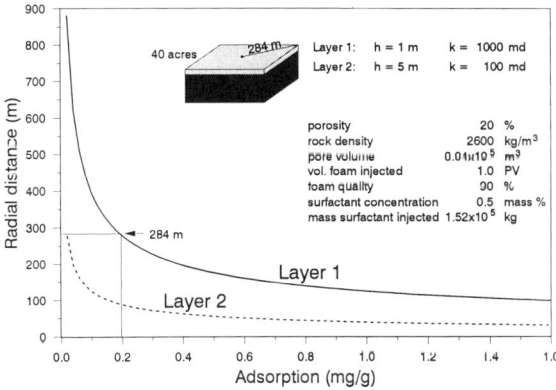

Figure 8. Example of radial surfactant propagation in a layered reservoir.

Figure 6 indicates that many of the tested systems fulfill this requirement. Also, more than half of the surfactants in Figure 6 adsorb at less than 0.4 mg/g, and this adsorption results in a reasonable distance of propagation (200 m) in the high-permeability layer. The distances plotted in Figure 8 are conservative because flow in a reservoir that requires sweep improvement measures is not likely to be completely cylindrical but takes place through preferential flow channels. The calculations would also be more favorable for a linear (as opposed to radial) drive pattern.

The experimental data in Figure 6 indicate that many surfactant systems allow for sufficient surfactant propagation even at high-salinity conditions, but clearly, the proper choice of surfactant is essential for a successful foam-flood. The following sections outline some surfactant systems suitable for foam-flooding at high salinity, and give a detailed review of the adsorption properties of these surfactants.

Mechanisms of Surfactant Loss

The following mechanisms are usually major contributors to surfactant loss in an oil-free porous medium:

- adsorption at the solid–liquid interface
- precipitation
- chemical degradation

Chemical stability of the surfactant is of concern in steam-foam flooding, where temperatures that exceed 180 °C may be encountered (*3, 13–17*). Surfactants that have been found thermally stable at steam-flood temperatures include alkyl aromatic sulfonates and olefin sulfonates. A drawback of these surfactants, particularly the aromatic compounds, is their limited solubility in salt solutions. Even when a surfactant is completely soluble in the reservoir brine, surfactant-induced ion exchange may cause precipitation during flow through the porous medium (*7*). Micelles of ionic surfactants bind divalent ions in preference to monovalent ions and provide a medium that exchanges calcium with clays (*18–21*). This behavior leads to a change in brine composition with possible loss in surfactant solubility. Therefore, the surfactant concentration should be well below the solubility limit in order to prevent surfactant loss by precipitation during foam-flooding.

Surfactant solubility and chemical stability are more easily assessed and controlled by proper surfactant selection than adsorption at the solid–liquid interface. In principle, proper foam-flood design should completely eliminate surfactant loss caused by the first two mechanisms. The same cannot be said about adsorption. First, there will always be some interaction between the surfactant and the solid, and therefore, some adsorption. Second, a higher degree of uncertainty in the parameters that determine adsorption makes adsorption losses more difficult to predict. A petroleum reservoir is a complex system consisting of multicomponent solid and fluid phases. The composition of the solid and liquid phases and its variation within the same reservoir may affect adsorption, but this composition is often not well-defined. Some of the factors determining adsorption of foam-forming surfactants, data collected in an effort to gain a better understanding of adsorption mechanisms, and a discussion on methods to minimize adsorption are the main topics of this chapter.

The presence of oil is obviously a factor that influences surfactant loss. Several factors in addition to those listed previously may contribute to surfactant loss when oil is present:

- partitioning of surfactant into oil
- surfactant adsorption at the oil–water interface
- deactivation of surfactant by binding to crude oil asphaltenes
- coadsorption of surfactant and oil

Most foam-forming surfactants, particularly those suitable for high-salinity conditions, are very hydrophilic and do not partition into oil, eliminating the first of the listed mechanisms. The amount of surfactant adsorbed at the oil–water interface depends on the surface excess of surfactant at this interface and on the amount of interface present. Although the surface excess of surfactant at the oil–water interface can be estimated from interfacial tension data using the Gibbs adsorption equation, the amount of interface that is present is not easily accessible to measurement.

The amount of surfactant that may be deactivated by asphaltenes possibly can be comparable to the amount adsorbed at the solid–liquid interface (22), and this mechanism may thus contribute significantly to surfactant loss in reservoirs containing oil of high asphaltene content. However, the experiments described in reference 22 were carried out with a surfactant that partitions into oil. Because many foam-forming surfactants do not partition significantly, surfactant loss by complexation to asphaltenes is not expected to contribute substantially to surfactant loss.

Coadsorption of surfactant and oil on the solid may enhance surfactant adsorption by reducing electrostatic repulsion between surfactant head-groups in the adsorbed layer (23, 24) or may decrease surfactant adsorption by reducing the packing density of surfactant molecules in the adsorbed layer or by occupation of some of the adsorption sites on the solid surface by oil (23, 25). Furthermore, natural surfactants from crude oil may occupy some of the adsorption sites and thus act as sacrificial adsorbates by reducing adsorption of the injected surfactant (26). Because of the multitude of possible mechanisms, contradictory results regarding the effect of oil on surfactant loss have been reported. Surfactant retention has been found to increase (23, 24), decrease (23, 25, 27–31), or remain the same (26, 32, 33) in the presence of an oil phase.

When dealing with a foam, gas–liquid interfaces will be present in addition to solid–liquid and liquid–liquid interfaces. Surfactant adsorption at the gas–liquid interface is obviously required for foam formation and therefore cannot be considered a mechanism of surfactant loss. Because gas is always the nonwetting fluid, the presence of a gas phase is not expected to affect contact between the solid and the aqueous phase and is not likely to affect adsorption of a water-soluble surfactant at the solid–liquid interface. Limited data comparing surfactant adsorption from a foam with adsorption from a bulk liquid during flow through a sand pack have indicated that this is, indeed, the case (34). If surfactant adsorption at the gas–liquid interface were to affect adsorption at the solid–liquid interface, the effect would likely be a reduction in adsorption on the solid because of a reduced surfactant concentration in the bulk aqueous phase.

Surfactant Adsorption at the Solid–Liquid Interface

Mechanisms of Surfactant Adsorption: Theory. Adsorption of surfactant at the solid–liquid interface can take place by a number of different mechanisms. Accordingly, the free energy of adsorption (ΔG_{ads}) can be expressed as follows (35, 36):

$$\Delta G_{ads} = \Delta G_{elec} + \Delta G_{cov} + \Delta G_{c-c} + \Delta G_{c-s} + \Delta G_{H} + \Delta G_{solv}$$

where the different free energy terms arise from electrostatic interactions (ΔG_{elec}), covalent bonding (ΔG_{cov}), van der Waals and hydrophobic (chain–chain) interactions (ΔG_{c-c}) between surfactant hydrophobes in the adsorbed phase, nonpolar chain–solid interactions (ΔG_{c-s}), hydrogen bonding (ΔG_{H}), and solvation and "desolvation" of adsorbate and adsorbent species (ΔG_{solv}). In an EOR application, the relative contribution of each term to overall adsorption is determined by parameters such as the following:

- Solid: rock type (sandstone or carbonate), mineral composition (clays, other minor or trace minerals), wettability, surface charge, and specific surface area
- Solvent: brine salinity, and brine composition

- Surfactant: surfactant type, surfactant composition, and surfactant solubility
- Temperature

Surfactant adsorption mechanisms are strongly dependent on the type of solid and are often classified accordingly. Adsorption on polar, practically insoluble solids, such as quartz, silica, and kaolinite, is dominated by electrostatic and cohesive chain–chain (hydrophobic) interactions (*35–39*). Adsorption probably takes place first by surfactant–solid electrostatic interaction or ion exchange, with the surfactant's polar group oriented toward the solid surface. As the surfactant concentration increases, surfactant molecules in the adsorbed phase aggregate by hydrophobic interactions between surfactant hydrophobes. Models that have been proposed for this process include Fuerstenau's "hemimicelle" hypothesis (*37*), Harwell's "admicelle" hypothesis (*40*), and Rupprecht's model of small surface aggregates (*39*). Surfactant aggregation leads to the formation of a bilayered structure or small surfactant aggregates with interpenetrating hydrocarbon chains. Molecules in the adsorbed phase are arranged so that surfactant polar groups are in contact with the solid and the bulk liquid, and surfactant hydrocarbon chains are in contact with each other. Therefore, hydrocarbon chain–water contact is minimized.

Surfactant adsorption on saltlike minerals, such as calcite and dolomite, is a more complex process and is less understood than adsorption on oxide surfaces. These minerals are relatively soluble and when in contact with an aqueous medium develop an interfacial region of complex composition (*41–43*). In addition to the two mentioned mechanisms of adsorption, covalent bonding, salt formation between surfactant and lattice ions at the solid surface, ion exchange of surfactant with lattice ions, and surface precipitation have been suggested as adsorption mechanisms (*36, 43–47*). The dissolution products of sparingly soluble minerals may interact with the surfactant, precipitate or adsorb at the solid surface, or lead to mineral transformations that affect surface composition and electrochemical properties (*46, 48–52*). All these factors can be expected to influence surfactant adsorption.

Petroleum reservoirs can exhibit the full range of wettabilities from water-wet to oil-wet (*53*). Adsorption of crude oil heavy ends modifies solid surface properties and is thought to change reservoir wettability toward more oil-wet. Surfactant adsorption on hydrophobic surfaces takes place by hydrophobic interactions between surfactant hydrocarbon chains and the solid surface (*35, 54–58*). At low surfactant concentrations, surfactant molecules are oriented parallel to the surface. As the surfactant concentration increases, hydrophobic interactions between surfactant hydrophobes become significant. The surfactant molecules become oriented vertically to the surface with the polar groups toward the aqueous phase.

Surfactant–solid and surfactant–surfactant hydrophobic interactions lead to minimization of solid–water and surfactant-chain–water contact and are energetically favorable. Unlike hydrophilic surfaces, hydrophobic surfaces do not lead to significant structuring of interfacial water, and the interfacial water is displaced from the surface relatively easily by the surfactant molecules. Consequently, surfactant adsorption on hydrophobic surfaces has often been found to be higher than adsorption on the corresponding hydrophilic surfaces (*39, 54, 56, 57, 59–62*), provided aqueous phase salinity is low.

The solvent properties of the aqueous phase for the surfactant affect surfactant adsorption as much as the properties of the solid. Petroleum reservoirs can vary widely in terms of brine composition, and if a pool has been waterflooded, salinity gradients may exist within the same pool. Brine salinity and ionic content strongly affect surfactant adsorption by influencing surfactant solubility and solid surface electrochemical properties. Increasing the brine ionic strength increases surfactant adsorption (*7, 10–12, 21, 33, 39, 63–69*) by decreasing surfactant solubility (*60, 64*) or by reducing electrostatic repulsion between surfactant head-groups in the adsorbed layer (*39*). A decrease in adsorption with increasing salt concentration has also sometimes been observed (*36, 70*) and has been attributed to competition between ionic surfactants and inorganic ions for the surface.

When the surfactant and the solid are oppositely charged, compression of the electric double layer by increased electrolyte concentration decreases the electrostatic potential for surfactant adsorption and thus reduces adsorption (*64*). When the surfactant and the solid are of like charge, electric double-layer compression increases surfactant adsorption by allowing the surfactant to access the surface more easily (*66*). Although monovalent ions lead to double-layer compression that affects only the magnitude of the solid surface charge, multivalent ions may specifically adsorb into the Stern layer and cause electrokinetic charge reversal (*36, 42, 70–74*). For a surfactant and solid of like charge, solid surface charge reversal caused by multivalent counterions will increase surfactant adsorption by electrostatic interaction. The presence of multivalent ions has been found to increase surfactant adsorption (*7, 11, 12, 33, 65, 66, 75*) by an amount that is larger than ionic strength effects alone.

Adsorption has sometimes been related to surfactant solubility or relative hydrophobicity (*60, 76–79*). The solubility of anionic surfactants increases with increasing temperature, decreasing salinity and divalent ion concentration, and decreasing surfactant molecular weight. Accordingly, surfactant adsorption has been found to increase with decreasing temperature (*10, 34, 80–83*), increasing salinity (*7, 10–12, 21, 33, 39, 63–69, 82*) and divalent ion concentration (*7, 11, 12, 33, 65, 66, 75*), and increasing surfactant molecular weight (*76, 79, 84, 85*). In general, surfactants that contain a phenyl group are less water-soluble than those that do not. Accordingly, adsorption of aromatic surfactants has been found higher than adsorption of aliphatic surfactants (*7, 34, 82*).

The surfactant-type will obviously also affect adsorption levels. Anionic surfactants have been most commonly used in EOR, because most sandstone reservoirs are thought to carry a negative surface charge that minimizes surfactant adsorption by surfactant–solid electrostatic interactions. Nonionic surfactants have also sometimes been described in the EOR literature. Adsorption of these surfactants is thought to take place by hydrogen bonding (*36, 79*). Some recent studies indicate that cationic and amphoteric surfactants show potential in foam-flooding (*2, 11, 12, 86*). Adsorption of amphoteric surfactants has rarely been studied but is thought to take place by interactions of the solid with both the negative and the positive group in the surfactant molecule (*11, 12, 87, 88*).

The Surface Charge of Reservoir Solids. Adsorption of ionic surfactants is strongly dependent on solid surface charge. Extensive data on the surface-charge properties of minerals is available in the mineral flotation literature (*71*). Measurements of solid surface charges are usually restricted to aqueous media of low ionic strength, typically up to 10^{-2} mol/L total dissolved solids concentration. Brines of much higher salinity (up to about 4 mol/L) and of complex composition are encountered in petroleum reservoirs. Because the measurement of electrophoretic mobility at high electrolyte concentrations is experimentally difficult or impossible, data on the surface-charge properties of reservoir solids in typical reservoir brines are not readily available in the literature. The work of Schramm et al. (*74*), who measured electrophoretic mobilities of a number of reservoir solids in a variety of brines at salinities up to 0.4 mol/L, is an exception. Isoelectric points of typical reservoir solids have been compiled from the literature in Table V.

The oxides of silicon and aluminum (e.g., quartz, alumina, and kaolinite) develop a pH-dependent surface charge by dissociation of the surface hydroxyls:

$$-\text{Si}-\text{OH}_2^+ \underset{}{\overset{H^+}{\rightleftarrows}} -\text{Si}-\text{OH} \underset{}{\overset{OH^-}{\rightleftarrows}} -\text{Si}-\text{O}^-$$

acid pH　　　　　　　　　　　　　　basic pH

Isoelectric points of quartz determined by different investigators range from pH 1.5 to 3.7 (Table V). The surface charge of kaolinite arises from two different mechanisms: A pH-dependent charge due to the dissociation

Table V. Isoelectric Points of Some Reservoir Solids

Solid	Isoelectric Point	Aqueous Medium	Ref.
Quartz	3.7	water, 10^{-4}–10^{-2} M NaCl	70
	1.5–3.7	not specified	89
	2–3.7	not specified	90
	2.7	water	91
	2.3	10^{-3} M NaClO$_4$	71
	2.4	water	87
	2	10^{-3} M NaCl	92
	1.5	not specified	93
	<2–3.5	10^{-3}–0.4 M NaCl and reservoir brine	74
Berea sandstone	3–5	10^{-3}–10^{-1} M NaCl	94
	4–5	10^{-3}–0.4 M NaCl and reservoir brine	74
Kaolinite	<3.5	not specified	95
	3.3–4.6	not specified	96
	<4	not specified	97
	4	0.6×10^{-3}–2×10^{-3} M NaCl	98
	<3	10^{-3} and 10^{-2} M NaCl	99
	4	10^{-4} M NaCl	99
	4.5–5	not specified	52
	5–6	not specified	100
	<1.5	10^{-3} M NaCl	92
	5	not specified	62
	3.5	not specified	93
	<2	10^{-3}–0.4 M NaCl and reservoir brine	74
Calcite	8–9.5	water	41
	10.8	not specified	101
	7–7.8	water	102
	>12.5	not specified	71
	5.4[a]	water, 10^{-2} M KCl	43
	8.2	2×10^{-3} M NaClO$_4$	103
	10.1	2×10^{-3} M NaCl	104
	4[a]	10^{-3} M KCl	105
	9.5	water	106
	10.5	water, 2×10^{-3} M KNO$_3$	49
	<6	water	107
	10.5	10^{-3} M NaCl	92
	7[a]	2×10^{-3} M NaClO$_4$	108
	8.2	water	64
	<7–8.6	5×10^{-3} M NaCl, 5×10^{-3} M NaCl + 10^{-3} M NaHCO$_3$	73
	10	10^{-2} M KCl	109
	7–9.6	water, 10^{-3}–10^{-2} M NaCl	110
	<7	10^{-3}–0.4 M NaCl and reservoir brine	74

Table V. *Continued*

Solid	Isoelectric Point	Aqueous Medium	Ref.
Dolomite	3^a	not specified	111
	<7	water, 10^{-3}–10^{-2} M KCl	42
	>12	water, saturated $CaSO_4$	71
	7	not specified	100
	9.5	not specified	62
	<7	10^{-3}–0.4 M NaCl and reservoir brine	74

^{a}Extrapolated.

of surface hydroxyls at the crystal edges, and a permanent negative charge of the crystal faces that arises from isomorphous substitutions in the crystal lattice (*97, 112*). Depending on solution pH, the face and edge charges may be of equal or opposite sign. The net isoelectric point (iep) of kaolinite can range from a pH lower than 1.5 to a pH of 6 (Table V). Both quartz and kaolinite thus carry a negative surface charge at neutral pH, and this charge leads to relatively low adsorption of anionic surfactants by electrostatic interaction with the solid.

The following surface-charge generation reactions have been suggested for calcite and dolomite upon preferential dissolution of lattice ions (*42, 43*):

$$\text{surface}-CO_3^- + H_2O \rightleftarrows \text{surface}-CO_3H + OH^-$$
$$\text{surface}-CO_3H + H^+ \rightleftarrows \text{surface}-CO_3H_2^+$$

$$\text{surface}-Me^+ + H_2O \rightleftarrows \text{surface}-MeOH + H^+$$
$$\text{surface}-MeOH \rightleftarrows \text{surface}-MeO^- + H^+$$

where Me is either calcium or magnesium. The relatively high degree of solubility of calcite and dolomite results in the formation of different ionic species in solution and (through preferential dissolution, adsorption, surface precipitation, or mineral transformation) at the solid–liquid interface (*41, 42*). Measured zeta potentials are extremely sensitive to sample preparation and experimental conditions (*41, 73, 103, 104*). Therefore, measurements of the isoelectric point of calcite in water or dilute electrolyte have yielded much more divergent results than for quartz and kaolinite. Measured isoelectric points range from pH 4 to 11 (Table V). Some investigators have found calcite to be positively charged throughout the pH range studied, but others have found it negatively charged throughout. Similarly, the isoelectric point of dolomite can range from pH 3 to 10 (Table V). Because the results that are reported diverge, it is difficult to predict the solid surface charge of carbonates, particularly in complex brines of relatively high ionic strength.

In mixed mineral systems, dissolution–precipitation equilibria may lead to mineral transformations that can influence surface electrochemical properties *(48–50, 52)*.

Electrophoretic mobilities of quartz and clays, both isolated from Berea sandstone, and of calcite and dolomite in three different brines are shown as a function of pH in Figure 9 *(74)*. These results are unique in that they were obtained with brines of higher ionic strength than are usually used in the measurement of solid surface charge. All three brines have the same ionic strength (0.406 mol/L) but differ in composition. The reservoir brine contains significant levels of divalent cations, which are mostly Ca^{2+}. Electrophoretic mobilities at pH 7, also taken from reference 74, are listed in Table VI.

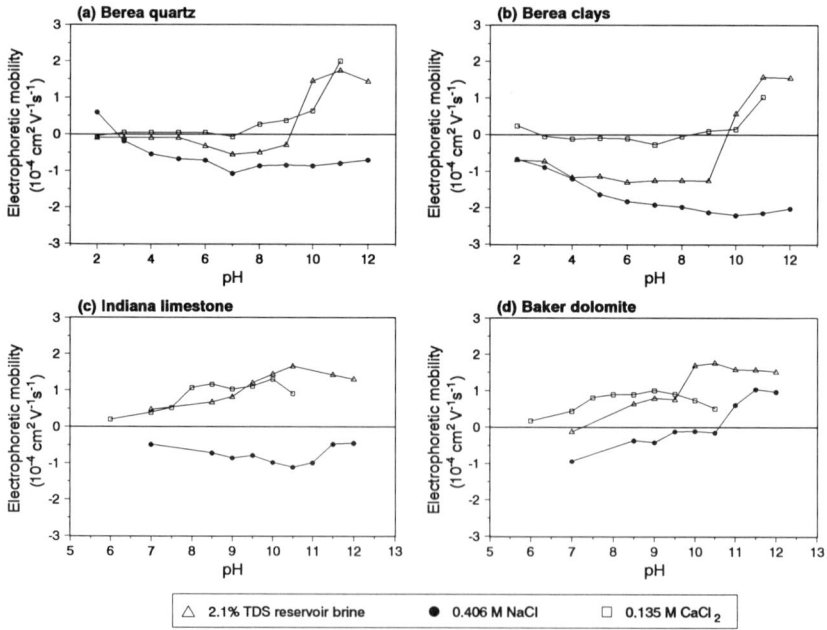

Figure 9. Electrophoretic mobilities of rock particles as a function of pH and brine ionic content. The ionic strength of all brines is 0.406 mol/L. (Reproduced with permission from reference 74. Copyright 1991 Elsevier Science Publishers.)

Table VI. Electrophoretic Mobilities of Some Reservoir Rock Particles at pH 7 in Brines of Different Composition and Constant Ionic Strength (0.406 mol/L)

Solid	NaCl Brine	Synthetic Reservoir Brine	$CaCl_2$ Brine
Berea sandstone	−1.30	−0.62	+0.21
Clays from Berea sandstone	−1.92	−1.25	−0.27
Quartz from Berea sandstone	−1.07	−0.55	−0.07
Baker dolomite	−0.94	−0.13	+0.44
Indiana limestone	−0.49	+0.47	+0.39

NOTE: Electrophoretic mobility is in the unit (10^{-4} cm^2 V^{-1} s^{-1}).
SOURCE: Reference 74.

The isoelectric point of both oxides (Berea quartz and clays) is at acidic pH (Figure 9), in agreement with literature data (Table V). Increasing the divalent cation concentration at constant ionic strength changes the electrophoretic mobilities toward less negative values. A second point of zeta potential reversal is observed at basic pH and has been attributed to adsorption of metal hydroxides on the solid surfaces (*71, 113*). The carbonates are more strongly affected by aqueous phase composition: A complete reversal of surface charge from negative to positive is observed in the presence of divalent cations (Figure 9). The addition of divalent cations to the aqueous medium has often been found to change the surface charge of quartz (*70, 71, 74*), calcite (*43, 71, 73, 74, 102, 103, 105, 114*), and dolomite (*42, 74*) toward less negative or to positive values. The change in surface charge has been attributed to adsorption of the divalent cations into the Stern layer (*70, 72*) or, with carbonates, to preferential dissolution of carbonate ions in the presence of excess Ca^{2+} or Mg^{2+} in the aqueous medium. This trend is also clearly reflected by the data in Table VI. All five solids carry the most negative surface charge in the NaCl brine and the least negative, or most positive, surface charge in the $CaCl_2$ brine. In any of the three brines, the electrophoretic mobilities at pH 7 exhibit the following trends (Table VI):

Berea clays < quartz < dolomite < limestone
most negative   least negative

The trends observed in the surface charges of these solids have bearing on surfactant adsorption, as will be evident.

Adsorption of Foam-Forming Surfactants: Experiments. This section discusses the adsorption behavior of a number of foam-forming surfactants under a wide variety of conditions. Surfactant adsorption data that appear in the petroleum literature are often difficult to compare because they are measured by different methods (batch or core-flood experiments), and experimental conditions may vary widely and are frequently not completely specified. The data discussed in this section, which are taken from references 7–12, 34, and 82 and some unpublished results, constitute the most extensive set of adsorption data for commercial foam-forming surfactants that were measured in a consistent manner by performing core-floods.

A core-flood for adsorption determination consists of injecting a measured volume of surfactant solution containing a nonadsorbing tracer into a brine-saturated core and collecting effluent fractions at the core outlet. Chemical analysis of the effluent samples allows the calculation of an adsorption level based on material balance considerations and also results in a set of effluent profiles for the surfactant and the tracer. In addition to the material balance, adsorption is evaluated by matching experimental effluent concentrations from the core-floods with a convection–dispersion–adsorption numerical model. The model parameters then allow calculation of a complete adsorption isotherm.

Examples of experimental and simulated effluent profiles and the adsorption isotherm based on the simulated surfactant profile are shown in Figure 10. For the data discussed in this chapter, adsorption was modeled using the surface excess formalism (*8–10, 115*), or in some cases, the Langmuir adsorption model (*8, 9, 34, 82*) as discussed in detail in these references. The model used to calculate the adsorption isotherm in Figure 10 assumes that surfactant adsorption takes place from the monomer

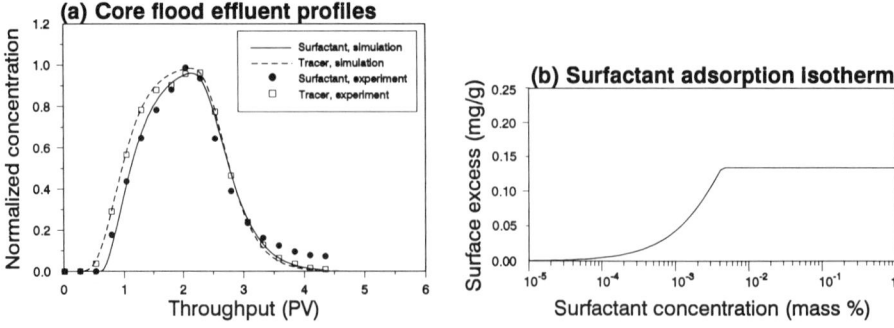

Figure 10. Example of experimental and simulated effluent concentrations from a core-flood and the surfactant adsorption isotherm calculated from the best-fit adsorption model parameters.

phase but not from the micelles. Hence, adsorption reaches a plateau at surfactant concentrations that exceed the critical micelle concentration.

In order to facilitate comparison between different systems in the discussion that follows, adsorption levels are taken from the adsorption isotherms at a fixed surfactant concentration of 1 mass %. Adsorption levels of surfactants typically used in foam-flooding are usually at or near their plateau at concentrations of this magnitude and are only weakly dependent on surfactant concentration.

Several abbreviations are used for surfactant names appearing in the figures discussed in this section. These are explained in the Appendix, which also lists trade names, suppliers, and chemical names for the surfactants. When the brines used in the adsorption experiments contain a single salt, they are identified by the salt they contain and its concentration. Brines for which only the total dissolved solids (TDS) concentration is specified are reservoir brines that contain mixtures of salts. The composition of these brines can be found in the references given. When concentrations are expressed as a percentage, mass percent is implied.

Dependence of Adsorption on Temperature. Figures 11 and 12 show the temperature dependence of adsorption for several foam-forming surfactants on sandstone or unconsolidated sand. Physical adsorption is an exothermic process and is expected to decrease with increasing temperature. This trend is observed for the anionic surfactants (Figures 11a and 12): Adsorption decreases up to an order of magnitude when the temperature is raised from 50 to 150 °C. In contrast, adsorption of the amphoteric surfactants is affected very little by temperature and may even show a slight increase with temperature in some cases (Figure 11b). An increase in adsorption with temperature has sometimes been taken as an indication of chemisorption (36).

Temperature gradients exist in a reservoir that has been subjected to steam-flooding, and therefore, a knowledge of the dependence of surfactant adsorption on temperature is important in the evaluation of steam-foam processes. Low adsorption levels in high-temperature or swept zones are beneficial to process performance because gas mobility reduction or gas blockage in the swept zones is desired.

Dependence of Adsorption on Brine Salinity and Divalent Ion Content. Brine salinity and composition probably constitute the primary criteria for selecting surfactants for foam applications. Many Canadian pools that are being flooded with hydrocarbon solvents contain near-saturated formation brines. Some of these pools have been waterflooded with fresh water, and therefore, salinity gradients exist. In addition, the majority of hydrocarbon-miscible floods in Canada are conducted in carbonate formations that contain formation waters with high levels of hardness.

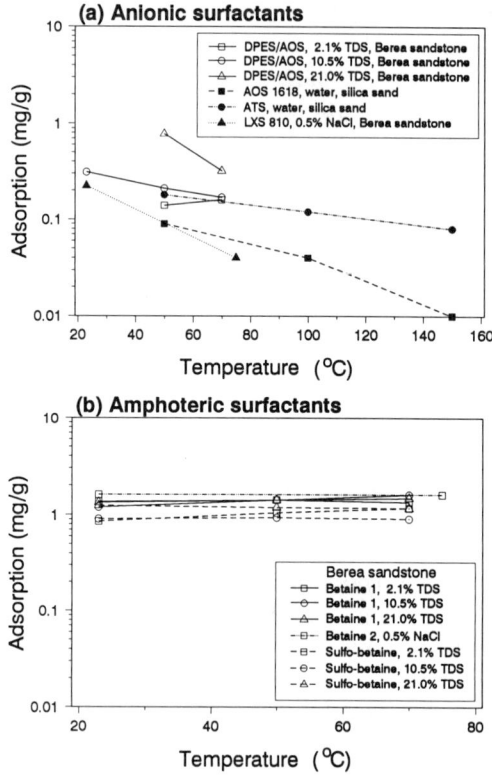

Figure 11. The dependence of surfactant adsorption on temperature, measured in Berea sandstone or silica sand. Adsorption levels were obtained using the surface excess model (7–10).

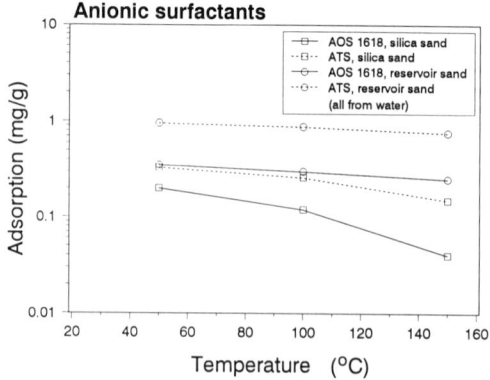

Figure 12. The dependence of surfactant adsorption on temperature. Adsorption levels were obtained using the Langmuir model (8, 9, 34, 82).

The dependence of adsorption of some anionic and amphoteric surfactants on brine salinity is shown in Figure 13. The measurements represent several different solids. A property that distinguishes three of the surfactants (diphenyl ether disulfonate–α-olefin sulfonate (DPES–AOS) mixture, betaine, and sulfobetaine) from those more commonly described in the literature dealing with laboratory experiments and field applications of foams is their complete solubility and excellent foaming characteristics in brines of extremely high salinity (2). Adsorption of these surfactants was measured at salinities up to 21 mass % TDS using synthetic reservoir brines of increasing overall salinity and constant ionic composition. These brines contained significant amounts of divalent cations, mostly Ca^{2+} (10–12). Two of the surfactants, an α-olefin sulfonate (AOS) and an internal olefin sulfonate (IOS) have limited solubility in brine. Adsorption measurements with these surfactants were carried out

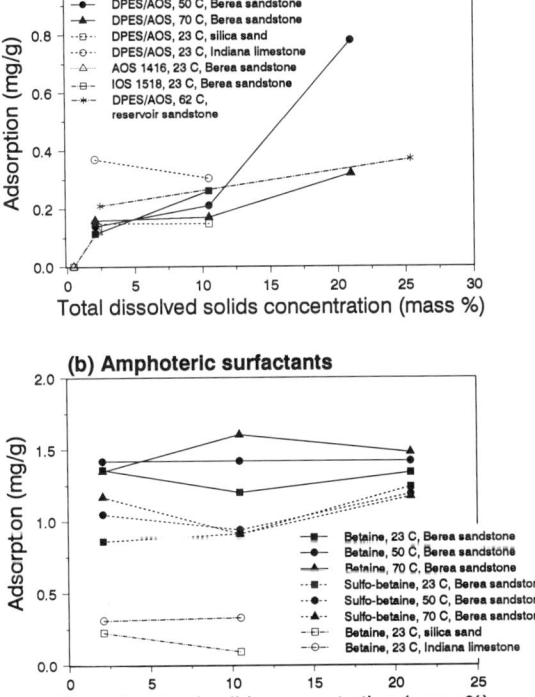

Figure 13. The dependence of surfactant adsorption on brine salinity (7, 10–12, and Mannhardt, K., Novosad, J. J., Petroleum Recovery Institute, unpublished data).

in sodium chloride brines and were restricted to relatively low salt concentrations (7).

With one exception (DPES–AOS on limestone), the anionic surfactants exhibit an increasing trend in adsorption with increasing salt concentration (Figure 13a), a result that is consistent with literature data. This result can be attributed to three effects: surfactant solubility, surfactant–solid electrostatic interactions, and reduced electrostatic repulsion between surfactant head-groups in the adsorbed layer in the presence of excess counterions.

An increase in electrolyte concentration reduces the solubility of anionic surfactants in the aqueous phase and increases their tendency to accumulate at the solid–liquid interface. An increase in temperature offsets the loss in solubility to some degree: For the DPES–AOS on Berea sandstone, the slopes of the lines in Figure 13a decrease as the temperature increases, and this finding lends support to the hypothesis that surfactant adsorption is related to surfactant solubility. Adsorption of surfactants that are less salt-tolerant than the DPES–AOS, such as the AOS and the IOS, increases much more steeply with salinity. Both surfactants adsorb negligibly at salinities of 0.5 mass % NaCl, but adsorb similarly to the DPES–AOS at a salinity of 2.3 mass %. At moderate salinities (on the order of 3 mass %), these surfactants precipitate, which severely limits their applicability to foam-flooding in many reservoirs that are currently being flooded with hydrocarbon solvents.

Electric double-layer compression due to increased electrolyte concentration may reduce surfactant adsorption when the surfactant and the solid are oppositely charged (64), but may increase adsorption when the surfactant and the solid are of like charge (66). At the conditions of the adsorption experiments of Figure 13, most likely the sandstone carries a negative surface charge, and the limestone is positively charged. Hence, there is a decrease in adsorption of the anionic DPES–AOS on limestone and an increasing trend in adsorption of all anionic surfactants on sandstone when the brine salinity is increased. Increasing the overall salinity of the synthetic reservoir brine also increases the divalent ion concentration. This increase results in a less negative surface charge at higher overall salinities, in turn resulting in increased surfactant–solid electrostatic attraction, and hence in increased adsorption of anionic surfactants on sandstone.

The tested amphoteric surfactants (betaine and sulfobetaine) are highly salt-tolerant and are excellent foamers. Adsorption of these surfactants is less dependent on salinity than adsorption of anionic surfactants, and the trends are not monotonic (Figure 13b). Adsorption of amphoteric surfactants may proceed by a complex interplay of mechanisms involving electrostatic and complexation mechanisms of both the cationic and the anionic group in the surfactant molecule (12, 87, 88). The trends in adsorption are due to a combination of effects that cannot easily be resolved into contributing mechanisms. The amphoteric surfactants clearly do not follow the general trend of decreasing adsorption with increasing solubility. These surfactants are highly soluble, but they adsorb more strongly on sandstone than the anionic surfactants. Probably adsorption of amphoteric surfactants proceeds by different mechanisms than adsorption of anionics.

The effect of divalent cations on surfactant adsorption is shown in Figure 14, which provides a comparison of adsorption levels on several solids measured in sodium chloride brine with those measured in brines containing sodium chloride and divalent cations. The ionic strength of all brines is constant at 0.403 mol/L, thus ionic strength effects are eliminated. Evidently, the dependence of surfactant adsorption on divalent ions varies with the type of surfactant and rock. In most cases, adsorption is increased by the presence of divalent cations. Adsorption of the sulfobetaine is less sensitive to divalent cations than adsorption of the betaine and the anionic surfactants. Adsorption of three surfactants on dolomite is not influenced very strongly by divalent cations.

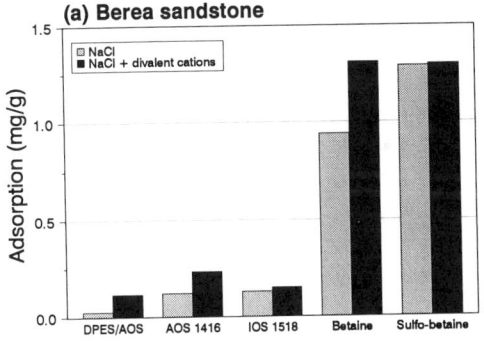

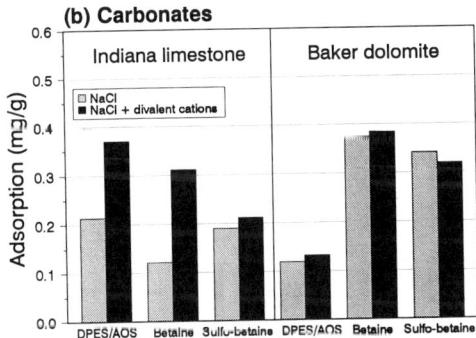

Figure 14. The dependence of surfactant adsorption on divalent cations, measured in brines of ionic strength 0.403 mol/L (7, 11, 12).

A previous section of this chapter showed that divalent cations cause the surfaces of reservoir solids to become less negatively charged or to reverse their charge to positive. Electrostatic attraction between the surface and anionic surfactants is thus increased. Alternatively, anionic surfactants may complex to divalent cations to form, for example, $CaRSO_3^+$, resulting in a cationic species that is electrostatically attracted to negatively charged surfaces (where R represents the surfactant hydrophobe)(66). Similarly, adsorption of cationic surfactants on carbonates is reduced by divalent cations (86, 116). With amphoteric surfactants, the trends are not as easily explained. Divalent cations can have opposing effects on amphoteric surfactant adsorption (12). Changing the surface charge toward more positive values will decrease electrostatic attraction between the surfactant's cationic group and the surface, but will increase adsorption through the surfactant's anionic group by electrostatic or complexation mechanisms. Depending on the magnitude of these two effects, the presence of divalent cations may then influence adsorption in either direction.

Dependence of Adsorption on Rock Type. Table I shows that gas injection EOR projects are being conducted in sandstone and carbonate pools. Hydrocarbon- and CO_2-miscible projects are run largely in carbonate reservoirs. With the exception of several studies that report adsorption levels of EOR surfactants on carbonates (4, 11, 12, 24, 33, 62–64, 86), the petroleum literature has dealt almost exclusively with anionic, and sometimes nonionic, surfactant adsorption on sandstones, because most studies have been carried out with surfactants used in low-tension flooding. These surfactants are not considered suitable for application in carbonate reservoirs because of their low salinity and hardness tolerance. Foam-forming surfactants suitable for high-salinity environments include amphoteric surfactants (2). The adsorption behavior of this surfactant type has also rarely been studied (10–12, 87, 88).

Figure 15 shows adsorption levels of three foam-forming surfactants suitable for high-salinity environments on four different solids representative of reservoir materials (12). To account for the differing specific surface areas of the rocks, adsorption is expressed in terms of the "packing density" of surfactant molecules on the solid surfaces. The data in Figure 15 show that surfactant adsorption is strongly dependent on the type of solid. Provided that foam performance is similar for all three surfactants, the proper matching of surfactant to reservoir in terms of adsorption is a crucial step in the optimization of process performance and economics.

Adsorption of both amphoteric surfactants on sandstone and dolomite is significantly higher than adsorption of the anionic surfactant on these rocks. On limestone, all three surfactants exhibit intermediate adsorption levels that are quite similar among the three surfactants, particularly in sodium chloride brine.

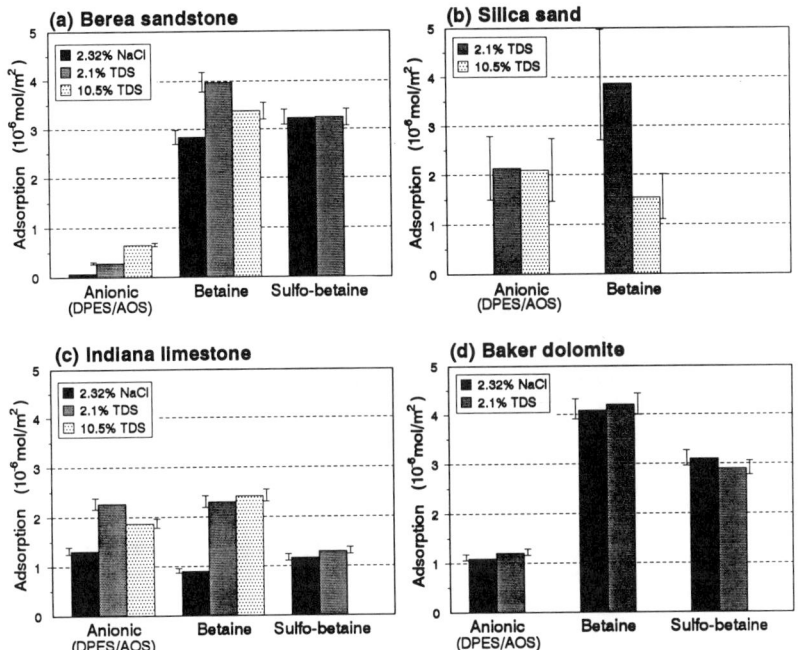

Figure 15. The effect of rock type on surfactant adsorption. (Reproduced with permission from reference 12. Copyright 1992 Elsevier Science Publishers.)

Clays are considered detrimental to EOR processes that are based on the injection of chemicals, such as foam-forming surfactants, because clays provide a large amount of surface area for adsorption. Table VII shows a comparison of specific surface areas of some clays (*97, 117, 118*) and of the solids used in the adsorption experiments of Figure 15 (*12, 119, 120*). Figure 15 allows comparison of adsorption levels in Berea sandstone, which consists mainly of quartz and 6–8% clays, with adsorption on clean quartz sand.

When normalized to unit surface area, the adsorption density of the anionic surfactant is higher on quartz than on Berea sandstone because quartz carries a more positive surface charge than the clays (The clays provide most of the surface area for adsorption in Berea sandstone). If it is assumed that the betaine adsorbs on sandstone at least in part by its cationic group, then the lower adsorption density of the betaine on quartz than on Berea sandstone can also be attributed to electrostatic interactions. Matrix grains of the size encountered in typical reservoir rocks have low specific surface areas. Accordingly, the absolute amount of surfactant adsorbed or the amount adsorbed per unit mass of rock is lower for a clean sand than for a sand containing clays (*12, 34, 82*). Therefore, the

Table VII. Specific Surface Areas of Some Reservoir Rocks and Clays

Solid	Specific Surface Area (m^2/g)
Clays	
Kaolinite	9–23
Smectite	35–110
Illite	40–110
Clorite	14–42
Reservoir rocks	
Berea sandstone	0.8–1.2
Indiana limestone	0.37–0.54
Baker dolomite	0.22–0.29
Quartz sand[a]	0.16

[a]Average grain size was 99 μm.
SOURCE: References 12, 119, and 120 for reservoir rocks, and references 97, 117, and 118 for clays.

high surfactant losses associated with reservoir rocks of high clay content can be attributed largely to the clays' specific surface area.

Possible adsorption mechanisms for the systems in Figure 15 have been discussed in detail in previous publications (11, 12). Adsorption of anionic and amphoteric surfactants is related to solid surface charge in Figure 16 (12). The data clearly indicate an increasing trend in adsorption of the anionic surfactant with an increasingly more positive surface charge, which is consistent with an electrostatic mechanism of adsorption (Figure 16a). The solid surface becomes more positive either when the rock is changed from sandstone to dolomite to limestone in a specific brine, or when divalent cations are added to the brine for any specific rock type.

A similar trend does not exist for the amphoteric surfactants (Figures 16b and 16c). Amphoteric surfactants may adsorb either by their anionic or by their cationic group, depending on solution pH and the surfactant's and the solid's isoelectric points (87, 88). The high adsorption levels of both amphoteric surfactants on sandstone suggest that adsorption takes place primarily by electrostatic attraction between their cationic group and the negatively charged sandstone surface. Considering the complex surface chemistry of carbonates, amphoteric surfactants in combination with carbonate surfaces make for an exceedingly complex system. Depending on aqueous-phase composition, the carbonates may carry a negative or a positive surface charge. The amphoteric surfactants may then be electrostatically attracted by either of their ionic groups. In addition, adsorption

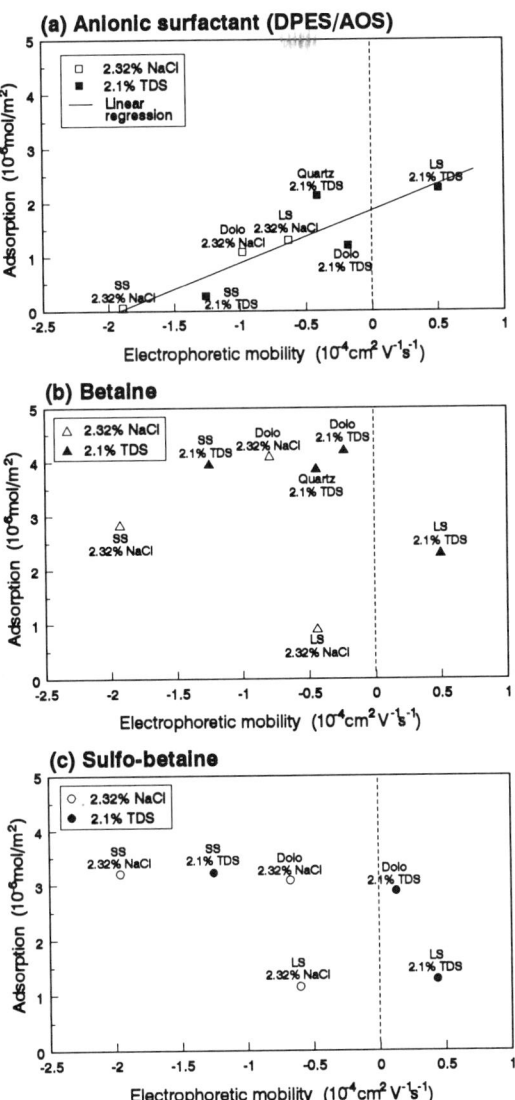

Figure 16. The effect of solid surface charge on adsorption of anionic and amphoteric surfactants. Key: SS, Berea sandstone; LS, Indiana limestone; and Dolo, Baker dolomite. (Reproduced with permission from reference 12. Copyright 1992 Elsevier Science Publishers.)

may be enhanced by complexation of the surfactants' anionic group with lattice cations or divalent cations in solution (12).

Surfactant adsorption generally causes a change in the electrophoretic mobility of a solid. Electrophoretic mobilities of the solids used in the adsorption measurements of Figure 16 were measured before and after equilibrating them with surfactant solutions. In Figure 17, the change in solid surface charge caused by adsorbed surfactant (surface charge of solid with adsorbed surfactant minus surface charge of solid without adsorbed surfactant) has been plotted against the surface charge of the clean solid (*12*). Adsorption of the anionic surfactant causes an increasingly more negative change in surface charge for solids with a more positive surface charge. Similarly, adsorption of the cationic surfactant causes a positive change in surface charge, as would be expected from electrostatic interactions. With both amphoteric surfactants, the change in surface charge depends on the type of solid. Sandstone surfaces become less negative with adsorbed surfactant, and carbonate surfaces become more negatively charged. These trends seem to indicate cationic behavior on sandstone surfaces, and more anionic behavior on carbonates. Amphoteric surfactants evidently adsorb by a complex interplay of mechanisms that cannot be resolved into contributing factors from the available data.

Displacement of the plane of shear caused by adsorbed surfactant may contribute to the changes in electrophoretic mobility plotted in Figure 17. For example, a small negative change in electrophoretic mobility (corresponding to a reduction of mobility from a positive value to zero) is

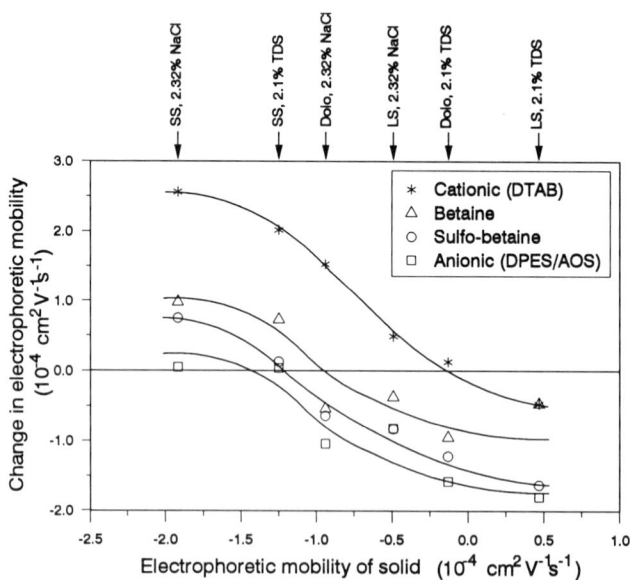

Figure 17. Change in particle electrophoretic mobility with adsorbed anionic, cationic, or amphoteric surfactant. Key: SS, Berea sandstone; LS, Indiana limestone; and Dolo, Baker dolomite. (Reproduced with permission from reference 12. Copyright 1992 Elsevier Science Publishers.)

indicated for the cationic surfactant–limestone–2.1% TDS brine system. However, because the trends observed in most of the data obtained with the anionic and the cationic surfactant follow the expected trends, the effects in Figure 17 probably are dominated by changes in surface charge caused by the adsorbed surfactant rather than by a shift in the plane of shear.

Effect of Wettability and Residual Oil on Adsorption. The surfactant adsorption data discussed in the preceding sections were derived from experiments carried out in clean, oil-free rocks. The presence of an oil phase may cause additional mechanisms of surfactant loss and may also alter system wettability toward more oil-wet by adsorption of petroleum heavy ends on the rock surfaces. Hydrocarbon- or CO_2-miscible flooding may promote wettability alteration by deposition of crude oil heavy ends. Because both EOR processes are candidates for foam-flooding, the effect of wettability on adsorption of foaming surfactants is of some interest. As has been discussed, surfactant adsorption mechanisms differ on hydrophobic and hydrophilic surfaces. In addition, oil may coat at least part of the solid surfaces in an oil-wet system and prevent a water-soluble surfactant from accessing the surface. On the other hand, an oil film coating oil-wet solid surfaces may provide a significant amount of oil–water interface where the surfactant may adsorb. This section discusses experimental data illustrating the effects of oil and system wettability on surfactant loss by adsorption.

Surfactant adsorption was measured in clean and in wettability-altered Berea sandstone cores. The wettability-altered Berea sandstone cores were prepared by adsorbing asphaltenes, which were precipitated with pentane from a heavy crude oil and then dissolved in methylbenzene, in the sandstone cores. After treatment with the asphaltene in methylbenzene solution, the methylbenzene was displaced from the cores with nitrogen, and the cores were dried. Surfactant adsorption was measured by surfactant-flooding clean and asphaltene-treated rocks that contained either no oil or waterflood residual oil. Because changes in wettability may depend on crude oil composition, three different oils (dodecane, a light crude oil, and a heavy crude oil) were used. The heavy oil contained 12 mass % asphaltenes, and the light crude oil had a low asphaltene content (0.2 mass %). Adsorption levels of an anionic and an amphoteric surfactant (DPES–AOS and betaine), both suitable for foam-flooding at high salinity conditions, were measured. Both surfactants are highly water soluble, and phase behavior studies ruled out partitioning into oil as a mechanism of surfactant loss.

To characterize the systems in which adsorption was measured, the wettability of clean and asphaltene-treated quartz crystals and Berea cores was assessed by several criteria. Water-advancing contact angles, measured with all three oils against clean quartz crystals immersed in brine, ranged from 32 to 56°, a result implying water-wet conditions. With asphaltene-treated quartz crystals, dodecane gave a water-advancing contact angle of 142°– 163°, a result indicating a change in the wettability of quartz to oil-wet (*121*). Figure 18 shows relative permeabilities to water at residual oil saturation, measured by waterflooding clean and asphaltene-treated Berea cores that contained different oils. The low end-point relative permeabilities to water in the clean cores indicate that water-wet conditions prevailed regardless of which oil was present. The significantly higher end-point relative permeabilities in the asphaltene-treated cores signify a shift toward more oil-wet behavior (*122, 123*).

Waterflood recoveries were similar in the clean cores with any of the oils, and recoveries were higher in the asphaltene-treated cores (Figure 19). This kind of behavior has been associated with mixed-wettability systems (*124–126*), defined by Salathiel (*125*) as systems in which the small pores are water-wet, and the larger pores form continuous oil-wet channels that allow efficient displacement of oil by water. All three criteria used for wettability assessment (contact angle, end-point relative permeabilities, and waterflood recoveries) indicate more oil-wet conditions after asphaltene treatment.

Figure 20 contains adsorption levels measured in the water-wet and mixed-wet sandstones in the absence and in the presence of residual oil. Similar trends are observed for the anionic and the amphoteric surfactant. Adsorption levels in the water-wet cores are essentially the same in the absence of oil and in the presence of any of the three oils. In a water-wet system, the solid surfaces are surrounded by water films. As long as water-wet conditions prevail, the aqueous surfactant solution is in contact

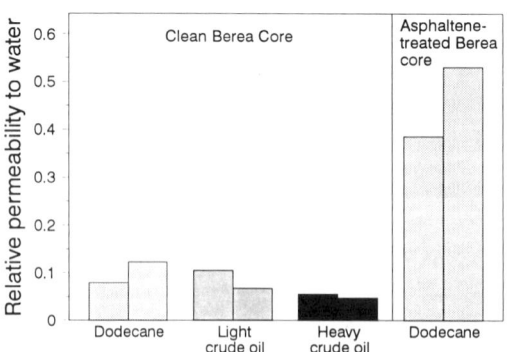

Figure 18. Relative permeability to water at residual oil saturation in clean and wettability-modified Berea cores containing different oils, calculated relative to the effective permeability to oil at irreducible water saturation.

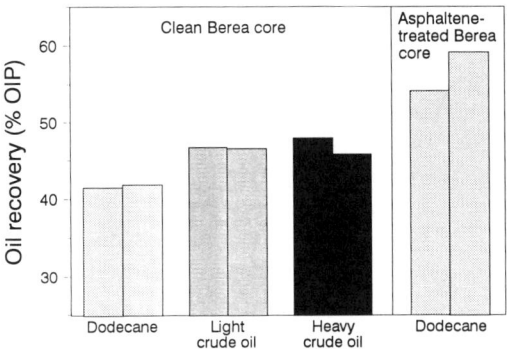

Figure 19. Waterflood recoveries from clean and wettability-modified Berea cores containing different oils.

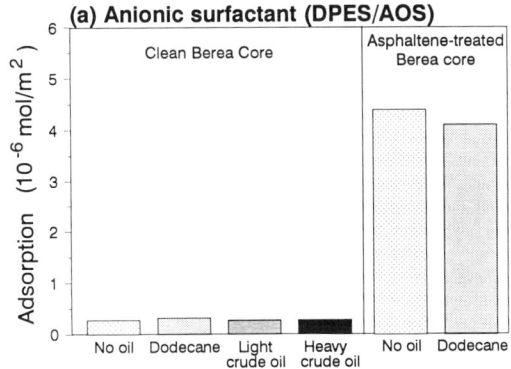

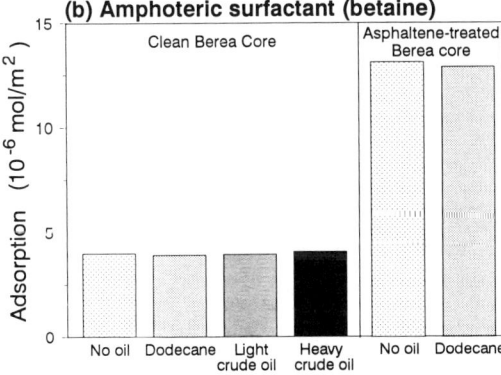

Figure 20. Surfactant adsorption levels, expressed as packing densities on the solid surface, in clean and wettability-modified Berea cores containing different oils.

with the solid surfaces, and adsorption is unaffected by the presence of an oil phase.

In the mixed-wettability cores, adsorption is affected only slightly by the presence of oil. The compensating effects of a decrease in the accessible solid–water interface and an increase in oil–water interface could be the reason for the relatively small effect. The most notable effect of the asphaltene treatment is the substantially higher adsorption density on the more oil-wet rock. Surfactant adsorption is higher on hydrophobic surfaces than it is on hydrophilic surfaces, a finding that is consistent with literature data.

The adsorption levels in Figure 20 are expressed as a packing density in the adsorbed phase or as the amount adsorbed per unit surface area. In addition to changing system wettability, the asphaltene treatment reduced the specific surface area of the Berea sandstone from 1 to 0.2 m^2/g. A reduction in specific surface area caused by asphaltene adsorption also was observed by Clementz (*127*), who attributed the decrease in surface area to blockage of the microporosity and interlamellar spaces in the clays. Although the more oil-wet rock has a higher affinity for the surfactants (Figure 20), it also provides less surface area for surfactant adsorption. The absolute amount of surfactant adsorbed or the amount adsorbed per unit mass of solid may be lower on the asphaltene-treated than on the clean sandstone, as with the amphoteric surfactant (Figure 21). Therefore, a change in wettability toward oil-wet that is caused by deposition of crude oil heavy ends strongly affects surfactant adsorption but is not necessarily detrimental to foam-flooding.

Minimizing Surfactant Adsorption

This section discusses three approaches that may be used to minimize surfactant adsorption: matching surfactant type to specific reservoir rock type based on surfactant ionic character and solid surface charge, application of surfactant mixtures, and sacrificial adsorbates (*128*).

Most reservoir rocks have a relatively small specific surface area, on the order of 1 m^2/g (Table VII), when compared to materials used in adsorption columns in the chemical industry that have specific surface areas of several hundred square meters per gram. The enormously large solid surface present in oil reservoirs comes from the sheer size of the reservoir volume. It cannot be eliminated, and therefore at least some adsorption of surface active materials is inevitable.

Even a cursory look at the consumption of chemicals illustrates the importance of minimizing surfactant adsorption (*128*). A small section of a densely drilled reservoir typical of older oil reservoirs that are prime

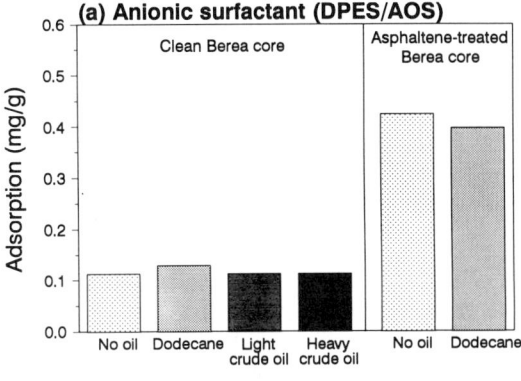

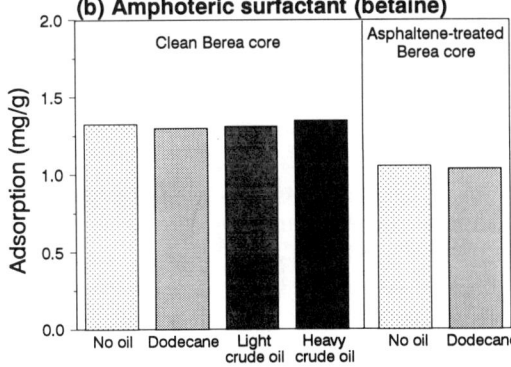

Figure 21. Surfactant adsorption levels, expressed as mass of surfactant lost per unit mass of rock, in clean and wettability-modified Berea cores containing different oils.

candidates for tertiary oil recovery processes is typified by a five-spot pattern of an injector and four producers that cover an area of 10 hectares. Assume a thickness of the oil-bearing zone of 5 m, a rock grain density of 2.6 g/cm^3, a porosity of 0.2, and a remaining oil saturation of 0.4. This brings the mass of reservoir rock to 1,040,000 metric tons. Even at what may appear to a chemist or a chemical engineer to be a nominal adsorption level of 1 mg of surfactant per gram of rock, 1,040 metric tons of surfactant would be adsorbed at the solid–liquid interface. This corresponds to consuming 26 kg of surfactant for 1 m^3 of potentially recoverable crude oil. At $4.00/kg of surfactant, $104 would be spent on the surfactant for each cubic meter of oil remaining in place. This example demonstrates that the value of the recovered oil would just cover the cost of the injected surfactant. Of course, additional costs are encountered during surfactant or foam injection. It is thus imperative to reduce surfactant adsorption by at least 1 order of magnitude to have any chance of economic success.

Sweep improvement processes, such as foam-flooding, have an economic advantage in that the injected surfactant is not expected to permeate throughout the whole volume of the reservoir. It is sufficient to sweep only the zones in which the injected gas is fingering through. Such zones are typically generated when fluids with high mobilities, such as hydrocarbon solvent, CO_2, or steam, are injected, or are inherently present in highly heterogeneous or fractured reservoirs.

Matching Surfactants to Specific Reservoir Solids. As mentioned, EOR field applications that use gas or solvent injection are being conducted both in sandstone and in carbonate reservoirs (Table I). These two types of rock have vastly different bulk composition (Table VIII) and surface electrochemical properties (Figure 9). Not unexpectedly, adsorption levels of different types of surfactant on these different rock types can be very different (Figure 15), and this difference has a significant impact on surfactant and foam propagation.

As an example, some of the adsorption levels in Figure 15 will be applied to the layered reservoir in Figure 8. The anionic DPES–AOS adsorbs at 0.11 mg/g on sandstone from 2.1% TDS reservoir brine. At this adsorption level, the surfactant can propagate 369 m in the upper, high-permeability layer. The betaine is comparable to the anionic surfactant in terms of gas mobility reduction. However, it adsorbs at 1.3 mg/g on sandstone from the same brine and would only propagate 109 m in the high-permeability layer. In a limestone, on the other hand, the betaine would travel 223 m compared to 205 m for the anionic surfactant.

Measurements of mobility-reduction factors showed that the surfactants for which adsorption levels are shown in Figure 15 are equally effective in generating mobility-control foams in porous media. Selection of

Table VIII. Mineral Composition (Weight%) of Rocks Used in Surfactant Adsorption Measurements

Mineral	Berea Sandstone	Indiana Limestone	Baker Dolomite
Quartz	85–90	1	trace
Feldspar	3–6		
Dolomite	1–2		98
Calcite		99	
Kaolinite	5–7	trace	trace
Illite	1	trace	trace
Chlorite and illite–smectite	trace		

SOURCE: References 12 and 120.

the most cost-effective surfactant then hinges on properly matching surfactant and rock in order to minimize adsorption. In terms of adsorption, the following surfactant selection criteria can be derived from the data in Figure 15: The anionic surfactant is the best choice of the three surfactants tested for foam-flooding in sandstone and dolomite reservoirs. In limestone reservoirs, the sulfobetaine is the surfactant of choice. In the presence of divalent cations, which are present at relatively high concentrations in carbonate reservoirs, the sulfobetaine exhibits the lowest adsorption level on limestone. Its adsorption is also the least sensitive to divalent cations.

Adsorption levels of ionic surfactants are determined to a large degree by surfactant–solid electrostatic interactions. Figure 22 shows the frequency of adsorption levels measured with surfactants and solids of like charge and with surfactants and solids of opposite charge. The data represent those measurements from Figure 6 for which information on solid surface charges was available. On the basis of Figure 17 and the discussion on the dependence of surfactant adsorption on rock type, the am-

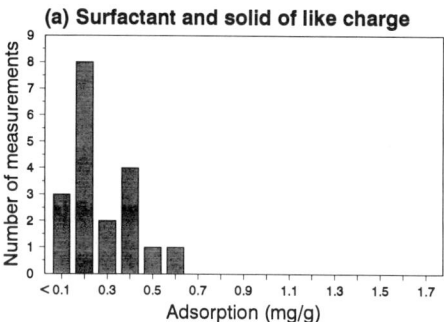

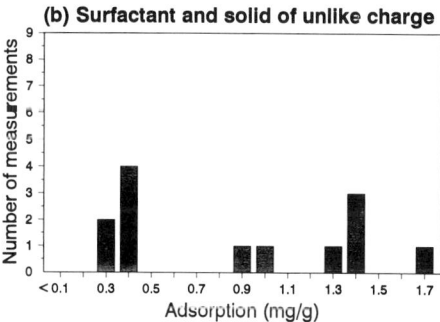

Figure 22. Distribution of adsorption levels measured with surfactants and solids of like and of opposite charge.

photeric surfactants were assumed to behave like cations on sandstone and silica and like anions on carbonates.

Clearly, adsorption levels for surfactants and solids of like charge cluster around the lower values. For oppositely charged surfactants and solids, there is more spread in the data because of the complex adsorption behavior of amphoteric surfactants. However, adsorption levels are clearly higher than for surfactants and solids of like charge. A knowledge of surfactant and solid electrochemical properties can thus aid in the proper selection of a surfactant for a specific reservoir. However, making assumptions about solid surface charges can be dangerous. Sandstone surfaces are often assumed to be negatively charged, and limestone surfaces are assumed to be positively charged. The data in Figure 9 show that these assumptions are not necessarily correct, particularly in the high-salinity and high-hardness brines encountered in many petroleum reservoirs. Solid surface charges depend not only on the type of rock but are extremely sensitive to pH, salinity, and hardness.

Surfactant Mixtures. Another promising approach to minimizing surfactant adsorption is through the formulation of surfactant mixtures. The mechanism of adsorption reduction in surfactant mixtures is based on a unique property of surface active substances: the formation of micelles (*129–134*).

The pseudophase-separation model for surfactant solutions (*84, 132, 135*) states that a surfactant solution above its critical micelle concentration (CMC) consists of two pseudophases in equilibrium with each other: singly dispersed surfactant monomer molecules and micelles. When the surfactant solution is in contact with a solid, the adsorbed phase constitutes a third pseudophase (*40*). Strong experimental evidence suggests that surfactant adsorption takes place from the surfactant monomer phase but not from the micelles (*84*), behavior that leads to competition of both the micelles and the adsorbed phase for surfactant molecules from the monomer phase. Adsorption from a surfactant solution above the CMC then depends not only on the affinity of the surfactant for the solid surface but also on its tendency to form micelles. If a mixture can be formulated such that at least one of the surfactants is incorporated into micelles preferentially over the adsorbed phase, then the micelles act as a sink for the surfactant and thus prevent it from being adsorbed.

In a solution containing a single pure surfactant, the monomer concentration and therefore the surfactant adsorption remain constant above the CMC because any additional surfactant is incorporated into micelles. In a surfactant mixture, the distribution of surface-active species in the monomer and micellar phases depends on their relative tendency to form micelles. The more hydrophobic components are incorporated into micelles preferentially, and the more hydrophilic components become enriched in the monomer phase. The composition of the phases and the monomer concentration of each surfactant continue to change with increasing overall surfactant concentration above the CMC (*129, 132, 136–141*). The monomer concentration of at least one of the surfactants may actually decrease with increasing overall surfactant concentration above the CMC, and this decrease will, in turn, lead to a decrease in adsorption.

An example of adsorption isotherms for a surfactant mixture with ideal mixing in the micelles is shown in Figure 23 (*134, 142*). Similarly shaped isotherms were modeled by Trogus et al. (*129*). The isotherm for one of the surfactants, the more hydrophobic one, exhibits a well-defined maximum. Ideally, the properties of the mixture would be such that overall adsorption from the mixture is lower than the sum of adsorption levels measured for each surfactant by itself (*129, 134*).

The potential of surfactant mixtures in lowering adsorption has been discussed in the literature (*129–134*). However, experimental work with surfactant mixtures has not advanced sufficiently to establish predictive capabilities for the formulation of surfactant mixtures with low adsorption levels. Two examples of experimental results obtained with commercially available foam-forming surfactants are shown next to illustrate the concepts.

Effluent profiles obtained from a core-flood performed with a mixture of two surface-active components (C_{12} and C_{18}) separated from a commercially available sulfobetaine are shown in Figure 24 (*115*). The points represent experimental data, and the lines were obtained by simulating the core-flood with a convection–dispersion–adsorption model that is based on the surface excess concept and takes into account monomer–micelle equilibrium (*115*). Because the mixture contains different homologues of the same surfactant, the ideal mixed micelle model

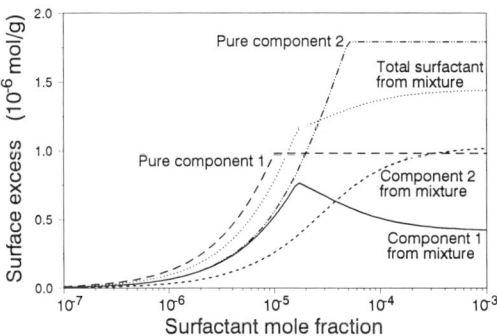

*Figure 23. Examples of calculated adsorption isotherms of two pure surfactants and their mixture (*134, 142*).*

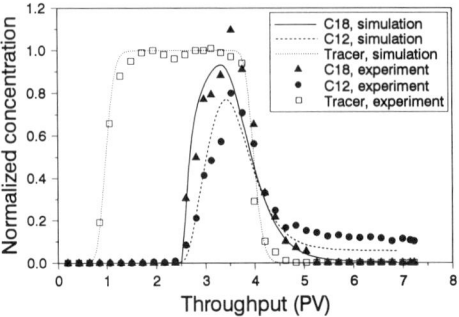

Figure 24. Effluent profiles of tracer and C_{12} and C_{18} components of a sulfobetaine from Berea sandstone. (Reproduced with permission from reference 134. Copyright 1991 Pergamon Press Ltd.)

was found to adequately describe this surfactant mixture (*115, 137, 143, 144*).

The C_{18} component is more hydrophobic than the C_{12} component. On the basis of considerations of solubility and batch adsorption measurements with homologous series of surfactants (*40, 76, 84, 145, 146*), one would expect the C_{18} component to adsorb more strongly and therefore to elute from the core later than the C_{12} component. The results in Figure 24 show just the opposite. Similar observations have been made by other authors (*130, 147, 148*). The explanation lies in the mixed micelle behavior of the surfactant mixture. Although the C_{18} component has a higher affinity for the solid surface, it also has a lower CMC and therefore a higher tendency to become incorporated into micelles, and this condition prevents it from being adsorbed more strongly.

The second example involves a mixture of two different types of commercial foam-forming surfactants: anionic and amphoteric (*7*). Unlike the mixture of the previous example, an anionic–amphoteric surfactant mixture probably does not follow ideal mixed micelle behavior (*138*). The results of three core-floods, performed separately with each surfactant and with a mixture of the two surfactants, are summarized as follows. The anionic surfactant adsorbs negligibly when used either by itself or when mixed with the betaine (at least at the low salinity used in these particular core-floods). Betaine adsorption is lowered by about an order of magnitude by mixing it with the anionic surfactant, from 1.7 down to 0.2 mg/g.

The first example illustrates that in micellar systems, the adsorption process may be dominated by monomer–micelle equilibrium. The second example provides experimental evidence that it is possible to lower adsorption by mixing surfactants.

When using surfactant mixtures, loss of the desirable properties of the mixture by chromatographic separation of the components during flow through the porous medium may pose problems. Much concern has been expressed about chromatographic effects in mixed surfactant systems, but available experimental data are insufficient to determine the severity of these effects. Some discussion in the literature suggests that chromatographic separation of components can be controlled by adjusting surfactant concentration, composition, and chemical structure (*130, 134, 147, 148*).

Sacrificial Adsorbates. The idea of preflushing a reservoir with cheaper chemicals in order to block the adsorption sites and reduce subsequent surfactant adsorption has been extensively studied. However, an inherent problem with all reservoir preflushes is that it is extremely difficult to place the sacrificial chemical in the same zones that will make contact with the surfactant that follows it. This problem can be attributed to a change in fluid mobilities caused either by the higher viscosity of the injected fluids or by the fact that residual oil is being mobilized and moved ahead of the surfactant bank. In fact, the function of a foam is precisely to change the flow pattern to previously unswept areas.

Effectiveness of sacrificial adsorbates is based on two basic assumptions: No significant desorption of the sacrificial agent takes place when the surfactant solution makes contact with the surfaces with the preadsorbed materials, and the sacrificial agent adsorbs on the same adsorption sites as the surfactant.

Information on the kinetics of adsorption and desorption of sacrificial agents is not readily available because these materials are typically complex mixtures of waste products such as lignosulfonates. In the example that follows, typical adsorption–desorption kinetics of surfactants are examined with the assumption that similar behavior may be exhibited by materials considered as sacrificial adsorbates. Assuming a rate equation for adsorption–desorption kinetics, the rate constants of adsorption and desorption can be estimated from surfactant effluent profiles measured during core-flooding (*10, 115*). For surfactants suitable for foam-flooding at high salinity, rate constants of adsorption were found to be of the order of 0.3 to 10 h^{-1}, and rate constants of desorption were of the order of 0.001 to 0.005 h^{-1}, assuming first-order kinetics (*7, 10–12*). Using the adsorption model described in reference 10 and rate constants of 1.0 h^{-1} for adsorption and 0.001 h^{-1} for desorption, the dependence of surface excess adsorption on time can be calculated during adsorption and desorption cycles as shown in Figure 25 (*128*). Although the rate constant of desorption is lower than the rate constant of adsorption by several orders of magnitude, Figure 25 indicates that significant desorption from the solid surface would occur over several days or weeks. If sacrificial materials desorb anywhere near the rates typical for surfactants, their usefulness

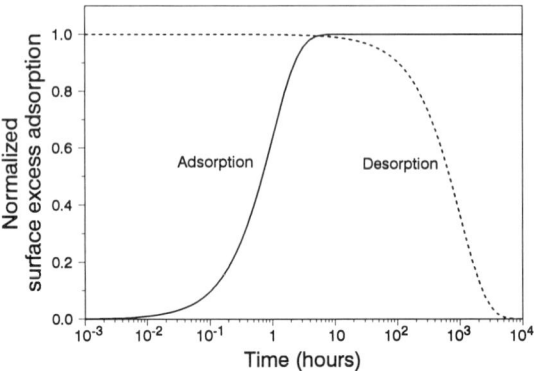

Figure 25. Illustration of the kinetics of adsorption and desorption, calculated assuming first-order kinetics and rate constants obtained from coreflood experiments. (Reproduced with permission from reference 128. Copyright 1991 Royal Society of Chemistry.)

would be severely limited in field applications where their protective action has to last for months or even years.

The second uncertainty associated with sacrificial adsorbates is illustrated by experiments in which several surfactant slugs were injected sequentially into the same Berea sandstone core. First, a slug of anionic surfactant (AOS 1416) was followed by brine injection and then by another slug of the same surfactant. The resulting effluent profiles are shown in Figure 26. Some surfactant adsorbs during the first surfactant flood, as indicated by the separation between the surfactant and tracer profiles. On the time scale of the experiment, the surfactant does not desorb appreciably during brine injection. Consequently, almost no additional adsorption occurs during the second surfactant flood: The surfactant and tracer profiles coincide. This kind of behavior is exactly what one would hope to achieve with a sacrificial agent. However, when the sequential slugs contain different chemicals, the situation changes drastically, as illustrated by the next core-flood.

Figure 26 also shows the effluent profile of a nonionic surfactant (alkylphenylethoxy alcohol) that was injected into the core containing the preadsorbed anionic surfactant. Material balance calculations indicated that the nonionic surfactant adsorbs as much as it would in a clean core despite the presence of the anionic surfactant on the solid surfaces. The anionic surfactant that was adsorbed during the first two floods is recovered completely.

The reverse sequence of core-floods is shown in Figure 27, where the nonionic surfactant is followed by another slug of the nonionic surfactant, and then by the anionic surfactant. As expected, adsorption of the non-

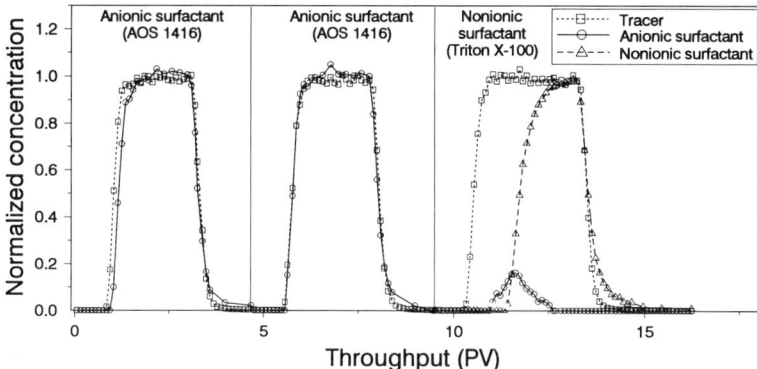

Figure 26. Effluent profiles obtained during sequential core-floods with an anionic and a nonionic surfactant.

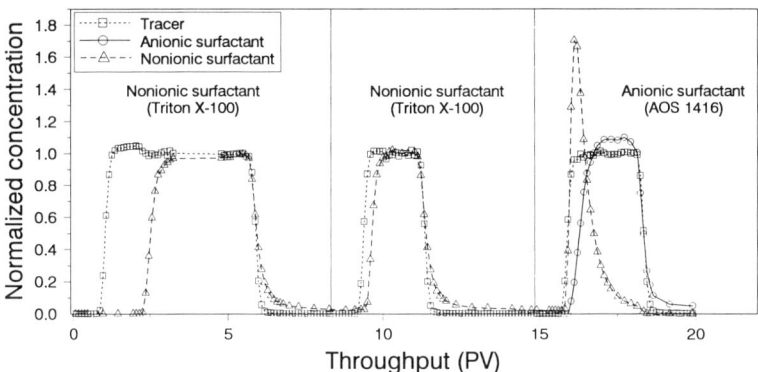

Figure 27. Effluent profiles obtained during sequential core-floods with an anionic and a nonionic surfactant, using the reverse sequence from Figure 26.

ionic surfactant is reduced during the second flood, because the rock contains preadsorbed nonionic surfactant. When the nonionic surfactant is followed by the anionic surfactant, 80% of the previously adsorbed nonionic surfactant is desorbed and recovered. The anionic surfactant adsorbs at a similar level as it would in a clean core. A similar sequence of core-floods with an anionic and an amphoteric surfactant was previously described (*128*).

The core-flood sequences in Figures 26 and 27 show that the two surfactants displace each other from the sandstone surface. Adsorption levels of both surfactants are unaffected by the presence of the other surfactant on the solid surfaces. Clearly, materials with the adsorption properties shown here would not be effective as sacrificial agents. However, one surfactant might be used as a desorbing agent for the other provided the chemical used as a desorbing agent is significantly cheaper than the primary surfactant.

From the previous discussion, it may be seen that many uncertainties are associated with the use of sacrificial adsorbates. Although preflushes with sacrificial adsorbates may show some improvement in chemical flooding processes (*149*), it remains to be demonstrated that the additional costs related to sacrificial materials can be fully compensated for by lowering the requirements for the primary surfactant.

Summary

Many hydrocarbon-miscible floods are run in reservoirs containing brines of extremely high salinity and hardness. Surfactants that may be used for mobility control foams at such conditions are commercially available. The effectiveness of foams generated with these surfactants was illustrated by way of representative mobility reductions factors measured in oil-free porous media.

One of the factors that determines foam propagation and foam-flood economics is surfactant loss in the reservoir, most importantly adsorption at the solid–liquid interface. Adsorption levels of foaming surfactants, mostly those suitable for high salinity conditions, cover a wide range and lead to vastly different distances of foam propagation. Therefore, selection of a surfactant with minimal adsorption levels for the reservoir conditions of interest is crucial.

The dependence of adsorption of some foaming surfactants on temperature, brine salinity and hardness, rock type, wettability, and the presence of a residual oil phase was illustrated using a large amount of measured adsorption data. Possible adsorption mechanisms were suggested. The experimental data together with the mechanistic information may provide guidelines for estimating surfactant losses to be expected during foam-flooding.

Three methods of minimizing surfactant adsorption were suggested: matching surfactant to reservoir conditions based on the information provided in this chapter, sacrificial adsorbates, and surfactant mixtures. Although much effort has gone into the formulation of sacrificial agents, considerable uncertainties are associated with the use of these materials. Surfactant mixtures provide a promising means of reducing adsorption. Theoretical and experimental information that sheds light on the mechanisms involved and demonstrates the effectiveness of some surfactant mixtures in reducing adsorption is available, but knowledge in this area has not advanced sufficiently to allow the formulation of mixtures with low adsorption levels.

List of Symbols

C	surfactant concentration
h	thickness of the layer into which foam is being injected
h_i	thickness of layer i
k	permeability
k_i	permeability of layer i
MRF	mobility-reduction factor
m_s	mass of surfactant injected
PV	pore volume
q	surfactant adsorption per unit volume of reservoir
Q	foam quality, or volume fraction of gas in a foam
r	radial distance
r_i	radial distance in layer i
R	radius of an injection pattern
V_f	pore volumes of foam injected
ΔG_{ads}	free energy change associated with adsorption at the solid–liquid interface
ΔG_{c-c}	free energy change due to chain–chain (hydrophobic and van der Waals) interactions between surfactant hydrophobes in an adsorbed surfactant layer
ΔG_{c-s}	free energy change due to nonpolar interactions between surfactant hydrophobes and a solid
ΔG_{cov}	free energy change due to surfactant adsorption by covalent bonding
ΔG_{elec}	free energy change due to surfactant–solid electrostatic interactions
ΔG_H	free energy change due to hydrogen bonding
ΔG_{solv}	free energy change due to solvation and "desolvation" of adsorbent and adsorbate
$\Delta P_{brine-gas}$	pressure drop measured across a core during simultaneous injection of brine and gas
$\Delta P_{surfactant-gas}$	pressure drop measured across a core during simultaneous injection of surfactant solution and gas (foam)
ϕ	porosity

Appendix. Abbreviations of Surfactant Names

Abbreviation	Trade Name	Supplier	Chemical Name
AOS 1416	Enordet AOS 1416	Shell	C_{14-16} α-olefin sulfonate
AOS 1618	Enordet AOS 1618	Shell	C_{16-18} α-olefin sulfonate
ATS	Suntech IV	Sun Refining	Alkyl toluenesulfonate
Betaine (same as Betaine 1)	Empigen BT	Albright & Wilson	Amide-modified alkylbetaine
Betaine 1	Empigen BT	Albright & Wilson	Amide-modified alkylbetaine
Betaine 2	Stepanflo 60	Stepan	Amide-modified alkylbetaine
DPES–AOS	Dow XS 84321.05	Dow	Diphenyl ether disulfonate and C_{14-16} α-olefin sulfonate mixture
DTAB		Aldrich	Dodecyl trimethylammonium bromide
IOS 1518	Enordet IOS 1518	Shell	C_{15-18} internal olefin sulfonate
LXS 810	Enordet LXS 810	Shell	C_{8-10} linear xylene sulfonate
Nonionic surfactant	Triton X-100	Rohm & Haas	Octylphenylethoxy alcohol
Sulfobetaine	Varion CAS	Sherex	Amide-modified sulfobetaine

Acknowledgments

The authors gratefully acknowledge the following companies for granting permission to include in this chapter data collected under private contract: Amoco Canada Petroleum Company, Calgary, Canada (Figure 6, Table IV); Chevron Canada Resources, Calgary, Canada (Figure 6, Table IV); Chevron Chemical Company, Calgary, Canada (Figure 6); and STATOIL, Stavanger, Norway (Figure 6).

References

1. Moritis, G. *Oil Gas J.* **1992**, *90(16)*, 51.
2. Novosad, J. J.; Ionescu, E. F. *Proceedings of the 38th Annual Technical Meeting of the Petroleum Society of CIM;* Canadian Institute of Mining, Metallurgy, and Petroleum: Calgary, Canada, 1987; paper 87–38–80.
3. Muijs, H. M.; Keijzer, P. P. M.; Wiersma, R. J. *Proceedings of the SPE/DOE Enhanced Oil Recovery Symposium;* Society of Petroleum Engineers: Richardson, TX, 1988; SPE/DOE paper 17361.
4. Lee, H. O.; Heller, J. P. In *Oil-Field Chemistry: Enhanced Recovery and Production Stimulation;* Borchardt, J. K.; Yen, T. F., Eds.; ACS Symposium Series 396; American Chemical Society: Washington, DC, 1989; pp 502–517.
5. Liu, P. C.; Besserer, G. J. *Proceedings of the 63rd Annual Technical Conference of SPE;* Society of Petroleum Engineers: Richardson, TX, 1988; SPE paper 18080.
6. Chad, J.; Matsalla, P.; Novosad, J. J. *Proceedings of the 39th Annual Technical Meeting of the Petroleum Society of CIM;* Canadian Institute of Mining, Metallurgy, and Petroleum: Calgary, Canada, 1988; CIM paper 88–39–40.
7. Mannhardt, K.; Novosad, J. J.; Jha, K. N. *Proceedings of the Annual Technical Meeting of the Petroleum Society of CIM;* Canadian Institute of Mining, Metallurgy, and Petroleum: Calgary, Canada, 1992; CIM paper 92–57.
8. Huang, A. Y.; Novosad, J. J. In *Fundamentals of Adsorption;* Liapis, A. I., Ed.; Engineering Foundation: New York, 1986; p 265.
9. Huang, Y. M.Eng. Thesis, University of Calgary, 1985.
10. Mannhardt, K.; Novosad, J. J. *Rev. Inst. Fr. Pet.* **1988**, *43*, 659.
11. Mannhardt, K.; Schramm, L. L.; Novosad, J. J. *SPE Adv. Tech. Ser.* **1993**, *1*, 212.
12. Mannhardt, K.; Schramm, L. L.; Novosad, J. J. *Colloids Surf.* **1992**, *68*, 37.
13. Handy, L. L.; Amaefule, J. O.; Ziegler, V. M.; Ershaghi, I. *Soc. Pet. Eng. J.* **1982**, *22*, 722.
14. Isaacs, E. E.; McCarthy, C.; Smolek, K. F. *Proceedings of the 2nd European Symposium on Enhanced Oil Recovery;* Éditions Technip: Paris, 1982; p 549.
15. Angstadt, H. P.; Tsao, H. *Soc. Pet. Eng. Reservoir Eng.* **1987**, *2*, 613.
16. Scharer, D. H.; Bright, D. B.; Merrill, C. L.; Shankar, P. K.; Bolsman, T. A. B. M. *Proceedings of the 76th Annual Technical Meeting of the American Oil*

Chemists' Society; American Oil Chemists' Society: Champaign, IL, 1985; p 2.
17. Marriott, R. C.; Kao, C. I.; Kristal, F. W. *Soc. Pet. Eng. J.* **1982,** *22,* 993.
18. Novosad, J. *Proceedings of the European Symposium on Enhanced Oil Recovery;* Elsevier Sequoia: Lausanne, Switzerland, 1981; p 101.
19. Hirasaki, G. J. *Soc. Pet. Eng. J.* **1982,** *22,* 181.
20. Hirasaki, G. J.; Lawson, J. B. *Proceedings of the 57th Annual Technical Conference of SPE;* Society of Petroleum Engineers: Richardson, TX, 1982; SPE paper 10921.
21. Lau, H. C.; O'Brien, S. M. *Soc. Pet. Eng. Reservoir Eng.* **1988,** *3,* 1177.
22. Kleinitz, W.; Rahimian, I. *Erdöl Erdgas Kohle* **1990,** *106,* 337.
23. Viswanathan, K. V.; Somasundaran, P. *Proceedings of the 6th International Symposiun on Oilfield and Geothermal Chemistry of SPE;* Society of Petroleum Engineers: Richardson, TX, 1982; SPE paper 10602.
24. Celik, M. S.; Shakeel, A.; Al-Yousef, H. Y.; Al-Hashim, H. S. *Adsorp. Sci. Technol.* **1988,** *5,* 29.
25. Lawson, J. B.; Dilgren, R. E. *Soc. Pet. Eng. J.* **1978,** *18,* 75.
26. Bae, J. H.; Petrick, C. B.; Ehrlich, R. *Proceedings of the Improved Oil Recovery Symposium of SPE;* Society of Petroleum Engineers: Richardson, TX, 1974; SPE paper 4749.
27. Gilliland, H. E.; Conley, F.R. *Oil Gas J.* **1976,** *74(3),* 43.
28. Malmberg, E. W.; Smith, L. In *Improved Oil Recovery by Surfactant and Polymer Flooding;* Shah, D. O.; Schechter, R. S., Eds.; Academic: Orlando, FL, 1977; p 275.
29. Celik, M.; Goyal, A.; Manev, E.; Somasundaran, P. *Proceedings of the 54th Annual Technical Conference of SPE;* Society of Petroleum Engineers: Richardson, TX, 1979; SPE paper 8263.
30. Novosad, J. *Soc. Pet. Eng. J.* **1982,** *22,* 962.
31. Austad, T.; Bjørkum, P. A.; Rolfsvåg, T. A.; Øysæd, K. B. *J. Pet. Sci. Eng.* **1991,** *6,* 137.
32. Smith, L.; Malmberg, E. W.; Kelley, H. W.; Fowler, S. *Proceedings of the 45th Annual California Regional Meeting of SPE;* Society of Petroleum Engineers: Richardson, TX, 1975; SPE paper 5369.
33. Lawson, J. B. *Proceedings of the 5th Symposium on Improved Methods for Oil Recovery of SPE;* Society of Petroleum Engineers: Richardson, TX, 1978; SPE paper 7052.
34. Maini, B. B.; Novosad, J. In *Surfactants in Solution;* Mittal, K. L., Ed.; Plenum: New York, 1989; Vol. 10, p 427.
35. Fuerstenau, D. W. In *The Chemistry of Biosurfaces;* Hair, M. L., Ed.; Marcel Dekker: New York, 1971; Vol. 1, p 143.
36. Somasundaran, P.; Hanna, H. S. In *Improved Oil Recovery by Surfactant and Polymer Flooding;* Shah, D. O.; Schechter, R. S., Eds.; Academic: Orlando, FL, 1977; p 205.
37. Gaudin, A. M.; Fuerstenau, D. W. *Trans. Am. Inst. Min. Metall. Pet. Eng.* **1955,** *202,* 958.
38. Somasundaran, P.; Fuerstenau, D. W. *J. Phys. Chem.* **1966,** *70,* 90.
39. Rupprecht, H.; Gu, T. *Colloid Polym. Sci.* **1991,** *269,* 506.
40. Harwell, J. H.; Hoskins, J. C.; Schechter, R. S.; Wade, W. H. *Langmuir* **1985,** *1,* 251.

41. Somasundaran, P.; Agar, G. E. *J. Colloid Interface Sci.* **1967**, *24*, 433.
42. Prédali, J.-J.; Cases, J.-M. *J. Colloid Interface Sci.* **1973**, *45*, 449.
43. Smani, M. S.; Blazy, P.; Cases, J. M. *Trans. Am. Inst. Min. Metall. Pet. Eng.* **1975**, *258*, 168.
44. Fuerstenau, M. C.; Miller, J. D. *Trans. Am. Inst. Min. Metall. Pet. Eng.* **1967**, *238*, 153.
45. Dobiáš, B. *Tenside Deterg.* **1976**, *13*, 131.
46. Ananthapadmanabhan, K. P.; Somasundaran, P. *Colloids Surf.* **1985**, *13*, 151.
47. Andersen, J. B.; El-Mofty, S. E.; Somasundaran, P. *Colloids Surf.* **1991**, *55*, 365.
48. Somasundaran, P.; Amankonah, J. O.; Ananthapadmabhan, K. P. *Colloids Surf.* **1985**, *15*, 309.
49. Amankonah, J. O.; Somasundaran, P. *Colloids Surf.* **1985**, *15*, 335.
50. El-Mofty, S. E.; Somasundaran, P.; Viswanathan, K. V. *Miner. Metall. Process.* **1989**, *6*, 96.
51. Hanna, H. S.; Somasundaran, P. *J. Colloid Interface Sci.* **1979**, *70*, 181.
52. Siracusa, P. A.; Somasundaran, P. *Colloids Surf.* **1987**, *26*, 55.
53. Cuiec, L. *Proceedings of the 59th Annual Technical Conference of SPE;* Society of Petroleum Engineers: Richardson, TX, 1984; SPE paper 13211.
54. Day, R. E.; Greenwood, F. G.; Parfitt, G. D. In *Chemistry, Physics and Application of Surface Active Substances;* Overbeek, J. Th. G., Ed.; Gordon and Breach Science: New York, 1967; Vol. 2, p 1005.
55. Connor, P.; Ottewill, R. H. *J. Colloid Interface Sci.* **1971**, *37*, 642.
56. Rupprecht, H. *Kolloid Z. Z. Polym.* **1971**, *249*, 1127.
57. Rupprecht, H.; Kindl, G. *Prog. Colloid Polym. Sci.* **1976**, *60*, 194.
58. Furlong, D. N.; Aston, J. R. *Colloids Surf.* **1982**, *4*, 121.
59. Aston, J. R.; Furlong, D. N.; Grieser, F.; Scales, P. J.; Warr, G. G. In *Adsorption at the Gas–Solid and Liquid–Solid Interface;* Rouquerol, J.; Sing, K. S. W., Eds.; Elsevier: Amsterdam, the Netherlands, 1982; p 97.
60. Noll, L. A.; Gall, B. L.; Crocker, M. E.; Olsen, D. K. *Surfactant Loss: Effects of Temperature, Salinity, and Wettability;* U.S. Department of Energy: Bartlesville, OK, 1989; NIPER–385, Coop. Agreement FC22–83FE60149.
61. Boneau, D. F.; Clampitt, R. L. *J. Pet. Technol.* **1977**, *29*, 501.
62. Yang, C.-Z.; Jao, W.-L.; Huang, Y.-H. *J. Pet. Sci. Eng.* **1989**, *3*, 97.
63. Somasundaran, P.; Hanna, H. S. *Soc. Pet. Eng. J.* **1979**, *19*, 221.
64. Al-Hashim, H. S.; Celik, M. S.; Oskay, M. M.; Al-Yousef, H. Y. *J. Pet. Sci. Eng.* **1988**, *1*, 335.
65. Bavière, M.; Bazin, B.; Noïk, C. *Soc. Pet. Eng. Reservoir Eng.* **1988**, *3*, 597.
66. Figdore, P. E. *J. Colloid Interface Sci.* **1982**, *87*, 500.
67. Hurd, B. G. *Proceedings of the Improved Oil Recovery Symposium of SPE;* Society of Petroleum Engineers: Richardson, TX, 1976; SPE paper 5818.
68. Lewis, S. J.; Verkruyse, L. A.; Salter, S. J. *Proceedings of the SPE/DOE 5th Symposium on Enhanced Oil Recovery of SPE and the Department of Energy;* Society of Petroleum Engineers: Richardson, TX, 1986; SPE/DOE paper 14910.
69. Glover, C. J.; Puerto, M. C.; Maerker, J. M.; Sandvik, E. L. *Soc. Pet. Eng. J.* **1979**, *19*, 183.
70. Gaudin, A. M.; Fuerstenau, D. W. *Trans. Am. Inst. Min. Metall. Pet. Eng.* **1955**, *202*, 66.

71. Ney, P. *Zeta-Potentiale und Flotierbarkeit von Mineralen;* Springer: Vienna, Austria, 1973.
72. Modi, H. J.; Fuerstenau, D. W. *J. Phys. Chem.* **1957,** *61,* 640.
73. Thompson, D. W.; Pownall, P. G. *J. Colloid Interface Sci.* **1989,** *131,* 74.
74. Schramm, L. L.; Mannhardt, K.; Novosad, J. *J. Colloids Surf.* **1991,** *55,* 309.
75. Gupta, S. P. *Proceedings of the 1st SPE/DOE Symposium on Enhanced Oil Recovery;* Society of Petroleum Engineers: Richardson, TX, 1980; SPE paper 8827.
76. Somasundaran, P.; Middleton, R.; Viswanathan, K. V. In *Structure/Performance Relationships in Surfactants;* Rosen, M. J., Ed.; ACS Symposium Series 253; American Chemical Society: Washington, DC, 1984; pp 269–290.
77. Somasundaran, P. *Adsorption from Flooding Solutions in Porous Media;* U.S. Department of Energy: Bartlesville, OK, 1988; DOE/BC/10848-10 (DE88001217) Contract No. AC19-85BC10848.
78. Somasundaran, P. *Adsorption from Flooding Solutions in Porous Media: A Study of Interactions of Surfactants and Polymers with Reservoir Minerals;* U.S. Department of Energy: Bartlesville, OK, 1989; DOE/BC/10848-15 (DE98000733) Contract No. AC19-85BC10848.
79. Trogus, F. J.; Sophany, T.; Schechter, R. S.; Wade, W. H. *Soc. Pet. Eng. J.* **1977,** *17,* 337.
80. Ziegler, V. M.; Handy, L. L. *Soc. Pet. Eng. J.* **1981,** *21,* 218.
81. Dilgren, R. E.; Owens, K. B. *J. Am. Oil Chem. Soc.* **1982,** *59,* 818A.
82. Novosad, J.; Maini, B. B.; Huang, A. *J. Can. Pet. Technol.* **1986,** *25,* 42.
83. Friedmann, F.; Chen, W.H.; Gauglitz, P.A. *Soc. Pet. Eng. Reservoir Eng.* **1991,** *6,* 37.
84. Scamehorn, J. F.; Schechter, R. S.; Wade, W. H. *J. Colloid Interface Sci.* **1982,** *85,* 463.
85. Gale, W. W.; Sandvik, E. I. *Soc. Pet. Eng. J.* **1973,** *13,* 191.
86. Tabatabai, A.; Gonzalez, M. V.; Harwell, J. H.; Scamehorn, J. F. *SPE Reservoir Eng.* **1993,** *8,* 117.
87. Gupta, S. C.; Smith, R. W. In *Advances in Interfacial Phenomena of Particulate/Solution/Gas Systems: Applications to Flotation Research;* Somasundaran, P.; Grieves, R. B., Eds.; AIChE Symposium Series 150; American Institute of Chemical Engineers, New York, 1975; Vol. 71, p 94.
88. Brode, P. F. III *Langmuir* **1988,** *4,* 176.
89. Parks, G. A. *Chem. Rev.* **1965,** *65,* 177.
90. Healy, T. W.; Fuerstenau, D. W. *J. Colloid Sci.* **1965,** *20,* 376.
91. Dobiáš, B. *Tenside Deterg.* **1972,** *9,* 322.
92. Wierer, K. A.; Dobiáš, B. *J. Colloid Interface Sci.* **1988,** *122,* 171.
93. Poirier, J. E.; Cases, J. M. *Colloids Surf.* **1991,** *55,* 333.
94. Sharma, M. M.; Kuo, J. F.; Yen, T. F. *J. Colloid Interface Sci.* **1987,** *115,* 9.
95. Street, N.; Buchanan, A. S. *Aust. J. Chem.* **1956,** *9,* 450.
96. Parks, G. A. In *Equilibrium Concepts in Natural Water Systems;* Gould, R. F., Ed.; Advances in Chemistry 67, American Chemical Society: Washington, DC, 1967; pp 121–160.
97. Wayman, C. H. In *Principles and Applications of Water Chemistry;* Faust, S. D.; Hunter, J. V., Eds.; John Wiley: New York, 1967, p 127.
98. Lorenz, P. B. *Clays Clay Miner.* **1969,** *17,* 223.

99. Williams, D. J. A.; Williams, K. P. *J. Colloid Interface Sci.* **1978**, *65*, 79.
100. Somasundaran, P.; Ramachandran, R. In *Surfactants in Chemical/Process Engineering;* Wasan, D. T.; Ginn, M. E.; Shah, D. O., Eds.; Surfactant Science Series 28; Marcel Dekker: New York, 1988; p 195.
101. Fuerstenau, M. C.; Gutierrez, G.; Elgillani, D. A. *Trans. Am. Inst. Min. Metall. Pet. Eng.* **1968**, *241*, 319.
102. Sampat Kumar, V. Y.; Mohan, N.; Biswas, A. K. *Trans. Am. Inst. Min. Metall. Pet. Eng.* **1971**, *250*, 182.
103. Mishra, S. K. *Int. J. Miner. Process.* **1978**, *5*, 69.
104. Le Bell, J. C.; Lindström, L. *Finn. Chem. Lett.* **1982**, *(6–8)*, 134.
105. Schulz, P., Dobiaš *Proceedings of the 15th International Mineral Processing Congress;* Societe de L'Industrie Minerale, Bureau de Recherches Geologiques et Minieres: France, 1985; p 16.
106. Pugh, R.; Stenius, P. *Int. J. Miner. Process.* **1985**, *15*, 193.
107. Smith, R. W.; Shonnard, D. *AIChE J.* **1986**, *32*, 865.
108. Rao, K. H.; Antti, B.-M.; Forssberg, E. *Colloids Surf.* **1988/89**, *34*, 227.
109. Oberndorfer, J.; Dobiáš, B. *Colloids Surf.* **1989**, *41*, 69.
110. Pierre, A.; Lamarche, J. M.; Mercier, R.; Foissy, A.; Persello, J. *J. Dispersion Sci. Technol.* **1990**, *11*, 611.
111. Prédali, J.-J.; Roubault, M. M. *C. R. Hebd. Seances Acad. Sci., Ser. D* **1967**, *265*, 477.
112. Swartzen-Allen, S. L.; Matijevic, E. *Chem. Rev.* **1974**, *74*, 385.
113. Hunter, R. J. *Zeta Potential in Colloid Science: Principles and Applications;* Academic, London, 1981.
114. Douglas, H. W.; Walker, R. A. *Trans. Faraday Soc.* **1950**, *46*, 559.
115. Mannhardt, K.; Novosad, J. J. *J. Pet. Sci. Eng.* **1991**, *5*, 89.
116. Saleeb, F. Z.; Hanna, H. S. *J. Chem. UAR* **1969**, *12*, 229.
117. Almon, W. R.; Davies, D. K. In *Clays and the Resource Geologist;* Longstaffe, F. J., Ed.; Mineralogical Association of Canada: Calgary, Canada, 1981; p 81.
118. Eslinger, E.; Pevear, D. In *Clay Minerals for Petroleum Geologists and Engineers;* SEPM Short Course Notes 22; Society of Economic Paleontologists and Mineralogists: Tulsa, OK, 1988.
119. Donaldson, E. C.; Kendall, R. F.; Baker, B. A.; Manning, F. S. *Soc. Pet. Eng. J.* **1975**, *15*, 111.
120. Churcher, P. L.; French, P. R.; Shaw, J. C.; Schramm, L. L. *Proceedings of the SPE International Symposium on Oilfield Chemistry;* Society of Petroleum Engineers: Richardson, TX, 1991; SPE paper 21044.
121. Anderson, W. G. *J. Pet. Technol.* **1986**, *38*, 1246.
122. Craig, F. F. Jr. *The Reservoir Engineering Aspects of Waterflooding;* Society of Petroleum Engineers of AIME: New York, 1971.
123. Anderson, W. G. *J. Pet. Technol.* **1987**, *39*, 1453.
124. Anderson, W. G. *J. Pet. Technol.* **1987**, *39*, 1605.
125. Salathiel, R. A. *J. Pet. Technol.* **1973**, *25*, 1216.
126. Richardson, J. G.; Perkins, F. M.; Osoba, J. S. *Pet. Trans. AIME* **1955**, *204*, 86.
127. Clementz, D. M. *Proceedings of the 3rd SPE/DOE Symposium on Enhanced Oil Recovery of SPE;* Society of Petroleum Engineers: Richardson, TX, 1982; SPE/DOE paper 10683.

128. Novosad, J. J. In *Chemicals in the Oil Industry: Developments and Applications;* Ogden, P. H., Ed.; Royal Society of Chemistry: Cambridge, England, 1991; p 159.
129. Trogus, F. J.; Schechter, R. S.; Wade, W. H. *J. Colloid Interface Sci.* **1979,** *70,* 293.
130. Trogus, F. J.; Schechter, R. S.; Pope, G. A.; Wade, W. H. *J. Pet. Technol.* **1979,** *31,* 769.
131. Scamehorn, J. F.; Schechter, R. S.; Wade, W. H. *J. Colloid Interface Sci.* **1982,** *85,* 494.
132. Scamehorn, J. F. In *Phenomena in Mixed Surfactant Systems;* Scamehorn, J. F., Ed.; ACS Symposium Series 311; American Chemical Society: Washington, DC, 1986; pp 1–27.
133. Minssieux, L. In *Oil-Field Chemistry: Enhanced Recovery and Production Stimulation;* Borchardt, J. K.; Yen, T. F., Eds.; ACS Symposium Series 396; American Chemical Society: Washington, DC, 1989; pp 272–288.
134. Mannhardt, K.; Novosad, J. J. *Chem. Eng. Sci.* **1991,** *46,* 75.
135. Shinoda, K. In *Colloidal Surfactants;* Shinoda, K.; Nakagawa, T.; Tamamushi, B.-I.; Isemura, T., Eds.; Academic: Orlando, FL, 1963; p 1.
136. Mysels, K. J.; Otter, R. J. *J. Colloid Sci.* **1961,** *16,* 462.
137. Clint, J. H. *J. Chem. Soc. Faraday Trans. 1* **1975,** *71,* 1327.
138. Rubingh, D. N. In *Solution Chemistry of Surfactants;* Mittal, K. L., Ed.; Plenum, New York, 1978; Vol. 1, p 337.
139. Holland, P. M.; Rubingh, D. N. *J. Phys. Chem.* **1983,** *87,* 1984.
140. Osborne-Lee, I. W.; Schechter, R. S.; Wade, W. H.; Barakat,Y. *J. Colloid Interface Sci.* **1985,** *108,* 60.
141. Kamrath, R. F.; Franses, E. I. In *Phenomena in Mixed Surfactant Systems;* Scamehorn, J. F., Ed.; ACS Symposium Series 311; American Chemical Society: Washington, DC, 1986; pp 44–60.
142. Mannhardt, K. M.Eng. Thesis, University of Calgary, 1988.
143. Lange, H.; Beck, K.-H. *Kolloid Z. Z. Polym.* **1973,** *251,* 424.
144. Osborne-Lee, I. W.; Schechter, R. S.; Wade, W. H. *J. Colloid Interface Sci.* **1983,** *94,* 179.
145. Somasundaran, P.; Healy, T. W.; Fuerstenau, D. W. *J. Phys. Chem.* **1964,** *68,* 3562.
146. Wakamatsu, T.; Fuerstenau, D. W. *Adv. Chem. Ser.* **1968,** *79,* 161.
147. Harwell, J. H.; Helfferich, F. G.; Schechter, R. S. *AIChE J.* **1982,** *28,* 448.
148. Harwell, J. H.; Schechter, R. S.; Wade, W. H. *AIChE J.* **1985,** *31,* 415.
149. Novosad, J. *J. Can. Pet. Technol.* **1984,** *23(3),* 24.

RECEIVED for review October 23, 1992. ACCEPTED revised manuscript March 16, 1993.

Near-Well and Oilwell Applications of Foam

8

Gas-Blocking Foams

Jan Erik Hanssen and Mariann Dalland

RF—Rogaland Research, P.O. Box 2503, Ullandhaug,
N—4004 Stavanger, Norway

Drawing mainly upon previous work by the authors, a review of key concepts in the use of foam for gas blockage in petroleum reservoirs is given. Demonstration experiments in radial and one-dimensional flow models that illustrate the use of foam barriers against gas coning are summarized. Observations and methods relevant for studying generation of gas-blocking foam are given. A standard procedure for evaluating blockage performance and a simple analytical model of a foam-generating displacement are proposed. From a review of sensitivity studies, conditions for obtaining gas blockage by foam are identified. Characteristic features of the gas-blocking state of foam in porous media are discussed that cover saturations, gas transport through foam, and loss of gas blockage. Two field cases are summarized.

FLUID FLOW MECHANISMS IN THE PRESENCE OF FOAM are exceedingly complex and may be sensitive to the experimental conditions as well as the prior saturation history of the system (1). In most studies of foam in porous media, the steady-state experimental mode has been used. This mode consists of injecting gas and surfactant solution at fixed flow rates or fixed fractional flow and measuring the resulting pressure gradient when production rates and saturations no longer change over a period of time. With strong foam, steady-state pressure gradients may be much higher than those in real reservoirs. Frequently measured values in core-floods are on the order of 10 bar/m even at reservoir rates (2–4), while typical gradients in North Sea reservoirs are on the order of 0.1 bar/m except in the immediate region near the wellbore. Because of the high flow resistance created by foam and its strongly nonlinear response to flow parameters, the laboratory case of externally imposed flow rates may not be representative for a real oil reservoir. The question has been raised whether a steady state of foam flow is attained at all in the field (5).

Specifically, what happens when gas enters a zone in the reservoir where foaming agent is present, but where the driving pressure of the gas is fixed? And what happens to a foam at steady state when the pressure drop across it becomes too small to support continued flow?

The purpose of this chapter is to describe an alternate mode of characterizing foam for reservoir application with emphasis on production—well treatments where the gas-blocking properties of foam, rather than its viscous flow effects, are paramount. The results should also be of relevance to other foam processes where gas flow is primarily to be blocked, such as the diversion of injected gas or steam and the sealing of leaks in gas storage reservoirs.

We start by an application example that highlights the use of foam to counteract excessive gas–oil ratios in oil production. We will review model experiments that are designed to match a novel production process that exploits the gas-blocking properties of foam. This review is followed by a description of relevant experiments and observations and a summary of our sensitivity studies on gas-blocking foam that aims to identify critical parameters in selecting foamers for use in the mentioned process. Next, some experimental results intended to lay a framework for a mechanistic interpretation of the gas-blocking state are summarized. The chapter is finished by quoting a few field trials where the gas-blocking properties of foam have been exploited. Most of the data reviewed concern aqueous foams, but some results on the nonaqueous foams that have been our main focus during the last few years are included.

Foam To Reduce Gas–Oil Ratios in Production Wells

Background. High-GOR (gas–oil ratio) production is a common problem for oil reservoirs that also contain free gas. Gas tends to be produced in preference to oil because of its higher mobility. This condition may cause reduced oil production rates, loss of drive energy, loss of recoverable oil (due to trapping by gas), and problems with fluid processing. Foam offers the potential of selectively reducing gas permeability and therefore is a means of reducing this problem, as was recognized at an early date (6, 7).

A foam whose purpose is to block gas flow should ideally be formed in all locations where gas breakthrough may occur and should remain essentially stagnant after its formation, maintaining the greatest possible gas mobility reduction for the longest possible period. These requirements are different from those for so-called mobility control foams.

We advocate the classification of foam processes from a practical operational viewpoint, that is, as treatments of gas-injection wells or oil-production wells, rather than by the presumed mechanisms of action of the foam and its macroscopic flow effects. This chapter primarily discusses producer treatments.

Demonstration. One mechanism of high-GOR production is known as gas coning, in reference to the cone-shaped region of high gas saturation, as illustrated in Figure 1. This figure also illustrates two ways to model coning in the laboratory: in a one-dimensional vertical core or pack, or in a radial geometry using a packed sector model.

From the known practical benefits to such wells of geological barriers, it was envisaged that foam could be effective as a means of reducing gas coning if it could be made to form an in situ barrier between the oil and the gas cap. Experiments were designed to test the creation of a barrier of gas-blocking foam and measure its effect on production using sector models.

Radial Flow Models. The sector models were constructed of transparent acrylic material that was packed with glass beads and initially saturated with a model oil in the lower part and gas in the upper part. Wall effects were found to be insignificant, and gas–liquid transition zones were of negligible height (1–2 cm). Further model properties are listed in Table I. The models were equipped with individual production–injection "perforations" in contact with specific intervals of the formation. All sector model experiments were conducted at room temperature and pressures below 1.5 bar absolute. It was possible to visually discern the presence of gas, injectant (with the aid of a dye), and model oil in the porous medium. With the aid of a strong backlight, the fluid that was present at the wall in a given location was also seen to be present across the entire model cross-section.

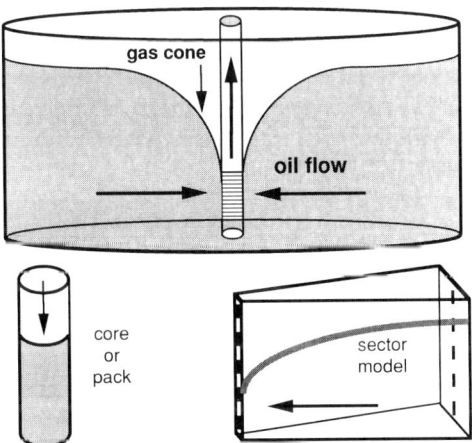

Figure 1. Gas coning. Idealized field case (above); radial laboratory model (below, right); and one-dimensional laboratory model (below, left).

Table I. Sector Model Experiments Used To Study Gas-Blocking Foam Barriers

Data	Model 1	Model 2
Length × height (cm)	40 × 20	80 × 40
Width, narrowest / widest (cm)	2.6 / 0.35	2.1 / 0.35
Pore volume (cm^3)	440	1460
Particle diameter range (μm)	80–105	70–110
Pack permeability (darcy)	8.5	5.5
Angle spanned (degrees)	3.22æ	1.29æ
Model oil	150 g/L NaCl brine	Exxon Isopar L
Model oil density (g/cm^3)	1.097	0.775
Model oil viscosity (cp)	1.21	1.42
Injectant solution	2% Perlankrol FN65[a] in distilled water	2% FC 740[b] in n-decane
Injectant density (g/cm^3)	1.00	0.730
Injectant viscosity (cP)	1.0	0.70
Critical rate to gas coning (cm^3/min)	1.2–1.3	<2.5
Theoretical critical rate (8) (cm^3/min)	2.1	1.0

NOTE: All data refer to standard room temperature of 23 °C and atmospheric pressure.
[a] Lankro Chemicals, United Kingdom.
[b] 3M Company, United States.

First, coning was successfully demonstrated in the sector models. Sharp, well-defined cones were formed, and steady-state flow of gas and oil was reached. The critical flow rate to gas coning was measured (Table I) and found to be in fair agreement with theoretical values (8) in view of the approximate nature of the theory. As expected, high gas–oil ratios were measured during oil production at supercritical rates, and the incremental gain in oil rate of producing at a higher pressure drop, ΔP, was very small at high ΔP.

Next, selective placement of foaming-agent solution between the in situ oil and the gas cap was demonstrated by formulating the injectant solution to have a density between those of the model formation fluids and injecting it at a low rate of 1.0 cm^3/min, so that gravity effects dominated the displacement, as shown in Figure 2. A foamer slug of 0.10 pore volume was found to achieve a penetration depth, R/r_e, of 0.6 in about 6 days, where R is the radius to the tip of the injectant zone and r_e is the drainage radius of the well. Such density-controlled placement is recommended for maximum later efficiency of the foam barrier (9, 10). However, placement by viscous-dominated flow (e.g., during injection of premade foam) may also, in some situations, produce a treatment effective against high GOR. In the latter case, gas coning is reduced by a high saturation of nearly immobile gas in the area near the wellbore (11).

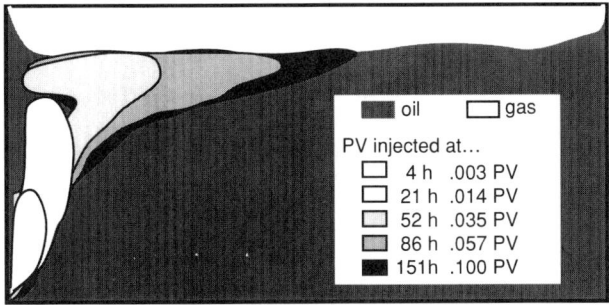

Figure 2. Placement of foaming-agent solution under gravity control as observed in sector Model 2. Injection from left to right through perforated lower 25% of wellbore height.

Generation of a barrier of gas-blocking foam was accomplished by returning the sector model to production after a horizontal surfactant solution layer was in place. First, a cone of colored surfactant solution formed and excess injectant was produced back. Subsequent downward movement of gas was significantly delayed by the formation of a foam front trailing the surfactant cone. After some time, a foam cone having essentially the same shape as the previously observed gas cone formed, and gas breakthrough was delayed or hardly noticeable. Production rates could now be increased to provide, in Model 1, oil rates of 8 to 10 cm^3/min (6–8 times the critical rate) with no or very low production of associated gas.

The effect of the gas-blocking foam on oil-production performance can be conveniently summarized by graphing the oil rate as a function of total rate as shown in Figure 3. The "strong foam" is that of Model 1 and represents target performance. The curve labeled "weak foam" is for the actual hydrocarbon foamer whose placement is shown in Figure 2. Its poor performance relative to the target points out the challenges of developing strong gas-blocking foams in nonaqueous systems.

In Model 1, a foam-barrier regeneration experiment, sensitivity tests of gas blockage to penetration depth (radial extent of the injectant layer), and injectant-layer thickness measurements were performed. For regeneration, a gas-blocking foam was shut in for 9 days during which time the drainage of liquid was observed as collection of liquid at the bottom of the cone. When production was resumed at the same ΔP, the foam came back at its previous gas-blockage efficiency within minutes. This result shows that enough foam remains intact or is regenerated by gas entering the region of drained liquid to maintain its efficiency as a gas-blocking agent.

The sensitivity tests revealed no dependence on penetration depth in a range of R/r_e between 0.5 and 0.9 and only a weak sensitivity to

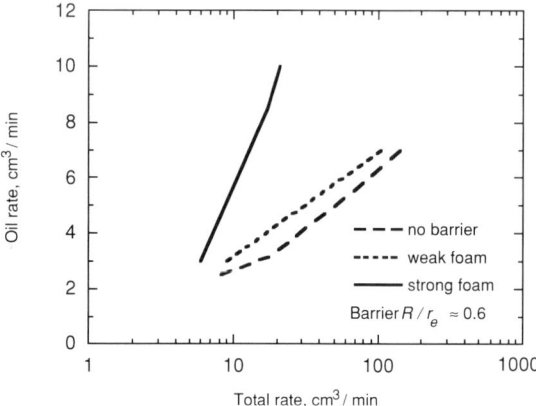

Figure 3. Measured production in sector models with gas coning in the absence and presence of barriers of gas-blocking foam.

injectant-layer thickness. There was a response in GOR to an oil-rate increase in going from 8 to 10 cm^3/min liquid for a slug of 0.02 pore volumes (PV), while the same GOR was measured at both rates for a 0.10 PV slug treatment (at the same penetration depth). More details about the results in sector model 1 were published previously (*12*).

One-Dimensional Flow Models. The generation of gas-blocking foam was also studied in vertical packs of glass beads with a permeability of 8.5 darcy using the same fluid system as in radial Model 1 but at 0.50% surfactant concentration. A slug of surfactant solution was placed near the top of the pack, and the rest was saturated by a higher-density liquid. Gas was next allowed to enter the pack at fixed inlet and outlet pressures that corresponded to an overall pressure gradient of 1 bar/m and a mean pressure of 5 bar. Foam was generated as the gas displaced the foamer solution downward, which in turn displaced the heavier liquid. Both displacements were piston-like. The foamer slug was continually shrinking at the expense of the growing foam, as illustrated by Figure 4.

The surfactant slug length was varied. Slugs of 0.2 and 0.1 PV that corresponded to a length in the pack of 37 and 17 cm, respectively, produced an excess of surfactant out of the pack and maintained slug integrity. The amount of surfactant solution consumed in generating foam was approximately constant, and a slug of 0.04 PV, a mere 7 cm in length, was just sufficient to fill the 2-m long column with foam.

In contrast to steady-state experiments, foam generation in the mode described caused no flow of foam out of the pack, although the strongly reduced gas flow rate showed that foam was definitely present in the

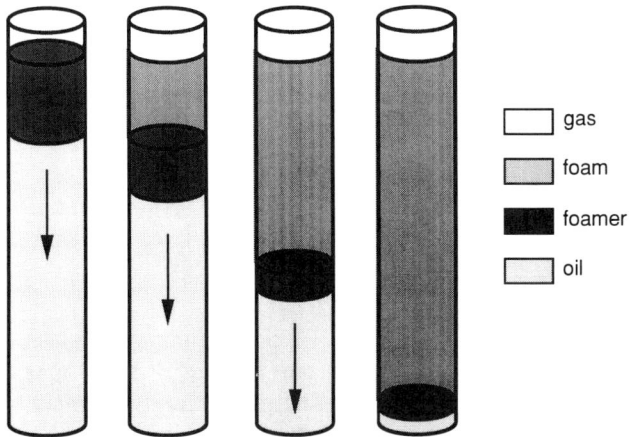

Figure 4. Generation of gas-blocking foam in vertical 2-m long packs of 8-darcy glass beads. Volume of surfactant slug not to scale.

pores. Only gas, and for a limited time after foam had reached the outlet, some surfactant solution, was seen to exit the porous medium. Using the viscosity of free gas, the ratio between gas permeability through foam and the absolute permeability, k_{gf}/k, was typically 10^{-4} to 10^{-3} at the initial pressure gradient, which is the pressure drop divided by length, $\Delta P/L$, of order 1 bar/m. The overall gas saturation, S_g, after liquid had ceased to flow varied only between 0.935 and 0.966, with a mean of 0.952 in six experiments where surfactant slug length, mean pressure, and differential pressure in foam generation was varied. The essentially invariant saturation is in good agreement with results for other systems used in characterizing the gas-blocking state of foam in a porous medium (*13, 14*).

Measurement of Foam Gas-Blockage Performance

In developing a gas-blockage process that uses foam, a test procedure was needed that represents foam generation and its effects on fluid flow in a manner mechanistically relevant to the process described previously. This experiment should distinguish clearly between the in situ gas-blockage performance of various systems, yet be simple enough to enable screening of a large number of products. Bulk-foam test methods (*15*) were ruled out after an introductory study showed bulk stability to be of no predictive power for gas blockage in porous media (*16*). A correlation might also not have been expected, because the properties of a confined foam depend heavily on the interaction between the lamellae and the pore walls (*3*). A simple porous-medium test was desirable and would, in addition, be useful for sensitivity and mechanistic studies. Steady-state methods cannot be used to characterize foams for gas blockage, because such experiments are performed under boundary conditions that are very different from those in gas blockage (*7, 17*). Therefore, a custom test protocol was defined on the basis of the vertical-pack experiments described combined with some reports from the early foam literature (*7–19*). The apparatus and procedure used are briefly described, and some examples of foam performance data follow.

Apparatus and Procedure. Packed columns, cores, or slim tubes (packed, coiled-steel columns) were mounted in a flow apparatus of the general design shown in Figure 5. The principle is the same for experiments at low and high pressure: Gas displaces foamer solution from a porous medium (containing appropriate desired residual saturations) at fixed inlet and outlet pressures. Various glass-bead fractions giving permeabilities from 1.6 to 30 darcy have been used to pack water-wet, highly repeatable model porous media. Sand may also be used. Transparent columns and tubing were used up to a pressure of 20 bar to aid in observing the displacement. Liquids were fed through gas-driven floating-piston cells, or (for establishing residual saturations) positive displacement

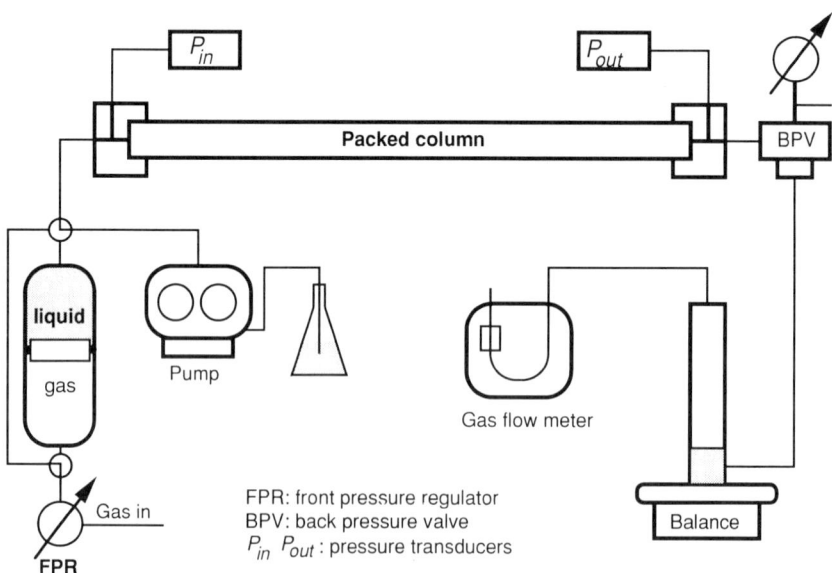

Figure 5. Generic apparatus used for studying gas-blocking foams. (Reproduced with permission from reference 13. Copyright 1993.)

pumps. Gas was fed at fixed pressure through a spring-loaded inlet pressure regulator. Effluent passed through a gas dome loaded back-pressure regulator. Pressures were measured at the column inlet and outlet. Overall pressure gradients were calculated as $\Delta P/L$. Actual, more detailed apparatus diagrams were given elsewhere (20, 21).

Pore volumes were measured, and average fluid saturations calculated by material balance. Absolute permeability, k, was measured to aqueous phase. In experiments with oil, the water-saturated column was flooded with oil to irreducible water saturation, S_{wi}, then with water to residual oil saturation to water, S_{orw}, and finally water was displaced miscibly by surfactant solution. (Without oil, the first two steps were omitted). Foam generation was accomplished by introducing gas into the porous medium (avoiding a pressure shock) at an inlet pressure, P_{in}, set to give a desired pressure drop ΔP. Production, pressures, and front movement were monitored. The foam was left to reach apparently stable flow, which could take hours or days, or never occur, depending on the system and conditions. If no consistent increase in gas rate and no measurable liquid production was observed during a time interval of a few hours, the average pressures and gas rates during that period were taken as stable.

From this point on, the experiment could be continued in two different manners. The system could be maintained at its original ΔP (foam-generation ΔP) in order to follow any long-term changes. Alternately, it could be subjected to a step-wise increase in ΔP, in which case the front pressure was increased to achieve a next higher overall pressure gradient, $\Delta P/L$ higher by ~1 bar/m. If a new stable flow state was reached, the pressure was increased again. This cycle of ΔP increase, observation, and measurement was repeated until the rate no longer became stable, or the pressure limit was reached. Stable gas rates were recalculated to mean pressure, and the apparent gas permeability through foam, k_{gf}, was calculated at each prevailing ΔP using Darcy's law and the viscosity of free gas. Relative gas mobility was calculated as k_{gf}/k.

Gas-Blocking Foam Generation. In a base case where distilled water was displaced by nitrogen from a long pack at fixed inlet and outlet pressures (corresponding to $\Delta P/L$ of 1 bar/m or less in order to match reservoir gradients), the expected unstable flood with an early breakthrough was observed. Adding as little as 0.1% foaming agent slowed down the front as a result of foam formation and established a piston-like front after some distance of travel. Figure 6 illustrates experiments with a sodium alkyl ethoxylated sulfate, Sulfotex RIF. (For convenience, surfactants are referred to in the following by trade names. Structures and manufacturers are listed in Table II. All concentrations quoted are percent weight by volume of active material). Adding 0.50 or 2.0% foamer

328   FOAMS: FUNDAMENTALS & APPLICATIONS IN THE PETROLEUM INDUSTRY

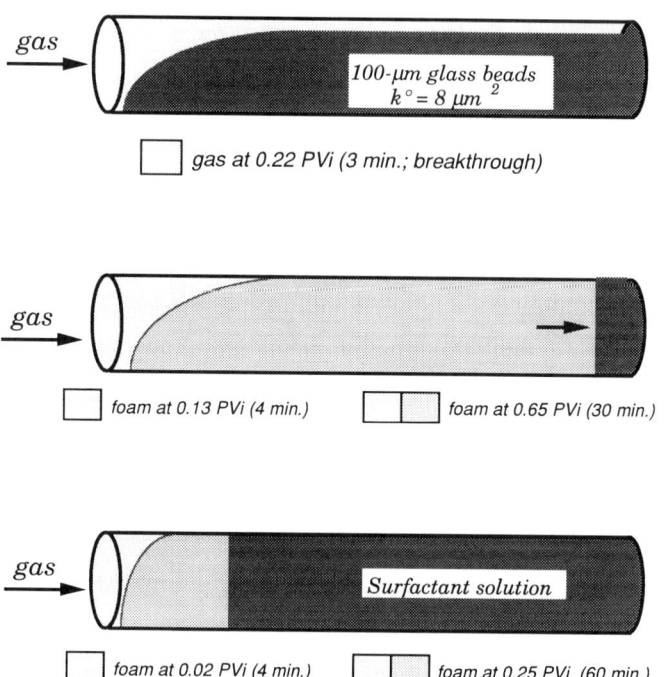

Figure 6. Typical displacement fronts observed in 8-darcy glass-bead packs during gas–aqueous phase displacements (from left to right). Top, no surfactant; middle, 0.10% Sulfotex RIF; and bottom, 0.50% Sulfotex RIF.

Table II. Surfactants Used

Trade Name	Structural Description	Maker
Perlankrol FN65	Na alkylphenyl 5,5-ethoxysulfate	Lankro
FC 740	fluorinated alkyl esters	3M Co.
Sulfotex RIF	Na alkyl ethoxysulfate	Henkel KG
Hostapur SAS	Na *sec*-alkyl sulfonate	Hoechst AG
Elfan OS 46	α-olefin sulfonate	Servo b.v
Ampholan B171	alkylcarboxybetaine	Lankro
Fenopon CD 128	NH_4 nonyl 2-ethoxysulfate	GAF Corp.
Fluowet OTN	fluoroalkylethoxy alcohol	Hoechst AG
FC 742	fluorinated alkyl esters	3M Co.

caused the displacement front to stabilize at a shorter distance from the inlet and continue propagating even more slowly, indicating the generation of a still stronger foam. In many cases, the foam generated initially is so strong that the displacement front stops entirely after a few centimeters of travel and shows no observable motion for weeks (22).

The foam-generating displacements also differed from gas–water displacements in their behavior at breakthrough. Whereas a long period of two-phase flow followed gas breakthrough in the surfactant-free case, only a brief period of foam production followed after the front had reached the outlet at 0.1%, and no foam at all appeared when using 0.5 and 2.0% surfactant. Instead, a prolonged period of liquid drainage occurred, approaching a low ultimate liquid saturation of order 0.05. The "irreducible" water saturation, S_{wi}, to gas flood in these bead packs is of order 0.10. This behavior is typical for foams at the gas-blocking state. At this state, which is transient but can be remarkably persistent, only gas is observed to enter and exit the porous medium, in contrast to the flowing steady state where foam can actually be seen flowing out of a core. If the ΔP applied to foam at the gas-blocking state is increased, the gas rate will typically increase rapidly for a short time and then stabilize at a new gas-blocking state. This action can frequently be repeated for a number of pressure steps.

Production history for a foam generated in a displacement of surfactant solution by gas at fixed pressures, and then subjected to a series of stepwise ΔP increases, is shown in Figure 7. An initial trickle of gas appears some time after the foam front has reached the outlet, and reaches a constant flow rate at a time that coincides with the overall water saturation, S_w, stabilizing at ~0.05 as production of liquid phase ceases. The vertical portions of the dotted line occur at the time of each differential pressure increase. In this case, the gas rate stabilizes at five successive gas-blocking states. The sequence of gas-blocking states defines a ΔP-dependent permeability such as that shown in Figure 8 for three repeat experiments. The rather broad error band is ascribed to the random nature

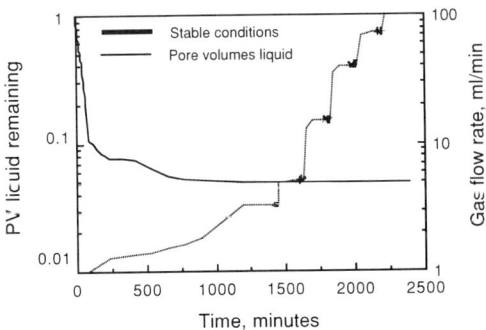

Figure 7. Primary data for 0.50% Hostapur SAS foam in a 100-cm pack of 8-darcy beads. Conditions: $P_{out} \simeq 6$ bar, $T = 21\ °C$, distilled water, and no oil. ΔP increased at times indicated by vertical portions of the gas-rate curve. (Reproduced with permission from reference 13. Copyright 1993.)

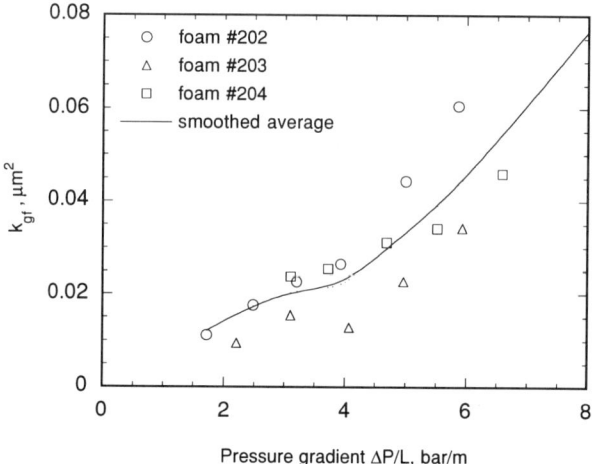

Figure 8. Repeatability of gas-blockage data for 0.50% Hostapur SAS foams in 100-cm packs of 8-darcy beads at 26 °C and no oil.

of initial gas fingering and foam generation and to inaccurate measurement of pressures and rates.

Figure 9 illustrates typical order-of-magnitude differences in foam performance data obtained by this method (20). A data set from the vertical columns of Figure 2 is included, and this demonstrates that the effect of orientation is insignificant in this case. The data set in the lower right corner of the figure represents target performance at applicable conditions.

A Simple Model of Pressure-Driven Foam Generation.

In systems favorable to generation of gas-blocking foam, the formation of a distinct displacement front after some distance of travel was noted by Raza (7). The near piston-like nature of the remaining displacement process, illustrated in Figure 10, allows a simple analysis that can provide information about the growing foam without the need for advanced in situ measurement techniques. This analysis is based exclusively on what might be termed the external data for the displacement: The liquid production rate, inlet and outlet pressures, and the position of the front observed in a transparent pack.

Assuming the observed presence of gas near the wall of the transparent pack to be identical to the pressure front (easily verified by observing with a backlight), the overall ΔP can be split between the section behind the front, where foam is assumed to be present everywhere, and the section ahead of the front where only surfactant solution is present.

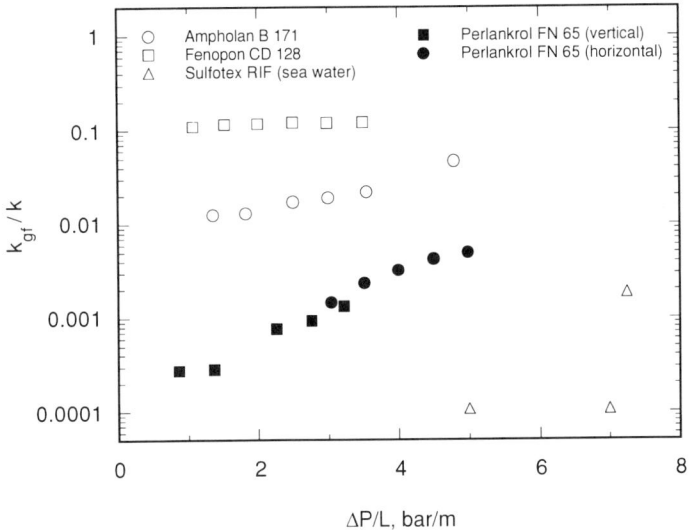

Figure 9. Gas-blockage performance of foams in 200-cm long 8-darcy bead packs at 0.50% foaming agent solution in distilled water except as indicated. Conditions: no oil, 21 °C, and initial outlet pressure $P_{out} \simeq 6$ bar.

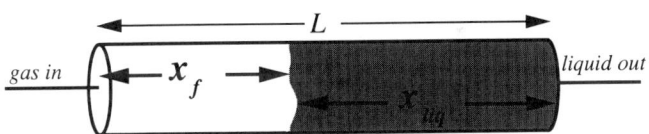

Figure 10. Typical front behavior in fixed-pressure gas–surfactant solution displacement in systems favorable to the creation of a gas-blocking foam. (Reproduced with permission from reference 14. Copyright 1993.)

Darcy's law can then be applied separately to each section. Ignoring gas compressibility,

$$\Delta P = \Delta P_f + \Delta P_{liq} \tag{1}$$

$$Q_{liq,out} = \frac{\Delta P_{liq} \, k \, A}{(1 - x_f) \, L \, \mu_{liq}} = Q_{g,in} = Q \tag{2}$$

$$\lambda_f = \frac{Q \, (x_f L)}{\Delta P_f \, A} \tag{3}$$

$$S_{gf} = 1 - \frac{x_f - V_{p,liq}}{x_f} \tag{4}$$

Here, ΔP_f and ΔP_{liq} are the pressure drops across the foam and the liquid, respectively. $Q_{liq,out}$ and $Q_{g,in}$ are the liquid rate out of the system and the gas rate into the system, respectively. A is the cross-sectional area of the porous medium, L is its length, and x_f is the fraction of the length containing foam. $V_{p,liq}$ is the produced liquid measured in pore volumes. Gas saturation in foam, S_{gf}, (foam quality) and foam mobility, λ_f, are averaged over the growing foam section. μ_{liq} is the viscosity of the foamer solution. Foam mobility is used here to avoid any assumptions of permeability or viscosity with regard to the growing foam.

Thus, we can construct an approximate model of a foam-generating displacement shown in Figure 11. The pressure and saturation profiles are shown for three points in time t_1, t_2, and t_3. Visual observations of a uniform saturation behind the front, which are confirmed by saturation measurements using microwave absorption (14), support drawing horizontal dashed lines for water saturation, S_w, in the graph on the right. The level of S_w behind the front is known at any time from the produced volume and the front position (eq 4). In the left graph, the pressure profiles are given by eq 2 for the liquid-containing portion of the pack, shown by solid lines, and the value for $x/L = 0$ is known. At long times, the pressure profile also is uniform. Because our experiments provide no support for any other shape, straight lines are drawn for P from the inlet to the observed foam front position $x_f(t_i)$. Fair agreement was reported when comparing this simple picture with measured saturation and pressure profiles (14).

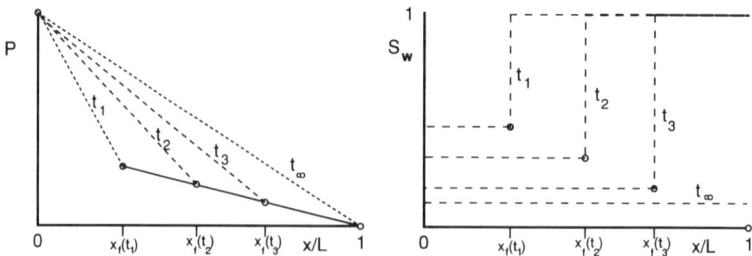

Figure 11. Constructed pressure (left) and water-saturation profiles (right) for a growing strong foam in a piston-like, pressure-driven displacement. (Reproduced with permission from reference 14. Copyright 1993.)

Critical Parameters

General. The study of a foam process and the selection of foaming agents for laboratory or field use must be done at an appropriate set of conditions, because foam properties are widely reported to depend on the mode of experiment and the state of the porous medium prior to foam generation. Thus, knowing what parameters influence the foam performance in a critical way is essential. For product screening, identification of the parameters that are not necessary to include is equally important, because this may allow more tests to be done in a shorter time. In our work on gas-blocking foams, we have studied the sensitivity of gas blockage to the following sets of parameters:

1. Those that are determined by the reservoir and system: temperature, salinity, oil (presence and type), pressure level, and gas type.

2. Those that depend on the experiment, or vary significantly within the reservoir: ΔP during foam generation, concentration, permeability, and length.

The most important factor that has not been varied is the wettability of the porous medium, because up to now, our studies have been limited to water-wet unconsolidated packs and outcrop sandstones. A few foam studies have been published on cores made artificially oil-wet (23, 24) that show a strong and rather complex influence of wettability on foam properties. We recommend the use of native-state reservoir cores, if available, for wettability–foam studies, because the reagents often used in making cores oil-wet may have unknown effects on foam if trace amounts remain in the cores. Besides, the effects of mixed or intermediate wettability may be of more interest to field applications than those of oil-wetting. Meanwhile, our work with nonaqueous foams has revealed an interesting challenge in that effective gas blockage by alcohol (25) and hydrocarbon foams can occur in water-wet packs. Because residual water is present in these experiments, the foamed organic solvent must be the nonwetting liquid phase.

Adsorption has not been explicitly studied for gas-blocking foams but of course must be taken into account for any chemical considered for reservoir application. Depending somewhat on how far into the reservoir the foamer is intended to move in a given situation, other work on foaming-agent adsorption (26, 27) is also considered to be applicable for the present systems. In our screening tests, in order to compare foamers at an equal basis, we have tried to eliminate adsorption loss by using low-adsorbing media and to ensure presaturation by overflushing with surfactant solution prior to foam generation.

Differential Pressure. At steady state, the mobility of foam has been found to vary with flow rates with a minimum velocity existing for generation of strong foam by snap-off (*28*). A corresponding dependence was sought in pressure-driven experiments but not found within a range of pressure gradients corresponding to field flow rates. A minimum overall pressure gradient for creating gas-blocking foam may still exist but must be below 0.05 bar/m. If the same relations apply to field cases, then a wide range of reservoir pressure gradients would favor generation of gas-blocking foam.

Temperature–Concentration–Salinity. As with bulk foams and foams flowing at steady state, temperature is detrimental to foam gas-blocking ability, as seen in Figure 12 for a nonaqueous foam. Temperature effects on foaming systems have been much studied because of their application in steam-floods. For aqueous foams, it appears to be possible to tailor a surfactant for good foam performance at any oil field temperature by varying its structure and hydrophile–lipophile balance (*29*). Much the same can be said about salinity with foamers available for a wide range of brine salinities including divalent ions. Gas blockage increases up to a certain limit and possibly levels-off at some concentration of foamer. This feature is reported in a different paper (*13*) that also shows that the concentration at which the plateau-level performance is reached depends on salinity for the system studied. Because a gas-blocking foam is generally regarded as a more static system than a flowing foam (*14, 17*) with fewer

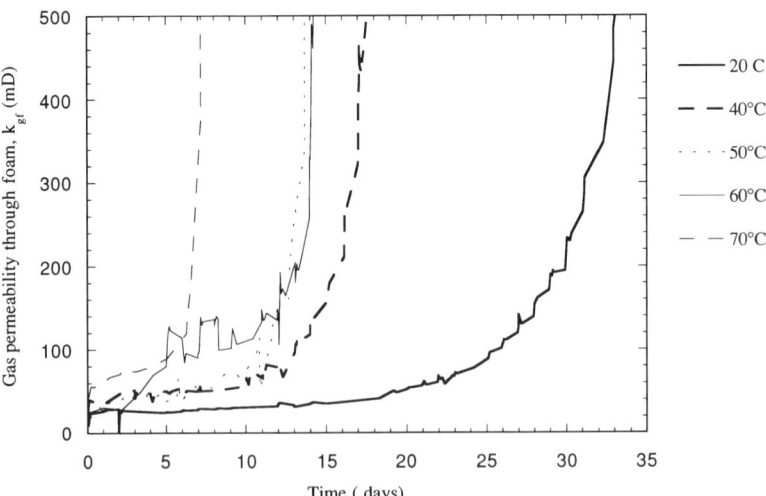

Figure 12. Gas-blocking ability of kerosene foams with 1.0% FC740 at various temperatures using 200-cm packs of 8-darcy beads.

generation events, somewhat higher concentrations may be required for effective blockage than for high mobility reduction in steady-state flow in order to stabilize the lamellae against breakage for longer times. An interesting observation from the data in Figure 13, which combine temperature and concentration effects for an aqueous fluorosurfactant and a conventional foamer, is that the loss of blockage efficiency by increased temperature can be be successfully compensated by increasing surfactant concentration.

Oil. As for other applications of foam in porous media, the presence of oil is also important to gas-blocking foams because it may impair the ability of foam to reduce gas mobility in situ. A foam to be used in blocking gas around an oil well must withstand prolonged exposure to oil and should thus be "intrinsically" oil tolerant. This necessity is in contrast to other applications where "dynamic metastability" of the foam has been claimed to be sufficient because the foam is assumed to flow through the pores and will have its lamellae in contact with oil for only brief periods (*30*). Oil–foam interaction has been a main focus of our work on aqueous gas-blocking foams and is published separately. In the present context, however, because the topic is reviewed elsewhere in this book, only the main conclusions are given.

1. Gas blockage by foam in a porous medium may be more strongly impaired by oil, even at low residual saturations, than is the bulk

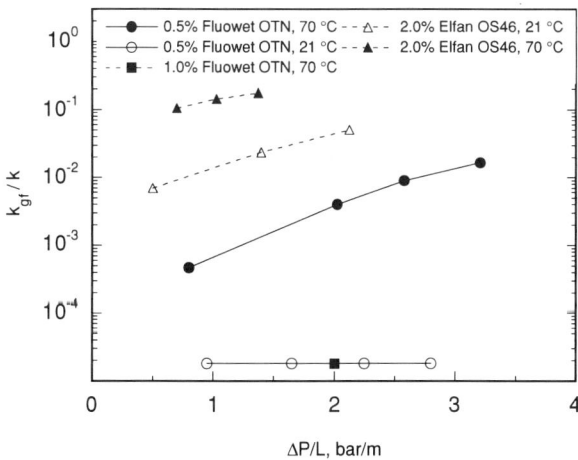

Figure 13. Foam gas-blockage performance at indicated temperatures in the presence of a North Sea crude at S_{or} using 200-cm packs of 8-darcy beads at $P_{out} \simeq 6$ bar. (Reproduced with permission from reference 13. Copyright 1993.)

stability of the same foams. With one set of model systems, oil was actually found to improve bulk stability of foam, while its effect on gas blocking ability was always negative (*31*).

2. Some foams are unaffected by the presence of oil (*20, 21*). Most nonfluorinated aqueous foamers studied, however, are strongly affected and commonly do not reach the gas-blocking state at residual oil.

3. The same physicochemical parameters appear to govern the interaction between foam and oil in gas-blocking and in flowing systems. However, as commonly evaluated from bulk equilibrated systems, the spreading and entering coefficients do not capture the full picture of the oil tolerance of a given foam as only two out of three "regimes" are predicted correctly (*32*).

For our laboratory work with aqueous foams, oil was generally present at S_{orw}. This value was chosen because it can be achieved in a reproducible manner. Oil saturation is largely unaffected by foam generation, because most foamers do not reduce the interfacial tension between oil and water, σ_{ow}, sufficiently to mobilize residual oil. Nonaqueous foams, conversely, are always evaluated at irreducible water, to capture any sensitivity to water (which, until now, has not been observed) as well as to maintain the porous medium in contact with its preferred wetting liquid.

Permeability. The effect of absolute permeability on foam gas-blockage performance is clearly seen from Figure 14. The same trend has been found for a different system, one in which foam performance at increasing pressure and prolonged exposure to initial pressure drop are measured (*21*). Bernard and Holm (*18*) measured relative gas mobility reduction in media from 0.3 to 100 darcy and found it to be at its greatest for a 7-darcy sand-pack for those tests conducted at a fixed pressure drop. A similar optimum could also exist in our data but would have to be below 1.6 darcy. For reservoirs with typical permeabilities on the order of 1–10 darcy, which includes a number of candidates for application of foam barriers in production wells, gas blockage would thus be favored. However, the result for the highest permeabilities indicates that gas blockage may be difficult to achieve in fractures, at least with the systems used here.

An explanation of the observed permeability dependence should be sought in the balance of lamellae generation, flow, and destruction in pores of different sizes such as that advanced by Gillis and Radke (*33*). Briefly, their picture is one where the very smallest pores contain only liquid, the medium-sized pores contain predominantly trapped lamellae, the larger fraction contain moving lamellae, and the very largest pores (if present) are occupied by continuous gas.

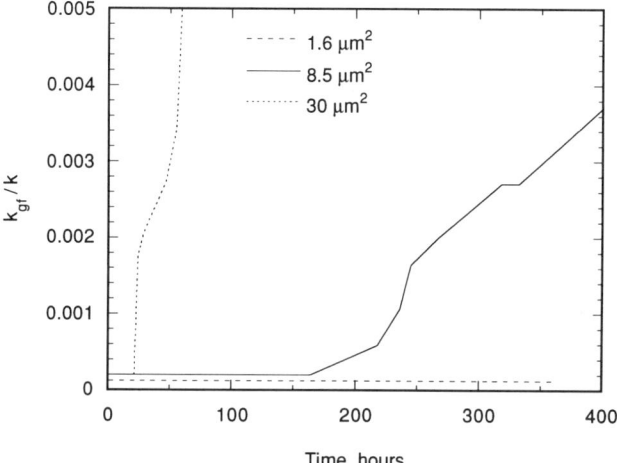

Figure 14. Gas blockage performance at prolonged exposure to original $\Delta P/L$ in 200-cm bead packs of indicated permeability. Conditions: 1.0% Fluowet OTN in sea water at S_{or}, $\Delta P/L \simeq 1$ bar/m, $T = 70\,°C$, $P_{out} \simeq 6$ bar, and crude oil. (Reproduced with permission from reference 13. Copyright 1993.)

Porous Medium Length. A foaming system was studied in packs of 22-, 47-, and 97-cm length, with all other parameters, including the initial $\Delta P/L$, held constant. A clear trend was found of gas blockage strength increasing with length. The results are shown in Figure 15. Clearly, a short pack does not allow formation of a gas-blocking foam. Other work on aqueous foams (21) has shown the same trends (including tests at prolonged exposure to the initial ΔP) in media of 24, 66, and 200 cm; the difference between 66 and 200 cm is much smaller than that between 24 and 66 cm.

The length dependence of gas-blocking foams, which was also observed by other investigators when using the fixed-pressure mode of foam generation (14, 25, 34), must be interpreted mechanistically. It is properly termed a length–pressure drop scaling effect. When the overall pressure gradient $\Delta P/L$ is kept constant between packs of different length, the inlet pressure will be lower in the shorter packs. Because the capillary inlet pressure of the smaller pores is the same for the same type of beads, gas cannot enter as many of these (or enters them only down to a larger radius), so that foam generation in them is not initiated to the same degree. Thus, in a fixed-pressure displacement, less driving force, that is, pressure, is available to generate strong foam in a short pack than in the long pack. Accordingly, we observe the liquid production rate (a coarse indicator of the foam generation rate) starting to decline at a lower liquid saturation the shorter the pack (25).

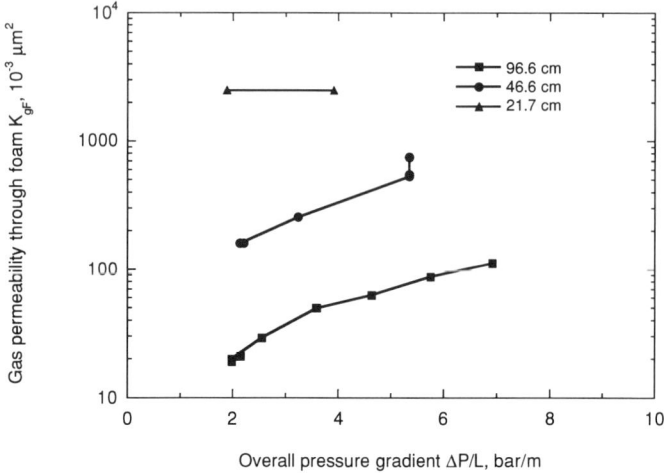

Figure 15. Gas-blockage performance of 2-methyl-1-propanol foams in 8-darcy bead packs of indicated length at 21 °C with $P_{out} \simeq 6$ bar. Foam generated at a constant overall pressure gradient $\Delta P/L \simeq 2$ bar/m. (Redrawn from data in reference 25.)

The differences are particularly striking in porous media that are so short that the initial gas tongue breaks through before its mobility has been significantly reduced by the growing foam. Gas flow through the tongue then becomes self-accelerating as the local gas saturation and effective permeability increase, possibly helped by a capillary end effect, retarded only by the initial weak foam. If no liquid is supplied to the gas tongue area, lamellae coalesce, reducing the flow resistance further, and forming continuous-gas channels, and gas rates soar. Even if a high local liquid saturation remains outside of the gas tongue, as there was for the short- and medium-length foams in Figure 15, strong gas-blocking foam could not be created because the original tongue took most of the gas flow.

Thus, the effect of using porous media that are too short in a constant-pressure foam generation is to inhibit the growth of a strong foam. This inhibition of growth is not expected to occur in steady-state experiments, because the effect of the initial foam in the case of fixed injection rates will be to increase the pressure drop and thus initiate the "cascading" process of gas entering and generating lamellae in gradually smaller pores, which again serves to increase the local pressure drop as necessary to maintain the imposed flow rate (*3*).

Our length-effect observations agree with a few observations in the early foam literature (*35*) that blockage or excessive pressure gradients was obtained in cores of more than 50- to 100-cm length. Steady-state foam flow could be obtained only in shorter media, which were then selected for laboratory tests. Other investigators have used cores only a few centimeters long to facilitate reaching the flowing steady state (36). In general, no rationale is offered for not considering scaling experimental conditions or results toward those of the reservoir.

Pressure Level and Gas Type. The pressure level and type of gas used might be expected to be critical for achieving effective gas blockage. Early studies indicated that gas blockage was a phenomenon achievable primarily at low pressure and conditions (18). In fact, sandstone cores are excellent media for reaching the gas-blocking state even in the presence of oil and natural gas and at reservoir pressure and temperature. As shown by Figure 16, porous-medium length appears to be more important for achieving gas blockage than either the pressure, the nature of gas, or the type of porous medium.

Foam studies should not be done at atmospheric outlet pressure (37) because this introduces bubble expansion effects. We have found unpredictable differences between gas-blocking foams generated at atmospheric outlet pressure and outlet pressures of 5–6 bar (13). All our subsequent foam performance tests have been conducted at an elevated back pressure, even if reservoir conditions are not desired. We also attempt to limit the ratio between pressure drop and overall pressure, $\Delta P/P$, to less than 20%

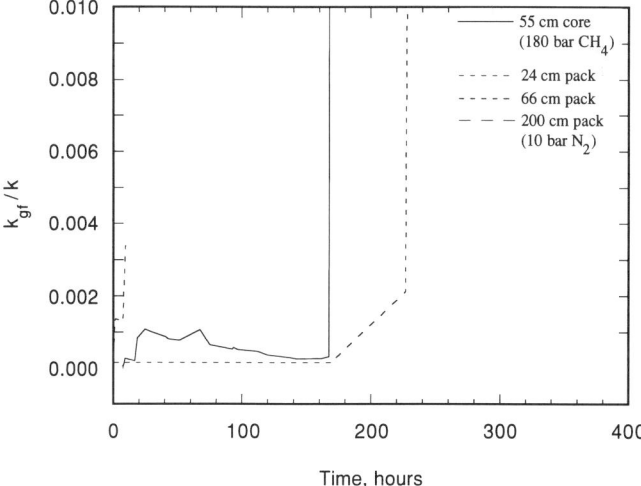

Figure 16. Gas-blockage performance of 1.0% Fluowet OTN foams in bead packs of indicated length at screening conditions, and in Bentheimer sandstone at reservoir conditions. Permeability is 1.4–1.8 darcy, temperature is 70 °C, and residual crude oil is present (except, 24-cm pack: 8 darcy and 21 °C).

so as to minimize bubble expansion. The reason for considering the type of gas together with the pressure level is that the differences in solubility, viscosity, and interfacial tensions, for example, for CH_4 and N_2, become significant only at reservoir conditions. At screening-test pressures, there probably is little difference between the gases and so N_2 can be used for added safety.

Summary: Conditions for Obtaining Gas Blockage. Thus, we arrive at the following set of critical parameters, in order of importance, for characterizing gas-blocking foams in the laboratory:

- porous medium length
- presence of oil
- temperature
- permeability

Porous-medium wettability is expected to also be critical but has not been studied. On the basis of our data, it appears not to be essential in screening to employ reservoir pressure, reservoir gas, and (if the rock in question is preferentially water-wet) reservoir material. This conclusion is supported by the results of screening aqueous foams: About 40 systems were evaluated for gas blockage at conditions defined by the critical parameters, and only four were found to be effective (20). However, of these, two (possibly three) were later shown to be equally effective at full reservoir conditions including sandstone, live oil, and methane gas, (21). A similar increasing degree of success has been missing in some other screening efforts, notably where the use of reservoir rock and high pressure, but not oil, have been prioritized from the start (38).

Of course, when a set of promising products has been evaluated for use in a given field and foam process type, qualification at field conditions is needed. However, we caution against introducing this too early in a project. A problem particular to foams is the use of composite reservoir cores to make porous media of sufficient length to achieve gas blockage, because the borders between core sections are likely to act as snap-off and termination sites, introducing strong and unpredictable end effects and possibly reducing the effective length of the composite core to that of the individual sections. Sand-packs (possibly made from crushed core material) are preferred, although in our experience, it is extremely important to keep a good quality packing by employing a prestressed pack. In any case, the use of composite cores in order to qualify gas-blocking foam performance at reservoir conditions should be accompanied by measured in situ saturations and pressure profiles to validate the data. Our work, as reviewed here, further shows the risk of unproductive product search to be even greater when bulk tests are used as prescreening, because the foams that actually perform best in porous medium tests frequently are not the most stable bulk foams (*16, 32*). However, because bulk foam tests differ in their design and interpretation, the design of a bulk test that is predictive of porous-medium performance should not be entirely ruled out.

The characteristic conditions for generating gas-blocking foam are thus summarized by Table III, and are compared with those from the literature for achieving steady-state foam flow.

Gas Blockage. Gas blockage, characterized by gas flowing at greatly reduced mobility through a foam-filled porous medium at near-constant saturation, is incompatible with current models of flowing foam. In this section, a few observations will be presented to give a basis for understanding and modeling with the ultimate goal of becoming able to describe also this aspect of foam in reservoir simulations. Our belief is that the differences between flowing and gas-blocking foams are essentially those of different boundary conditions and that the same mechanisms are operating on the pore level.

The characteristics of the foam-generating displacement of surfactant solution by gas was discussed. The two most important phenomena in this regard are the self-sharpening fronts (Figure 6) and liquid ceasing to flow as a separate phase after a period of slow drainage. The following describes the situation at the gas-blocking state.

Saturation. As already mentioned for the vertical bead pack demonstration experiment, gas-blocking foams can become exceedingly dry. Some results for aqueous foams are summarized as follows. With eight different Hostapur SAS foams in 8-darcy bead packs at ambient temperature without oil, overall aqueous-phase saturations from 0.04 to 0.06 (±0.01) pore volume were obtained (*22*). No additional liquid production was observed during several days of blockage, when up to 100 pore volumes of gas flowed through at mobilities reduced by 2–4 orders of magnitude. Visual observation at the onset of gas blockage showed surfactant solution to be distributed uniformly throughout the foam-filled

Table III. Conditions for Gas-Blocking and Steady States of Foam in Porous Media

Parameter	Gas Blockage	Steady-State Flow
Foam generation mode	Gas displaces surfactant	Coinjection
Boundary conditions	ΔP fixed	Gas and liquid rates fixed
Persistent flow state	Gas or nothing flows	Foam flows
Typical $\Delta P/L$ at ~1 m/day	0.1–1 bar/m	Often ≥ 10 bar/m
S_w in foam-filled medium	$\leq S_{wc}$	$\geq S_{wc}$
Porous-medium length	Long (>0.5–1 m)	Any (often short)
Conc. for strong foam	Typically $\geq 0.5\%$	0.01 to $\geq 1\%$

pack. Values for S_w of 0.03–0.05 during gas blockage were measured for three other aqueous model foams in 1-m long bead packs (*13*).

In the presence of oil at high temperature, it is harder to obtain accurate saturation data, but three experiments on 0.5% Fluowet OTN in sea water with crude oil at 70 °C in 8-darcy beads gave S_w between 0.04 and 0.08 (*21*). Comparable residual nonwetting-phase saturations for the same 8-darcy medium in the absence of foam were 0.13 for the residual water saturation to gas, S_{wrg}, in two gas floods and an average of 0.11 for residual water saturation to oil, S_{wro}, in 24 oil floods with a standard deviation of ±0.02 (*20*). Values of S_w below "irreducible" were also observed by other investigators (*39, 40*) in studying foams at comparable conditions. In contrast, steady-state foam flow is consistently associated with S_w values above connate (*2–4, 18, 34*).

The cause of these low saturations is probably that the slow drainage of liquid, which in the generation of a gas-blocking foam typically continues from breakthrough until the time that gas flow stabilizes, may proceed very far (further than in a flowing foam) because most of the lamellae are stagnant. Capillary pressure suction will continue to drain liquid from a lamella until a saturation has been attained that corresponds to a balance between capillary pressure, P_c, and the disjoining pressure, π, of a given foam film. Because the stability criteria for static and flowing lamellae are not identical (*41*), a stagnant foam is not necessarily broken even at conditions when flowing foam would collapse by the overall P_c exceeding the critical capillary pressure, P_c^* (*42*).

In sandstones, values for S_w closer to connate (0.18 to 0.21 in 1.3-darcy Bentheimer) were obtained with fluorosurfactant foams, although gas blockage was as strong as in bead packs of comparable permeability (*21*). Because most of the water in such cores is presumed to occur in the smallest pores, which hold no foam (*33*), it probably means that the actual gas-blocking foam is also of a similar dry nature in the rock.

Gas Transport. The large gas mobility reduction associated with gas-blocking foam and the nonlinear response of gas mobility to increased differential pressure are not as expected for continuous-gas foams (*43*), which generally are reported to reduce gas mobility by factors of 5–10 (*28, 44*). In their pioneering work on modeling foam behavior in porous media, the Shell researchers noted that for gas to flow in a discontinuous-gas foam, lamellae must be forced through the pore network. In our results, the gas-blocking state of foam is consistently associated with seeing no lamellae or foam exiting the porous medium or collecting near the outlet because a uniform liquid saturation distribution has been measured by microwave attenuation (*45*).

The fact that foam at the gas-blocking state reduces gas mobility by several orders of magnitude and simultaneously shows a strongly non-

Newtonian response to applied gas differential pressure effects must imply that only a few or no continuous gas channels exist at low ΔP, but that more such channels may open in response to an increased gas ΔP. Hysteresis data (14) show that they may also close again if ΔP is reduced, provided enough foamable liquid is present to allow regeneration of the pore-blocking lamellae. Figure 17 compares typical examples that cover our range of gas-blocking foam results to a set of literature data by Prieditis (40). Prieditis' work also included foams conducting only gas flow, but these were generated in steady state and then depleted of surfactant by switching-off liquid flow. This procedure resulted in foams with permanently open gas channels and Newtonian rheological behavior. The only significant difference between the two sets of data is in the mode of foam generation.

From our discussion of length effects, the main difference between gas-blocking and flowing foams is that gas-blocking foams are generated under conditions that do not allow sustaining the high local pressure gradient required to keep bubble trains moving. Conditions for trapping and mobilization of foam lamellae, which are long known (6), have been rather well characterized by the work of Rossen (46, 47). Further, indications are that a major portion of the initially generated lamellae are preserved in a pressure-driven experiment (14, 17). Thus the trapped fraction can approach 1. However, because gas may, in fact, transport through a gas-blocking foam in large volumes over long times, a part of the pore volume must conduct gas flow. Because the mobility reduction is much stronger

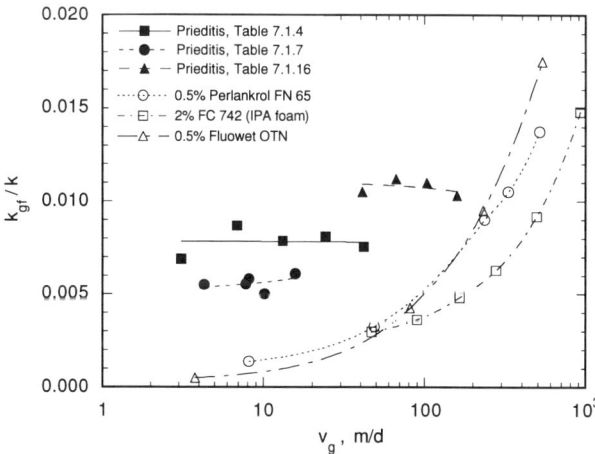

Figure 17. Relative gas mobility as a function of gas velocity, at zero liquid rate, of foams generated by coinjection in 2.9- and 10-darcy bead packs (40) and in the pressure-driven mode in 8-darcy bead packs (13, 20, 25). (Reproduced with permission from reference 14. Copyright 1993.)

than expected for a continuous-gas foam (43), we propose a network model of the gas-blocking state, as illustrated in Figure 18. In such a model, in contrast to the conventional view of a foam with continuous gas, flow paths may be open to gas in a statistical sense only, and may close again in response to regeneration of lamellae at key sites. A simple equation can be derived to describe the response of gas flow rate, or k_{gf} through foam, to increased ΔP. Increasing the "load" on the network will lead to one or a few lamellae at sites that may open large parts of the network to flow; hence creating a nonlinear dependence on pressure drop. Percolation concepts have previously been used (48, 49) with some success in modeling other aspects of foam in porous media.

Gas-blocking foams typically display a nonlinear response to applied pressure. The frequently observed yield point and the power-law dependence of gas flow on ΔP above the yield point suggest a percolation approach to modeling. A square, two-dimensional network model with nonlinear element characteristics:

$$q = 0 \qquad \text{for } \Delta p < \Delta p_t \qquad (5)$$

$$q = \sigma_1(\Delta p - \Delta p_t) \qquad \text{for } \Delta p > \Delta p_t$$

with conductances σ constant for all elements, and the microscopic pressure thresholds, Δp_t, uniformly distributed within a range 0 to 1, has been

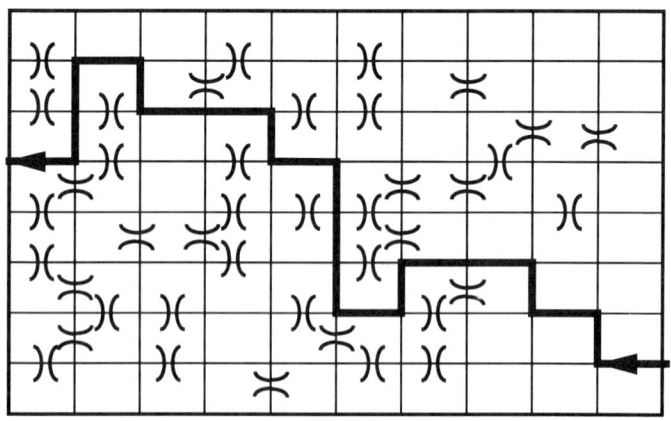

Figure 18. A bond percolation model of gas flow in a pore network containing foam lamellae blocking pore throats. Line intersections represent pore bodies. (Reproduced with permission from reference 14. Copyright 1993.)

calculated to have a macroscopic response to applied ΔP (50) given by

$$Q = 0 \qquad \text{for } \Delta P \leq \Delta P_T \qquad (6)$$

$$Q = \sigma_1 (\Delta P - \Delta P_T)^\alpha \qquad \text{for } \Delta P > \Delta P_T$$

The power exponent, α, is approximately 2 for a wide pore size distribution, σ_1 is a constant conductance, and ΔP_T is the macroscopic threshold pressure into a regime of power-law dependence on reduced pressure. The resemblance of this model to our problem of stagnant foam subjected to an increasing gas pressure is readily seen.

The theory predicts that, in related data sets, the gas-blockage performance of different foams (expressed as k_{gf} should fall on a common straight line when plotted as a function of $(\Delta P - \Delta P_T)^\alpha / \Delta P$.

Figure 19 shows experimental gas-blockage data for some model foam systems (31). Foams were generated in pressure-driven displacements of surfactant solution by gas. A gas-blocking state was reached with only gas entering and exiting the pack. At this state, the pressure drop was increased in steps and the stabilized Q_g and ΔP were measured at each successive gas-blocking state. In the present context, the data are

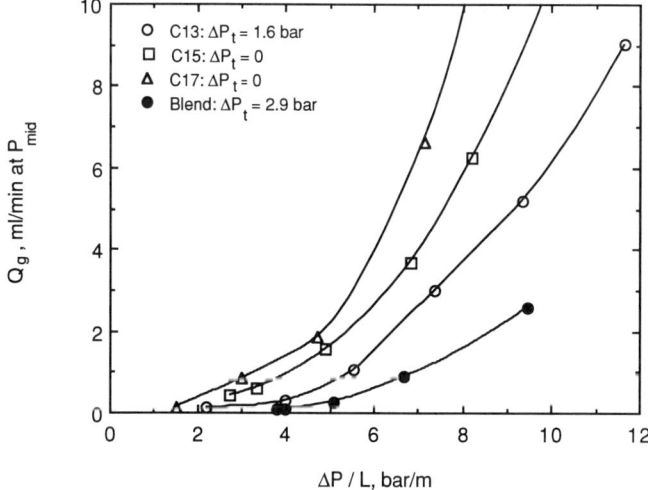

Figure 19. Data for foams from 0.5% Hostapur SAS carbon number fractions in 100-cm 8-darcy bead packs at 21 °C with $P_{out} \simeq 7$ bar and no oil. Threshold pressures are indicated. (Reproduced with permission from reference 14. Copyright 1993.)

presented as flow rates at mean pack pressure. Figure 20 shows the same data replotted with k_{gf}/k as a function of the normalized and reduced pressure drop in accordance with the network model. The indicated threshold pressures were found by second-order polynomial curve fits on the Q versus ΔP data and the best-fit power exponent of 2.5 was found by trial and error. The data fall on one line except two points for C_{17}, which may be outside the power-law regime. The linear correlation coefficient is 0.93. Similarly, a data set for the same surfactants in the presence of oil ($S_{or} \sim 0.05$) was also generalized by a corresponding $(\Delta P - \Delta P_T)^\alpha / \Delta P$ function with essentially the same value for the power-law exponent.

Loss of Gas Blockage. In steady-state flow of foam, the major mechanism of foam decay is dynamic capillary suction coalescence *(41, 51)*. This process may occur in earlier stages of an experiment with gas-blocking foam but is expected to be less frequent later in the experiment.

Evaporation. Some observations on breakdown of gas blockage by evaporation, which has not previously been considered as a decay mechanism, are reported here.

In the mentioned experiment recording microwave saturation profiles *(14)*, data were also taken during breakdown of gas blockage at prolonged exposure to initial ΔP. This particular foam was not a very efficient gas-blocking agent. Liquid saturation, after having reached a uniform level throughout the porous medium, continued its initial decline until reaching uniformly a value of ± 0.015. Then, zero liquid was recorded at the first measurement location that indicated a dry section had developed from the

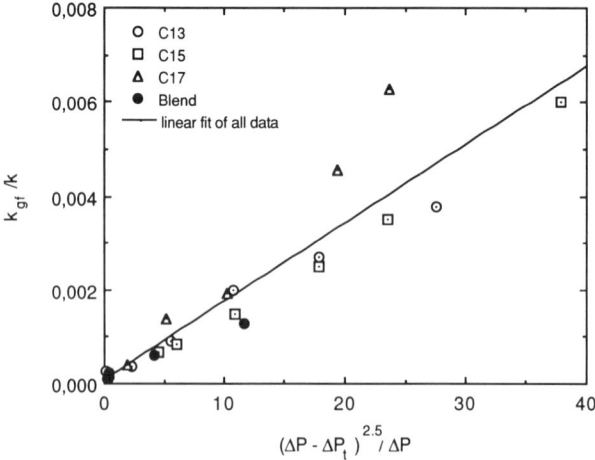

Figure 20. Fit of data in Figure 19 using relationship derived from network model. (Reproduced with permission from reference 14. Copyright 1993.)

inlet and continued to grow with the passage of more gas through the pack. An evaporation loss was expected, because in this experiment, gas is not preequilibrated with the foamed aqueous phase. The rising gas rate as the foam was drying agrees with the length dependence of gas blockage (*13, 25, 52*) because the growing dry section in effect makes the foam shorter.

Foam decay by evaporation will occur whenever a flowing gas phase is not in equilibrium with the liquid constituting the foam films. For the microwave experiment of reference 14, comparing the mass balance (excluding water produced in the vapor phase) with the in situ saturations, shows that the gas did not become saturated in water even after stripping essentially all the water from the first half of the pack. The poor mass transfer indicates that most of the liquid is effectively "shielded" from the flowing gas by existing in trapped foam. This trend is found in all effective gas-blocking foams characterized. Because one cannot assume equilibrium, evaporation from a gas-blocking foam is difficult to model. Rates of evaporation are influenced by the presence of a surfactant film and will depend on the type and concentration of surfactant as well as on the thickness and structure of the liquid film. The effect of evaporation on gas-blockage performance of a strong foam is illustrated by Figure 21.

The more persistent of the two foams shown clearly resulted from bringing the gas to partial saturation in water vapor prior to injection. Because of liquid dropout problems as gas passes through the porous

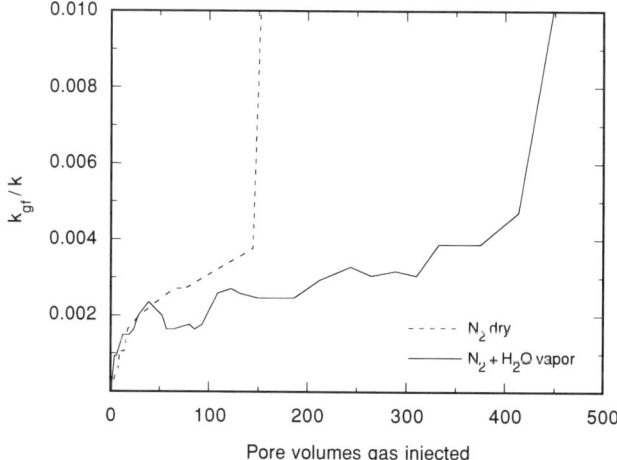

Figure 21. Gas-blockage performance of 1.0% Fluowet OTN foams in 200-cm packs of 8-darcy beads. Conditions: nitrogen gas, dry or partially saturated with water vapor as indicated; $T = 70\ °C$; $P_{out} \simeq 6\ bar$; $\Delta P/L \simeq 2\ bar/m$; *sea water; and* S_{or} *of crude oil. (Reproduced with permission from reference 14. Copyright 1993.)*

medium, a presaturator is not used in our standard test procedure. It follows that the measured persistence of a gas-blocking foam, whether expressed as blockage time or pore volumes of gas transported, is useful in ranking foams at comparable conditions, but does not scale to field conditions. For the purpose of using foam as a gas-coning barrier, evaporation is not likely to limit foam stability in the reservoir because the foam is created by equilibrated gas. However, if foam is applied near the injection points of a nonequilibrated gas, the importance of evaporation as a foam decay mechanism should be evaluated.

Diffusion. Coalescence by diffusion from smaller to larger bubbles remains as a potential decay mechanism for stagnant confined foam in contact with an equilibrated gas. In a typical gas-blocking foam, the large bubbles (which estimated to be on the order of 10 pore bodies per lamella), small areas of bubble contact, and modest pressure gradients all contribute to increase diffusion times drastically compared to bulk foams. Indications are that characteristic diffusion times may well be on the order of weeks or months (*14*). This finding agrees with the observed very long persistence times for gas-blocking foam.

Field Examples

Few cases are described in the literature where producing wells have been treated with foam, injector treatments being more common. An early patent (*53*) describes how a well in the Pennsylvanian sandstone formation, which produced at high GOR and low oil rates, was treated with foaming-agent solution preceded and followed by brine. After the well was put back on production, oil flowed at the original rate and a low GOR until gas suddenly broke through after 4 days. The improvement was not due to oil and gas segregating in the cone area during shut-in, because that was not observed in a previous long shut-in period. This example shows that foam probably was formed in situ by the field gas moving toward the perforations, and foam was able to block gas flow. The effect was short-lived because, in our interpretation, (1) gravity-controlled placement was not possible, and gravity would, in fact, have been a negative influence in this case, because the heavier (aqueous) foamer solution would sink toward the bottom of the oil zone; and (2) the foam used was oil-sensitive and would be expected to break after a short time even if it had been optimally placed.

Another producing-well treatment against high GOR, in this case more likely due to gas cusping (flowing along bedding planes) rather than coning as in Figure 1, was reported at Prudhoe Bay (*54*). Two unsuccessful attempts were first made to create foam in situ by injecting surfactant and surfactant chased by gas. These failures could be ascribed to gravity segregation causing the (aqueous) surfactant or foam to be formed mainly where gas was not flowing. A third trial, where preformed foam was injected, was effective in reducing GOR and increasing oil rates for about 40 days. Most likely, the effect of the foam treatment was limited to a few meters into the well and, in fact, may have been mainly to seal a leaking set of perforations. The gas-blocking action of foam was in any case demonstrated.

Conclusions

A gas-blocking state of foam may be reached in a porous medium whenever gas is allowed to displace a foaming-agent solution at fixed differential pressure. At appropriate conditions, foams may be generated that are able to reach a nearly constant saturation in the liquid forming the foam films. The subsequent flow of foam and surfactant solution out of the porous medium may be reduced to zero, and the flow of gas may be lowered for extended periods.

For gas blockage to be achieved, the porous medium must exceed a certain minimum length that must be greater than the length of travel for the displacement front to become piston-like. Gas blockage of a strength equal to that in model porous media at laboratory conditions is readily achieved in sandstone cores containing reservoir fluids at reservoir conditions provided that the media length and permeability are comparable.

Increasing the surfactant concentration enhances the gas-blocking ability of the resultant foam at least up to a certain level, which corresponds to a multiple of the critical micelle concentration. The presence of oil at saturations as low as 0.04 can completely destroy the gas-blocking ability of some foams, but others are less affected. In the range 20–70 °C, higher temperature is also detrimental to the gas-blocking ability of foam in that only some surfactants at higher concentrations are able to block gas efficiently at higher temperature. The pressure and the pressure drop during displacement of surfactant solution by gas do not seem to have a critical influence on gas blockage.

An experimental method that may include all parameters found critical for achieving gas blockage has been defined and has allowed evaluation of the gas-blockage efficiency of foams in porous media. Recommendations for product selection are given.

The gas-blocking state does not fit the common conceptual models of foam in porous media, which have been constructed largely on the basis of steady-state studies. Gas-blocking foams can maintain strongly reduced mobility despite the passage of hundreds of pore volumes of gas at liquid saturations often lower than those associated with the critical capillary pressure for the existence of stable moving lamellae. The gas-blocking foams display a nonlinear response to increased flow rate, yet do not appear to contain lamellae moving over macroscopic distances. An alternate description proposed is one of essentially all the foam lamellae being trapped and of gas percolating through this network of stagnant lamellae in a number of paths increasing over time and with increased differential pressure. Starting points for modeling the generation of gas-blocking foam by fixed-pressure displacement, the transport of gas through a stagnant foam and the loss of gas blockage caused by evaporation are sketched.

Foam in the gas-blocking state can be used to treat oil wells producing at excessive GOR. The foam can be formed in situ by the field gas moving toward the perforations and impair further coning of gas.

Acknowledgments

We acknowledge with thanks the support of Norske Shell and Statoil for RF's foam-block process development projects from 1985 to 1992, and to the national research and development program RUTH for continuing support of our work on foam processes. J. E. Hanssen is also grateful to the Royal Norwegian Research Council for Science and the Humanities (NAVF) for a Ph.D. project grant. Thanks are due to RF management for supporting the preparation of this chapter.

List of Symbols

A	area (m^2)
GOR	gas–oil ratio (m^3/m^3)
k	absolute permeability [darcy ($\cdot$ 1.013× 10^{-12} = m^2)]
k_{gf}	permeability of gas through foam [darcy ($\cdot$ 1.013× 10^{-12} = m^2)]
L	length (m)
P	pressure
ΔP	pressure drop [bar ($\cdot$ 0.1 = MPa)]
$\Delta P/L$	overall pressure gradient (bar/m)
PV	number of pore volumes
R	radius from well to tip of injectant (m)
r_e	drainage radius of well (m)
S	fractional saturation

t	time (sec)
V	volume (m³)
$x, x/L$	length fraction

Greek

α	power exponent
λ	mobility (darcy/cp)
μ	viscosity [cp ($\cdot\, 10^{-3}$ = kg/m)]
σ	interfacial tension [dyn/cm (= mN/m)]

Superscripts and Subscripts

*	critical
c	capillary
f	foam
g	gas
in	inlet
liq	liquid
o	oil
out	outlet
p	produced
r	residual
t, T	threshold
w	water

References

1. Hanssen, J. E. In *SPOR Monograph: Recent advances in Improved Oil Recovery Methods for North Sea Sandstone Reservoirs;* Skjæveland, S. M.; Kleppe, J., Eds.; Norwegian Petroleum Directorate: Stavanger, Norway, 1992; pp 277–283.
2. De Vries, A. S.; Wit, K. *SPE Reservoir Eng.* **1990,** *5,* 185–192.
3. Persoff, P. J.; Radke, C. J.; Pruess, K.; Benson, S. M.; Witherspoon, P. A. *SPE Reservoir Eng.* **1991,** *6,* 365–372.
4. Radke, C. J.; Ettinger, R. A. *SPE Reservoir Eng.* **1992,** *7,* 83–90.
5. Brigham, W. E.; Castanier, L. M.; Liu, D. *Proceedings of the 60th SPE California Regional Meeting;* Society of Petroleum Engineers: Richardson, TX, 1990; paper SPE 20071.
6. Marsden, S. S.; Albrecht, R. A. *SPE J.* **1970,** *10,* 51–55.
7. Raza, S. H. *SPE J.* **1970,** *10,* 328–336.

8. Skjæveland, S. M.; Papatzacos, P.; Høyland, L. A. *SPE Reservoir Eng.* **1989**, *4*, 495–502.
9. Ekrann, S. *Proceedings of the 6th European Symposium on Improved Oil Recovery;* European Association of Petroleum Geoscientists and Engineers: Zeist, The Netherlands, 1991; pp 167–178.
10. Hanssen, J. E.; Ekrann, S. U.S. Patent 4 903 771, 1990.
11. Ekrann, S.; Hanssen, J. E. *Proceedings of the 13th Collaborated Project on Enhanced Oil Recovery Symposium and Workshop;* Schramm, L. L., Ed.; International Energy Agency; Petroleum Recovery Institute: Calgary, Canada, 1992.
12. Hanssen, J. E.; Rolfsvåg, T. A.; Dalland, M.; Corneliussen, R. *Proceedings of the 5th European Symposium on Improved Oil Recovery;* European Association of Petroleum Geoscientists and Engineers: Zeist, The Netherlands, 1989; pp 737–746.
13. Hanssen, J. E. *J. Petroleum Sci. Eng.* **1993**, *10*, 117–133.
14. Hanssen, J. E. *J. Petroleum Sci. Eng.* **1993**, *10*, 135–156.
15. Bikerman, J. J. *Foams;* Springer-Verlag: Berlin, Germany, 1973.
16. Hanssen, J. E. *Proceedings of the 6th SPE/DOE Symposium on Enhanced Oil Recovery;* Society of Petroleum Engineers: Richardson, TX, 1988; paper SPE/DOE 17362.
17. Schechter, R. S.; Sanchez, J. M.; Monsalve, A. *Proceedings of the 61st Annual Technical Conference;* Society of Petroleum Engineers: Richardson, TX, 1986; paper SPE 15446.
18. Holm, L. W.; Bernard, G. G. *SPE J.* **1964**, 267–274.
19. Göktekin, A. *Erdöl-Erdgas Z.* **1968**, *84*, 208–226.
20. Hanssen, J. E.; Dalland, M. *Proceedings of the 7th SPE/DOE Symposium on Enhanced Oil Recovery;* Society of Petroleum Engineers: Richardson, TX, 1990; paper SPE/DOE 20193.
21. Hanssen, J. E.; Dalland, M. *Proceedings of the 6th European Symposium on Improved Oil Recovery;* Stavanger, Norway, 1991; pp 95–104.
22. Meling, T. M.S. Thesis, Rogaland University Center, 1989.
23. Sanchez, J. M.; Hazlett, R. D. *Proceedings of the 64th SPE Annual Technical Conference and Exhibition;* Society of Petroleum Engineers: Pichardson, TX, 1989; paper SPE 19687.
24. Holt, T.; Kristiansen, T. S. *Proceedings of the 5th Annual SPOR Seminar;* Norwegian Petroleum Directorate: Stavanger, Norway, 1990.
25. Hanssen, J. E.; Haugum, P. *Proceedings of the SPE International Symposium on Oil-Field Chemistry;* Society of Petroleum Engineers: Richardson, TX, 1991; paper SPE 21002.
26. Mannhardt, K.; Novosad, J. J. Chapter 7, this book.
27. Holt, T.; Kristiansen, T. S. *Proceedings of the 5th European Symposium on Enhanced Oil Recovery;* European Association of Petroleum Geoscientists and Engineers: Zeist, The Netherlands, 1989.
28. Radke, C. J.; Ransohoff, T. C. *SPE Reservoir Eng.* **1988**, *3*, 573–585.
29. Muijs, H. M.; Keijzer, P. P. M.; Wiersma, R. J. *Proceedings of the 6th SPE/DOE Symposium on Enhanced Oil Recovery;* Society of Petroleum Engineers: Richardson, TX, 1988; paper SPE/DOE 17361.
30. Radke, C. J.; Manlowe, D. J. *SPE Reservoir Eng.* **1990**, *5*, 495–502.
31. Hanssen, J. E.; Meling, T. *Prog. Colloid Polym. Sci.* **1990**, *82*, 140–154.

32. Dalland, M.; Hanssen, J. E.; Strøm Kristiansen, T. *Proceedings of the 13th Collaborated Project on Enhanced Oil Recovery Symposium and Workshop;* Schramm, L. L., Ed.; Petroleum Recovery Institute: Calgary, Canada, 1992.
33. Radke, C. J.; Gillis, J. V. *Proceedings of the 65th SPE Annual Technical Conference and Exhibition;* Society of Petroleum Engineers: Richardson, TX, 1990; paper SPE 20519.
34. Chou, S. I. *Proceedings of the 66th Annual SPE Technical Conference and Exhibition;* Society of Petroleum Engineers: Richardson, TX, 1991; paper SPE 22628.
35. Minssieux, L. *J. of Pet. Technol.* **1974**, 100–108.
36. Heller, J. P.; Lee, H.-O. *SPE Reservoir Eng.* **1990**, *5,* 193–197.
37. Heller, J. P.; Kuntamukkula, M. S. *Process Prod. R&D* **1987**, *26,* 318–325.
38. McPhee, C. A.; Tehrani, A. D. H.; Jolly, R. P. S. Proceedings of the 6th SPE/DOE Symposium on Enhanced Oil Recovery; Society of Petroleum Engineers: Richardson, TX, 1988; paper SPE/DOE 17360.
39. Holm, L. W. *SPE J.* **1968**, *8,* 359–369.
40. Prieditis, J. PhD Thesis, University of Houston, 1988.
41. Radke, C. J.; Jménez, A. I. In *Oil-Field Chemistry: Enhanced Recovery and Production Stimulation;* Borchardt, J. K.; Yen, T. F., Eds.; ACS Symposium Series 396; American Chemical Society: Washington, DC, 1989; pp 461–479.
42. Khatib, Z. I.; Hirasaki, G. J.; Falls, A. H. *SPE Reservoir Eng.* **1988**, *3,* 919–926.
43. Falls, A. H.; Hirasaki, G. J.; Patzek, P. W.; Gauglitz, D. A.; Miller, D. D.; Ratulowski, T. *SPE Reservoir Eng.* **1988**, *3,* 884–892.
44. Friedmann, F.; Jensen, J. V. *Proceedings of the 56th SPE California Regional Meeting;* Society of Petroleum Engineers: Richardson, TX, 1986; paper SPE 15087.
45. Hanssen, J. E. PhD Thesis, Rogaland University Center, 1992.
46. Rossen, W. R. *J. Colloid Interface Sci.* **1990**, *136,* 1–38.
47. Rossen, W. R. *J. Colloid Interface Sci.* **1990**, *139,* 457–468.
48. Rossen, W. R.; Gauglitz, P. A. *AIChE J.* **1990**, *36,* 1176–1188.
49. Chambers, K. T. MS Thesis, University of California, Berkeley, 1990.
50. Moe, K. MS Thesis, Rogaland University Center, 1989.
51. Radke, C. J.; Chambers, K. T. In *Interfacial Phenomena in Petroleum Recovery;* Morrow, N. R., Ed.; Surfactant Science Series; Marcel Dekker: New York, 1991; pp 191–255.
52. Chou, S. I. Proceedings of the SPE/DOE 7th Symposium on Enhanced Oil Recovery; Society of Petroleum Engineers: Richardson, TX, 1990; paper SPE/DOE 20239.
53. Heuer, G. J., Jr.; Jacocks, C. L. U.S. Patent 3 368 624, 1968.
54. Bain, G. F.; Kuehne, D. L.; Krause, R. E.; Lane, R. H. *Proceedings SPE/DOE Eighth Symposium on Enhanced Oil Recovery;* Society of Petroleum Engineers: Richardson, TX, 1992; paper SPE/DOE 24191.
55. Hanssen, J. E.; Jakobsen, K. R.; Meling, T. *Prog. Colloid Polym. Sci.* **1990**, *81,* 264–265.

RECEIVED for review December 1, 1992. ACCEPTED revised manuscript April 26, 1993.

9

Foams for Well Stimulation

David J. Chambers

Canadian Fracmaster Ltd., No. 1700, 355 Fourth Avenue S.W., Calgary, Alberta T2P 3J2, Canada

> *Foam stimulation fluids use N_2 or CO_2 dispersed within water, acid, alcohol–water mixtures, or hydrocarbon liquids to increase the productivity of petroleum reservoirs. Foam stimulation fluids that are properly designed and applied can result in more productive reservoirs through reduction of possible damaging fluids, improved fluid cleanup, and in certain instances reduced fluid leakoff and improved proppant-carrying capabilities. Foam fluids, as with any stimulation fluids, are not necessarily the most suitable fluid to use in every instance. Foam stimulation fluids have a specific purpose, and achieving the maximum potential of foamed fluids requires thoughtful design and careful application. Specific knowledge of the properties and limitations of foam, fracturing procedures, and reservoir conditions is essential to design a proper foam stimulation treatment.*

STIMULATION IS A COMMON PROCEDURE used to increase the productivity or injectivity of wells. Initially, or during the life of a well, the productivity or injectivity may be lower than what is expected of a well in a particular reservoir. The economic potential of a well can be increased using sound stimulation practices.

Stimulation of a well is accomplished by increasing the flow capacity near the wellbore and allowing formation fluids to enter the wellbore more easily. When low permeability exists, when formation is damaged, or when altered native reservoir characteristics around the wellbore exist, near wellbore stimulation may be beneficial.

Not all wells are good stimulation candidates. Reservoirs that have depleted pressures or low hydrocarbon volumes should be considered as too high a risk to get a return on investment for a stimulation treatment.

The types of stimulation treatments available on the market today are almost as varied as the number of formations that produce hydrocarbons. Stimulations can be categorized into four primary types: hydraulic fracturing, acidizing, soaking–squeezing solvents or gases, and stress or explosive fracturing.

Hydraulic fracturing and acidizing are the most common of the four stimulation methods. Foamed fluids are used in both hydraulic fracturing and acidizing, and thus will be the focus of this chapter.

Hydraulic fracturing increases well productivity by creating a highly conductive channel from the wellbore to some distance within the reservoir. The highly conductive channel is created by pumping a fluid into the formation at rates greater than the fluid leak-off rate. Fluid pressure in the wellbore will increase until the tensile stress holding the rock together is exceeded. At this point, a fracture is created in a direction perpendicular to the plane of least resistance. The fracture will continue to extend radially away from the wellbore until the fracturing fluid in the fracture leaks off into the formation at the same rate as the fluid being pumped into the well or the fracture reaches a barrier.

The width of the created fracture is maintained by using proppants that are mixed into the fracturing fluid during the stimulation process. The proppants prevent the fracture channel from closing after the treatment has been completed, and thus create a conductive path to assist the movement of hydrocarbons from the reservoir to the wellbore.

Generally, fracturing fluids use either water or hydrocarbon as the base fluid. Acid fracturing, however, uses acid as the base fluid. In acid fracturing, acid is pumped into the formation without a proppant. The fracture is created in similar manner to that of hydraulic fracturing; however, the acid chemically reacts with the reservoir rock by etching the formation face. After the formation closes, the etched channels continue to exist, and thus allow hydrocarbon to more easily flow into the wellbore. Acid fracturing is used on carbonate formation rock only. Matrix acidizing differs from hydraulic acid fracturing by the manner in which it stimulates a well. Matrix acidizing is used to remove damage around the immediate wellbore caused by drilling, completion, production, work-over, or kill fluids. Matrix acid stimulations do not penetrate as deeply into a formation as hydraulic acid fracturing because of the rapid chemical reaction between the acid and the formation.

Historically, the fracture stimulation fluids have been water or hydrocarbon-based. With respect to acid stimulation, the base fluid has been hydrochloric acid. As fluid technology developed, stimulation fluids have become more complex. However, the pursuit of the ideal stimulation fluid is an ongoing process. An ideal fluid would incorporate all of the desirable features of several stimulation fluids into one particular fluid. A few of the desirable features are the following:

1. low fluid loss
2. sufficiently high effective viscosity

3. low friction pressures
4. temperature stability
5. shear stability
6. low formation damaging effects
7. low residue deposition
8. good posttreatment viscosity breaking characteristics
9. good posttreatment cleanup and flow-back behavior
10. low cost

The utilization of gases entrained within the stimulation fluid enabled many of these features to be achieved. During the 1950s, hydrocarbon gases were used to enhance well cleanup after matrix acid squeezes. Subsequently, N_2 and later CO_2 were introduced as energizers for well stimulation. Presently, N_2 and CO_2 are being used exclusively, because other readily available gases such as air and hydrocarbon gases pose a potential danger to people and equipment during the treatment.

Grundman and Lord in their Society of Petroleum Engineers paper entitled "Foamed Stimulation" (1), reported that the first use of foam as a stimulation fluid was in 1968. Foamed stimulating fluids became more widespread in the early 1970s, as a result of the research of Mitchell, Blauer, and others. Initially, foams were considered to be the ultimate stimulation fluid. They were thought to have all the desirable properties of a stimulation fluid as is common to the introduction of new stimulation fluids. As the use of foams increased during the 1970s, the following advantages became apparent:

1. exceptional flow-back and cleanup
2. reduced liquid volume
3. good proppant-carrying capabilities

Other highly regarded characteristics of stimulation fluids such as low fluid loss and high viscosity were not to be as exceptional as initially thought (2).

Perhaps the petroleum industry's most widely recognized advantage to using foam as a stimulation fluid is its ability to cleanup the liquid load fluid used to perform the fracture. Upon completion of the fracture treatment, the fracture fluid, after it has "broken", should ideally be completely removed from the well. This step ensures that no residual polymers, fines, or unbroken gel will hinder the conductivity of the fracture. Also, a quick cleanup in order to bring the well back onto production is important. Fracture cleanup is accomplished by back-flowing the fluid from the formation to the surface. As the surface pressure is released, the gas within the foam expands, the foam stability is diminished, and the gas–liquid interface weakens. Further release of pressure at the surface will cause a considerable expansion of the gas. As the volume of gas is large in comparison to the volume of the liquid, the foam will flow back as a mist (1).

Throughout the 1970s, foamed fluids were considered to be the most acceptable method to treat shallow, low-pressure gas wells, because of the reduced liquid volume and excellent flow-back capabilities. Further developments such as fluorocarbons made it possible to foam hydrocarbon-based fluids. Thus fracturing in both oil and gas reservoirs was made available; however, gas well stimulation was more common. Advancements in gelling and gel-breaking chemicals, developments of mechanical concentrators, and improved blending processes permitted massive high-rate foam fracturing treatments.

As with the energized fluids, foams first incorporated N_2 as an energizer. The introduction of CO_2 enabled the industry to stimulate wells deeper than previously allowed by N_2 foams, and to use the benefits of a foamed fluid. Foams that use gaseous N_2 for the internal phase have lower hydrostatic pressures in comparison to CO_2 foams that use liquid CO_2 as the internal phase. For this reason, the surface treating pressures for N_2 foams are greater than for CO_2 foams in order to achieve the same bottomhole fracture pressure, provided that tubular friction pressure does not offset the gain by the hydrostatic pressure. As a result, the surface treating pressures for N_2 foams may exceed the maximum allowable rating for the wellhead and surface equipment prior to reaching the required formation fracturing pressure in deep wells. A third-generation foam stimulation fluid employs both nitrogen and carbon dioxide. The resulting foam combines the advantages of CO_2 foams with those of N_2 foams.

Foams are characterized in many different ways. Stability, quality, and texture describe different characteristics regarding foams. Generally, foam quality is most often used to categorize foams that are used in the service industry. However, other foam characteristics and their relationship with each other can be important.

Foams are stable mixtures of fluid and gas; the fluid is the external phase, and the gas is the internal phase (Figure 1). A surfactant used to impart stability to the mixture concentrates at the gas–liquid interface to reduce the surface tension and form stable lamellae.

The static stability of a foam refers to the ability of the foam to resist bubble breakdown. The foam's static stability can be quantified by measuring its half-life. This is the time required at static conditions for the foam to drain half of its liquid volume. As parameters such as type of stabilizer, containment pressure, or foam generation process change, the stability will also change. Foam half-life is not a direct measurement of stability. Variations of foam stability will occur under different conditions.

Foams are best defined by their quality, which is the percentage of gas contained within the foam. Mitchell foam quality, which is more wide-

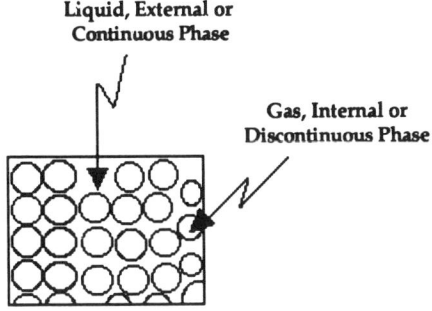

Figure 1. Cross-sectional diagram of a foam.

ly recognized in the petroleum industry, is the volume percentage as gas and is mathematically described as (2–4):

$$\text{foam quality} = \frac{\text{gas volume}_{(T,P)}}{\text{gas volume}_{(T,P)} + \text{liquid volume}} \times 100\% \quad (1)$$

Another foam quality that is not used as commonly as the Mitchell quality is "slurry quality". Slurry quality (Q_s), or total internal phase volume, is defined as the ratio of the volume of gas (V_G) plus the volume of proppant (V_P) to the total slurry volume ($V_G + V_P + V_L$) (3–5).

$$Q_s = \frac{V_G + V_P}{V_G + V_P + V_L} \quad (2)$$

This quality is often used when determining rheological properties of foams when proppant is present (6) (Figure 2).

Monodispersed foams were initially considered to exist only in the range of qualities between 52% and 96%. For foam qualities below 52%, no stability was achieved because of what was first believed to be a lack of bubble interaction (7). Polydispersed bubble interaction occurs at regions higher than 54% (2). For qualities greater than 96%, the foam inverts to a liquid in gas emulsion. The solid's suspension and viscosity characteristics disappear as a liquid mist is created (2). As the foam quality increases, the bubbles become more closely packed. Deformation of monodispersed foam bubbles begins at 74% quality and is higher for polydispersed foams. The effective viscosities of foam according to Blauer et al. (7) increase and approach the asymptote at 96%. For qualities above 52%, the foam begins to exhibit bubble interactions. The measured

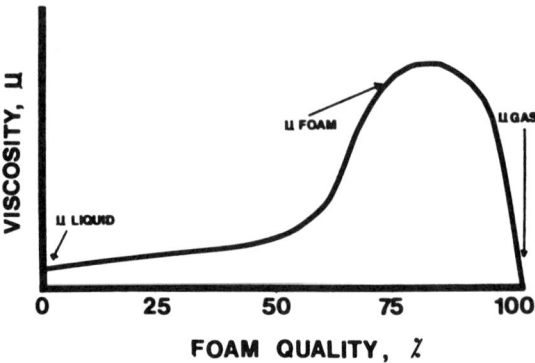

Figure 2. A plot of foam quality vs. foam viscosity. (Reproduced with permission from reference 6. Copyright 1975 Canadian Institute of Mining, Metallurgy, and Petroleum.)

viscosities of foams differ depending upon bubble size, bubble uniformity, external phase viscosity, and shear rate during measurement.

Improvements in chemical research and development have increased the stability of foams by improving gellants and stabilizers. Thickening of the continuous phase increases the difficulty for gas bubbles to coalesce. Although relatively high qualities are required to achieve stability, qualities of less than 52% can maintain dispersion of the gas phase. Further improvements in foam stability can be accomplished by cross-linking the polymers in the aqueous phase. To date, foams have been established with qualities less than 40% (5). The use of low-quality foams is advantageous if viscosity or fluid loss control is a primary concern. If the formation sensitivity or fluid flow-back is important, low-quality foams are not desirable because of the reduced gas volume and increased fluid volume.

Foam exhibits favorable fluid loss control in low-permeability formations, 10 millidarcy (md) or less. However, in naturally fractured or medium-permeability formations, 10–100 md, the foamed fluids leak-off at higher rates (2, 6, 8). Conventional stimulation fluids have more desirable leak-off coefficients in naturally fractured or medium-permeability reservoirs than foams (6). Leak-off control in low-permeability reservoirs is controlled by the viscosity of the external phase according to Harris (8). Formations that have high permeabilities, 100 md or greater, generally do not benefit substantially from being stimulated unless the area near the wellbore has been damaged. In such cases, the primary objective is to remove the wellbore damage by use of a small stimulation, and therefore leak-off is not of great concern. As far as stimulation is concerned, the incremental increase in production of high-permeability reservoirs would not justify the expenditure of the treatment, and so high-permeability formations are poor stimulation candidates.

Foamed Fluids

Foam fluids are essentially two-phase fluids that consist of an inner phase and an outer phase. The inner phase is generally gaseous, usually N_2; however, it can be liquid or dense vapor as is the case when CO_2 is used. The external phase is primarily composed of a saline–water mixture with either a surfactant or gellant depending upon the viscosity and stability requirements. Other external fluids commonly used are either hydrochloric acid or alcohol–water mixtures. Diesel fuel, reformates, or other hydrocarbon-based solvents can also be used as external-phase fluids but require N_2 as the energizer. Carbon dioxide and hydrocarbons produce a single-phased fluid, because CO_2 is very soluble in hydrocarbon liquids.

Foamed fluid characteristics, to a large extent, depend upon the nature of the external phase, but the importance of the internal phase can not be underrated. This section of the chapter will examine each of the internal phases of oil field stimulation foams.

N_2-Foamed Fluids. N_2 has been used in oil field stimulation work since 1961. In the mid 1970s, foamed fracturing fluid using N_2 was developed. More recent uses of N_2 foams in well stimulation involve well clean-outs and treatment diversions.

The N_2 used for well stimulation is transported to the well as a liquid in a cryogenic transport tank. The tank that contains the N_2 is kept at a temperature of -196 °C and a pressure of 30 kPa. The liquid N_2 is pumped with a cryogenic positive displacement pump into a heat exchanger where it is transformed from a liquid into a gas by raising the temperature (Figure 3). The dry gas is pumped into the liquid or slurry treatment line downstream from the liquid pumpers. At this point, the N_2 is commingled with the treating fluid and creates a foam that is pumped downhole.

As a result of the addition of N_2, the foam injection rates are generally 2 to 5 times those of the liquid pump rates. The following expression relates to the volume change (E_f) to the foam quality (Q) at given temperature (T) and pressure (P).

$$E_f = \frac{1}{1 - Q_{(T,P)}} \qquad (3)$$

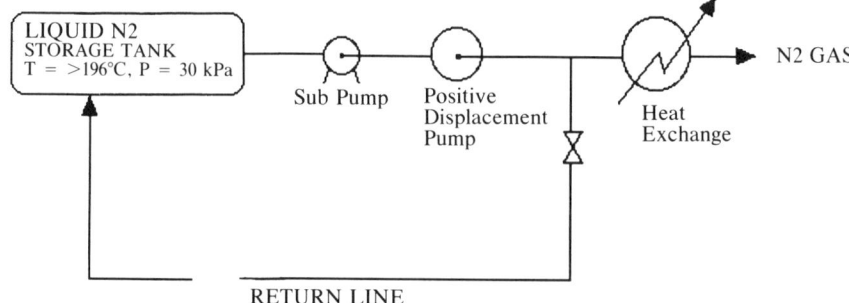

Figure 3. A N_2 pumper schematic.

A foamed fluid with a downhole quality of 75% may, for example, require a surface quality of 80%. (Foamed stimulation treatments are always stated in terms of downhole quality; surface quality is dependent upon zone, friction pressure, depth, and treating pressure.) A closed system is necessary to inject N_2 because it is a gas. Therefore, the sand can only be added to the liquid portion of the foamed fluid. In this situation, the proppant, which is added to the liquid portion of the foamed fluid stream, must be added at 2000 kg/m^3 in order to achieve a downhole concentration of 400 kg/m^3. This condition illustrates that maximum downhole concentrations depend upon the downhole foam quality and the maximum proppant concentrations that the blender is capable of pumping.

Slurries with high proppant concentrations are difficult to achieve with any foam stimulation fluid. Heavier loading of gellants, cross-linkers, and sand intensifiers, or reducing foam qualities, and increasing the sand concentrations can slightly increase downhole slurry proppant concentrations. Increasing the proppant concentrations with any of these methods may deteriorate the desirable properties of foam fluids. For instance, increasing the fluid viscosity by using higher gellant loadings or cross-linkers can increase the potential for formation damage because of an increase in gellant residue, plugging both the proppant pack permeability and the surface area of the fracture face. As well, increasing the proppant density by reducing foam quality will reduce the flow-back capabilities of the fracture fluid. This reduction of flow-back may further increase the risk of damage to fluid-sensitive formations by increasing the contact time of the stimulation fluid in the fracture. In designing the treatments, the benefits gained by increasing the proppant concentrations must be weighed against the negative aspects, such as increased potential of formation damage and reduced flow-back capabilities.

Because of its abundance in the atmosphere and its general inertness, N_2 gas is one of the most useful and environmentally safe gases in the oil industry. The abundance of N_2 in the atmosphere enables it to be isolated by liquefaction and fractional distillation. The extraction process is relatively simple, making N_2 economical as well as relatively safe.

N_2 gas behaves in a similar manner as an ideal gas. However, at higher pressures and temperatures, the gas deviates from ideal, and equation 4 can be used to determine the density of the foam. The gas is relatively light, weighing approximately 28 g/mol. Because the gas is fairly light and the foamed fluid is composed of about 75% N_2, the column of foamed fluid in the well has a lower hydrostatic pressure than other fracture fluids including CO_2 fracture foams. For that reason, nitrified fluids are more beneficial in treating shallow low-pressure reservoirs.

A procedure for calculating the density of foam at a given depth is as follows. This procedure uses a method that Sage and Lacey (9) developed to calculate gas compressibilities. The first step is choosing an incremental depth over which the densities will be calculated. The smaller the increment, the more accurate the calculation; however, the more steps are necessary to reach completion over the same depth. Temperature is calculated using $T = 10\ °C + 3\ °C/100$ m. The compressibility (z) can then be calculated as follows:

$$z = 0.021AP^2 + 0.145BP + C \qquad (4)$$

where P is pressure in kPa (initially surface treating pressure), and A, B, and C are constants that are dependent on temperature in Kelvin. The compressibility factor for N_2 is shown in Figure 4. If P is less than 27,580 kPa (4000 psi) then:

$$A = (1.67939 \times 10^{-9}) - [1.1204 \times 10^{-9}(T)] +$$
$$[2.6043 \times 10^{-12}(T^2)] - [2.0687 \times 10^{-15}(T^3)]$$

$$B = -0.0003122 + [1.52784 \times 10^{-6}(T)] - [1.7399 \times 10^{-9}(T^2)]$$

$$C = 1$$

If P is greater than 27,580 kPa but less than 55,160 kPa (4000 psi < P < 8000 psi) then:

$$A = 0$$

$$B = 0.00022817 - [7.31 \times 10^{-7}(T)] + [7.452 \times 10^{-10}(T^2)]$$

$$C = 0.0956 + 0.0045(T) - [4.86 \times 10^{-6}(T^2)]$$

Compressibility Factor For Nitrogen

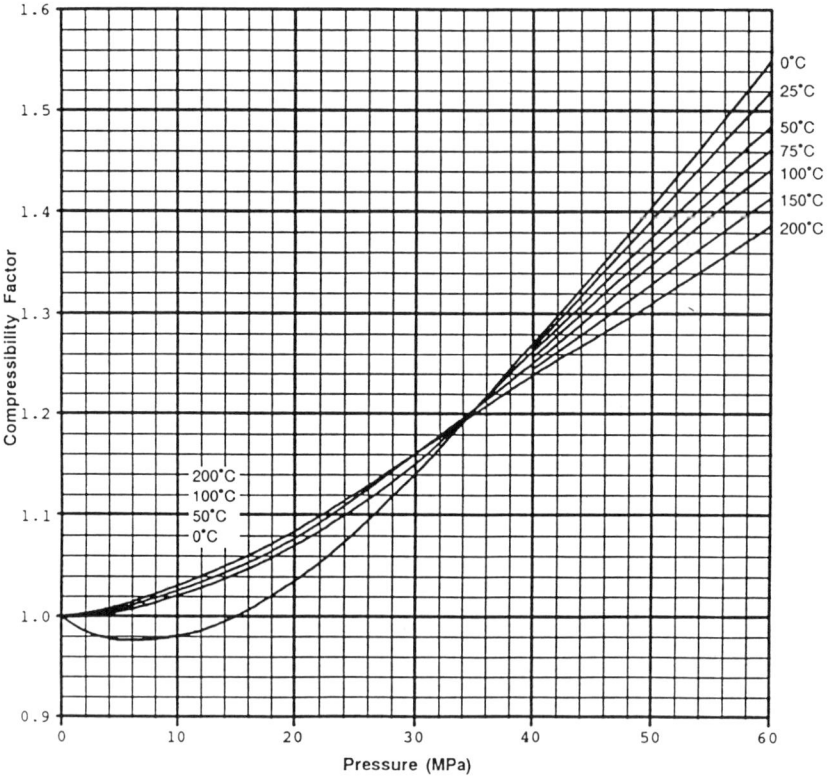

Figure 4. A plot of the compressibility factor for N_2 at various temperatures and pressures (developed from equation 4).

If P is greater than or equal to 55,160 kPa (8000 psi) then:

$$A = 0$$

$$B = 0.00022043 - [6.336 \times 10^{-7}(T)] + [5.8644 \times 10^{-10}(T^2)]$$

$$C = 0.1573 + 0.0044(T) - [4.536 \times 10^{-6}(T^2)]$$

After the compressibility factor has been arrived at, the N_2 space factor can be determined. The N_2 space factor is the volume of N_2 gas at a particular temperature and pressure that will occupy one standard cubic meter at standard conditions. The space factor can be determined as follows:

$$\frac{1}{V_{STP}} = \frac{2.84P}{zT} \tag{5}$$

The quality of the foam (Q) is then calculated:

$$Q = \frac{N_2 \text{ ratio}}{N_2 \text{ ratio} + \text{space factor}} \tag{6}$$

Because the foam quality and liquid ratio can be calculated, the density of the foam can be determined:

$$\rho_{foam} = Q(\rho_{N_2}) + (1 - Q)(\rho_{fluid}) \tag{7}$$

where Q is foam quality, ρ_{N_2} is the density of the N_2, and ρ_{fluid} is the density of the fluid (liquid) portion.

Example 1: Hydrostatic Pressure Calculation. The following is an example of the foam density calculation. Because the calculation is an iterative process, the example will be in tabular form in order to account for each interval. The example has a few characteristics that should be noted. First, the pressure column is in absolute rather than gauge pressure; therefore, standard atmospheric pressure must be added to the pressure. Also, foam rate and foam velocity have been calculated but are not necessary for this example. The surface conditions at which the foam was pumped were as follows:

- N_2 ratio, 700.0 m³/m³
- fluid density, 1000.0 kg/m³
- surface pressure, 25.0 MPa
- depth, 1500.0 m
- equivalent i.d., 139.7 mm
- bottomhole pressure (B.H.P), 30.97 MPa
- fluid rate, 2.0 m³/min

CO_2-Foamed Fluids. Several studies have compared N_2 and CO_2 foams (10–13). The results of the studies concluded that both types of foams exhibit very similar behaviors. This result is interesting because

Example 1. Foam Simulation

Depth (m)	Temperature (K)	Max. B.H.P. Pressure (MPa)	z	Space Factor (m^3/m^3)	Decimal Fraction N_2 (m^3)	Decimal Fraction Fluid (m^3)	Foam Density (kg/m^3)	Foam Rate (m^3/min)	Foam Velocity (m/min)
33.33	284.00	25.00	1.10	229.13	0.75	0.25	462.49	8.11	529.12
66.67	285.00	25.15	1.10	229.35	0.75	0.25	462.83	8.10	528.73
100.00	286.00	25.30	1.10	229.57	0.75	0.25	463.16	8.10	528.35
133.33	287.00	25.45	1.10	229.79	0.75	0.25	463.49	8.09	527.98
166.67	288.00	25.61	1.10	230.00	0.75	0.25	463.81	8.09	527.62
200.00	289.00	25.76	1.11	230.20	0.75	0.25	464.12	8.08	527.26
233.33	290.00	25.91	1.11	230.41	0.75	0.25	464.43	8.08	526.91
266.67	291.00	26.06	1.11	230.61	0.75	0.25	464.74	8.07	526.57
300.00	292.00	26.21	1.11	230.80	0.75	0.25	465.03	8.07	526.23
333.33	293.00	26.36	1.11	231.00	0.75	0.25	465.32	8.06	525.90
366.67	294.00	26.52	1.11	231.18	0.75	0.25	465.61	8.06	525.58
400.00	295.00	26.67	1.12	231.37	0.75	0.25	465.89	8.05	525.26
433.33	296.00	26.82	1.12	231.55	0.75	0.25	466.16	8.05	524.95
466.67	297.00	26.97	1.12	231.73	0.75	0.25	466.43	8.04	524.65
500.00	298.00	27.12	1.12	231.90	0.75	0.25	466.70	8.04	524.35
533.33	299.00	27.28	1.12	232.07	0.75	0.25	466.96	8.03	524.06
566.67	300.00	27.43	1.12	232.24	0.75	0.25	467.21	8.03	523.78
600.00	301.00	27.58	1.12	233.25	0.75	0.25	468.73	8.00	522.08
633.33	302.00	27.74	1.12	233.32	0.75	0.25	468.84	8.00	521.96
666.67	303.00	27.89	1.13	233.40	0.75	0.25	468.96	8.00	521.83

700.00	304.00	28.04	1.13	233.48	0.75	0.25	469.07	8.00	521.70
733.33	305.00	28.20	1.13	233.55	0.75	0.25	469.19	7.99	521.57
766.67	306.00	28.35	1.13	233.63	0.75	0.25	469.30	7.99	521.44
800.00	307.00	28.50	1.13	233.70	0.75	0.25	469.41	7.99	521.32
833.33	308.00	28.66	1.14	233.78	0.75	0.25	469.53	7.99	521.19
866.67	309.00	28.81	1.14	233.86	0.75	0.25	469.64	7.99	521.06
900.00	310.00	28.96	1.14	233.93	0.75	0.25	469.76	7.98	520.94
933.33	311.00	29.12	1.14	234.01	0.75	0.25	469.87	7.98	520.81
966.67	312.00	29.27	1.14	234.09	0.75	0.25	469.99	7.98	520.68
1000.00	313.00	29.42	1.15	234.16	0.75	0.25	470.10	7.98	520.55
1033.33	314.00	29.58	1.15	234.24	0.75	0.25	470.22	7.98	520.43
1066.67	315.00	29.73	1.15	234.31	0.75	0.25	470.33	7.97	520.30
1100.00	316.00	29.88	1.15	234.39	0.75	0.25	470.45	7.97	520.17
1133.33	317.00	30.04	1.15	234.47	0.75	0.25	470.56	7.97	520.05
1166.67	318.00	30.19	1.15	234.54	0.75	0.25	470.68	7.97	519.92
1200.00	319.00	30.34	1.16	234.62	0.75	0.25	470.79	7.97	519.79
1233.33	320.00	30.50	1.16	234.70	0.75	0.25	470.91	7.97	519.66
1266.67	321.00	30.65	1.16	234.77	0.75	0.25	471.02	7.96	519.54
1300.00	322.00	30.81	1.16	234.85	0.75	0.25	471.14	7.96	519.41
1333.33	323.00	30.96	1.16	234.93	0.75	0.25	471.26	7.96	519.28
1366.67	324.00	31.11	1.17	235.01	0.75	0.25	471.37	7.96	519.15
1400.00	325.00	31.27	1.17	235.08	0.75	0.25	471.49	7.96	519.02
1433.33	326.00	31.42	1.17	235.16	0.75	0.25	471.60	7.95	518.90
1466.67	327.00	31.58	1.17	235.24	0.75	0.25	471.72	7.95	518.77
1500.00	328.00	31.73	1.17	235.32	0.75	0.25	471.84	7.95	518.64

there is a significant difference in the physical properties of the two gases and the field implementation procedures of both systems. CO_2 is pumped as a liquid, and N_2 is pumped as a gas.

CO_2 at atmospheric conditions exists as a colorless dense gas. When the gas is compressed and cooled, a liquid is formed. The CO_2 used in oil field stimulation applications is transported as a liquid at approximately -20 °C and 2 MPa.

The primary difference between N_2 and CO_2 foams is that N_2 is pumped as a gas, and CO_2 is pumped as a liquid. The introduction of liquid CO_2 to the external phase occurs under pressure, commingled with the sand-laden fracture fluid as does N_2. As the CO_2 is mixed with the external phase that is at approximately 20 °C, it will warm and expand. The CO_2 or internal phase will remain in a liquid state until the temperature reaches 31 °C (Figure 5) (*14*). After the CO_2 reaches 31 °C, it becomes a supercritical fluid. The heating may occur in the pipe as a result of friction, or it may occur in the formation as a result of conduction from the formation. The CO_2 will change phases (liquid to supercritical) as a result of the heating.

Foams are structured, two-phase fluids that are compressible in nature. CO_2 foam when initially mixed is a fluid that is composed of an aqueous liquid–liquid CO_2 mixture. Although a liquid–liquid, two-phase structured fluid is classically termed an emulsion, the end-use application

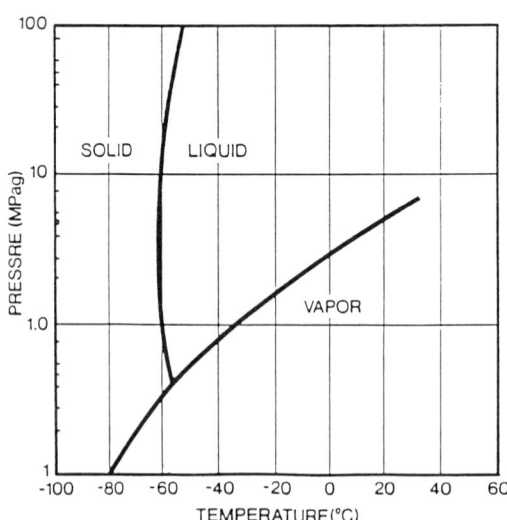

Figure 5. A three phase diagram for CO_2. (Reproduced with permission from reference 14. Copyright 1986 Canadian Institute of Mining, Metallurgy, and Petroleum.)

of the two-phase fluid is normally above the critical temperature of 31 °C, where CO_2 will exist only as a supercritical fluid. At typical formation depths, the internal phase is heated by the formation to form a gas that is dispersed within the external liquid phase. The fluid properties are not lost as the CO_2 is being converted from a liquid to a gas (8). Therefore, the term "foam" can be used to describe the two-phase fluids that use CO_2 as the energizer.

CO_2 exhibits significant solubility in both water and hydrocarbons, as shown in Figures 6 and 7. When CO_2 is dissolved in water, it ionizes to form a weak organic acid (carbonic acid) that buffers the pH of the aqueous system. The following is the resultant chemical reaction.

$$CO_2 + H_2O \rightarrow HCO_3^- + H^+ \quad K_a = 2 \times 10^4$$

The pH is reduced to approximately 3.2, and the system assists to control the swelling of formation clays (13). A potential drawback to the CO_2 fluid systems is the effect that the carbonic acid has on most gelling agents and metal cross-linkers. Acid hydrolysis of the gellant polymers oc-

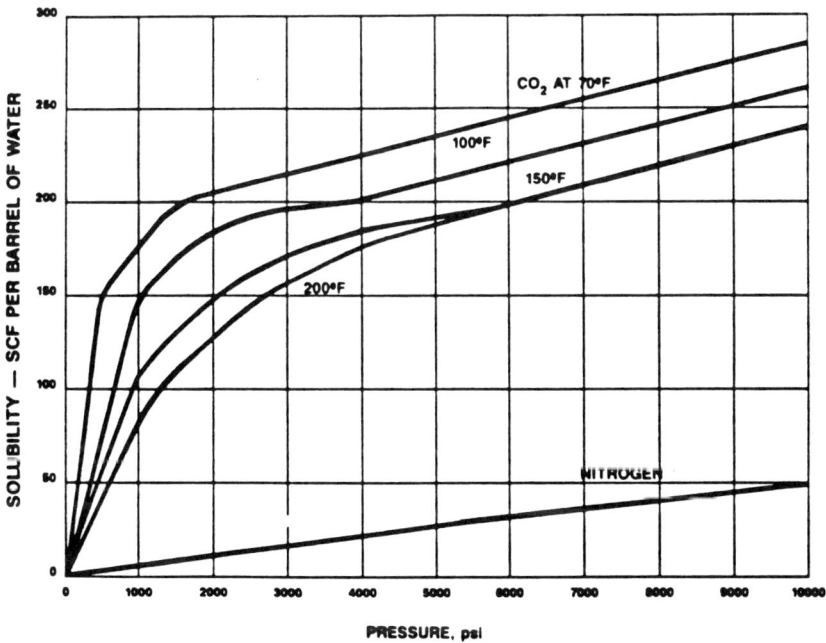

Figure 6. A plot of the solubility of CO_2 and N_2 in water. (Reproduced with permission from reference 5. Copyright 1985 Society of Petroleum Engineers.)

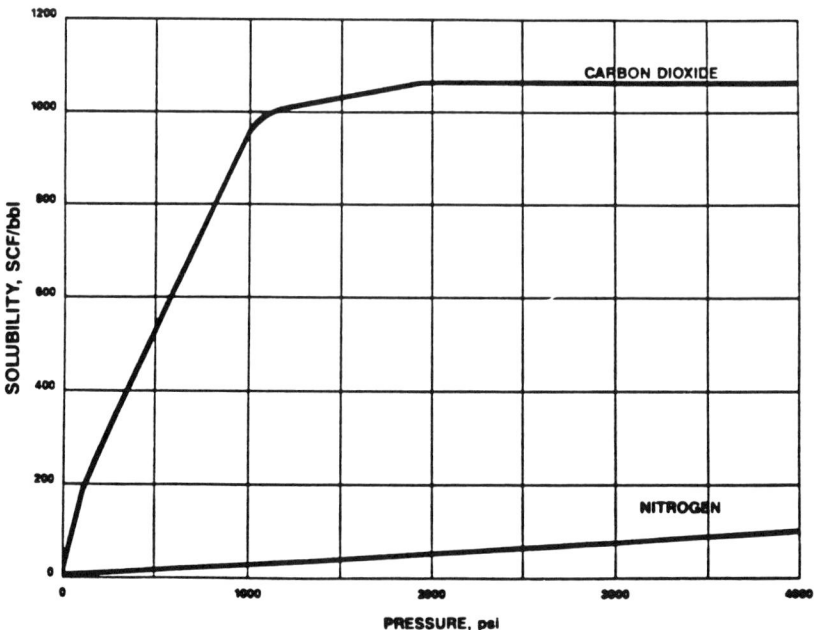

Figure 7. A plot of the solubility of CO_2 and N_2 in 40 °API Crude Oil. (Reproduced with permission from reference 5. Copyright 1985 Society of Petroleum Engineers.)

curs, and most gellants and some foamers are degraded. The viscosity of the aqueous phase and the foam stability is reduced. Therefore, the CO_2 foams generally require larger amounts of gellants or foamers to achieve stability (15).

To compare the effect of CO_2 on liquid-phase gellants and foamers, N_2 was used as the energizer. To simulate the acidity of CO_2 foam systems, the pH in the N_2 system was lowered to 3.2. Tests illustrated that several foamers and gellants that did not perform acceptably with CO_2 performed successfully with N_2. "These results suggest a chemical interaction, such as carboxylation, between the CO_2 or carbonic acid and some of the foamers" (15).

An advantage of using CO_2 as an energizer is that it dissolves in oil and to a lesser extent in water. As the solubility of CO_2 in crude oil increases, the viscosity of the oil decreases dramatically. Thus the ability for the crude oil to be produced is improved. An example of the viscosity reduction effects of CO_2 on different crude oils at different temperatures and constant pressure is shown in Table I (16).

Table I. Viscosities of Oil with CO_2 at Different Temperatures

CO_2 Ratio (m^3/m^3)	20 °API Oil at 21 °C	20 °API Oil at 49 °C	38 °API Oil at 29 °C
0	400	80	5.40
70	23	14	2.50
123	20	12	1.50
190	20	12	1.10

NOTE: All viscosity values are given in centipoises.

As illustrated by Table I, the solubility of CO_2 in fluids decreases with increasing temperatures, and the formation volume factor shows a corresponding reduction.

Table II compares the solubility of CO_2 in various well treatment fluids at various pressures (16).

The solubility of CO_2 in crude oil or formation water has a distinct advantage in fracturing operations. CO_2 is stored until fluid recovery begins, and as the pressure decreases, gaseous CO_2 breaks out of solution and expands to push fluids to the surface. These properties allow CO_2 to remain in the formation for several hours prior to flowing the well back. However, if the carbonated foamed fluid is left in formation for an extended period of time, then the CO_2 will dissipate.

Foamed fluid rheological properties are very responsive to changes in foam quality. Therefore, in order to achieve the desired treatment goals, accurate metering of the constituents of the foam is essential. The liquid CO_2 rate is measured using a low-pressure turbine metering device. To determine the volume of CO_2 required at the downhole conditions, a mass balance across the total system can be accomplished.

CO_2 is considered to be a nonideal gas. That is, it obeys the real gas law, $PV = znRT$. Using a mass balance, the mass remains constant both at

Table II. Solubility of CO_2 (m^3/m^3) at Various Pressures

Treatment Fluid	690 kPa	6890 kPa	13,790 kPa	27,580 kPa
Fresh water (38 °C)	3.5	26.9	30.8	33.8
Salt water (100,000 ppm – 38 °C)	2.3	19.1	22.5	24.6
Salt water (260,000 ppm – 38 °C)	1.1	9.4	11.2	12.2
Crude oil (38 °API – 30 °C)	8.0	181.4	190.3	190.3
Crude oil (20 °API – 49 °C)	6.2	73.5	123.5	123.9

standard and in situ conditions. Rearranging the equation to calculate mass and equating standard conditions to reservoir conditions allows us to calculate the volume of gas at standard conditions.

Where the volume is V, the temperature is T, and the pressure is P at standard conditions (SC), knowing $V_{SC} = 1$, $T_{SC} = 288.6$ K, $P_{SC} = 101.35$ kPa, the following equation can be derived.

$$V_{SC} = \frac{2.85 P_r V_r}{z_r T_r} \quad (8)$$

where P_r is the pressure of gas at reservoir conditions, V_r is the volume of gas at reservoir conditions, z_r is the compressibility of the gas at reservoir conditions, and T_r is the temperature at reservoir conditions.

The solubility of the CO_2 must be taken into account to include the dissolved volume of CO_2 that exists within the liquid portion of the foam. S_r is the solubility of CO_2 in the liquid phase of the foam, V_f is the average foam velocity, and the volume of CO_2 dissolved can be written as $S_r V_f (1 - Q)$. Depending on whether the external phase of the foam is aqueous or hydrocarbon-based, S_r can be determined by referring to Figure 6 or 7. Therefore, the total downhole volume of CO_2 required is expressed by

$$V_{CO_2} = \frac{2.85 P_r Q V_f}{z_r T_r} + S_r V_f (1 - Q) \quad (9)$$

Because volumetric rather than mass metering instruments are typically employed on actual treatments, conversion from standard condition gas injection rates to pumping condition liquid rates must be performed. Figures 8 and 9 can be used to estimate the pumping requirements at liquid conditions.

The compressibility of CO_2 can be determined from a compressibility chart that is available from several sources (17). Another more accurate method of acquiring compressibility data is explained in reference 18.

Manual calculations of these relationships would not be timely enough to keep pace with the varying surface treating pressures. Microprocessors enable the calculations to be completed as soon as the input parameters change. Therefore, more rapid changes in volumetric rates at surface can occur, and the bottomhole quality is kept constant. Bottomhole foam quality calculations do not generally encompass heat transfer from formations to the fluids.

Combined N_2- and CO_2-Foamed Fluids. A third-generation foam fluid uses both N_2 and CO_2 simultaneously as the internal phase in

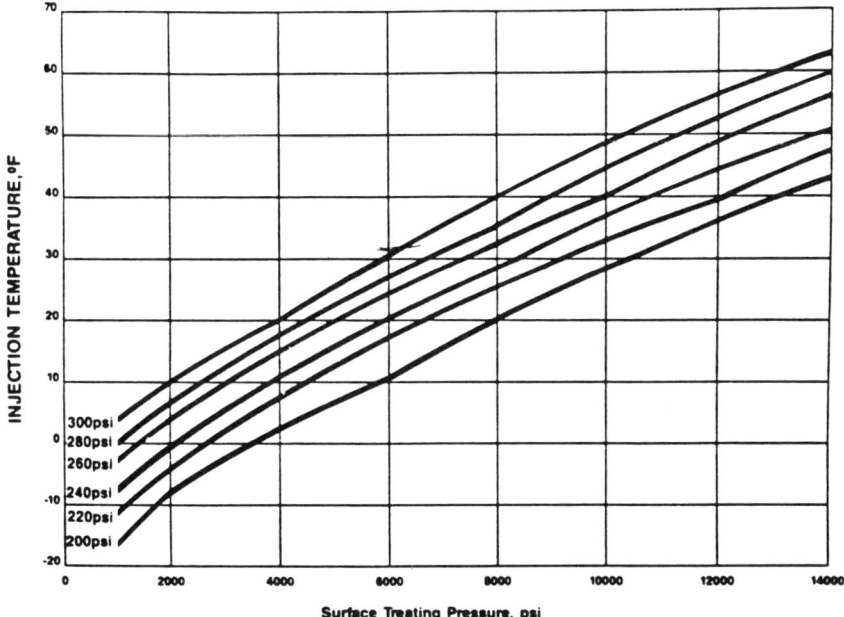

Figure 8. A plot of CO_2 injection temperature vs. pressure. (Reproduced with permission from reference 5. Copyright 1985 Society of Petroleum Engineers.)

an aqueous foam. The foam can reach qualities up to 85%, with the internal phase containing 35–50% CO_2 and 15–35% N_2. The external phase can be loaded up to 8.5 kg/m^3 of water-base gellant (19).

The use of combined gas foamed fluid is not an industry-wide standard as yet. The foam, which was developed and used primarily by a particular service company, is known as "binary foam". Binary foam is presently being scrutinized carefully by industry experts. Some believe that binary foams were introduced to avoid patent infringements and conclude that no advantage is gained. Others maintain that the binary foams are advantageous because they provide quicker cleanup and achieve higher stimulation rates.

Combined foams contain less CO_2 than conventional CO_2 foams. The volume of CO_2 that is removed is replaced by N_2 gas. The reduction of CO_2 in combined foams as compared to conventional CO_2 foams allows for reduced friction pressures and lowered hydrostatic pressures. Combined foams can be run at lower treating pressures or higher rates at the same treating pressure as conventional CO_2 foams (20). As well, clay swelling does not appear to be affected by the reduced amount of CO_2. A sufficient quantity of CO_2 exists to maintain a low pH. A low pH is bene-

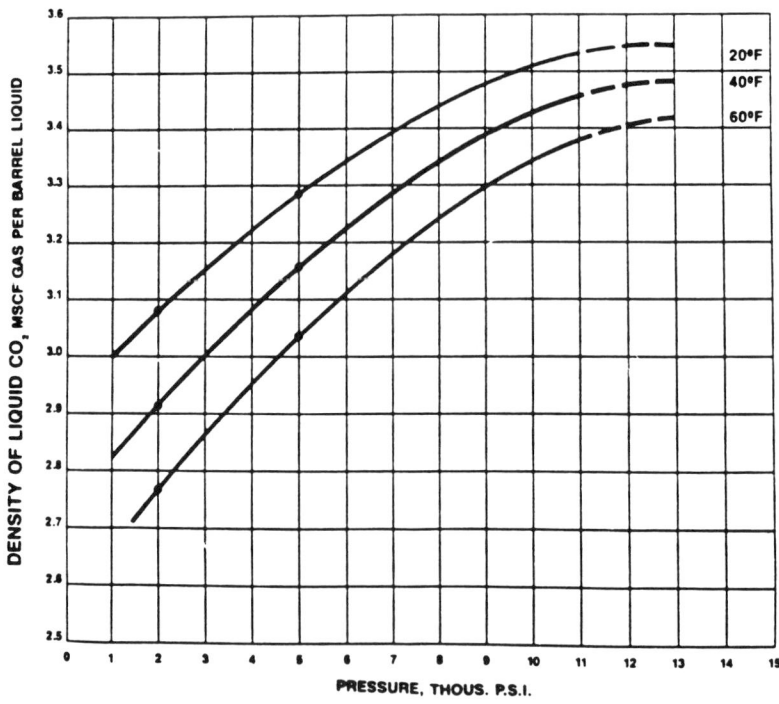

Figure 9. A plot of the gas yield of liquid CO_2. (Reproduced with permission from reference 5. Copyright 1985 Society of Petroleum Engineers.)

ficial when stimulating formations with water-sensitive clays. Clays will be inhibited from swelling in a low pH environment (4, 20, 21).

Combined gas foam and single gas-phase foam fracs benefit gas reservoirs that are underpressured and have low permeabilities. The down side of the combined foam stimulation treatments is the quantity of gas equipment required and the complexity of the data treatment center. N_2 and CO_2 tanks and pumpers are required on location; this extra equipment could increase the cost substantially. The increase may be significant enough to effect the economics of the stimulation. Field setups are accomplished as if separate N_2 and CO_2 foam treatments are being completed. The monitoring equipment must be able to calculate and record downhole and surface foam qualities, surface treating pressure, slurry densities, CO_2 flow rate, N_2 flow rate, slurry rate, and slurry temperature. As a result of the complexity of adding two gases to create a single foamed fluid, further advancements in the monitoring equipment would be necessary. The computer equipment must monitor the constantly changing surface pressure and account for the changing compressibilities of the two gases. The changing gas compressibilities and the fluctuation in the complex computer monitoring equipment must be taken into account in calculating bottomhole pressures for this type of foamed stimulation.

Third generation foam fluids are still in the infancy stages. Although some benefits may be gained when using the foam, more research and field implementation is required to gain proper economic evaluation.

Foamed Acid Fluids

Acidizing has been used for many years as a means of increasing the productivity of petroleum-bearing carbonate reservoirs. Because of its low cost and abundance, the primary acid used to stimulate carbonate reservoirs has been hydrochloric acid. Acetic and formic acids are used occasionally; however, their primary use is as an additive rather than a base acid.

Foamed Acid Fracturing. The goal of acid fracturing is to create a fracture within the reservoir and to etch the sides of the fracture face. Etching of the fracture face with acid dissolves a portion of the fracture face and allows a conductive path to be created after the fracture has closed. The further the live acid can reach into the fracture, the longer the conductive channel will increase well productivity.

One of the difficulties with acid fracturing is delivering the live acid to the fracture tip without it leaking off into the formation or spending. Hydrochloric acid is a strong acid, and the mass transfer is often the limiting factor in the reaction with limestone (*22*). Consequently, the rate of spending is dependent upon the injection rate. At low injection rates, the acid penetrates only a short distance into the fracture before consumption. This condition causes the limestone in the area near the wellbore to be dissolved excessively and prevents deep stimulation.

The key to a successful acid fracture stimulation is to control the fluid leak-off (*23*). Most fluid loss additives control leak-off by building up a low-permeability filter cake against the fracture face, thereby creating a wall-building mechanism. The three mechanisms that govern fluid loss as described by Howard and Fast (*24*) are

- C_I viscosity and relative permeability effect
- C_{II} reservoir fluid viscosity–compressibility effect
- C_{III} wall-building effect

The following equations have been derived for these three mechanisms (24).

$$C_I = 1.483 \times 10^{-3} \left[\frac{k_f \Delta P \phi}{\mu_f}\right]^{\frac{1}{2}} \qquad (10)$$

Where k_f is the permeability of reservoir rock to foam fracture fluid, ΔP is the differential pressure, ϕ is porosity, and μ_f is the viscosity of the foam fluid.

$$C_{II} = 1.183 \times 10^{-3} \Delta P \left[\frac{k_r \phi C_r}{\mu_r}\right]^{\frac{1}{2}} \qquad (11)$$

Where k_r is the permeability of reservoir rock to the reservoir fluid, C_r is the coefficient of compressibility of the reservoir fluid, and μ_r is the viscosity of the reservoir fluid.

$$C_{III} = 0.0164 \frac{m}{A} \qquad (12)$$

Where m is the slope of the plot of fluid loss versus the square root of time, and A is cross-sectional area.

The composite fluid leak-off coefficient (C_T) is given by the equation:

$$\frac{1}{C_T} = \frac{1}{C_I} + \frac{1}{C_{II}} + \frac{1}{C_{III}} \qquad (13)$$

Foam does not exhibit wall-building characteristics, and the effects of formation fluid viscosity are negligible compared to the viscosity of foam under shear conditions (25, 26). For those reasons, foam is viscosity controlled. Thus

$$C_T = C_I = 1.483 \times 10^{-3} \left[\frac{k_f \Delta P \phi}{\mu_f}\right]^{\frac{1}{2}} \qquad (14)$$

Methods of controlling foamed acid leak-off by creating a wall-building mechanism are possible. Alternating stages of cross-linked gelled water and foamed acid increase the wall-building coefficient which influences the leak-off control mechanism. Reducing the fluid loss in this manner enables greater fracture extension.

Foamed acid fluids were developed to control accelerated leak-off without the addition of fluid loss additives. Foams are clean fluids and therefore less damaging. Because foams are not wall-building fluids, leak-off control is not affected by fracture face erosion because of the presence of acid.

Investigations to determine the leak-off control mechanisms of foam have shown (26–29) that the effective permeability of a porous medium is greatly reduced in the presence of foam. Some basic assumptions were used during the testing to determine the leak-off control mechanisms of foamed fracturing fluids. The first assumption was that the liquid or continuous phase moves freely, and permeability reduction is a function of the liquid saturation. The other assumption was that the gas or discontinuous phase flows only by rupture and reformation of the foam film. The resistance of foam to flow through porous media is a function of the stability of the foam.

The investigations (26–29) illustrated the difference between reactive and nonreactive foams. During the tests, core permeabilities ranged from 0.5 to 5.0 md. Fluid loss of the nonreactive foam was approximately half of the reactive foam, although the stability of each did not show a significant difference. This result suggests two possible scenarios. The first is that the foamed acid is destabilized in its reaction with limestone, and this destablization causes greater fluid loss of the gas phase. The second is that the permeability is increased as the acid dissolves the limestone.

Field results are inconclusive as to the mechanism that controls fluid loss of the reactive foams. In certain situations, improvement in productivity has been gained by using foamed acid instead of conventional acid treatments. However, many field applications of foamed acid show no stimulation benefit over conventional acid treatments.

Laboratory tests comparing conventional and foamed acid leak-off were completed in reference 26.

The reactive foam exhibited good fluid-loss properties; however, the gas phase of the foam was lost to the rock matrix at an extremely rapid rate. In an actual treatment stimulation scenario, the rapid loss of gas would result in the foam quality being depleted at a short distance into the fracture. The treatment would then essentially be reduced to a conventional acid treatment with no real gain in production being accomplished.

The high level of N_2 gas loss severely limits the effectiveness of foamed acid fracturing. In an effort to improve fluid-loss control, various techniques and additives have been evaluated. Using foamed brine or a gelled water pad to precede the foamed acid was one of the best techniques used to control fluid loss. The nonreactive viscous pad creates fracture extension that allows the acid to reach deeper within the fracture before being spent. Longer, more conductive fractures are created enabling greater productivity from the reservoir. The use of hydrocarbon-based foams or brine preflushes containing fluid-loss additives do not improve fluid-loss control of the acid. Hydrocarbon-based gellants have a tendency to break foams, but brine preflushes won't retard the acid reaction.

Successful fracture acidizing treatments do not depend solely on good fluid-loss properties (30). Adequate fracture flow capacity must be established by the acid system used. The quantity of rock removed and the pattern in which it is removed from the fracture face are important. Fracture flow capacity is dependent on the nature of the rock and the characteristics of the acid, such as acid type, volume, and concentration. Other factors that lead to increased fracture flow conductivity include foam quality and pumping rate. As long as the foam is stable, foamed acid of any quality increases the fracture flow conductivity when compared with non-foamed acids. As well, increases in total treatment pumping rates achieve better fracture conductivities because of greater acidized fracture lengths.

Foamed stability affects acid-etched fracture flow capacity in the same way that it affects fluid loss control. By increasing foam stability, a substantial increase in fracture flow conductivity will occur. By increasing the foam stability, the pattern of rock removal is more effective without overetching the fracture face (30). Using lower quality foams, gelling the reaction phase and increasing the viscosity of the foam achieve maximum fracture flow capacity.

Foamed Matrix Acidizing. Matrix acidizing is a stimulation treatment used to remove damage near the wellbore without creating a fracture. The process involves the injection of a reactive fluid into the porous medium at a pressure below the fracturing pressure. The fluid dissolves some of the porous medium and consequently increases its permeability.

Matrix acidizing treatments differ from acid fracturing treatments. Acid fracturing treatments are completed only on carbonate reservoirs. Matrix acid treatments are generally completed on carbonate reservoirs, but on occasion sandstone reservoirs are also acidized. Acidizing sandstone formations result primarily in dissolution of permeability-damaging minerals rather than creating new flow paths, as is the case when acidizing carbonate formations. Fluid leak-off into sandstone formations is a primary method of allowing the acid to come into contact with the blockage, thus enabling its removal. Fully foamed matrix acid treatments in sandstone formations are unlikely as foams inhibit fluid loss, except when used for treatment diversion.

The ideal matrix acid treatment in carbonate formations is one in which the etched pattern consists of the smallest amount of etching close to the wellbore but with maximum penetration. Peters and Saxon state that the benefits of acid spending near the wellbore decrease very rapidly with the acid volume pumped. Their study indicates that productivity response increases with deeper acid penetration (*31*). Acidized permeability has little effect on productivity, and penetration is more important than the amount of calcium carbonate dissolved. The important aspect of matrix acidizing is to achieve a maximum depth of penetration rather than increasing the acid volume to dissolve large volumes of reservoir rock. A methods of increasing acid penetration is to retard the acid by using foam or other means.

In dolomite, acid spends differently than in limestone formations. The reaction of HCl with dolomite is rate-limited at formation temperatures under 50 °C. Wormhole development, which is common in limestone acidizing, is not characteristic in dolomite acidizing (*32*). As pores become connected during matrix acidizing operations, dolomites form caverns. The use of foamed acid ensures that the acid is spent in the primary channels and allows deeper penetration. Foams are good acid extenders. Deeper penetration can be achieved with foamed acid as opposed to an equal volume of nonfoamed acid.

One primary benefit of using a foamed acid in a matrix application is the use of the energy for faster cleanup of the undissolved fines created during the treatment. The benefits realized by using a foamed acid treatment fluid are very similar to those benefits gained from using nonreactive foamed fluids. Foamed acids use a smaller volume of acid than conventional acid treatments, and deeper penetration into the formation is attained. Thus a reduction in the volume of acid pumped would reduce the cost of products to attain the same productivity gain. However, equipment costs would then increase because of the use of gas transports and pumpers.

Foamed Diversion. Diverting agents are used in the oil industry as a method of improving matrix treatment fluid coverage across the zone of interest. The theory behind using diverting agents is that during a typical well treatment, the pumped fluids will choose the path of least resistance into the formation. Often, this tendency results in a poor treatment because the treatment fluid will be more readily accepted by the most permeable zone rather than the most damaged zone.

Fluid diversion is a technique designed to alter the treatment fluid injection profile. Because fluids choose the path of least resistance, diversion is a resistance problem. Darcy's law for linear flow can be used to define the parameters that govern this resistance.

$$q \propto \frac{k(\Delta P)A}{\mu L} \qquad (15)$$

where k is permeability, ΔP is the differential pressure, A is the surface area in the zone of interest, μ is the viscosity, and L is the length of the pipe.

The purpose of diversion is to alter the injection rate per unit area, q/A, so that each zone is accepting fluid at approximately the same rate. Therefore, the reservoir properties that cause the flow rate into the formation to vary are permeability, differential pressure, and length.

In mathematical terms, the injection rate per unit area is equal for all zones (*33*).

$$\frac{k_1(\Delta P_1)}{L_1} = \frac{k_2(\Delta P_2)}{L_2}$$

Should the ratios not be equal, it is necessary to consider using a method of fluid diversion. This inequality can result from

1. zones with differing permeabilities, either due to damage or resulting from natural causes
2. zones with differing formation pressures
3. reservoir fluids with differing compressibilities in different zones
4. reservoir fluids with differing viscosities in different zones
5. naturally fractured reservoirs
6. any and all combinations

The types of diverters vary extensively, from mechanical systems to chemical diverting techniques. The latest generation of diversion techniques involves pumping an immiscible mixture of two fluids such as emulsions or foams. Emulsions have not been used as extensively as foams as a diversion technique, because they are more difficult to clean up in low-pressured reservoirs. Foams, however, with similar properties to emulsions, tend to clean up very easily.

Foams can flow like liquids and remain motionless like a solid. Advantages of foams include their capability to be run over a wide range of pressures and temperatures, ease of application, and flow-back capabilities.

Foam properties vary with changes in pressure and temperature; thus foam qualities increase as they flow away from the wellbore. The property is advantageous when treating an interval with more than one zone having different pressures. In the lower pressure zone, the foam will have a higher quality and potentially more diverting effects.

Investigations (*34, 35*) to study the effects and limitations of brine and acid foams as diverting agents for nonfoamed acids developed several interesting conclusions. In several case studies of actual stimulated wells in which foamed diversion was used, foam was an effective diversion fluid that provided uniform treatment over the entire perforated interval. The most effective foams were those in which the foam quality exceeded 60%, and those in which the diversion duration lasted much longer with the higher quality foams. Alternating stages of foamed diverter and either nonfoamed or commingled acid were more effective than a single stage of foamed diverter. Alternating stages ensure that a more complete diversion occurs, and allows the tighter zones to receive more treatment fluid.

Brine foams in certain circumstances are more acceptable as diversion fluids than acid foams, as shown in Tables III and IV. The data illustrate that either high-porosity or high-permeability limestones reacted with the foamed acid to create poor diversion characteristics. Foam brines that will not react with the carbonate formation produced better diverting fluids.

Table III. Resistance Tests with Foamed and Nonfoamed Brine

| Foam Quality (%) | | | Brine | Apparent Foam | Plug Duration |
Q_1	Q_2	Q_3	k (md)	k/k_1	(min)
79.1	80.3	82.0	10.4	0.086	110
57.8	58.6	61.2	6.2	0.092	60
48.0	48.7	51.5	9.8	0.17	30
40.3	41.1	43.6	10.1	0.31	15

NOTE: k_1 was the permeability to nonfoamed brine after partial acidization. Q_1 is quality at point of mixing, Q_2 is quality at core inlet, and Q_3 is quality at core outlet.
SOURCE: Reproduced with permission from reference 34. Copyright 1991 Society of Petroleum Engineers.

Table IV. Resistance Tests with Foamed and Nonfoamed Acid

Nominal Foam Quality (%)	Brine k_1 (md)	Apparent Foam k/k_1	Plug Duration (min)
80	28.2	0.14	4.4
70	17.3	0.18	4.5
60	16.4	0.12	2.1
50	17.4	0.13	0.9
40	30.6	0.13	-2.6

NOTE: k_1 is the permeability to nonfoamed brine after partial acidization.

Thompson and Gdanski (*34*) also performed dual-core experiments to determine the best diversion method using foam, and the maximum permeability difference needed to achieve an equal flow rate through the core. Multiple diversion techniques were used, including foamed acid, multiple stages of foamed acid, and various qualities of foamed brine. The tests showed that foamed brine reduced flow rates better than foamed acid. Also, higher quality foamed brines were most effective. In order to effectively use foamed diversion fluids, the permeabilities of the zones of interest must be relatively similar. The limit on permeability differences is approximately a factor of 10. Otherwise, the more permeable zone will accept both the diversion and treatment fluids.

The use of foam as a diversion fluid is gaining wider acceptance throughout the petroleum industry. Although foams are not suited for all applications, they are very versatile. The specific foamed diversion technique that is used in a treatment design is dependent on individual well characteristics and the stimulation objective.

Design Considerations

Two important parameters that describe a foam are texture and stability. Although both are dominant factors in the determination of foam rheology, neither is stipulated but is assumed when designing a foam fluid treatment.

Texture. Foam texture is an important parameter that affects the rheology of the foam fluid. Texture of a foam is a means of classifying a foam according to its bubble size, shape, and distribution within the foam matrix. Texture is a description of the manner in which the gas bubbles are distributed throughout the liquid phase of the foam. This property not only influences the foam's rheology but also its fluid loss, proppant transport, and cleanup properties. The texture of a foam is a qualitative rather than a quantitative value, and therefore a number cannot be used to describe it; a physical description will be used. Factors that effect the texture of foams are quality, pressure, foam generating technique, and chemical composition.

Foams are fluids that depend on shear history. The texture of a foam will reach an equilibrium state at a particular shear rate. Finer textured, more dynamically stable foams are produced at high shear rates, higher pressure, and with higher quantities of surfactant (*36*). Reidenbach et al. (*11*) observed that at higher shear rates, finer more uniform bubbles were created. This information indicates that at downhole conditions during fracture stimulation when conditions of high pressure and shear are present, foams are finely textured with parallel-piped uniform bubbles.

Harris (37) determined by experimentation that fracturing foam fluid bubble sizes varied from 300 to 1200 μm with a size distribution varying by a factor of 10. Because of the narrow size distribution of fracturing foam bubbles and the small bubble size in relation to the fracture-flow passages, foams can be considered to be homogeneous. Density is a function of temperature, pressure, and quality.

The understanding that foams are pseudohomogeneous was a fundamental development that Blauer et al. (7) used to develop an expression for the effective viscosity of foams.

Surfactant types and concentrations have an important impact on foam texture and viscosity. In general, bubble sizes decrease with an increase in the concentration of surfactants; decreased bubble size increases the viscosity of the foam. Because foaming requires energy to create new surfaces and surfactants give lower surface tension, they will promote foaming and give more stable foams. Therefore, smaller bubble sizes require larger surfactant concentrations to maintain stability. Increasing gellant or cross-linking the external phase dramatically increases the foam viscosity, but texture is left relatively unaffected.

Stability. The static stability of a foam is the ability to resist bubble breakdown resulting from bubble collapse or coalescence. Foam instability can be caused by the drainage of liquid from the foam resulting from increased quality above 0.95 because of pressure reduction, heating, bubble rupture, or coalescence.

A method to quantify the stability of a foam is to measure its half-life. The half-life is the time required for the foam to drain half its liquid volume. The longer it takes to drain the liquid, the more stable the foam will be. A major factor that strongly affects the stability is the method of foam generation. In addition, the experimentally measured half-life is a function of the height of the foam column.

Those factors that were previously mentioned that produce finer-textured foams also produce more stable foams. Factors such as surfactant type, concentration, increasing pressure, and higher inputs of mechanical energy generate more stable foams. For higher temperatures such as those that exist downhole, dynamic foam stability relies upon surfactant type and concentration rather than the addition of thickeners (polymer stabilizers). It is not known what rates are necessary to maintain dynamic stability in fractures, or whether those conditions typically exist.

Half-life tests have shown that the use of gellants increases the static stability of foams, and the use of cross-linkers significantly lengthens half-lives (Figure 10). The figure illustrates the half-lives for different stabilized foams as a function of quality. The addition of gellants and cross-linkers dramatically increase both the viscosity and the static stability of the foam.

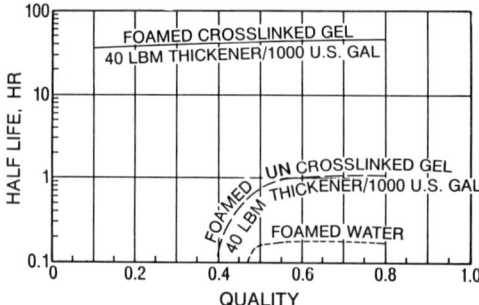

Figure 10. A plot of the half-life vs. quality of foamed fluids. (Reproduced with permission from reference 36. Copyright 1983 Society of Petroleum Engineers.)

Effects of Quality on Foam Viscosity

Stimulation fluid viscosity is valuable in determining fluid loss, proppant carrying capability, friction pressure drop, and to a lesser extent fracture width.

Because of the nature of foams, the rheological character of foam fluids is difficult to quantify. The viscosity of foams is primarily dependent upon foam quality and external-phase fluid viscosity. Stimulation fluids are subjected to pressure variations from surface to downhole conditions; therefore, foam quality and viscosity will change accordingly. In order to overcome the changing conditions experienced during stimulations, the foam fluids are designed to reach a specified quality at downhole conditions.

Blauer (7) described the effect of foam quality on viscosity by dividing the range of foam qualities into regions of different bubble interactions. The first region exists between 0 and 52% foam quality. Spherical bubbles are uniformly dispersed throughout, the bubbles are not in contact with one another, and flow is Newtonian. Above a quality of 52%, spherical bubbles are loosely packed in a cubic arrangement and contact one another during flow, resulting in an increase in viscosity. At a foam quality of 74%, the bubbles deform from spheres to parallel-pipe shaped during flow. Maximum foam viscosities occur in this range. Above qualities of 95%, foams are no longer stable, and a mist is formed.

A major factor that affects foam quality is the shear rate that is imposed upon the foam. In his Ph.D. dissertation, Mitchell (38) showed the relationship between viscosity, quality, and shear rate for Newtonian external foams (Figure 11). Two observations can be made from the diagram. The first is the absence of shear rate dependence on foams with qualities

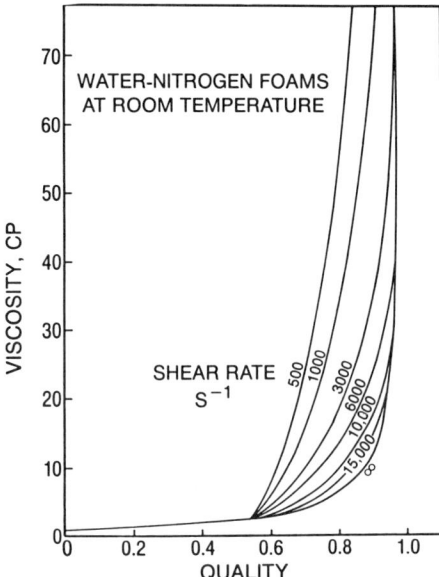

Figure 11. A plot that shows the effect of quality and shear rate on the viscosity of water–N_2 foams at room temperature.

below 55%. This result indicates that foams with qualities less than 55% are Newtonian fluids. Foams with qualities above 55% exhibit shear-thinning properties. That is, foams at a given quality have lower apparent viscosities at higher shear rates.

Foams that are composed of non-Newtonian external phase fluids do not behave similarly to Newtonian foamed fluids. Non-Newtonian foamed fluid viscosities are totally shear rate dependent and shear-thinning (Figure 12) (11).

Although these studies were completed using N_2 as the internal phase, Reidenbach et al. (11) found that similar results occurred using liquid CO_2 as the dispersed phase. The similarity between the resulting emulsion and N_2 foams illustrates that the composition of the dispersed phase is not important, and that the two-phase structure is dominant.

Foam Friction

Friction pressures and viscosities are two important physical characteristics of well stimulation fluids. Friction pressures, fluid hydrostatics and fracturing gradients determine the minimum surface pumping pressure necessary to maintain fracture growth. Often friction pressure is the de-

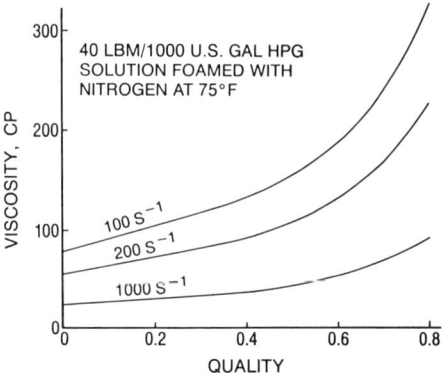

Figure 12. A plot that shows the effect of quality and shear rate on a gelled foam system. (Reproduced with permission from reference 11. Copyright 1983 Society of Petroleum Engineers.)

termining factor as to the downhole tubular configuration necessary to maintain a minimum surface pressure.

Certain methods of calculating friction pressures involve the use of pressure charts that are in the form of pressure drop per length of pipe versus the flow rate of the foam. These types of charts incorporate foam quality and tubular geometry but may neglect consideration of one or more of the parameters such as temperature, pressure, foam texture, polymer or surfactant type, and concentration. Each of these parameters is often unspecified but drastically affects the friction pressures of foams. Information derived from such charts should be taken only as a guideline.

Incremental numerical schemes have been developed with computers. Most models should take into account the numerous variables while pumping foams and therefore contribute more accurate pressure predictions. Generally, the service companies or companies that use foams incorporate or are planning to incorporate these models in their foam treatment designs.

Foam Fracturing Design Criteria

When designing a fracturing treatment, the proper choice of the fluid is very important. Because of the specific well constraints, selection of the correct fluid for the treatment is difficult. The characteristics are

- fluid loss
- compatibility with formation and in situ fluids

- rheology and friction
- proppant effects

Although foamed fluids are physically different than other types of stimulation fluids, their characteristics are important considerations in the design of fracture treatments.

Fluid Loss. As mentioned earlier in this chapter, foamed fluids have viscosity-controlled leak-off characteristics, as shown in equation 14. C_T, according to Blauer and Kohlhaas (40), will approximately equal C_I because C_{II}, the reservoir fluid viscosity coefficient, is 2 to 4 orders of magnitude greater than C_I, and C_{III} is nonexistent in foams. In clarifying that statement, cross-linked or heavily gelled foams will develop wall-building characteristics; in this case, C_{III} will be dominant. The fluid leak-off controlling mechanism for foams that are cross-linked or heavily gelled will have to be determined either by minifrac (a diagnostic treatment used to gather technical information pertaining to a particular reservoir) methods or in the laboratory. Foams that are lightly gelled or use a surfactant will not build filter cake, and therefore fluid leak-off will remain viscosity dominant; that is, C_T will be approximately equal to C_I.

Blauer and Kohlhaas (40), in their study to determine foam fluid loss characteristics, found that C_I for foams was much less than for gelled or untreated water. Consequently, the total fluid leak-off for foams is much less than that of gelled water, and foams are also more efficient. Assuming rates and injected volumes are the same, foam, which is more efficient fluid, will create fractures with greater area. This feature will result in higher productivity increases using foams instead of gelled water (40).

Fluid and Formation Compatibility. Foamed fluids are mixtures of gases and liquids. The volumetric ratio of gas is given by the foam quality for stimulation fluids. It is designed to be between 70 and 85%. The gas ratio remains high for several reasons and therefore, a reduced volume of liquid is available to contact the formation. For this reason, possibility of formation damage is reduced.

As stated in the previous section of this chapter, foamed fluids have low fluid loss in low-permeability formations. As a result, fracture fluid invasion of the formation is small, and consequently the reaction between the treatment fluid and the reservoir fluid or rock is minimized.

Because of the energized nature of foamed fluids, the amount of time that they are in contact with the formation is kept to a minimum. If allowed to sit for extended periods at bottomhole conditions, these fluids will lose the energy available from the gas. The gas dissipates into the formation. For this reason, the contact time of foam in a formation is minimized, and potential damaging effects are further reduced.

Foams are very compatible even with the most sensitive formations. The reduced liquid volumes, contact time, and fluid loss make foams one of the most versatile fluids for the stimulation of shallow, sensitive formations. References 6, 22, 25, 35, and 36 illustrate improved success of production by using foamed stimulation fluids.

Rheology and Friction. Rheology is an important criterion in the design of a stimulation treatment as several important design elements are affected by the rheological predictions. Friction pressures are dependent upon fluid rheology, and surface treating pressure is directly dependent upon the fluid friction pressure. Determinations of fracture widths and lengths and economical predictions from fracture simulators use the rheological predictions. Thus, accurate modeling of actual foam rheological properties enables prediction of the success of the fracture treatment and the surface treating pressures.

Rheology is the study of the deformation and flow of fluids. Four different models are used to characterize the flow of fluids: Newtonian, Bingham plastic, power law, and viscoelastic. In the characterization, models have been developed to relate the observed effects that shear rate has on foam. Several scientists who have studied foamed fluid rheology categorize foam into various models.

Mitchell (7), Blauer, and co-workers (7, 40) modeled foams as Bingham plastic. Foams were modeled as power law fluids by Patton et al. (41). A more rigorous model is the Herschel–Bulkley model, which combines both the Bingham plastic and Power law models (2, 11, 42, 43).

Mathematically the models are expressed as

$$\text{Bingham plastic: } \tau = \tau_{yp} + \mu_p \delta \tag{16}$$

where τ is the shear stress, τ_{yp} is the shear stress at yield point, μ_p is the plastic viscosity, and δ is the shear rate.

$$\text{power law: } \tau = K_f'(\delta)^{n'} \tag{17}$$

where K_f' is the consistency index of foam, δ is the shear rate, and n' is the shear rate exponent.

$$\text{Hershel–Bulkley: } \tau = \tau_{yp} + K_f'(\delta)^{n'} \tag{18}$$

By using the shear–stress relationship that the Herschel–Bulkley model follows, the pressure loss for foamed fluids in pipe can be determined, as shown in equation 18. The shear stress at the wall can be written as follows:

$$\tau_w = \frac{d \Delta P}{4L} \qquad (19)$$

where d is the inside diameter of the pipe, ΔP is the pressure differential, and L is the length of the pipe.

The shear rate can be written as

$$\delta = \left[\frac{8v}{d} \right] \qquad (20)$$

where v is the average fluid velocity and d is the inside diameter of the pipe.

Substituting these values into equation 18 gives the following equation:

$$\frac{d \Delta P}{4L} = \tau_{yp} + K_f' \left[\frac{8v}{d} \right]^{n_f'} \qquad (21)$$

The yield point (τ_{yp}) is a function of the foam quality. The relationships are as follows:

$$\tau_{yp} = 0.07Q \quad \text{for } Q \leq 0.6 \qquad (22)$$

$$\tau_{yp} = 0.0002 \exp 9Q \quad \text{for } Q > 0.6 \qquad (23)$$

The consistency index of the foam (K_f') is dependent upon the foam quality and the consistency index of the liquid phase (K_L') (42).

$$K_f' = K_L' \exp(C_I Q + 0.75 Q^2) \qquad (24)$$

where:

$$C_I = 4(n'_{24})^{1.8} \qquad (25)$$

and n'_{24} is the shear rate exponent of the liquid phase at 24 °C.

The bottomhole working temperature is generally much greater than 24 °C for which both n' and K_L' have been developed. Both n_t' and K_t' must reflect the effects that temperature has on viscosity. In reference 42, equations have been developed to reflect these temperature effects.

$$K_T' = K_{24}' \exp[(C_{II}Q - 0.018)(1.8T - 43)] \qquad (26)$$

where

$$C_{II} = \exp - [3.1 + 3n_{24}'] \qquad (27)$$

Equations 22–27 show that quality is an important factor in the determination of the rheological indices. In most fluids, an increase in temperature will decrease viscosity; however, the temperature-thinning effect is greater at low qualities than at high qualities. With increasing temperature, the flow behavior index increases at a greater rate with fluids at low-quality than with fluids at high quality. As well, the decrease in the consistency index is larger at low qualities than at high qualities. This result indicates that viscosities increase at higher qualities, and that the thinning effect of increasing temperature is more pronounced as foam qualities decrease. The difficulty in developing an expression for pressure loss of a fluid due to friction is increased when the fluid is foam. The expression must account for both turbulent and laminar flow. It must take into consideration both ungelled (Newtonian fluids) and gelled (pseudo-plastic fluids) foams.

The relationship in equation 21 is accurate for laminar fluid flow through other geometrics as well as pipe. If the shear stress at the wall of the pipe is greater than the yield point value, the flow is independent of geometry. This relationship is important when determining pressure drop through the fracture.

Equation 21 was developed assuming fluid flow is laminar. Laminar flow is believed to be the primary flow regime for fluids in the fracture. However, the flow regime of foam through wellbore tubulars may be either laminar, turbulent, or a combination of both (*11*).

Melton and Malone (*44, 45*) developed an expression for turbulent flow of a non-Newtonian fluid through tubulars that was modeled after the Bowen relationship. Reidenbach et al. (*11*) modified the relationship to account for the changing density of foam as it travels through the tubulars. Equation 28 was developed for non-Newtonian fluids; however parameters were developed to account for Newtonian fluid flow behavior. Table V shows the parameters for water.

$$\frac{\Delta P d}{4L} = A' \rho_f^{X'} d^b \left[\frac{8 V_f}{d}\right]^S \qquad (28)$$

Table V Scale-Up Equation for Turbulent Flow of N_2 Foams

External Phase	A'	X'	b	S
Water	1.28×10^{-5}	0.8	1.87	1.80
HPG				
1.2 kg/m^3	2.239×10^{-4}	0.37	1.37	1.37
2.4 kg/m^3	3.873×10^{-4}	0.31	1.31	1.31
3.6 kg/m^3	6.556×10^{-4}	0.27	1.27	1.27
4.8 kg/m^3	1.0×10^{-3}	0.23	1.23	1.23

SOURCE: Reproduced with permission from reference 11. Copyright 1983 Society of Petroleum Engineers.

The parameters A', b, and S were obtained from a plot of τ_w versus $8v/d$ in the turbulent region. A' incorporates the effects of fluid viscosity and density in the turbulent region, b accounts for changing diameter characteristics during turbulent flow, d is the inside diameter of the pipe, ρ_f is the foam density, X' is the density parameter that accounts for changes in density occurring as a result of changes in quality or pressure, V_f is the volume of foam, and S is indicative of the turbulence level and the deviation of fluid flow from Newtonian behavior.

Table V contains experimentally developed constants for water as the external phase fluid gelled with varying quantities of hydroxypropyl guar (HPG). The foam quality is 70%, but the effect of quality on the equation is minimal because of the density correction factor. Table V shows that water consistently produced higher results for X', b, and S. This result is indicative of the fact that gelled aqueous foam systems are drag-reducing fluids. Also in the gelled systems, the exponents b and S are identical, and this implies that the shear stress is a function of density and velocity only. Furthermore, X' is approximately equal to $S - 1$, this relationship correlates density changes resulting from changes in foam quality or pressure.

Steady-state flow is never achieved with foamed fluids; rather, the flow is dynamic. Foams flow dynamically because the pressure, which is continually changing, affects the viscosity, flow rate, and density of the foam at any given interval in the tubular. This problem can be accounted for by numerically integrating the mechanical energy balance equation from bottomhole to surface conditions.

In calculating the pressure drop due to friction for laminar flowing foam, equation 21 is used. By rearranging the equation to calculate $\Delta P/L$ and breaking the total depth of the tubulars into several increments, the pressure drop over each increment can be calculated. The summation of the pressure drop over each interval will be the total pressure drop due to the friction.

For turbulent fluid flow, equation 28 is used, and the incremental pressure drop is calculated over each interval. The total pressure drop is calculated in the same manner, by summing the individual pressure drops.

If there is no indication about the nature of the fluid flow, both the laminar and turbulent pressure drops must be calculated in the manner as mentioned in the previous paragraphs. The larger value of friction pressure is to be used.

Proppant Effects. The introduction of proppant into stimulation fluids further increases the difficulty in the prediction of pressure drop due to friction. The calculation used to determine friction pressures is further complicated by the addition of proppant. Harris et al. (46) developed a useful analytical method to account the effects in rheology by introducing a proppant.

The method assumes that the proppant is part of the internal phase, and the foam quality is effectively increased. The viscosity of the foam slurry is then estimated as being equal to the proppant-free foam, the quality being calculated using the internal-phase volume method shown in equation 2.

Example 2. Consider a N_2–water foam without proppant where the quality is 0.65. Enough sand is added to the foam to achieve a proppant concentration of 700 kg/m^3. The volume fraction of sand and gas can be calculated as follows.

Assume the proppant bulk density is 2650 kg/m^3.

$$V_P = \frac{700 \text{ kg/m}^3}{2650 \text{ kg/m}^3} = 0.26$$

because

$$V_P + V_G + V_L = 1$$

then

$$V_G + V_L = 0.74$$

V_P is the volume of proppant. The volume fraction of gas (V_G) and liquid (V_L) does not change; therefore, Mitchell's foam quality expression can be used to determine the volume fraction of the gas.

$$\frac{V_G}{V_G + V_L} = Q$$

$$\frac{V_G}{V_G + V_L} = 0.65$$

but

$$V_G + V_L = 0.74$$

then

$$V_G = 0.74(0.65) = 0.48$$

To determine the effect of proppant on friction pressure, an estimation can be made by keeping the A', X', b, and S parameters constant and adjusting the density and velocity terms in the Bowen equation 28. This step can be accomplished because these parameters are largely affected by the external-phase rheology that may be determined from turbulent studies of the proppant-free foam. Studies on N_2 foams support this method of friction pressure estimation; however, there have been no studies for CO_2 foams to date.

Implementation

Field Equipment Setup. The equipment setup for foam stimulation treatments depends upon the type of foam treatment being implemented.

During a basic nitrified foam fracture treatment, water from a tank is mixed continuously with proppant in a blender, at sand–water ratios increasing progressively to as high as 2400 kg/m^3 depending on whether the fluid is gelled or not. A foamer or surfactant is then added to the slurry and is proportioned at a ratio of 0.2–1.0%. The blender then acts as a primer pump and transfers the slurry to the high-pressure pumpers. High-pressure N_2 gas is manifolded into the slurry to generate the foam downstream of the frac pumpers. KCl, gellants, or other chemical treating agents may be added to the base water if required to improve its compatibility with reservoir rock or formation fluids. Figure 13 schematically illustrates this process.

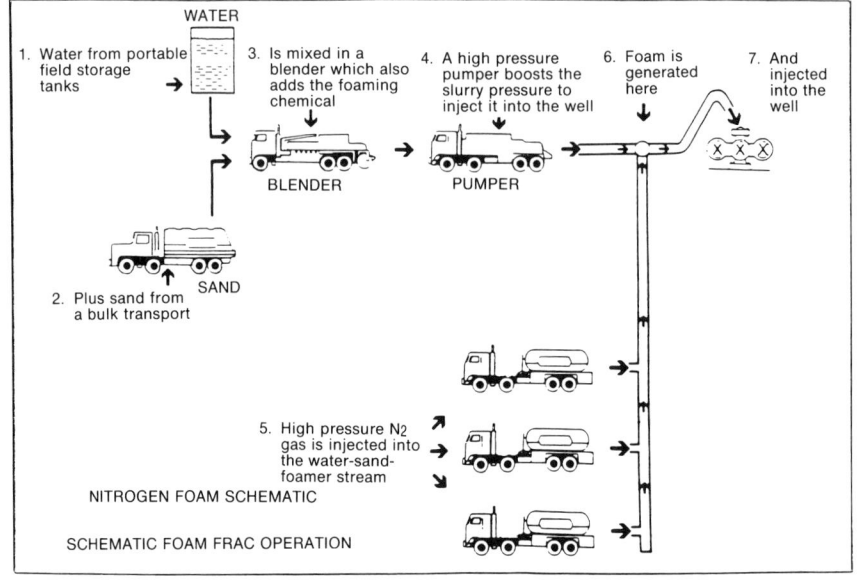

Figure 13. A diagram that shows the N_2 foam fracture equipment setup.

The blender and high-pressure pumpers are generally truck-mounted equipment conventionally used in oil field servicing applications. The N_2 gas is created from liquid N_2 that is passed through a high-pressure positive displacement pump followed by a heat exchanger. In the heat exchanger, the N_2 goes through a phase change from a liquid at -196 °C to a gas at approximately 20 °C as shown in Figure 3.

The nitrified foam is generated at the manifold where the fluid slurry joins together with the dry N_2 gas. At this point, the foam volume is between 2 to 5 times that of the slurry volume. The foam is then directed downhole.

CO_2 foams are implemented slightly different than N_2 foams because CO_2 foams use different equipment than N_2 foams. CO_2 foams use CO_2 in the liquid phase or dense vapor, depending on temperature and pressure, rather than the gaseous phase such as nitrified foams. The CO_2 is transported to location in the liquid phase at approximately -20 °C and 2000 kPa. The CO_2 transports contain small positive displacement pumps that supply CO_2 to the frac pumpers.

The slurry or liquid stream is similar to that of nitrified foam fracturing setups. Water or the desired fluid is mixed with sand continuously at the blender, which also adds gellants, emulsifiers, breakers, or other chemicals necessary to achieve the desired viscosity.

The blender then primes the high-pressure pumpers that are in parallel with the CO_2 high-pressure pumpers. The slurry and CO_2 streams are manifolded together. At this point, an emulsion is formed with CO_2 being the internal phase. The CO_2 will remain in a dense vapor or as a liquid, depending upon temperature and pressure (Figure 5).

To ensure that the correct proportions of fluid, sand, and CO_2 are mixed to achieve the proper downhole qualities, rates, pressures, and densities are logged, and adjustments are made automatically on computer-controlled treatments, as seen in Figure 14.

Acid foam treatments can be implemented using either CO_2 or N_2. The main difference is that no proppant is run with acid foams and the liquid phase is acid rather than water, alcohol, or hydrocarbons. Because of the nature of the corrosive fluid, safety considerations are more stringent during acid treatments.

Field Monitoring. The interpretation of bottomhole pressure has become an important factor in the stimulation of oil and gas wells since the work of Nolte and Smith was published (47).

The use of real-time bottomhole treating pressure enables the qualitative analysis of fracture propagation. Treatment design parameters such as fluid loss coefficient, fracture height, length, width, and closure time can also be determined with pressure decline analysis.

Bottomhole pressures can be recorded with mechanical devices such

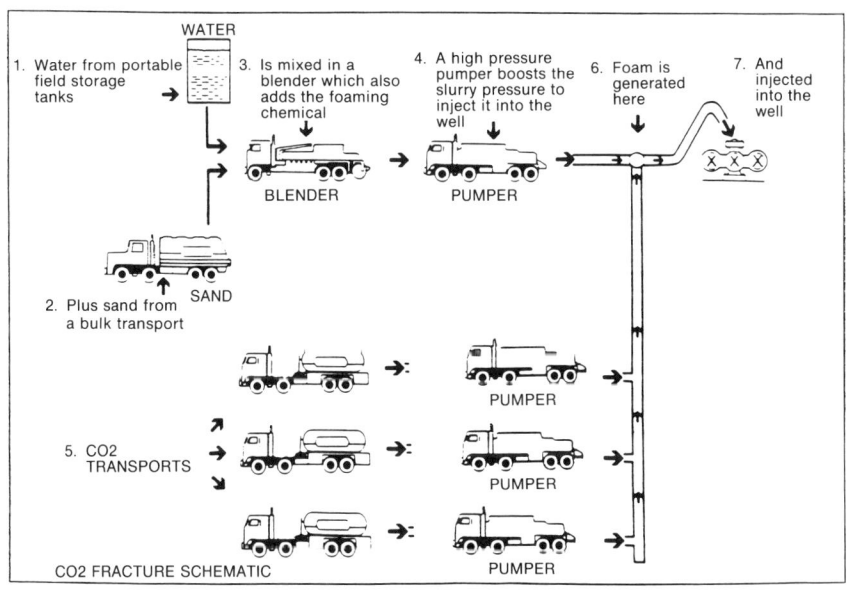

Figure 14. A diagram that shows the CO_2 foam fracture equipment setup.

as a reference string or with wire line tools. The use of these devices may escalate the cost of the work-over to the point of making the treatment uneconomical, or tubular restrictions may result in higher treatment pressures.

Implementation of an on-site computer control unit eliminates the need for mechanical systems, and enables real-time computation of bottomhole pressures. Monitoring techniques and pressure calculation methods vary extensively throughout the service industry. However, the measurable parameters such as surface treating pressure, liquid and gas injection rates, proppant concentration, and slurry temperature are the parameters that must be used to determine the bottomhole pressures.

Determination of bottomhole pressures for conventional fracture fluids such as gelled oil or water are relatively simple compared to foam bottomhole pressure calculations. However, when compressible fluids such as foams are used, more rigorous calculations are required to determine bottomhole pressures. Foam density, foam quality, and injection rates are functions of temperature and pressure.

The physical properties of foams are changed by both the fluctuations in the surface rates of components and the temperature and pressure gradients the foam experiences as it travels down the wellbore and into the formation. Bottomhole treating pressures (BHTP) can be calculated as follows:

$$BHTP = STP + HH - FP - PFP \qquad (29)$$

The only variable that can be measured directly is the surface treating pressure (STP). The hydrostatic head (HH) and friction pressures (FP) must be calculated from two measured parameters, fluid injection rate and proppant concentration, and the perforation friction pressure drop (PFP) can be determined from the physical properties of the fluid.

The hydrostatic head calculation can be completed in a similar manner, as in Example 1. If proppant is added to the foam, the hydrostatic pressure can be determined for each interval by knowing the proportion of gas, liquid, and proppant for each interval. The fractions of each component can be determined in a similar manner to Example 2. The hydrostatic pressure can be calculated by the summation of the hydrostatic pressure for each increment over the entire depth of the well.

Calculation of friction pressure drop using both equations 21 and 28 will account for both laminar and turbulent flow. These calculations are also done over each increment. The total pressure drop due to friction is then the sum of the incremental pressure drops over each interval divided by the total length. Because there will be two different answers, the larger will be used.

In fracturing, the pressure drop across the perforation is relatively insignificant in comparison to the pressure drop through the tubulars and the hydrostatic head. Perforation friction pressure drop is generally assumed to be between 1 and 2 MPa. In certain circumstances, the perforation friction pressure is important. For those instances, the following equation can be used.

$$\Delta P = \frac{q^2 \rho}{N^2 d^4} \tag{30}$$

Where q is the flow rate, ρ is the bulk density, N is the shear rate exponent, and d is the inside diameter of the pipe in meters.

The use of the bottomhole treating pressures enables the fracture parameters to be calculated more accurately. As well, the continuous or dispersed phases can be altered as the treatment progresses in order to achieve maximum fracture propagation. The more accurate the prediction of the bottomhole pressures, the more reliable the data can be to create the best possible stimulation treatment.

Field Safety. The objective of well stimulation is to use highly pressurized fluids to improve the productivity of petroleum reservoirs. This objective will not be achieved if the productivity is improved at the expense of the workers or equipment safety. The well stimulation sector of the petroleum industry is a hazardous sector that not only involves the use of highly pressurized fluids, but also includes fluids that may be flammable or corrosive or contain radioactive tracers. Foams that use compressed gas may be hazardous if the gas is allowed to expand quickly. Reservoir fluids also pose a danger to field personnel and equipment, because the fluids are generally flammable, combustible, or extremely poisonous, as is the case with H_2S.

To maintain safety-conscious personnel, safety must be a priority with the well service company's management as well as with the oil company's personnel. Neither the service company nor the oil company should ignore unsafe practices.

All stimulation treatments pose potential hazards. Because energized fracturing fluids involve compressed gases that are more explosive during a quick reduction in pressure than nonenergized fluids, foam treatments can be more hazardous than other stimulation treatments. Therefore, extreme caution should be exercised when conducting foam or energized stimulation treatments. Field personnel should be aware that compressed gases are involved with these treatments.

Typically, safety begins with the field personnel. All well-site operations require the use of safety equipment, such as fire-retardant clothing and coveralls, hard hat, and steel-toed boots. Safe practices, however, require more than the use of safety apparel. Each person on site must be aware of the treatment procedure and the potential hazards involved in the treatment, and know how to protect oneself and others in the vicinity should an unforeseen circumstance occur.

To increase safety awareness of all personnel on site, a pretreatment safety meeting is essential. During these meetings the following are discussed

1. treatment procedures
2. pressure limitations
3. safety hazards, such as unusual chemicals and flammable liquids
4. emergency procedures
5. H_2S escape routes (if applicable)
6. the assignment of emergency tasks to particular experienced personnel

The next step to ensure a safe treatment is to pressure-test all of the equipment and surface lines that are to be used during the treatment. The surface lines are pressure-tested with the same fluid that will be pumped through them during the treatment, for example, N_2 or CO_2 through gas lines, water through water lines. The pressure at which the lines are tested should be equal or exceed the maximum treating pressure of the treatment but less than the maximum allowable pressure rated for the lines.

Upon completion of the pressure test, the treatment is commenced with all personnel on site fully aware of the potential hazards, escape routes, safety procedures, and pressure limitations.

Conclusion

Foam stimulation fluids provide viable alternatives to stimulate low-pressure, fluid-sensitive gas wells where permeabilities are less than 5 md. Foams have been successfully used on shales, tight sandstones, and carbonates. Some interesting applications of foam fracturing have been on very sensitive formations and coal-bed methane fracturing.

Many sensitive formations have had a history of damage, or in some cases, wells have had complete loss of production when conventional fracture fluids were used. Foamed fluids have shown dramatically successful productivity increases on these formations.

Presently, there are only a few constraints on foam. Economics limits the use of foams on deeper, high-pressure gas reservoirs because of the pressure limitations. Another limitation to foamed stimulation fluids, in particular nitrified foams, is the relative permeability problems associated with their use in oil reservoirs above the bubble point pressure. The N_2 gas, which is slowly soluble in crude oil, will create a trapped-gas saturation. The relative permeability effect will be slowly removed as the N_2 becomes more soluble in oil. CO_2 foams do not have as dramatic an effect because N_2 foams on oil reservoirs because CO_2 is very soluble in oils.

High-permeability reservoirs may be difficult to stimulate using foams because the fluid losses to leak-off can be high. This problem can be alleviated to some degree with the use of gellants, cross-linkers, or other fluid-loss control additives. Productivity gains using foamed fluids may not be as high on higher permeability formations because of the foams' inability to achieve long fracture extension. Therefore, the use of foams may be limited in these formations.

Recent gains using foamed fluids result from the improvements in on-site quality control. More accurate monitoring equipment, the ability to adjust input parameters during the treatment, and real-time calculations of bottomhole pressures and qualities enable more accurate rheology calculations, and thus allow for high-efficiency stimulations and improved productivity gains.

Foams provide a viable alternative as a stimulation fluid in the well servicing industry. The application of foamed fluids must be suited to the particular characteristics of the reservoir in order to maximize the economic potential of the treatment. As long as foam treatments are implemented correctly in suitable reservoirs, the treatments will be successful and economic.

Table of Conversions

Multiply	By	To Obtain
Barrel	1.589×10^{-1}	Cubic meters
Barrel	42.0	Gallons
Centipoise	1.0×10^{-3}	Pascal · second
Cubic feet	2.832×10^{-2}	Cubic meters
Feet	3.048×10^{-1}	Meters
Feet/minute	3.935×10^{-2}	Meters/second
Gallons	3.785×10^{-3}	Cubic meters
Inches	2.54×10^{-2}	Meters
Pounds	4.54×10^{-1}	Kilograms
Pounds/gallon	120	Kilograms/cubic meter
Pounds/square inch	6.895	Kilopascal
Pounds/1000 gallons	1.198×10^{-1}	Kilograms/cubic meter

List of Symbols

A	surface area of the zone of interest
A'	constant in equation 28 for turbulent flow, Bowen's model
b	diameter exponent in equation 28, dimensionless
BHTP	bottomhole treating pressure, equation 29.
C_I	fluid-loss coefficient for fracture fluid in total fluid-loss equation [m/(min)$^{1/2}$]
C_{II}	fluid-loss coefficient for reservoir fluid in total fluid-loss equation [m/(min)$^{1/2}$]
C_{III}	fluid-loss coefficient for fracturing fluid in total fluid-loss equation [m/(min)$^{1/2}$]
C_2	parameter in foam consistency index, equations 20 and 21
C_3	parameter in foam consistency index, equations 20 and 22
C_R	coefficient of compressibility of reservoir fluid
C_T	total fluid loss coefficient [m/(min)$^{1/2}$], equation 14
d	inside pipe diameter (m)
E_f	expansion factor
FP	friction pressure
g	acceleration of gravity, 9.81 m/s^2 (32.2 ft/s^2)
g_c	gravitational dimensional constant
k	permeability (md)
k_f	permeability of reservoir rock to foam fracture fluid (md)
k_r	permeability of reservoir rock to reservoir fluid (md)
K_{24}'	foam consistency index of liquid phase at 24 °C (75 °F), (Pas$^{n'}$)
K_L'	proportionally constant, equation 48
L	length of pipe (m)
m	slope of plot of fluid loss vs. square root of time (cm/min$^{1/2}$
N	shear rate exponent
N_{24}'	shear rate exponent of liquid phase at 24 °C
ΔP	differential pressure (kPa)
ΔP_f	foam differential pressure (kPa)
PFP	perforation friction pressure, equation 29.
P_r	pressure at reservoir conditions (kPa)
P_{SC}	pressure at standard conditions (kPa)
Q	foam quality (Mitchell quality)
Q_s	foam slurry quality
q	flow rate (m^3/min)

S_r	solubility of CO_2 in liquid phase of foam
STP	surface treating pressure (kPa), equation 29
T_r	temperature at reservoir conditions (°C)
T_{SC}	temperature at standard conditions (°C)
V_{CO_2}	CO_2 volume (m^3)
V_d	volume fraction of dispersed phase (m^3)
V_f	volume of foam (m^3)
V_G	volume of gaseous phase
V_L	volume of liquid phase
V_P	volume of proppant
V_r	volume of foam at reservoir conditions (m^3)
V_{SC}	volume of foam at standard conditions (m^3)
v	average fluid velocity (m/min)
v_f	average foam velocity (m/min)
X'	density parameter
z_r	compressibility of gas at reservoir conditions

Greek

δ	shear rate (s^{-1})
μ	viscosity (cp)
μ_a	apparent viscosity (cp)
μ_c	viscosity of continuous phase (cp)
μ_d	viscosity of dispersed phase (cp)
μ_e	effective viscosity (cp)
μ_f	viscosity of foam fluid (cp)
μ_p	plastic viscosity (cp)
μ_r	viscosity of reservoir fluid (cp)
ρ	bulk density (kg/m^3)
ρ_f	foam density (kg/m^3)
ρ_L	density of liquid (kg/m^3)
ρ_r	relative density (kg/m^3)
τ	shear stress (kPa)
τ_w	shear stress at the wall (kPa)
τ_{yp}	yield point (kPa)
ϕ	porosity

References

1. Grundman, S. R.; Lord, D. L. *Foam Stimulation;* Production Operations Symposium; Society of Petroleum Engineers: Richardson, TX, 1981; paper SPE 9754.

2. King, G. E. *Foam and Nitrified Fluid Treatments: Stimulation Techniques;* Distinguished Lecture Program; Society of Petroleum Engineers: Richardson, TX, 1985; paper SPE 14477.
3. Mack, D. J.; Harrington, L. J. *Oil Gas J.* **1990,** *July 16,* 49–58.
4. Toney, F. L.; Mack, D. J. *Proceedings of the Production Operations Symposium;* Society of Petroleum Engineers of AIME: Richardson, TX, 1991; paper SPE 21644, pp 23–33.
5. Garbis, S. J.; Taylor, J. L. *Proceedings of the Production Operations Symposium;* Society of Petroleum Engineers of AIME: Richardson, TX, 1985; paper SPE 13794, pp 27–40.
6. Bullen, R. S.; Bratrud, T. F. *Proceedings of the Petroleum Society of The Canadian Institute of Mining 26th Annual Meeting;* Petroleum Society of the Canadian Institute of Mining, Metallurgy, and Petroleum: Calgary, Canada, 1975.
7. Blauer, R. E.; Mitchell, B. J.; Kohlhaas, C. A. *Proceedings of the 44th Annual California Regional Meeting of the Society of Petroleum Engineers;* Society of Petroleum Engineers of AIME: Richardson, TX, 1974; paper SPE 4885.
8. Harris, P. C. *SPE Prod. Eng.* **1987,** *2(2),* 89–94.
9. Sage, B. H.; Lacey, W. N. *Monograph of API Research Project 37: Thermodynamic Properties of Lighter Paraffin Hydrocarbons and Nitrogen;* American Petroleum Institute: Washington, DC, 1950.
10. Neill, G. H.; Dobbs, J. B.; Pruitt, G. T.; Crawford, H. R. *Proceedings of the 38th Annual Fall Meeting of the Society of Petroleum Engineers of AIME;* Society of Petroleum Engineers of AIME: Richardson, TX, 1963; paper SPE 738.
11. Reidenbach, V. G.; Harris, P. C.; Lee, Y. N.; Lord, D. L. *SPE Prod. Eng.* **1986,** *1(1),* 31–41.
12. Harris, P. C. *Production Operations and Engineering Proceedings;* Society of Petroleum Engineers of AIME: Richardson, TX, 1990; paper SPE 20642, pp 271–278.
13. Phillips, A. M.; Couchman, D. D. *High Temperature Rheology of CO_2 Foam Fracturing Fluids;* Southwestern Petroleum Short Course; Western Company of North America, 1987.
14. Lancaster, G.; Sinal, M. L. *J. Can. Pet. Technol.* **1987,** *26(5),* 26–30.
15. Phillips, A. M.; Couchman, D. D.; Wilke, J. G. *Proceedings of SPE/DDE Low Permeability Reservoir Symposium;* Society of Petroleum Engineers of AIME: Richardson, TX, 1987; paper SPE/DDE 16916, pp 267–272.
16. Canadian Fracmaster Ltd., Calgary, Alberta, Unpublished Data, Revised 1990.
17. *CRC Handbook of Tables for Applied Engineering Science,* 2nd ed; CRC Press: Boca Raton, FL, 1973; p 64.
18. Soave, G. *Chem. Eng. Sci.* **1972,** *27,* 1197–1203.
19. Cameron, J. R.; Prud'homme, R. K. In *Recent Advances in Hydraulic Fracturing;* Gidley, J. L.; Holditch, S. A.; Nierode, D. E.; Veatch, R. W., Eds.; Monograph Series; Society of Petroleum Engineers: Richardson, TX, 1989; Vol. 12, pp 198–207.
20. "The Benefits of Running CO_2 and N_2 in the Same System"; Technical Manual Western Binary Foam System; Western Petroleum Services: Houston, TX, 1990.

21. Preprint of Reference 3.
22. Holcomb, D. L. *Proceedings of the Permian Basin Oil and Gas Recovery Conference;* Society of Petroleum Engineers of AIME: Richardson, TX, 1977; paper SPE 6376.
23. *Reservoir Stimulation;* Economides, M. J.; Nolte, K. G., Eds.; Schlumberger Educational Services: Houston, TX, 1987; pp 17-1–17-3.
24. Howard, G. C.; Fast, C. R. *Hydraulic Fracturing;* Monograph of the Henry L. Doherty Series; Society of Petroleum Engineers of AIME: Richardson, TX, 1970; Vol. 2, pp 34–40.
25. Petryk, R. P.; Goruk, B. W. *Proceedings of the 30th Annual Technical Meeting of the Petroleum Society of the CIM;* Petroleum Society of the Canadian Institute of Mining, Metallurgy, and Petroleum: Calgary, Canada, 1979; paper 79-30-37.
26. Scherubel, G. A.; Crowe, C. W. *Proceedings of the 30th Annual Technical Meeting of the Petroleum Society of CIM;* Petroleum Society of the Canadian Institute of Mining, Metallurgy, and Petroleum: Calgary, Canada, 1979; paper 79-30-36.
27. Holm, L. W. *Transactions* **1968,** *243.*
28. Bernard, G. G.; Holm, L. W.; Jacob, W. L. *Proceedings of the SPE Fall Meeting;* Society of Petroleum Engineers: Richardson, TX, 1965; paper SPE 1204.
29. Bernard, G. G.; Holm, L.W. *Soc. Pet. Eng. J.* **1964,** *4,* 267–274.
30. Ford, W. G. *Proceedings of the SPE 55th Annual Technical Conference and Exhibition;* Society of Petroleum Engineers of AIME: Richardson, TX, 1981; paper SPE 9385.
31. Peters, F. W.; Saxon, A. *Proceedings of the SPE Asia–Pacific Conference;* Society of Petroleum Engineers: Richardson, TX, 1989; SPE paper 19496.
32. Bernadiner, M. G.; Thompson, K. E.; Fogler, H. S. *Proceedings of the SPE International Symposium on Oilfield Chemistry;* Society of Petroleum Engineers of AIME: Richardson, TX, 1991; paper SPE 21035, pp 363–370.
33. Burman, J. W.; Hall, B. E. *Proceedings of the 61st Annual Technical Conference and Exhibition of the Society of Petroleum Engineers;* Society of Petroleum Engineers of AIME: Richardson, TX, 1986; paper SPE 15575, p 12.
34. Thompson, K. E.; Gdanski, R. D. *Proceedings of the SPE Eastern Regional Meeting;* Society of Petroleum Engineers of AIME: Richardson, TX, 1991; paper SPE 23436.
35. Kennedy, D. K.; Kitziger, F. W.; Hall, B.E. *Production Operations and Engineering Proceedings—SPE Annual Technical Conference and Exhibition;* Society of Petroleum Engineers of AIME: Richardson, TX, 1990; paper SPE 20621, pp 65–76.
36. Watkins, E. K.; Wendorff, C. L.; Ainley, B. R. *Proceedings of the 58th Annual Technical Conference;* Society of Petroleum Engineers of AIME: Richardson, TX, 1983; paper SPE 12027.
37. Harris, P. C. *SPE Prod. Eng. 1989, 4(3),* 249–257.
38. Mitchell, B. J. Ph.D. Dissertation, University of Oklahoma, Norman, 1970.
39. Sporker, H. F.; Trepess, P.; Valko, P.; Economides, M. S. *Production Operations and Engineering, Part 2—Proceedings of the SPE Annual Technical Conference and Exhibition;* Society of Petroleum Engineers of AIME: Richardson, TX, 1991; paper SPE 22840, pp 243–251.

40. Blauer, R. E.; Kohlhaas, C. A. *Proceedings of the 49th Annual Fall Meeting of the Society of Petroleum Engineers;* Society of Petroleum Engineers: Richardson, TX, 1974; paper SPE 5003.
41. Patton, J. T.; Holbrook, S. T.; Hsu, W. *Soc. Pet. Eng. J.* **1983,** *23,* 456–460.
42. Harris, D. C.; Reidenbach, V. G. *Proceedings of the 59th Annual Technical Conference and Exhibition;* Society of Petroleum Engineers of AIME: Richardson, TX, 1984; paper SPE 13177.
43. Cawiezel, K. E.; Niles, T. D. *Proceedings of the SPE Hydrocarbon and Evaluation Symposium;* Society of Petroleum Engineers of AIME: Richardson, TX, 1987; paper SPE 16191.
44. Melton, L. L.; Malone, W. T. *Soc. Pet. Eng. J.* **1964,** *4,* 56.
45. Bowen R. L., Jr. *Chem. Eng.* **1961,** *68,* 143.
46. Harris, P. C.; Klebenow, D. E.; Kundert, D. P. *Proceedings of the SPE Rocky Mountain Regional Meeting;* Society of Petroleum Engineers of AIME: Richardson, TX, 1988; paper SPE 17532.
47. Nolte, K. G.; Smith, M. B. *J. Pet. Tech.* **1981,** *33,* 1767–1775.

RECEIVED for review October 14, 1992. ACCEPTED revised manuscript July 12, 1993.

10

Role of Nonpolar Foams in Production of Heavy Oils

Brij B. Maini and Hemanta Sarma

Petroleum Recovery Institute 100, 3512 33 Street, N.W., Calgary, Alberta T2L 2A6, Canada

> *In this chapter, the properties of nonaqueous hydrocarbon foams will be reviewed and the effects of foam formation on flow of oil–gas mixtures in porous media will be discussed. A laboratory technique for investigating the role of foamy-oil behavior in solution gas drive is described, and experimental verification of the in situ formation of non-aqueous foams under solution gas drive conditions is presented. The experimental results show that the in situ formation of nonaqueous foam retards the formation of a continuous gas phase and dramatically increases the apparent trapped-gas saturation. This condition provides a natural pressure maintenance mechanism and leads to recovery of a much higher fraction of the original oil in place under solution gas drive.*

F OAMS IN WHICH THE LIQUID PHASE IS NONAQEOUS are encountered in many industrial processes, such as distillation of crude oil and production of polymer latex. In this chapter, we focus on the role of such foams in production of heavy oils under solution gas drive. The oil in such reservoirs contains dissolved gas at high pressure. During primary production, the oil is driven into the well by the initial high pressure of the reservoir. Under flowing conditions, the pressure is higher at the boundary of the area being drained by the well compared to the pressure within the wellbore. This difference between the pressure at the drainage boundary and the pressure within the wellbore is called pressure drawdown in reservoir engineering terminology, and is an important parameter in designing the optimum production strategy.

Naturally, the oil experiences decreasing pressure as it moves from the interior of the reservoir toward the wellbore and later from the bottom of the well toward the surface. The reduced pressure causes the dissolved gas to come out of solution. In most light-oil reservoirs, the released gas readily separates from the oil, and no foam is created. However, several heavy-oil reservoirs in Alberta and Saskatchewan show "foamy-oil" behavior; the released gas forms an oil-continuous foam that has the appearance of a chocolate mousse. This foam can be quite stable and may persist for several hours in open vessels. Several of these reservoirs also show anomalously high production. Both the rate of production and the total recovery under solution gas drive are much higher than what would be expected from measured reservoir parameters. History-matching the primary production from these wells often requires unrealistic adjustment of measured parameters, such as increasing the absolute permeability by an order of magnitude or increasing the trapped-gas saturation to a very high level (*1*). The need for increased trapped-gas saturation suggests that in situ foam formation may be involved.

Most likely, the in situ formation of an oil-continuous foam plays a role in the production of heavy oils under solution gas drive. Before discussing this role of oil-continuous foams, the similarities and differences between such nonpolar, nonaqueous foams and the aqueous foams will first be reviewed.

Foam Stability in Nonpolar Media

The basic theories of foam stability have been discussed in detail in the first two chapters of this book from the perspective of aqueous foams. Much of what has been said there applies equally to nonpolar foams. However, some important differences exist between the two types of foams, and the following discussion will attempt to clarify these differences.

As discussed in the preceding chapters, all foams are thermodynamically unstable, and given sufficient time will collapse into separate gas and liquid phases. Therefore, foam stability describes only the relative resistance of the foam to eventual phase separation. The collapse of foam is generally a two-stage process involving gradual thinning of liquid films followed by breakage of these films. The thinning of a foam lamella occurs under the influence of gravitational and capillary forces, and involves flow of the liquid from the central portion of the lamella toward its Plateau borders. Several factors play a role in providing foam stability by resisting the thinning and rupture of liquid films.

Role of Liquid-Phase Viscosity. As in any fluid flow process, the liquid viscosity offers resistance to flow and has a direct bearing on the rate of drainage of the liquid from foam films. The rate of film thinning will decrease as the liquid viscosity increases, and in the extreme case of very high viscosity (for example, the solidified films of latex foam), the resistance to flow can make the foam very stable.

The experimental evidence of the role of liquid viscosity in determining nonaqueous foam stability has been provided by several researchers. Brady and Ross (2) found that the foam stability of refined mineral oils increased linearly with kinematic viscosity. McBain and Robinson (3) showed that a high bulk liquid viscosity was often associated with high foam stability. Callaghan and Neustadter (4) measured the foam stability of crude oil foams and reported that the average lifetimes of such foams are almost linearly dependent on bulk viscosity.

Role of Adsorbed Surfactant Layer. Foams, irrespective of the nature of liquid and gas involved, require a third component for stabilization of thin films (lamellae) of the liquid. In the familiar case of aqueous soap films, this third component is the soap, a surface-active chemical that adsorbs at the gas–liquid interface and lowers the surface tension of water. The two effects, adsorption at the liquid surface and the depression of surface tension, are intimately linked and occur concomitantly. The adsorption is defined as the excess moles of solute per unit area of the liquid surface. In a binary system, this surface excess can be directly related to the lowering of surface tension by Gibbs adsorption equation:

$$\Gamma_2 = -\frac{1}{RT}\frac{d\gamma}{d \ln a_2} \qquad (1)$$

where Γ_2 is the excess moles of solute per unit area of the surface, R is the gas constant, T is the absolute temperature, γ is the surface tension, and a_2 is the chemical activity of the solute in the solution (5). The lowering of surface tension by adsorption of the surface-active chemical makes the liquid surface elastic against sudden expansion or contraction. When local areas of the liquid surface are suddenly expanded, the surface tension in the expanded area becomes temporarily large (compared to the surface tension of the undisturbed surface) because the surfactant molecules have not had sufficient time to become adsorbed at the surface. The greater surface tension in the expanded area exerts a pull on the surrounding areas of lower surface tension, and this pull causes a surface flow and contraction of the expanded surface. This surface elasticity is discussed further in Chapter 1.

This mechanism makes the foam lamella resistant to minor mechanical shocks. The degree of protection against breakage provided by this process depends on two factors: (1) the magnitude of surface tension reduction that determines the maximum pull that can be generated by the instantaneous surface tension gradient, and (2) the rate at which the solute migrates to the surface to destroy the increased surface tension. The migration rate of the solute toward the surface must be slow enough to enable the surface tension gradient to do its work before being destroyed. Therefore, the kinetics of adsorption at the surface is an important factor in determining the effectiveness of a surfactant in creating a stable foam.

The role of surface elasticity provided by the adsorbed layer of surfactant molecules remains the same in aqueous and nonpolar foams, however, its relative importance and effectiveness is different. A major difference lies in the magnitude of surface tension gradient that can be generated by the surface-active chemicals. Because the surface tension of pure water is very high (73 mN/m), many surfactants can lower the surface tension by more than 50 mN/m. The surface tension of nonpolar solvents is typically around 25 to 30 mN/m. Therefore, even the best surfactants can only provide a much smaller reduction in its value.

Another difference lies in the role of electric double-layer repulsion, which is often a key factor in stabilizing aqueous foams with ionic surfactants. The adsorption of ionic surfactant at the liquid surface leads to the formation of a charged surface and a diffuse layer of counterions. As the foam lamellae thin because of the drainage of liquid, these counterions begin to repel each other and retard further thinning. Because ionization is not possible in nonpolar solvents, this double-layer mechanism is not operative in nonpolar foams.

Because of the limited magnitude of surface tension gradients and absence of electric double-layer effects, the stabilization of foams in nonpolar liquids requires other ways of retarding the thinning of foam lamellae. These include the high liquid-phase viscosity that has been discussed earlier and increased surface viscosity because of presence of highly viscous or even rigid liquid-crystal films.

Role of Insoluble Surface Films. The importance of viscous liquid-crystal films at the interface between an nonpolar solvent and gas in stabilizing foams was studied by Friberg and co-workers (6–8). They found that the presence of a liquid-crystalline phase in a mixture of an organic solvent and a surfactant dramatically increased the "foamability" of the mixture. This improvement in the foam stability was not simply because of the increased viscosity of the mixture, but appeared to be caused by a specific surface action of the liquid-crystalline phase. The liquid-crystalline phase was found to form a thick layer of adsorbed phase at the liquid–gas interface. Thus the interface toward the gas phase became immobilized by the highly viscous or even rigid liquid-crystal. This immobilized interface had the obvious effect of severely retarding the liquid drainage from the foam. The reason for such preferential location of the liquid-crystalline phase at the surface toward the air was explained to be the lower surface tension of the liquid-crystal due to a higher concentration of the low-energy methyl groups in its surface.

Friberg et al. (6) suggested the following simple rules to serve as the starting point for selection of surfactants for producing hydrocarbon foams: (1) mutual solubility of the surfactant and the hydrocarbon, (2) no solubility of the surfactant in water, and (3) solubility of the water in the surfactant to form a liquid-crystal. They noted that in addition to the formation of lamellar liquid-crystal phase, the foam stability also requires the surface tension of the liquid-crystal phase to be lower than the surface tension of the isotropic surfactant solution in the hydrocarbon solvent. Therefore, solvents with higher surface tension are more likely to produce stable foams by this mechanism. The mechanism responsible for foam stability in such systems is the induced surface activity caused by the formation of a new association structure (liquid-crystal) in which the low-energy methyl groups are preferentially exposed at the surface.

Role of Incipient Phase Change. Ross (9) investigated the relationship between foaminess of a fluid mixture and the position of its composition on the phase diagram. He reported that the foaminess of a solution increases as its composition moves closer to a phase boundary. The reason for this observed behavior is related to structural changes in the solution prior to phase separation. He also noted that adsorption is the precursor of imminent phase separation in the sense that the surface offers a region for partial segregation of molecules prior to a more complete segregation into a separate bulk phase. Therefore, higher levels of adsorption occur at concentrations close to a phase boundary. Because adsorption is a measure of surface activity, compositions closer to a phase boundary are likely to display stronger interfacial phenomenon associated with surface activity. Ross and Nishioka (10) measured the foam stability of binary mixtures of 2,6-dimethyl-4-heptanol and 1,2-ethanediol as a function of composition and temperature. They found that the maximum foam stability occurred at a temperature and composition near that of the critical point in the single-phase region. They also noted that the foam stability decreases sharply at the onset of two-phase separation; the newly formed conjugate liquid acts as a defoamer for its foamy conjugate.

Experimental Investigation of Foamy-Oil Flow in Porous Media

As mentioned earlier, heavy oil produced by solution gas drive often displays marked foaminess in wellhead samples. This feature is not surprising because the two key factors needed for nonpolar foam stability are present in the heavy-oil system. The viscosity of the liquid phase (heavy oil) is high enough to retard drainage of liquid films by capillary forces. More important, plastic surface films, most likely stabilized by high molecular weight porphyrins, have been observed in such crude oil systems. The wellhead foam that had the appearance of chocolate mousse is generated by the liberation of dissolved gas, due to reduced pressure, that fails to coalesce and escape the liquid. A large part of this transition, from the high-pressure state in which the gas remains dissolved in oil to atmospheric pressure condition in which most of the gas comes out of the solution, occurs before the oil enters the production well. Clearly, the gas is evolved within the reservoir. The obvious question raised by this scenario is whether or not the released gas forms a nonaqueous foam before entering the production well, and if such a foam is formed in situ, what effect such foam formation has on the oil-production behavior of the reservoir.

The previous chapters of this book presented the fundamentals of foam transport in porous media and discussed the use of aqueous foams for reducing the mobility of gases in oil recovery operations. The experimental and theoretical considerations presented in Chapters 3, 5, 6, and 8 show that the flow behavior of an aqueous foam is markedly different from that of a nonfoaming water–gas mixture: The formation of a foam greatly reduces the mobility of the gas phase without appreciably changing the water mobility. Therefore, the flow behavior in porous media of a foam-forming oil–gas mixture is likely to be very different from that of a nonfoaming oil–gas mixture, and the mobility of the gas phase could be significantly reduced by foam formation, but the oil mobility remains unaffected.

The objective of the experimental study described next was to examine whether the in situ formation of a foam occurs in primary production of heavy oils by solution gas drive. A simple apparatus was designed to conduct primary depletion experiments in the laboratory. The presence of foam within the porous matrix was inferred from the observed production and pressure-drop behavior of the system.

Equipment and Materials. *Experimental Apparatus.* The apparatus used for simulated primary depletion experiments is shown, schematically, in Figure 1. In the actual oil field situation, primary depletion involves flow of oil from the reservoir into a production well under the driving force provided by expansion of the initially compressed oil and by the release of the initially dissolved gas. The flow geometry in the field is cylindrical, and this condition makes the flow velocity inversely proportional to the distance from the axis of the well. Because of the increasing flow velocity with decreasing distance to the well, a large fraction of the total change in the pressure occurs within a few meters of the wellbore. Such cylindrical geometry would be very difficult to duplicate in the laboratory even if one was attempting to look at only the region within 1 or 2

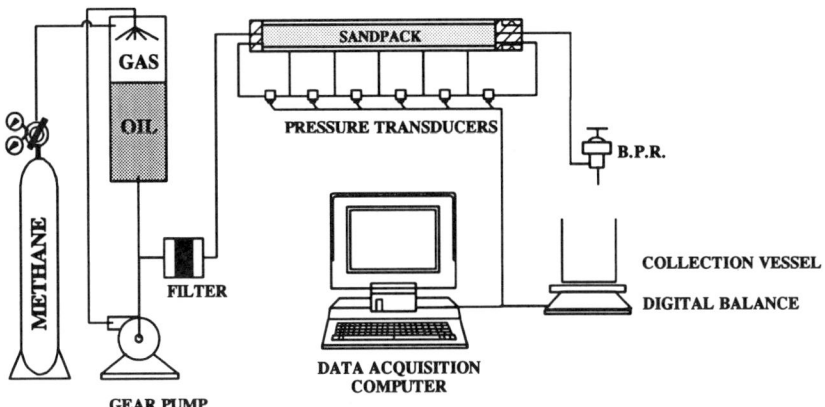

Figure 1. Schematic diagram of the experimental apparatus.

m of the wellbore. Consequently, a linear geometry was used in the laboratory experiments. Experiments were conducted in a 2-m long, radially stressed, sand pack. The sand pack was equipped with six intermediate pressure taps for continuous monitoring of the pressure distribution within the sand during the flow experiments. The dimensions of the sand pack and the properties of the porous medium are listed in Table I.

Recombined oil (also referred to as "live oil") was prepared by saturating the oil with methane gas in recombination equipment connected to the inlet end of the core holder. Produced fluids were collected into a graduated cylinder placed on an electronic balance for measuring the oil production rate. An automated data acquisition system was employed for recording of the oil-production rate and the differential pressures in each segment during the flow experiment.

Heavy Oils. The two heavy oils were Lloydminster heavy oil supplied by Mobil Oil Canada and Lindbergh heavy oil supplied by PanCanadian Petroleum Limited. The Lloydminster had a density of 981 kg/m^3 at

Table I. Properties of the Porous Medium

Parameter	Value
Length (m)	2.0
Cross-sectional area (m^2)	16.1×10^{-4}
Sand size (μm)	74–105
Porosity (fraction)	0.33
Pore volume (mL)	1062
Permeability (μm^2)	3.35
Confining pressure used (MPa)	14.0

15 °C, an asphaltene weight fraction of 0.1411, and a viscosity at 20 °C of 8580 mPa·s. The Lindbergh had a density of 992 kg/m^3 at 15 °C, an asphaltene weight fraction of 0.2067, and a viscosity at 20° C of 10,500 mPa·s. Of the two crude oils used, the Lindbergh oil was more viscous (10,500 mPa.s), and contained about one and a half times higher asphaltene than the Lloydminster heavy oil. The asphaltene contents of the crude oils were measured by precipitation with pentane.

Preparation of Recombined ("Live") Oil. Prior to the start of flow experiments, each oil sample was cleaned of its suspended materials. The oil was then recombined with methane gas in the recombination equipment at a pressure of 4.83 MPa. During this recombination process, the solution gas–oil ratio (SGOR) was monitored periodically, and when a constant SGOR was established, the oil was considered to be fully saturated. This usually took about 5–6 days of continuous mixing. Properties of recombined oils are reported in Table II.

Porous Medium. Ottawa sand was used as the porous medium for flow experiments. The porosity and absolute permeability to the water were determined prior to the start of flow experiments (Table I), and the core was then resaturated with "live oil" by displacing the water. During the resaturation process, almost 98% of the water was displaced. This value of irreducible water saturation (2%), although much lower than the field values, is not exceptional in laboratory tests of this nature.

Procedure. The sequence of flow experiments for each oil started with the core saturated with the "live oil" at saturation pressure, 4.83 MPa. The pressure at the inlet end was always maintained at 4.83 MPa. A back-pressure regulator (BPR) was used to maintain constant pressure at the outlet end, and this pressure was reduced in steps of approximately 0.34 MPa, starting from the saturation pressure. The term "pressure drawdown" will be used in the following discussion to denote the pressure difference between the inlet end and the outlet end of the sand pack. The pressure drawdown reaches its maximum value when the pressure at the outlet end becomes equal to the atmospheric pressure.

Table II. "Live Oil" Properties

Oil	Density (kg/m^3)	Viscosity $(mPa·s)$	Initial Soln. Gas–Oil Ratio (m^3/m^3)	Core Avg. S_g at Maximum Drawdown (%)
Lloydminster	968.0	3007.5	14.8	13.8
Lindbergh	978.0	3970.2	14.2	14.8

At each setting of the outlet pressure, the steady-state pressure profile and the production rate were measured. Therefore, for each oil, 10–12 flow experiments were conducted. End-point average gas saturations (S_g) were estimated after the flow test at the highest pressure drawdown and are presented in Table II.

Results and Discussion

During each flow experiment, the oil production and pressure drop in each segment of the porous medium were monitored and recorded as functions of time to ascertain whether the steady-state flow had been established and to analyze the flow mechanisms in the porous medium. A constant slope in the cumulative oil production versus time plot and a constant-pressure gradient in all segments over several hours (2–3 h) were considered a positive indication of the steady-state flow. When such steady state was attained for a particular pressure drawdown, the back-pressure regulator was reset for the next higher pressure drawdown. In this way, flow experiments were carried out at progressively higher pressure drawdown.

The effects of the pressure drawdown on the oil-production rate are shown in Figure 2. Both the experimental data and theoretical calculations based on the Darcy's law are shown. Initially the oil-production rate increased linearly with the increasing pressure drawdown across the porous medium, in agreement with Darcy's law for single-phase flow through porous media. At low drawdown pressures, the dissolved gas remained largely in solution, and therefore, the oil was flowing as a

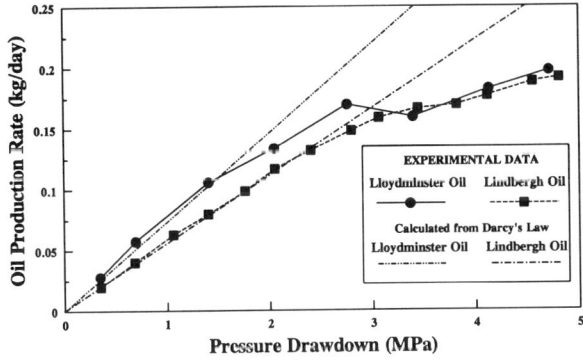

Figure 2. Comparison of experimental oil production rate with theoretical calculation based on Darcy's law.

single-phase fluid. However, as the pressure drawdown was increased beyond a certain pressure (1.4 MPa), the increase in the production rate was no longer linear, and the rate of increase in the production rate declined. This behavior is obviously caused by the evolution of dissolved gas at higher pressure drawdown, which results in two-phase flow within the porous medium.

Thus the production versus drawdown behavior was consistent with the conventional picture of solution gas drive and revealed no surprises.

Pressure Distributions in the Sand-Pack. Profiles that show the pressure distribution in each flow experiment are presented in Figures 3 and 4. Each pressure profile in Figures 3 and 4 represents the pressure distribution in the porous medium at a particular pressure drawdown when the steady-state flow was attained. A general trend among these pressure profiles is obvious. The pressure distribution in the porous medium remained linear when the drawdown pressure was below a certain pressure, beyond which the pressure distribution started becoming increas-

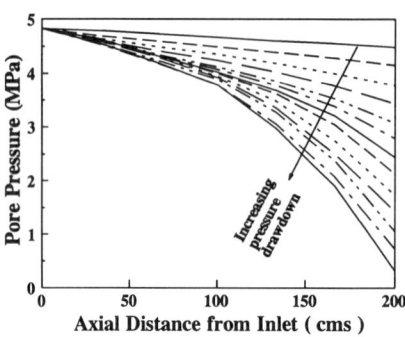

Figure 3. Experimental pressure profiles at various pressure drawdowns in the Lindbergh system.

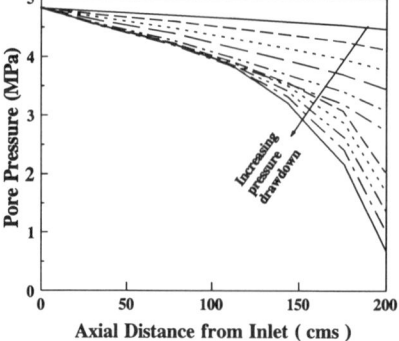

Figure 4. Experimental pressure profiles at various pressure drawdowns in the Lloydminster system.

ingly nonlinear. This trend is also consistent with the conventional theory of solution gas drive and confirms the oil production rate behavior discussed earlier. Thus the effect of pressure drawdown on oil production rate and the pressure profiles within the sand did not reveal any unusual behavior. However, the recorded history of pressure gradients within individual core segments showed an unusual behavior.

Pressure Gradient in Individual Segments. As the drawdown was progressively increased, the nonlinearity of the pressure profile became more severe, a result suggesting that more and more gas was being liberated and was hindering the flow of oil. The oil–gas mixture being produced from the core was mostly in the form of an oil-continuous foam. However, occasionally, a large slug of gas was produced without any oil. Such gas production was accompanied by dramatic changes in the pressure profile within the sand pack. These happenings and the corresponding mechanisms within the porous medium can be better illustrated with the help of Figures 5–8, which show the differential pressure in individual segments and the cumulative oil production as functions of time.

A comparison of the pressure-drop patterns presented in Figure 5 and 6 demonstrates that the drawdown pressure significantly influenced how the pressure in each individual segment was distributed and balanced under different drawdown pressures. For example, at the pressure drawdown of 1 MPa, the segmental pressure drop remained constant, and this pressure drop was accompanied by a steady oil production rate as implied

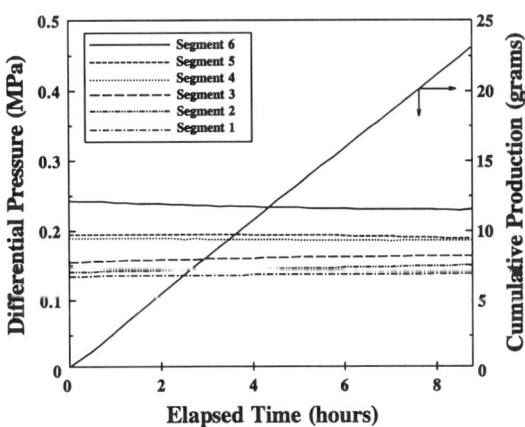

Figure 5. Change in the pressure drop across different core segments and cumulative oil production with time at 1.0-MPa drawdown pressure in the Lindbergh system.

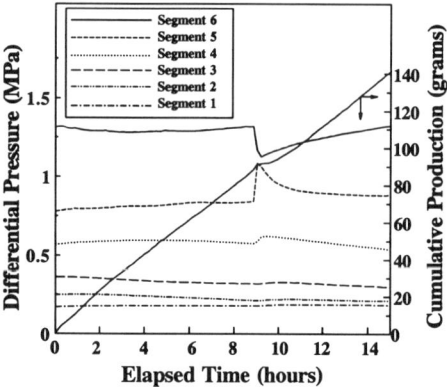

Figure 6. Change in the pressure drop across different core segments and cumulative oil production with time at 3.4-MPa drawdown pressure in the Lindbergh system.

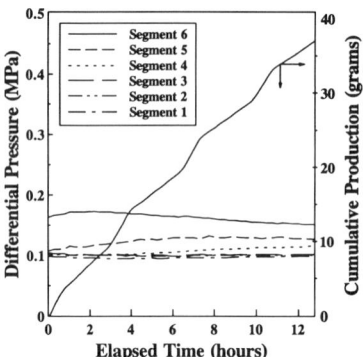

Figure 7. Change in the pressure drop across different core segments and cumulative oil production with time at 0.7-MPa drawdown pressure in the Lloydminster system.

by the corresponding cumulative oil production curve. At the higher pressure drawdown of 3.4 MPa, an unexpected cyclic behavior became evident in the segmental pressure-drop history. The pressure drop in the segment closest to the outlet showed a sudden sharp decline, and the pressure drop in the upstream section showed a sharp increase. This change in the pressure distribution was accompanied with production of a large volume of gas at the outlet with very little oil production. This behavior occurred with both oils at high drawdown pressures and was reproduced in several different tests.

The pressure-drop profile for Segment 6 and the observed production

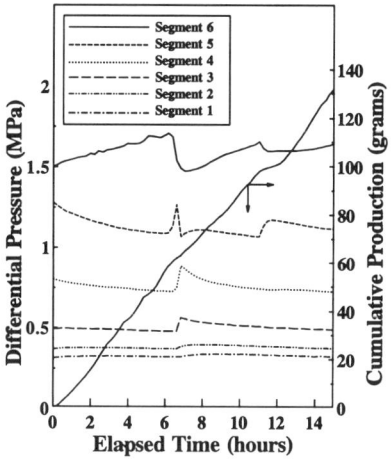

Figure 8. Change in the pressure drop across different core segments and cumulative oil production with time at 4.2-MPa drawdown pressure in the Lloydminster system.

behavior suggest that with continued accumulation of gas in this segment, the pressure drop continued to increase until the gas saturation reached some critical value. Then the gas was produced at the outlet end in sudden bursts. This gas production was simultaneously accompanied by a sharp and immediate decrease in the pressure drop in Segment 6 and a lower oil production rate. But this free flow of the gas was soon arrested, and the next cycle of increasing pressure drop in this segment started.

The core average gas saturation at the end of the high drawdown experiments with the Lloydminster and the Lindbergh heavy oils was around 15% (Table II). Because the local pressure was lowest in the last segment, probably this segment contained a large part of this gas saturation. The pressure profiles presented in Figures 3 and 4 also support this contention. For example, during the horizontal flow for the Lloydminster oil (Figure 5), the profile at the maximum pressure drawdown indicated the linearity in the first three segments and nonlinearity for the last three segments. This finding suggests that the gas was evolved primarily from the last three segments.

Moreover, even among these last three segments, the gas saturation would be highest in the segment closest to the outlet. This condition was inferred from the progressive increase in the pressure drop recorded for these segments. For example, a pressure drop of 0.62 MPa was recorded in Segment 4, and in Segments 5 and 6, they were 0.99 and 1.58 MPa, respectively. Thus the pressure drop in the last segment (Segment 6) was almost equal to the sum of pressure drops in Segments 4 and 5. And because the magnitude of the pressure drop is a measure of the evolved gas saturation, it can be argued intuitively that the last segment alone could have contained about half of the total gas saturation in the porous medium. This result would imply that the gas saturation was around 40% in the last section.

The foregoing discussion implies that although a very high gas saturation existed in the sand pack, the gas was not flowing as a continuous phase. This situation strongly suggests a foam-type behavior. If the oil–gas mixture forms a foam, the gas can remain discontinuous even at very high saturations. Because foam was being produced at the outlet, it is most likely that an oil-continuous foam was present in the sand pack. The foam-flow hypothesis can explain the observed behavior. The foam forces the gas to flow as a discontinuous phase even at high gas saturations. However, being a nonpolar foam in which the high viscosity of the liquid is a major factor in providing stability, the increasing gas saturation due to reduced pressure makes the liquid lamellae thinner, and beyond a system-dependent critical gas saturation the foam becomes unstable and locally collapses to form a continuous gas saturation. When generated, the free gas saturation rapidly depletes the pressure in the region in which it exists. This explains the sudden decline in the pressure drop in the last section and the readjustment of pressure drops in other sections. However, the continuous gas flow does not continue for long. Because of the rapid depressurization in the gas phase, the surrounding oil that was at a higher pressure begins to foam and moves into the space occupied by the freely flowing gas. This action cuts off the free gas flow, and a new cycle of gradual pressure drop increase in the last section starts again.

Blowdown Experiments. Although the flow experiments described capture some of the essential elements of the flow of heavy oils under solution gas drive, they have some limitations. The flowing ratio of oil and total gas (dissolved plus free) is forced to remain fixed at the injected-gas–oil ratio. Also, the experiments do not provide an estimate of the recovery factors obtainable under solution gas drive. A different type of experiment was needed to estimate the total recovery potential of solution gas drive. This experiment started with the sand pack at maximum "live oil" saturation. The pack was allowed to blow down to atmospheric pressure through the outlet end, and the inlet end remained closed. Figure 9 shows the recovery and pressure-drop behavior. More than 20% of the original oil was recovered in this primary depletion experiment. This value is surprisingly high for the viscous oil system and suggests that the critical gas saturation was much higher than what would be measured by an external gas drive experiment. Typically, the external drive experiments in such systems show the critical gas saturation to be less than 5%. Therefore, this experiment also suggests that a mechanism is present in

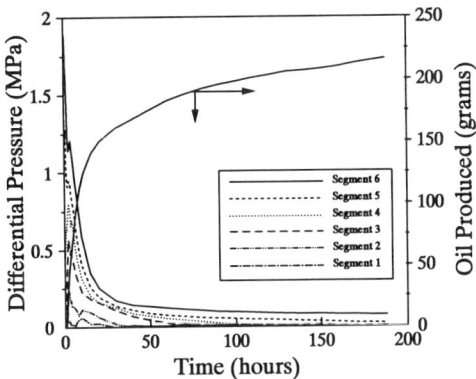

Figure 9. Change in the pressure drop across different core segments and cumulative oil production with time during the blowdown experiment with the Lloydminster system.

heavy-oil systems to increase the critical gas saturation. We suggest that this mechanism is the formation of a oil-continuous foam.

Summary and Conclusions

Results of the experimental study suggest that the formation of an oil-continuous foam may be involved in flow of heavy oil under solution gas drive. Such foam formation can be very beneficial for increasing the oil recovery. It delays the formation of a continuous gas phase, and thereby acts as a natural pressure-maintenance mechanism. In terms of the conventional solution gas drive theory, it serves to greatly increase the apparent trapped-gas saturation.

Relating this to the anomalous production behavior of Lloydminster type heavy-oil reservoirs under solution gas drive, the foam-flow hypothesis can explain the high primary recovery factors. However, it does not explain the high rate of production. Evidently, a different mechanism is responsible for the high rates. Most probably, an increase in the effective permeability caused by sand dilation is involved. The foam-flow mechanism also explains why some of these prolific producers under primary production may show poor response to steam stimulation. At elevated temperatures, the foam is likely to be less stable. Therefore, the gas phase would become continuous at much lower gas saturations. Consequently, the benefit of foamy-oil may not be available at steam temperatures.

Acknowledgments

We gratefully acknowledge the valuable contribution of Fausto Nicola and Jon Goldman in conducting the experiments. We also thank PanCanadian Petroleum Limited and Mobil Oil Canada for providing the crude oils.

References

1. Loughead, D. J.; Saltuklaroglu, M. *Proceedings of the Ninth Annual Heavy Oil and Oil Sand Technology Symposium;* University of Calgary: Calgary, Canada, 1992; paper 9.
2. Brady, A. P.; Ross, S. J. *J. Am. Chem. Soc.* **1944,** *66,* 1348.
3. McBain, J. W.; Robinson, J. V. *Natl. Advis. Comm. Aeronaut. Tech. Notes* **1949,** *1844.*
4. Callaghan, I. C.; Neustadter, E. L. *Chem. Ind.* **1981,** 53.
5. Gibbs, J. W. *The collected works of J. W. Gibbs;* Longmans, Green: London, 1928; Vol. 1, p 119.
6. Friberg, S. E.; Sun W. M. In *The Structure, Dynamics and Equilibrium Properties of Colloidal Systems;* Bloor, D. M.; Wyn-Jones, E., Eds.; Kluwer Academic: Hingham, MA, 1990; pp 529–539.
7. Friberg, S. E.; Concepcion, S. *Langmuir* **1986,** *2,* 121–126.
8. Friberg, S. E.; Wohn, C.-S; Greene, B.; Gilder, R. V. *J. Colloid Interface Sci.* **1984,** *101,* 593.
9. Ross, S. In *Interfacial Phenomenon in Apolar Media;* Eicke, H.-F.; Parfitt, G. D., Eds.; Marcel Dekker: New York, 1987; pp 1–39.
10. Ross, S.; Nishioka, G. In *Foams;* Akers, R. J., Ed.; Academic: London, 1976; pp 17–31.

RECEIVED for review October 14, 1992. ACCEPTED revised manuscript April 29, 1993.

Foams in Surface Facilities

11

Bituminous Froths in the Hot-Water Flotation Process

Robert C. Shaw[1], Jan Czarnecki[1], Laurier L. Schramm[2], and Dave Axelson[2]

[1]Research Department, Syncrude Canada Ltd., 10120 17 Street, Edmonton T6P 1V8, Alberta, Canada
[2]Petroleum Recovery Institute 100, 3512 33rd Street, N.W., Calgary, Alberta T2L 2A6, Canada

A special kind of nonaqueous foam known as bituminous froth is produced during the application of the hot-water flotation process to Athabasca oil sands, a large-scale commercial application of mined oil sands technology. These froths are multiphase, composed of oil, water, gas, and solids, and form an interesting kind of petroleum industry foam. This chapter presents a review of the occurrence, nature, properties, and treatment of bituminous froths.

Oil Sands

Oil sands are essentially beds of sand containing a heavy hydrocarbon: bitumen. Bitumen is chemically similar to conventional oil but has comparatively high density (low API gravity) and high viscosity. Although a number of definitions for bitumen have been used by various authors, a reasonable set of definitions is given in Table I, based on the United Nations Institute for Training and Research (UNITAR) sponsored attempts to establish definitions (*1–3*).

Oil sand deposits are present in many locations around the world, and they comprise an oil capacity of some 1440 to 2100 billion barrels (*4, 5*), most of which is contained in Canada and Venezuela (almost as much as the world's total discovered medium- and light-gravity oils). The major world deposits are fairly similar and occur along the rim of major sedimentary basins in either fluviatile or deltaic environments (*4–10*).

Table I. Table of UNITAR-Sponsored Established Definitions for Bitumen

Hydrocarbon	Viscosity Range (mPa·s at deposit temperature)	Density Range (kg/m³ at 15.6 °C)
Crude oils	<10,000	<934
Heavy crude oil	<10,000	934–1000
Extra heavy crude	<10,000	>1000
Bitumen–tar	>10,000	>1000

Of the Canadian deposits, the largest, Athabasca, is at least 4 times the size of the largest conventional oil field, Ghawar, in Saudi Arabia (5). Of the Athabasca's estimated 600 billion barrels of bitumen, about 60 billion barrels could be recovered by surface mining of the oil sand followed by beneficiation to separate the oil. Currently two commercial plants are producing synthetic crude oil from the Athabasca deposit. In these operations, the oil sands are first mined, and the bitumen is extracted by a hot-water flotation process, which produces a bituminous froth. After breaking the froth, the separated bitumen is subsequently upgraded by refinery-type processes to produce synthetic crude oil. In order to understand the nature of the froths produced, the nature of oil-sand structure will be reviewed first, and then the flotation process from which froths are produced will be examined.

Oil-Sand Structure. In oil-sand processing, the composition and structure of oil sands are thought to have a great impact on the way flotation must be operated. Although the oil sands resemble conventional oil deposits in many respects, there are important differences. The bulk composition of a given oil sand and some of its physical properties are largely determined by its geological origin.

In the shallow regions of the Athabasca deposit, the oil is in a drainage basin that was filled in with sediments as it was alternately flooded by a sea (estuarial environment) and then rivers (fluvial environment), so that a number of distinct depositional environments of estuarine and marine sediment occurred (10–15). As a result, the oil-bearing sands have great variability in their compositions and properties.

The oil, in mobile form, probably migrated into the sediments and then altered in place because of some combination of the following (9, 10, 13, 16):

- Lighter components may have evaporated or migrated away from the bulk oil.
- Lighter components may also have dissolved somewhat into the pore water.

- In shallow depths, oxidation may have occurred (increasing the average molecular mass and viscosity of the oil).
- Anaerobic bacteria may have degraded the lighter hydrocarbons.

Any combination of these factors would have resulted in a residuum of the heavier components, hence bitumen.

The mineral fraction is made up mostly of quartz grains, with smaller amounts of heavy minerals, feldspar grains, mica flakes, and clays (*17–19*). The clays in this deposit are predominantly illite and kaolinite with some chlorite. In general, the oil-bearing sands are fine grained to very fine grained (62.5–250 μm) with only a small percentage of fines (*20, 21*). The oil-bearing sands have fairly high porosities (25–35%) compared with the 5–20% porosity of most petroleum reservoirs (*17*). The high porosity is achieved by having a low occurrence of mineral cements, which is why the deposits are classified as oil sands rather than sandstones (*18*). The Athabasca oil sands with the highest oil content have oil saturations of about 18% w/w (36% v/v) and water saturations of about 2% w/w (4% v/v) (*18*).

This picture provides a connection to the structure of oil sands. The arrangement of phases in oil sands is largely determined by the volume fractions of bitumen, water, and quartz sand. The quartz sand forms the bulk of the material, with either the bitumen (in rich oil sands) or the water (in lean oil sands) forming the continuous phase. This structure results in a simple relationship between oil content (grade) and particle size distribution (*18, 20, 22, 23*). The oil content of oil sands decreases as the median diameter of the particles decreases and as the fraction of particles in the clay-size range increases.

For the most part, the mineral grains in Athabasca oil sand are water-wet, although an examination of literature citations that claimed to support this conclusion turned up a surprising lack of evidence. Such citations can ultimately be traced to the works of Clark and Ball (*24–26*). Clark (*24*) advanced the postulate that water is present as a film around the sand particles that are in turn surrounded by the oil. This postulate is based on microscopic observations of freshly cut oil-sand samples that yielded bright, clean (wet) exposed quartz particles as the oil envelopes were ruptured and sheared away. Early papers by Ball (*25, 26*) cite Clark's findings together with the observation that dry heating did not cleanly separate oil whereas boiling in water did. It was inferred that in heating, the water film evaporated, and left oil to attach to the solids so that a residue of oil was left after the bulk had flowed away.

An NMR study of oil sand (*27*) found evidence for not only bulk water between sand grains, but also water bound directly to mineral surfaces. These results are consistent with the theoretical proposal by Takamura (*28*) that connate water exists as pendular rings around sand-grain contact points and also as roughly 10-nm-thick films on sand-grain surfaces. Contrary points of view also exist (e.g., ref. 29). The paradigm of water-wet solids in the Athabasca oil sands remains a reasonable yet little examined postulate.

The usual representation of oil-sand structure is shown in Figure 1, where the bitumen is not in direct contact with the mineral phase, but instead is separated by at least a thin film of water. The separation of oil from solids by a water film is widely held to be the characteristic difference between Athabasca oil sand and oil sand from other oil-sand deposits in the world (e.g., California, New Mexico, or Utah). These other oil-sand deposits are thought to consist of oil-wet solids. That is, the bitumen occurs in direct contact with the mineral grains. The significance of the distinction is that the "oil-wet" oil sands are considered to be more difficult to beneficiate using hot-water flotation because of the difficulty in dislodging bitumen from an oil-wet surface.

Flotation Applied to Oil Sands. Although many different separation methods have been tested, hot-water extraction and flotation has received the most attention and forms the basis for existing large-scale commercial plants. Descriptions of the commercial hot-water flotation processes have been given in the technical and patent literature (*8, 9, 22, 30–35*). The oil sand is surface-mined after removal of as much as 30 m of overburden material (Figure 2). The mining of the ore body, which is about 60 m thick, is accomplished by large draglines or bucket wheel excavators. About 2 tons of oil sand of 10% bitumen content are needed to yield 1 barrel of oil (after extraction).

The hot-water flotation process is used to separate the bitumen from the associated minerals by exploiting differences in their bulk and surface properties. A commercial process layout is shown in Figure 3. First the oil sand is "conditioned" by slurrying it with water in rotating horizontal drums called tumblers. Heat (viscosity reduction) and shear are employed to overcome the forces holding oil-sand lumps together. In this ablation, successive layers of each lump are warmed and sheared off until everything is suspended. Besides stirring to maintain a state of suspension, a number of other things must happen in this "conditioning" step. The bitumen has to be separated from the solids, which make up about 70% of the slurry, and prepared for separation from the aqueous phase. Several things are done to accomplish this preparation. Steam is added to raise the exit temperature to 80 °C. Air is not sparged in directly but becomes worked in to aerate the bitumen by inclusion of about 30% v/v gas (*22*). Sodium hydroxide is added to raise the solution pH. The amounts of the reagents added are typically in the proportion: oil sand–water–NaOH (20%) = 1:0.19:0.0012 by mass. A continuous flow of slurry is maintained.

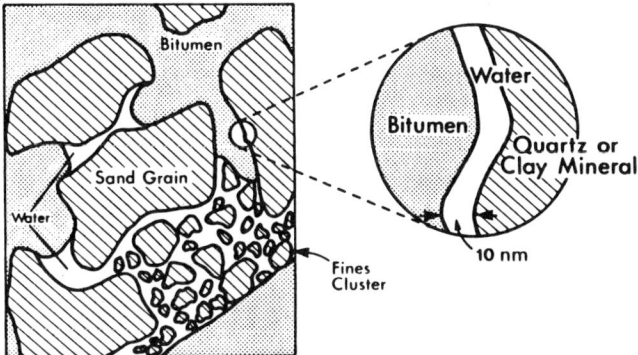

Figure 1. A model for the arrangement of the phases in Athabasca oil sands. (Reproduced from reference 28. Copyright 1982.)

The mined oil sand being treated at this stage is neither homogeneous nor uniform. Figure 2 shows a section of core from the Athabasca deposit. Although the mining and handling process (Figure 3) causes the various layers to be broken up and mixed (resulting in a more uniform feed), strong variations in feed quality persist. These variations necessitate constant adjustment to maintain optimal processing conditions. Sodium hydroxide and slurry water addition are the process control variables. An appreciable time is required to achieve good distribution of the bitumen, minerals, and reagents and to allow chemical and surface reactions to occur. In 5–10 min, a quasi-steady state is reached, probably not full thermodynamic equilibrium, and the slurry is discharged from the tumblers onto vibrating screens and washed with hot spray water to remove solid materials and undigested oil-sand lumps. This step may also provide additional air entrainment and thus promote further aeration of bitumen.

The diluted (flooded) slurry contains about 7% aerated bitumen droplets, 43% water, and 50% suspended solids. It is pumped to large primary separation (flotation) vessels. The vessels are maintained in a quiescent condition to facilitate the flotation of the aerated (lower density) bitumen droplets as well as the settling of the heavier (coarse) solids to the bottom. What is floated can be in some sense considered an aggregated colloid, rather than a dispersed one, and consequently high shear rates must be avoided to obviate destruction of the aggregates. The slurry is retained in these vessels for about 45 min to allow floating and settling and to provide adequate time for aqueous film drainage from the froth. To maximize the flotation and sedimentation processes, viscosity and density in the central (middlings) vessel region are kept low by adjusting the flood water addition and middlings removal rates. Because the process is continuous, the presence of fine minerals (e.g., clays) makes the vessel suscep-

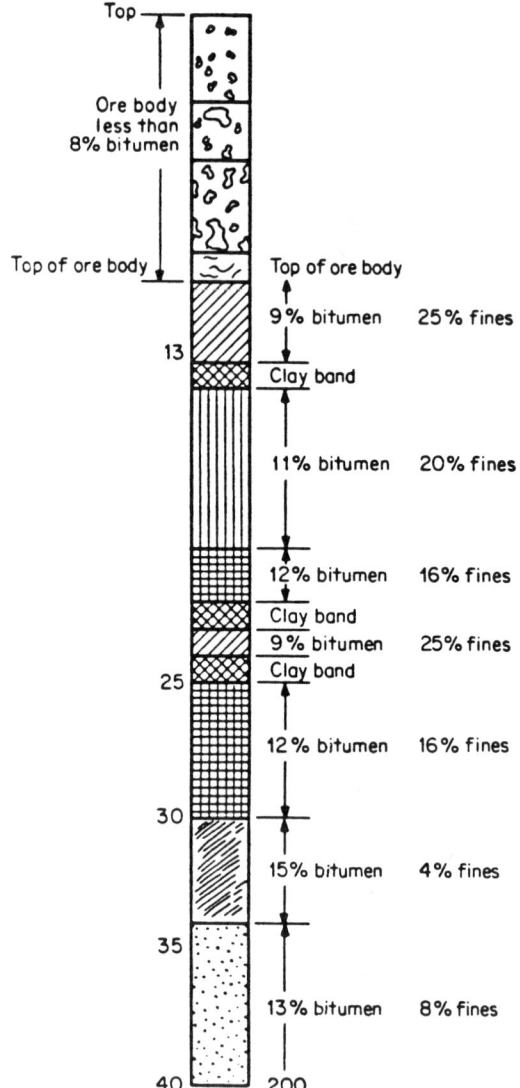

Figure 2. A typical cross-section of a core hole from the Mildred Lake area of the Athabasca oil-sands deposit. (Reproduced from reference 35. Copyright 1984 McGraw-Hill.)

tible to buildups in solids concentration that can increase the viscosity (22, 36, 37). Viscosity also is controlled by the regulation of flood water addition and middlings withdrawal rates.

At the surface, the rising aerated bitumen droplets gather to form a froth layer. A moderate depth of froth is necessary to allow some back-

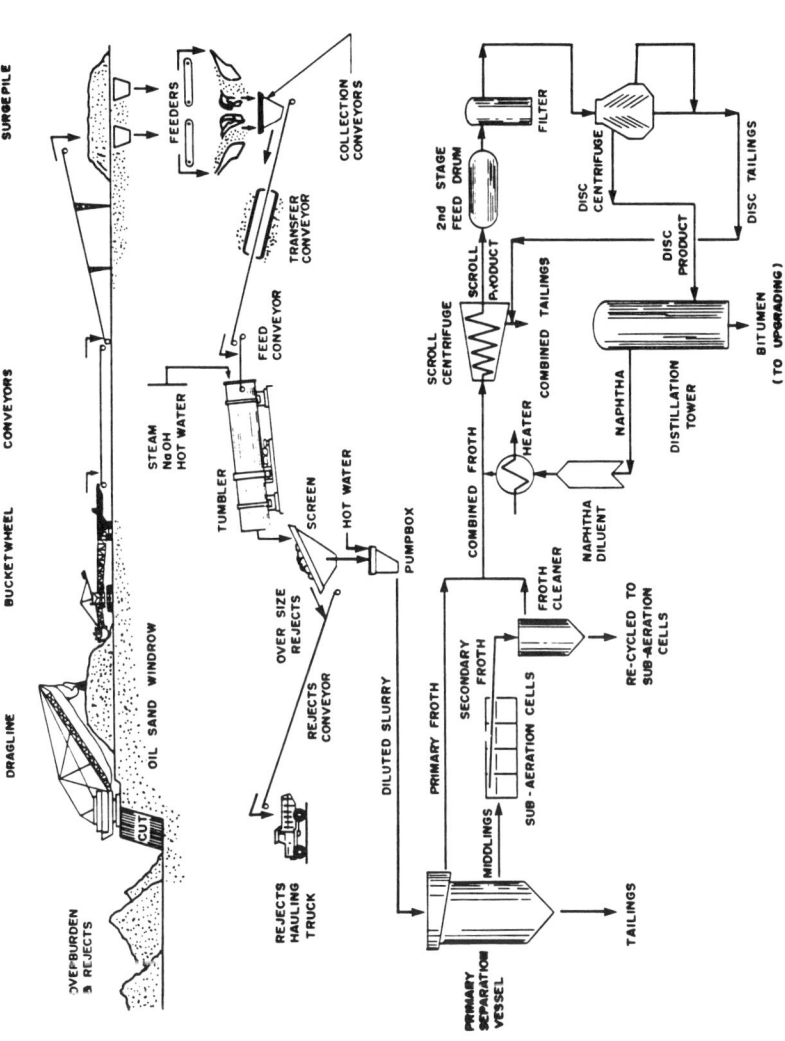

Figure 3. Diagram of a commercial oil-sands mining and hot-water flotation process. (Reproduced from reference 44. Copyright 1980.)

drainage to take place with release of nonfloated particles that have been unavoidably entrained to some extent between the droplets. The froth layer is removed mechanically from the surface and flows away in a launder. A good quality froth might consist of 66% bitumen, 9% solids, and 25% water.

The smaller suspended solids, which do not settle rapidly, and the smaller and poorly aerated bitumen droplets, which do not float rapidly, are all drawn off in a slurry from the middle of the vessel (middlings). The bitumen droplets in middlings have either too little air content or have diameters that are too small for rapid enough flotation. The middlings stream contains enough bitumen that it is pumped to a secondary (scavenging) flotation circuit for additional bitumen flotation. Conventional flotation cells that employ vigorous agitation and air sparging are used to cause further bitumen aeration and flotation. This step results in a secondary froth that is much more contaminated with water and solids than is the primary froth and typically contains 24% bitumen, 17% solids, and 59% water. Other variations of this process are also practiced. For example, in one commercial operation the middlings stream and primary tailings stream are combined and pumped to a special tailings oil recovery (TOR) flotation circuit. The middlings from this TOR vessel are then pumped to scavenging circuits, as discussed in the text, and the TOR froth is recycled into the flooded slurry that is fed into the primary separation vessels. The TOR tailings are combined with the tailings from the scavenging circuits.

The coarse rapidly settling solids in the primary separation vessel are kept in motion by mechanical rakes at the bottom of the vessel and are drawn off from the bottom as a slurry (primary tailings). The tailings from the primary and secondary flotation processes are combined and transported to a tailings pond. Some of the supernatant water from the pond can be recycled into the process; the remaining fine solids and water undergo a slow consolidation into sludge.

The primary froth contains air and is therefore compressible and more viscous than bitumen (38). To make it easier to pump, it is deaerated in towers by causing it to cascade over a series of cones, flowing against the upward flow of steam. The froth from the secondary flotation is "cleaned" in stirred thickeners to remove some of the water and solids, and then deaerated. After deaeration, the primary and secondary froths are combined into a single feed for a froth treatment process. This deaerated froth contains about 65% oil, 25% water, and 10% solids.

A froth treatment process is used to remove water and fine solids from the froth (8, 9, 30, 32, 35). The froth is first diluted with heated naphtha in about 1:1 volume ratio. The diluted froth is then centrifuged in scroll centrifuges (at about 350 × g) to remove coarse solids (greater then 44 μm). The froth is then filtered and pumped into disc-nozzle centrifuges. Here, higher gravity (G-force) (about 2500 × g) is used to remove essentially all of the remaining solids and most of the water. The diluted bitumen product is then ready for stripping of the naphtha and for upgrading into synthetic crude oil.

The Production of Froth. In the conditioning step, bitumen becomes separated from the sand and mineral particles because of a combination of the effects of mechanical shear and disjoining pressure (Figure 4a). The disjoining pressure in the aqueous films that separate the bitumen and solids is increased by increasing electrostatic repulsion caused by a combination of ionizing surface functional groups and the adsorption of anionic surfactant molecules at the interfaces (39). Next, bitumen–air attachment occurs, but this mechanism is not entirely clear. The process conditions that most favor bitumen–solids separation, that is, a high degree of electrostatic repulsion due to charged surfactant molecules at the interfaces, also will act in opposition to gas–bitumen attachment because the gas bubbles also acquire a surface charge of the same sign (40).

On the other hand, in mineral flotation, where there is gas–solid attachment without filming, such electrostatic repulsion is not as important a factor as are the inertia effects when the particles and droplets are larger than 10–40-μm diameter. It is possible for bitumen droplets to attach to gas bubbles and form bubble droplet pairs or aggregates, as in mineral flotation, but for the most part a balance of interfacial tensions in the system favors filming of the bitumen around the gas bubbles. Figure 4b and the photographs in Figure 5 show the spontaneous filming of bitumen around a gas bubble brought into contact with the solution–bitumen interface. These aerated bitumen globules are what float upwards in the flotation vessels to form froth.

Gas bubble–mineral particle attachment also occurs for the fraction of mineral particles that are not hydrophilic, so that some bubble–particle aggregates also float upwards and become incorporated in the froth. An important goal of the flotation process is to produce bituminous froth without entraining large amounts of solids. The entrained water and froth are later removed only with considerable difficulty, so the question of how to produce higher quality froth is of interest. Because of the higher quality, primary flotation froth does not need the difficult cleaning and is more highly valued than secondary froth. Thus, much flotation process optimization is directed at optimizing the primary froth yield in the process. This is why so many oil-sands patents are directed principally at the primary recovery aspect of the process (e.g., ref. 32). As a practical matter, the floated solids do become part of the nature of the froths and will be discussed further in this work.

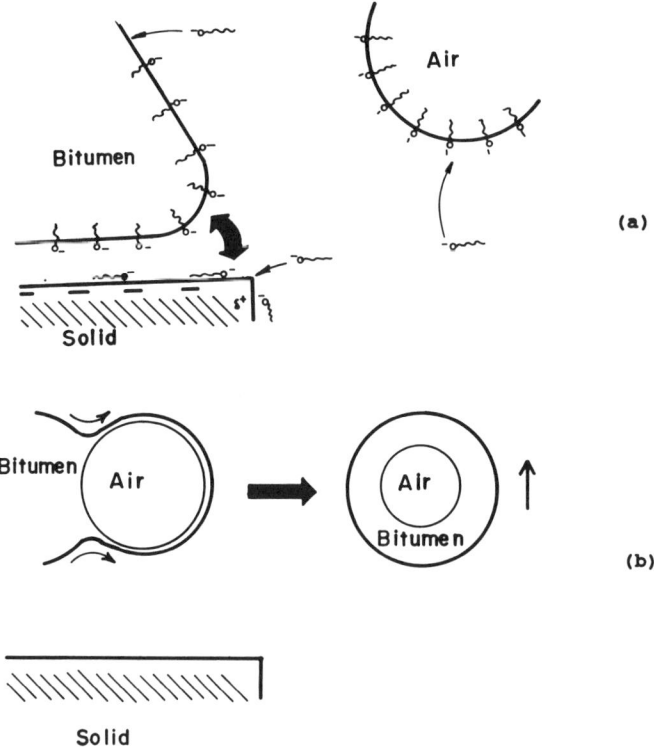

Figure 4. Illustration of two of the steps in the hot-water flotation process: a, the separation of bitumen from solids showing the adsorption of naturally produced surfactants; and b, the attachment and filming of bitumen around gas bubbles.

Physical Properties of Froths

Composition and Density. Density is of fundamental importance to any flotation process. The density of bitumen is much less variable and less dependent on temperature than on its viscosity. Thus, the density data reported by Ward and Clark (*41*) and Camp (*22*) shows slight temperature dependence for Athabasca bitumens. However, the density of froth is strongly influenced by the gas phase and also the entrained solid particles. Primary froth typically has a composition of 66% bitumen, 25% water, and 9% solids, and secondary froth typically contains 24% bitumen, 59% water, and 17% solids. Table II shows a series of freshly formed primary froth densities that vary from 0.7 to 0.9 g/cm^3 at about 70 °C (the experimental conditions are described later).

Figure 5. Photographic sequence in which an air bubble, on the tip of a capillary, is pushed down through an alkaline solution, at 80 °C, until it just touches a layer of bitumen that had been coated onto a silica surface. The bitumen spontaneously spreads over the surface of the bubble causing it to detach from the capillary and become engulfed. The presence of solid particles in the bitumen on the surface can be observed in the lower photo. (Photomicrographs by L. L. Schramm.)

Rheology. The viscosity of Athabasca bitumen in place is so high (about 1,000,000 mPa·s at reservoir temperature) that the bitumen in oil sand is practically immobile. This feature makes the bitumen difficult to displace in attempts at in situ recovery (18, 42). When mined and beneficiated, bitumen viscosity is reduced but still has some important influences including a contribution to froth viscosity.

Several studies of the rheological properties of Athabasca bitumen have been reported (38, 41, 43). For all practical purposes, Athabasca bitumen can be considered to be Newtonian in character except at quite low temperatures and very low shear rates (38, 43). In an early study of bitumen viscosity, Ward and Clark (41) found that the values vary with the type (location) of the host oil sand, the method by which the bitumen was extracted for study, and with temperature. Other studies (43–46) concentrated on the influence of gas saturation and pressure. For bitumen in the

Table II. Viscosities, Densities, and Compositions of Some Fresh Primary Froths by On-Line Measurements

Froth Composition (mass%)			Froth Viscosity (mPa·s at 56 s^{-1} shear rate)	Froth Density (g/cm^3)	Temperature (°C)
Oil	Water	Solids			
62.4	27.0	10.4	1.9×10^3	0.89	69
63.4	31.8	4.2	2.2×10^3	0.84	70
70.3	24.8	4.0	1.7×10^3	0.81	70
66.3	28.1	4.3	1.7×10^3	0.81	70
61.3	27.9	11.2	2.5×10^3	0.91	70
57.2	34.3	8.4	1.7×10^3	0.85	71
65.2	21.7	12.2	3.5×10^3	0.81	60
61.1	27.8	9.8	3.1×10^3	0.78	67
64.8	23.6	9.8	2.4×10^3	0.80	67
62.0	22.5	13.9	2.3×10^3	0.84	66
62.4	25.8	10.4	4.0×10^3	0.74	60
65.8	20.5	11.5	2.9×10^3	0.82	61
62.3	27.1	9.9	1.3×10^3	0.92	79
58.1	33.4	8.2	6.9×10^3	0.70	65
67.8	21.0	9.6	4.3×10^3	0.75	65
63.5	28.4	6.1	1.1×10^3	0.79	74
65.3	27.6	5.3	2.1×10^3	0.87	64
62.8	30.6	5.5	1.3×10^3	0.89	74
60.8	31.8	4.7	4.7×10^3	0.89	64

Athabasca deposit, the temperature dependence of the viscosity is shown in Figure 6. The Athabasca bitumen rheology is not typical of that in other world locations. For example, the rheology of bitumen in the Utah deposits, which have much higher effective viscosities (48, 49), is more complex and exhibits markedly non-Newtonian behavior (47).

Very little information about the rheological properties of froths has been published because the froths are not very stable and collapse with time. Also, great care is needed because froths, which are a multiple dispersion of four phases, are very subject to phase rearrangements during the conduct of experimental measurements. Some approximate froth viscosities for commercial-scale froth production mentioned in the patent literature are

- Suncor, 7.5×10^3 mPa·s at 66 °C (50)
- Syncrude, 1.5×10^3 mPa·s at 71 °C (51)

Some measurements have been reported (38) for froths produced using a standard lab-scale hot-water process simulation test, known as the batch extraction unit (BEU), shown in Figure 7 (52). The batch extraction unit described elsewhere (53) simulates the commercial hot-water flo-

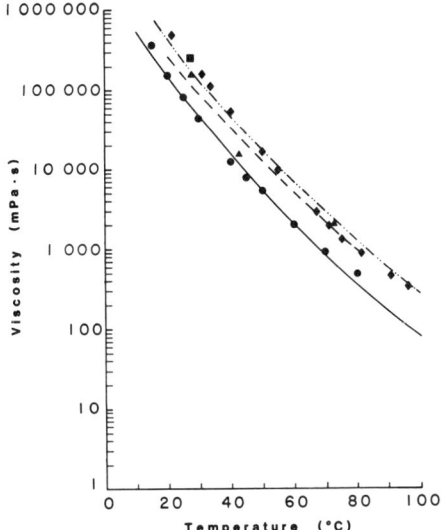

Figure 6. The viscosities of various Athabasca bitumens at different temperatures. (Reproduced from reference 44. Copyright 1980.)

tation process. Sanford (*54*) and others (*55–57*) showed that the batch test results establish trends that translate directly to the pilot scale.

From the batch extraction froths it was concluded that (*38*):

- Froths are non-Newtonian (shear-thinning) when freshly produced but collapse upon standing to eventually become Newtonian.
- Froth stability and also froth rheology exhibit varying time-dependencies, some have quite remarkable stability (no change in viscosity over 20 to 40 h), and some have marked instability (50% reduction in viscosity within 20 h).
- Fresh froth viscosities can be quite high compared to that of bitumen at the same temperature, primary froth having a viscosity of typically 1.8 to 3.0 × 10^3 mPa·s at about 80 °C (shear rate of 58 s^{-1}), compared to a bitumen viscosity of 495 mPa·s under the same conditions.
- The time dependence was surmised to be due to the escape of dispersed air as the froth collapsed.

On the basis of the batch tests, little or no rheological difference could be discerned between primary and secondary froths.

In an attempt to obtain more reproducible, more representative data for freshly produced froth, some on-line measurements for continuously produced primary froth have also been made. The processing was carried

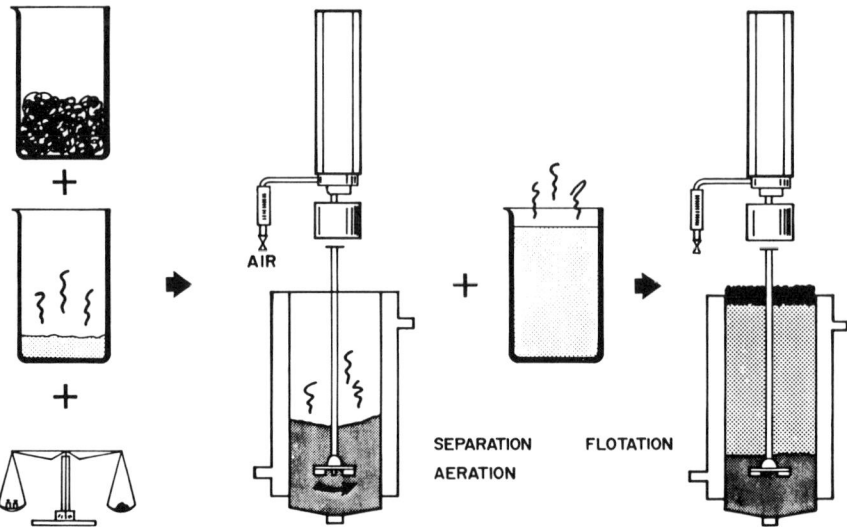

Figure 7. Basic steps involved in the standard batch hot-water flotation (batch extraction) test. (Reproduced from reference 52. Copyright 1989 Alberta Oil Sands Technology and Research Authority.)

out in a 2270-kg/h continuous pilot unit operated by Syncrude Canada Ltd., which is a scaled-down version of the commercial oil-sands plant (*30, 58*). The processed oil sand was obtained from the Syncrude Mildred Lake operation and found to be of marine depositional origin and contained 9.1% (mass) bitumen, 9.0% water, and 81.8% solids. Approximately 29.4% of the mass of solids was fines (defined as the percentage of solids less than 44 μm in diameter). The conditioning and flotation process flow sequences are basically those shown in Figure 3 (except for the mining and froth treatment segments). The circuit was operated under various conditions with respect to process control variables such as water and alkaline agent additions. During the process operations, samples of froth were periodically withdrawn; bulk densities were determined; and then the samples were assayed for bitumen, water, and solids content by standard methods (*59*).

Rheological measurements were made using a Haake Rotovisco rheometer (model RV12, Haake Inc., Saddle Brook, NJ) with concentric cylinder sensor system T-II and the model M500 measuring drive unit. A model PG142 programmer (Haake) was used to provide linearly increasing and decreasing shear rates. The unit was mounted at the top of the primary flotation vessel as illustrated in Figure 8. When the extraction circuit was operating at steady state, the measuring head was lowered so that the flow-through sensor became completely immersed in the froth layer.

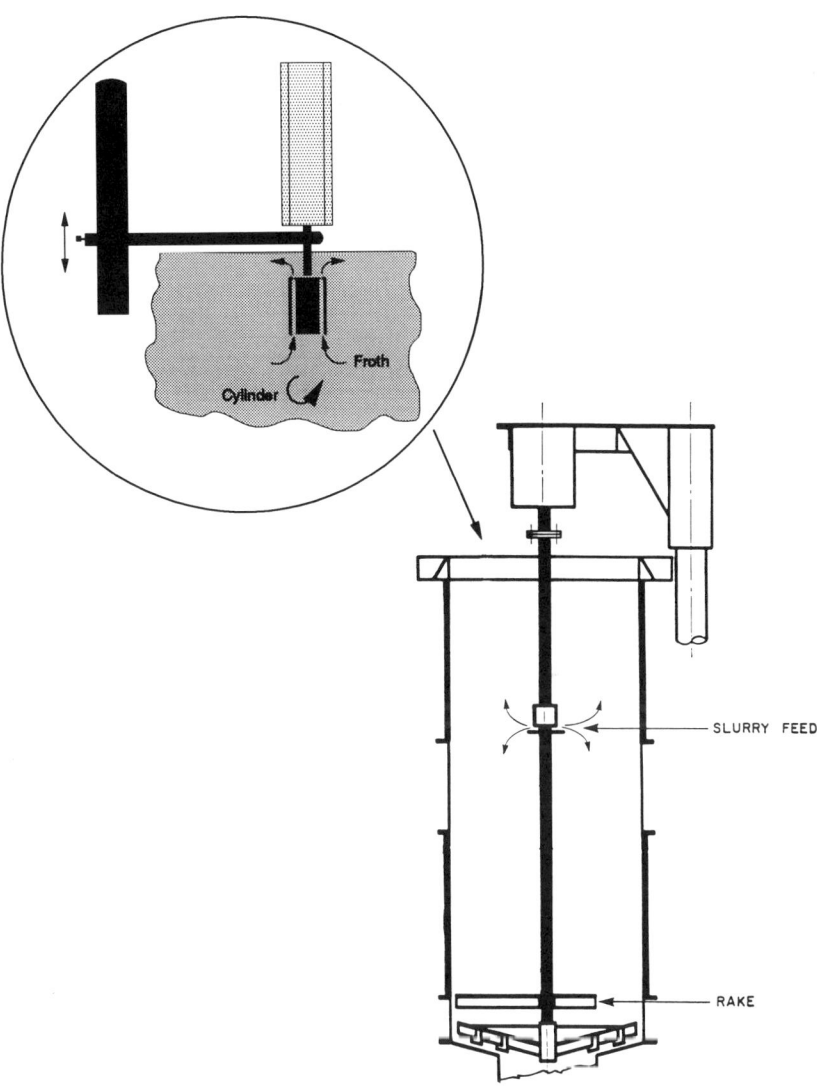

Figure 8. Drawing of the primary separation vessel from a 2270-kg/h continuous pilot plant, showing the mounting location for the froth rheometer. Inset: the rheometer's sensor system, including flow-through concentric cylinders.

The froth level was continuously checked to ensure that the sensor was always completely immersed in froth. Between measurements, the sensor was raised and reimmersed at different positions into the froth to provide fresh samples in each instance. The froth temperatures were measured directly at the same time as each measurement.

A typical rheogram for fresh primary froth is shown in Figure 9. As can be seen the froth is shear-thinning and thixotropic. At shear rates higher than about 100 s^{-1}, the apparent nature of the froth changed and yielded poorly reproducible but lower viscosities. This effect may have been due to a slip in the sensor system as the original viscosities were restored when the shear rate was reduced. Some viscosity results are shown in Table II, together with corresponding density and composition data. Whereas bitumen itself has a viscosity of about 930 mPa·s at 70 °C, the primary froths exhibited viscosities in the range 1700–4700 mPa·s at 58 s^{-1} at the same temperature, a factor of 2–5 times higher.

These results correspond to froths that have combined water and solids contents as high as 43%. At some higher water concentration, phase inversion would occur and cause a marked reduction in viscosity (60). At the temperatures studied (60–80 °C), froth viscosity is a strong inverse function of density because of different degrees of dispersed air content and also a strong inverse function of temperature. The strong influence of gas content is illustrated in Figure 10, where a strong influence of bulk density on viscosity is apparent.

Froth Structures

Froth Formation Models. The froth recovered from the primary separation vessel is a complex mixture of bitumen, water, inorganic solids,

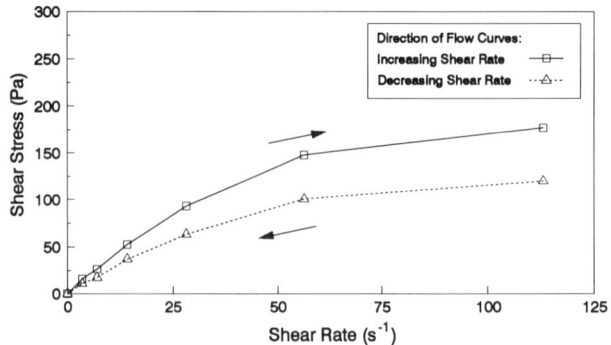

Figure 9. Flow curve for primary froth at 70 °C, measured in situ in the continuous pilot plant flotation vessel.

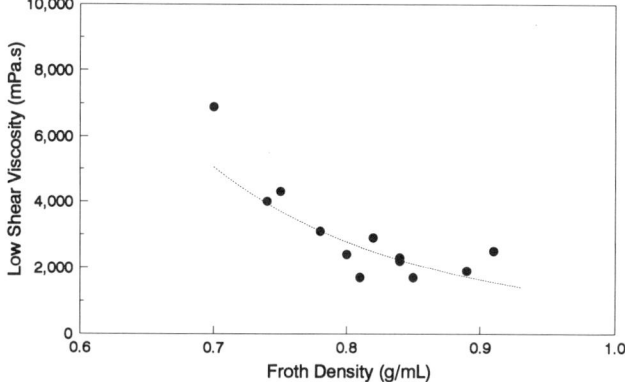

Figure 10. Bulk density dependence of low-shear froth viscosity. The viscosities were measured at a shear rate of 56 s^{-1}, in situ, for froths at temperatures of between 65 and 70 °C in the continuous pilot plant flotation vessel.

and air. The quality of the product is determined by the relative amounts of water and solids present in the material and the ease by which these constituents can be separated from the froth in downstream operations.

The mechanisms by which water and solids become associated with the froth is an issue of considerable debate. It is generally contended that most of the water associated with froth is due to the entrainment of middlings as the froth layer is formed in the primary separation vessel. The premise implies that froth water content should be related to the packing density of the aerated bitumen droplets as they collect at the interface (a function of droplet size distribution, deformability, froth residence time, and other aspects affecting the void fraction), and the thickness of the bitumen film enveloping individual air bubbles (a factor that defines the bitumen droplet density and determines the mass ratio of bitumen to interstitial middlings). Fine solids dispersed within the interstitial middlings phase (water-wet clays and silica) can account for some of the mineral matter present in the froth. Oil-sand minerals that are intrinsically hydrophobic and thus preferentially floated with the bitumen (heavy minerals, coal, and oil-wet clays) constitute another source of froth solids. Particle size measurements on solids in froth tend to display a bimodal distribution that serves to differentiate hydrophobic and hydrophilic minerals. Oil-wet particles are typically coarser than the solids suspended in the middlings directly below the froth interface. The particle size distribution of solids in the aqueous phase of froth agrees well with that predicted from settling velocities given the bulk residence time in the primary separation vessel.

The packing density of aerated bitumen droplets at the froth interface is a function of droplet size distribution, the flux of material impinging on the boundary, and the bitumen droplet coalescence rate. These three variables govern the time available for water drainage and the extent to which droplets can orient themselves relative to one another.

Under normal vessel loading conditions, the rate of bitumen coalescence is slow relative to the rate at which bitumen droplets collect at the interface. As a consequence, much of the occluded water cannot drain. Paths for drainage become exceedingly tortuous as the froth interface is continually replenished with aerated bitumen droplets.

Experimental studies undertaken by Schutte (61) to examine the partitioning of polar constituents of the bitumen to various process streams in the extraction circuit indicated that froth is enriched in asphaltenes. Schutte postulated that asphaltenes play a key role in promoting air attachment to bitumen. Further observations also indicated that the water and solids content of froth is directly related to the asphaltene content of the bitumen. Findings were rationalized in terms of the effect of asphaltenes on the rate of coalescence of aerated bitumen droplets at a froth interface. Schutte believed that rapid coalescence would inhibit the entrainment of water and solids and thereby favor good product quality. Experimentation with a model system using paraffin oil and process water indicated that the presence of asphaltenes substantially increases the time required for coalescence.

Droplet diameters and size distribution are related to slurry conditioning variables such as the nature of the oil sand and the chemicals used to process the ore. Houlihan (62) made extensive measurements of bitumen droplet diameters and rise velocities in the primary separation vessel of a continuous pilot-scale extraction circuit. He observed that droplet size distributions for a given feed are relatively narrow. Data obtained showed that, when processing a low-grade oil sand, the mean diameter of the droplets was approximately half that for a high-grade oil sand. This result is illustrated in the histograms (62) shown in Figure 11. For both types of oil sand that are processed at standard operating conditions, the bitumen droplets that float are for the most part spherical and consist of a thin film of oil enveloping an air bubble.

At low alkali addition levels, the bitumen was attached to air bubbles as discrete particles. This mode of bitumen attachment is also seen at reduced temperatures and is akin to the type of mechanism encountered in mineral flotation. At more reasonable alkali addition levels however, the bitumen will encapsulate the air bubble. The size of the droplets increases with increasing dosage of alkali. Bitumen droplet size was found to be invariant with depth in the separation vessel, and this invariance implies that bitumen droplets do not coalesce as they rise through the middlings phase.

Calculations can be performed to show the expected relationships between froth quality, bitumen droplet aeration, and packing density for

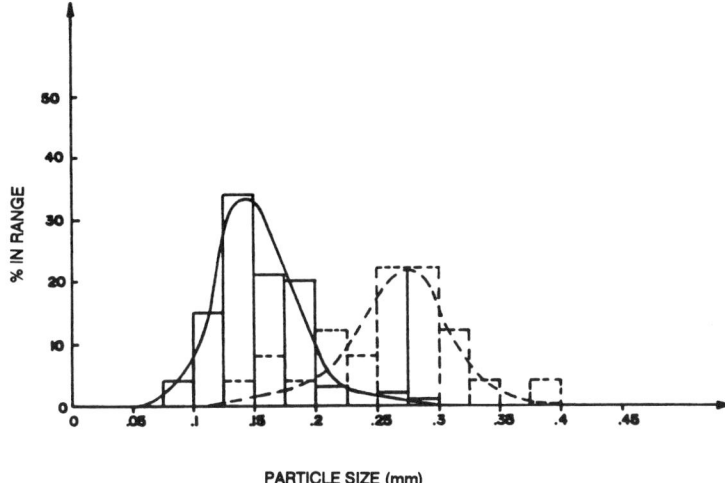

Figure 11. Size distributions of bitumen droplets formed when processing an average quality oil sand (broken curve) and a poor quality oil sand (solid curve). The particle sizes reported are for two-phase droplets consisting of a bitumen layer around a gas bubble. (Reproduced from reference 62, from Syncrude Canada Ltd.)

an ideal system of uniform droplet size. These calculations were conveniently undertaken by assuming the aerated bitumen droplets to be spherical and resistant to deformation. Further, droplets are of equal size and congregate in a hexagonal close-packed array with an occluded volume fraction of 0.74 (63). The interstitial fluid was assumed to be water. A density of 1 g/mL was used for both the water and bitumen. The mass of air within a droplet was considered negligible relative to the mass of bitumen.

A scaled representation of the thickness of bitumen layers corresponding to a range of droplet diameters and densities is provided in Figure 12. The general relationship between froth quality and aerated bitumen droplet density, which can be obtained for any collection of equal-sized droplets arranged in a hexagonal close-packed structure, is shown in Figure 13. Data indicate that froth with a high bitumen content is favored with increasing droplet density.

The linear correlation between aerated bitumen droplet density and bulk froth density for this ideal system is illustrated in Figure 14. The intercept of this line is equivalent to the assumed void fraction, and the slope represents the volume fraction of the material occupied by the aerated bitumen droplets. Experimental data for mean particle diameter and froth bulk density obtained under a variety of process conditions could

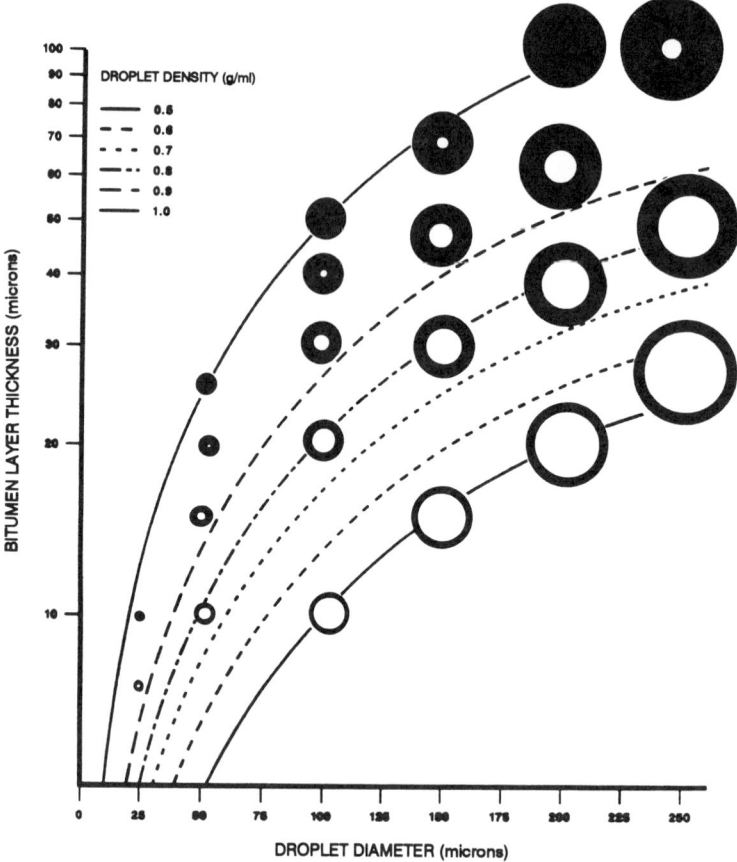

Figure 12. Illustration of the relative bitumen layer thicknesses of aerated bitumen droplets.

permit calculation of actual void fractions from the slope and intercept of a best-fit line.

Danielson (64) measured bitumen droplet densities and corresponding froth compositions for two oil sands of differing processibility. For one oil sand, the mode of the observed distribution of aerated bitumen droplet densities was 0.31 g/mL with the recovered froth found to contain 52.9% bitumen. For the other oil sand, the mode of the density distribution was 0.51 g/mL with a corresponding froth assay of 63.2% bitumen. Although these froth compositions are somewhat higher than those predicted from Figure 13, the trend toward improved froth quality with increasing droplet density is evident.

The comparison indicates that the selected packing density and as-

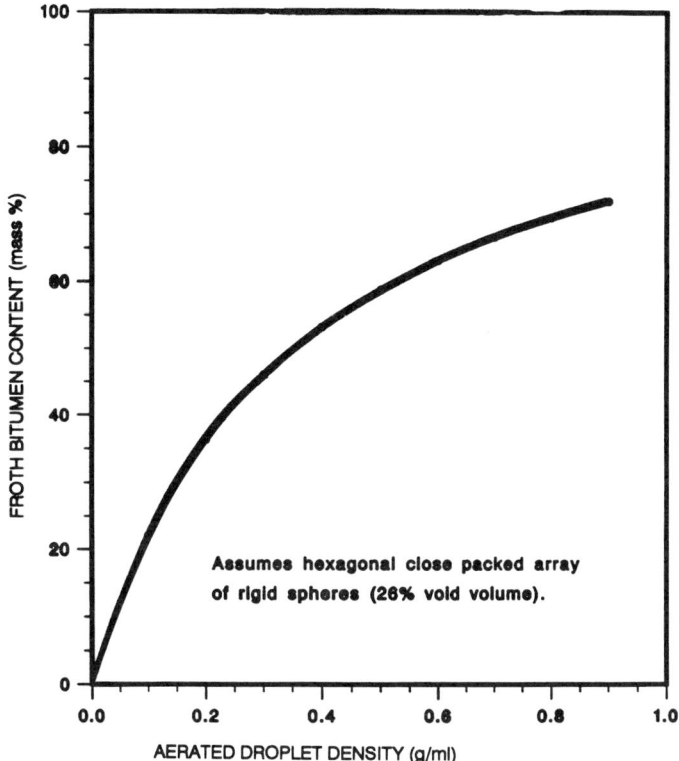

Figure 13. Relationship between froth quality and aerated bitumen droplet density.

sumption of uniform droplet size may be inadequate in representing the real system. Small-sized droplets that could fill the interstices of the larger droplets would give rise to increased bitumen concentration in the froth. A similar result is obtained upon increasing the packing density of the droplets through deformation and drainage of the interstitial fluid. The effect of varying packing density on froth bitumen content is shown in Figure 15.

Further refinements to this model of the froth formation process can be made by substituting a representative middlings composition for the interstitial fluid. In addition, the assay of bitumen films surrounding the air bubbles can be modified to account for the presence of hydrophobic solids and emulsified water.

When plotted on triangular coordinates, primary froth assays from commercial operation typically lie on a straight line and indicate an upper limit of 67.5% bitumen concentration. This observation adds credence to

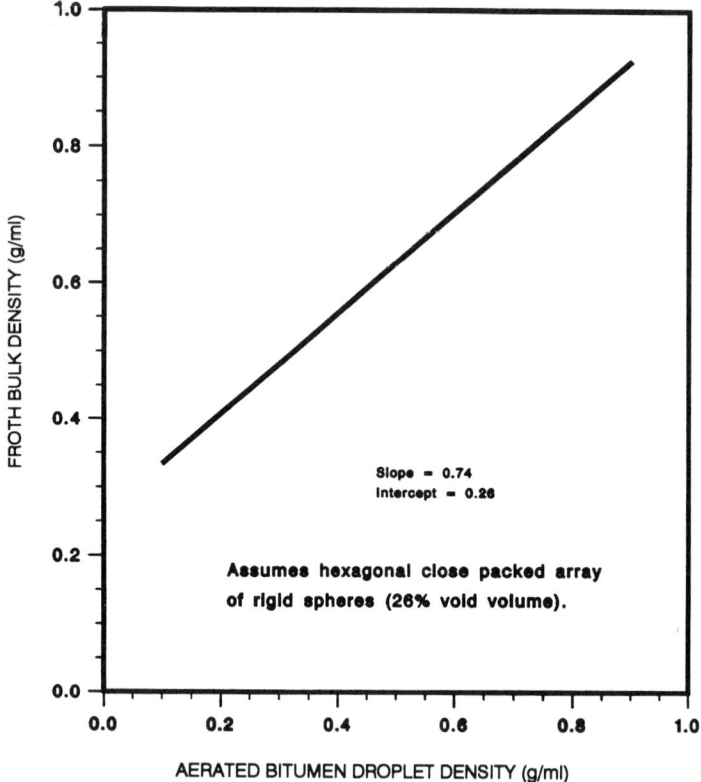

Figure 14. Relationship between bulk density and aerated bitumen droplet density.

the simple middlings entrainment model and provides some valuable insight regarding the nature and structure of froth. The data envelope, reproduced in Figure 16, suggest that froth can be considered a mixture of two components, one bitumen-rich and the other bitumen-lean. The composition of these two components is not precisely fixed. Results indicate however that the assays are restricted to the following linear ranges for bitumen to water to solids ratio:

- bitumen-rich component: from 91.5:0:8.5 to 67.5:24.5:8.0
- bitumen-lean component: from 45.0:47.0:8.0 to 0:92.5:7.5

The variability in froth composition reflects the complex nature of the slurry conditioning and phase-separation processes.

The concept that froth is a binary mixture containing bitumen-rich material and a largely aqueous phase (essentially middlings) was a step

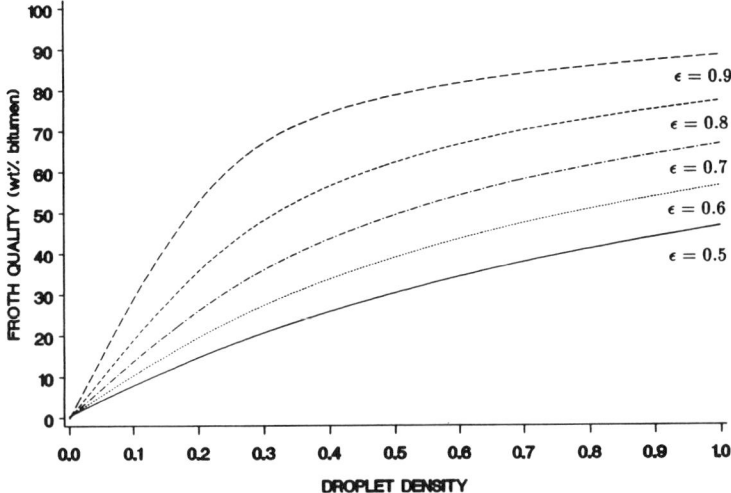

Figure 15. Froth quality as a function of packing density (ϵ) and aerated bitumen droplet density.

forward in describing the nature of froth. A more fundamental advance in understanding froth structure came from consideration of the volume fractions of the various components in froth prior to deaeration.

Viewing froth as a ternary mixture of air, middlings, and a bitumen-rich component with negligible water, Crickmore and Schutte (65) showed that at process temperatures of approximately 80 °C, the volume fraction of the bitumen-rich component, V_B, was essentially constant at 0.56. By difference, the sum of the volume fractions of the middlings and air, $V_M + V_A$, was 0.44. This finding is consistent with a physical picture of aerated froth as an expanded bed of dispersed spheres of the bitumen-rich component with the middlings and the air occupying the interstices. Data from froths formed at lower temperatures revealed that another distinct type of froth may be produced having a constant middlings volume fraction of 0.23 and consequently a volume fraction $V_B + V_A$ of 0.77.

Building on the assumptions that

- froth is composed of three constituents—a bitumen-rich component that contains some proportion of oil-wet solids, air, and middlings from the zone immediately beneath the froth layer

- froth is generated from droplets of the bitumen-rich component that have filmed over bubbles of air (these bitumen–air aggregates rise through the middlings continuous zone until the aggregates crowd together in the initial stages of the froth formation process)

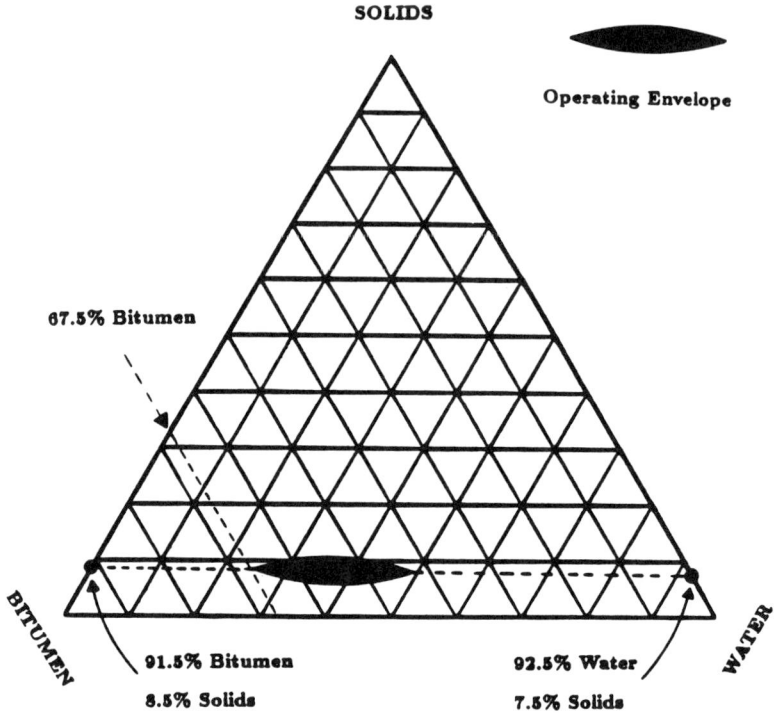

Figure 16. Bitumen, water, and solids content of froths produced from Syncrude's commercial hot-water flotation plant.

- there is no significant loss of either component in these aggregates to the surrounding middlings or ambient air

it is possible to describe a process whereby a dilute dispersion of aerated bitumen particles in the middlings is transformed into the material observed as froth.

If the ratio of the volume fraction of bitumen-rich component to that of air in the aggregates is denoted by G, then this value will be constant throughout the froth formation process. The only variable along the path from a dilute dispersion of aggregates to a froth is the relative volume fraction middlings. If G is greater than 2.67, the resultant froth is regarded as "stable" and has the component volume fractions:

$$V_B = 0.56 \quad V_A = \frac{0.56}{G} \quad V_M = 0.44 - \frac{0.56}{G}$$

If the aggregates have too much air (or conversely, insufficient bitumen-rich component) to yield a G value for a "stable" froth, the resultant material will have the component volume fractions:

$$V_B = \frac{0.77G}{(1+G)} \quad V_A = \frac{0.77}{(1+G)} \quad V_M = 0.23$$

A froth with this range of compositions is a "loose" froth typical of operation at temperatures below 80 °C. The evolution of a dilute dispersion of bitumen–air aggregates into either "stable" or "loose" froths is sketched in Figure 17. This triangular diagram, shows that G is the sole parameter that determines the path that a dispersion of aggregates undergoes in the formation of a froth. This diagram can also be used to predict the volumetric composition of froth when an insufficient residence time is allowed in the primary separation vessel. G will remain unchanged, but the composition will not reach the ultimate froth structure boundary formed by the lines $V_M = 0.23$ and $V_B = 0.56$.

The general model for froth formation can be amended to account for water dispersed in the bitumen-rich phase. Experimental investigations conducted by Chung et al. (66) indicate that as much as 16% of the water in the froth originates from slurry water used during oil-sand digestion. This water is present as an emulsion formed under the turbulent conditions experienced in the tumbler, pumps, and pipelines upstream of the primary separation vessel. The condensation of steam in contact with bitumen during the digestion step may also be a contributing factor.

Microscopic and Magnetic Resonance Imaging Studies. On the basis of microscopic studies, Swanson (67) proposed a model of froth that comprises a continuous phase of bitumen in which large water droplets are emulsified. The emulsified water is contained within "membranous-sacs" and is rich in submicrometer-sized mineral particles (Figure 18). The membranous interfacial film was probably formed from solids (mostly clays) that have an organic coating, which makes them oil-wetting or biwetted. As illustrated in Figure 18, the bitumen-coated gas bubbles rise to the surface of the flotation vessel where the bubbles collapse and yield the water droplets dispersed in bitumen. Similarly, such species coalesce within the froth layer as part of the emulsion inversion process to produce a bitumen continuous phase that might be expected to yield gas bubbles that have diameters just smaller than the original globules and also water droplets formed from the original thin films, whose droplets should be much smaller. In addition, there will be larger-sized water droplets that are entrained into the froth layer. These droplets are termed free water.

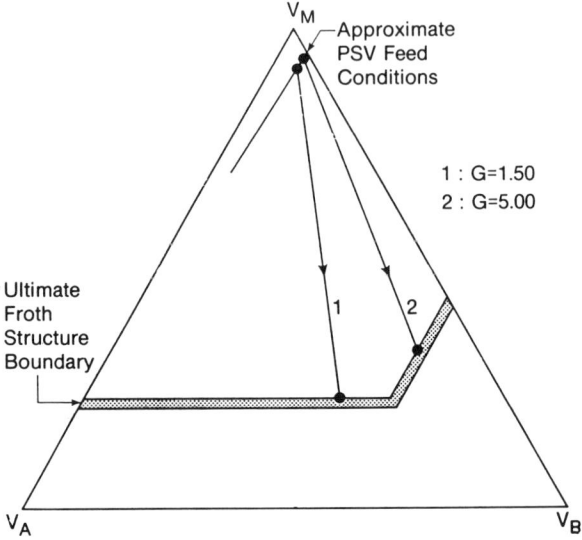

Figure 17. The structure of primary froth and the role of the variable G. G = 2.67 corresponds to the intersection of the two lines given by V_B = 0.56 and V_M = 0.23.

THE PROPOSED MECHANISM OF WATER-SAC FORMATION

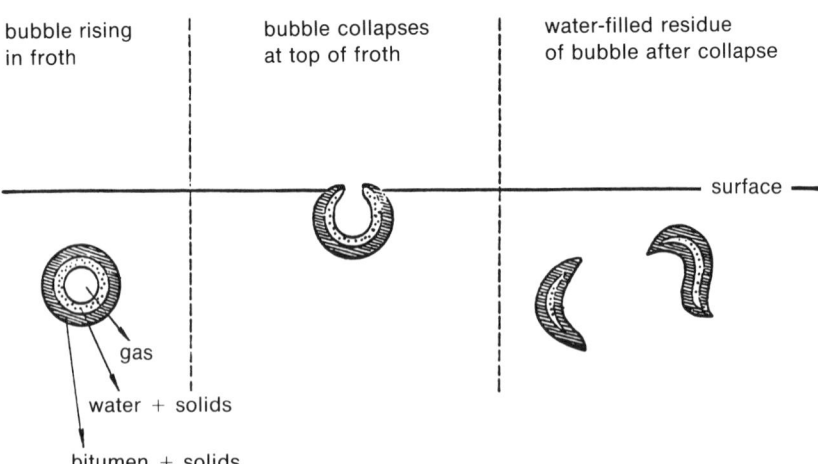

Figure 18. A mechanism of membranous emulsified water formation in froth, based on microscopic examinations (67).

This picture is consistent with the available results for dispersed gas bubble and water droplet sizes. Chung et al. (66) measured emulsified water droplet sizes in froth samples obtained from the previously described 2270-kg/h continuous pilot plant. After removing the large-size free water droplets, a second, smaller size fraction of water droplets was observed, which ranged up to about 18 μm. Houlihan (62) conducted a number of direct photomicrographic studies of rising bitumen droplets in the same 2270-kg/h continuous pilot plant. Under good processing conditions, the rising droplets were observed to be approximately spherical and had diameters of about 280 ± 140 μm (total range from 125 to 400 μm). Various other similar studies have always yielded droplet size ranges within the bounds 100 to 600 μm. The thickness of the bitumen film has been estimated to be about 30 μm, so that the droplet diameters of 100–600 μm can also be taken as a first estimate of the gas bubble sizes to be expected in froth. The case of Houlihan's measured mean droplet size of 280 ± 140 μm would then translate into an expectation of bubble sizes of about 216 ± 140 μm.

To probe further into the arrangement of the phases of froth nuclear magnetic resonance imaging, MRI, was used. In MRI, a sample to be imaged is placed in a homogeneous static high-magnetic field with a superimposed, time-dependent, linear gradient magnetic field. In this way, the total magnetic field strength depends on position in the sample. Resonance is induced with a radio frequency, just like in conventional NMR spectroscopy. In the imaging system, the position-dependent resonance frequencies and signal intensities allow the determination of the concentration, chemical environment, and position of any NMR-active nuclei in the sample. The instrument used was a Biospec-X 24/30 imaging system (Bruker Spectrospin Canada Ltd.) operating at 2.35 T, with a superconducting solenoid having a bore diameter of 30 cm.

Freshly produced froth samples for imaging were obtained from the previously described 2270-kg/h continuous pilot plant. Again, a good grade of oil sand was obtained from the Syncrude Mildred Lake operation, and the circuit was operated under good processing conditions so as to provide a thick and easily sampled layer of primary froth. During the process operation, samples of froth were obtained from different depths in the froth by inserting a brass cylinder into the froth layer, allowing froth to flow around it for a period of time, and then quickly filling the cylinder with liquid nitrogen. This procedure caused a layer of froth adjacent to the outer cylinder wall to freeze very quickly. The cylinder was then withdrawn, the frozen froth was chipped off, and the frozen froth samples were kept stored and transported in liquid nitrogen.

The frozen samples were placed in the imaging system and allowed to thaw. Starting when frozen, then immediately upon thawing of both water and oil phases, and again at intervals thereafter, MRI images were taken for a horizontal slice through the samples of 3.0-mm thickness. Figures 19–21 show MRI images for samples taken at the top, middle, and bottom of the froth layer, respectively. In all these images, the oil phase appears black and is clearly the continuous phase. The images marked "a" in these figures show the water droplets as white, and the images marked "b" show the gas bubbles as white. Table III shows estimates for the relative phase saturations for each sample as determined from image intensity histograms. Estimates for the mean droplet and bubble diameters are also shown.

The gas content is somewhat higher in the top of the froth layer, and the water content is somewhat higher toward the bottom, as would be expected. The high viscosity of the bitumen apparently prevents rapid

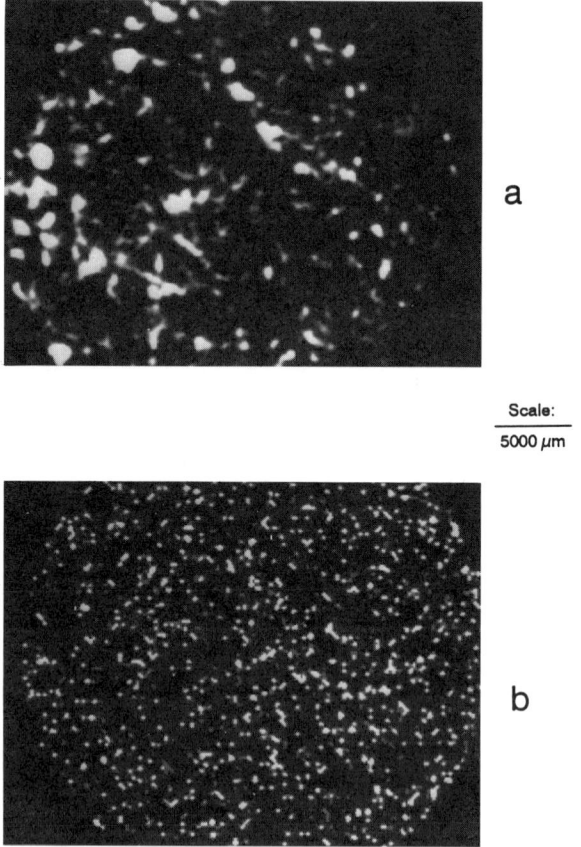

Figure 19. NMR images of froth sampled from the top of the froth layer in the continuous pilot plant flotation vessel. The images have been reconstructed to highlight the water droplets (a) and the gas bubbles (b).

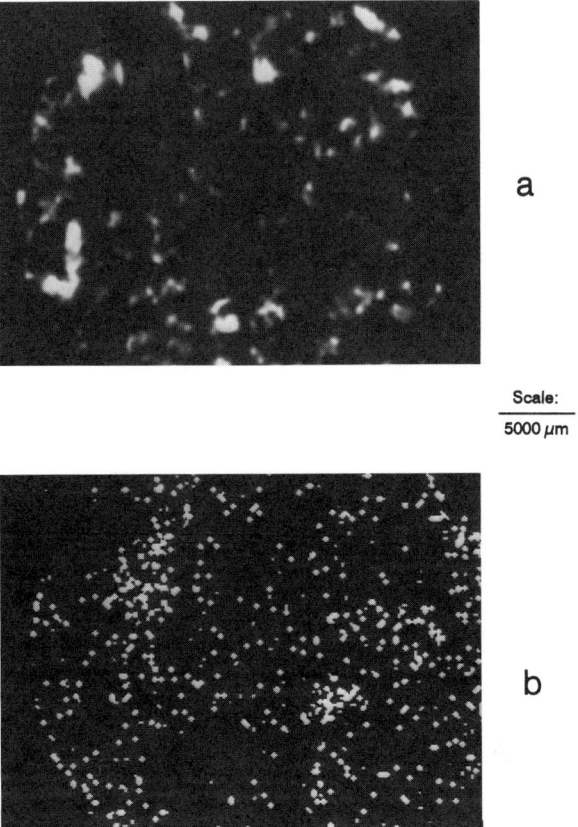

Figure 20. NMR images of froth sampled from the middle of the froth layer in the continuous pilot plant flotation vessel. The images have been reconstructed to highlight the water droplets (a) and the gas bubbles (b).

separation of the phases. A reference sample was obtained from the top of the froth layer at the same time as the other samples and was not frozen but rather kept at ambient temperature and then imaged at the same time as the frozen–thawed samples. Its oil–water–gas content was determined to be 71.0:18.6:10.4 volume %, very similar to the values reported in Table III.

As a further test, the thawed samples were retained at ambient laboratory temperature and reimaged after 26 days. Figure 22 shows MRI images for slices from the top, middle, and bottom froth layer samples. The water droplets appear as white, and the oil phase (and gas bubbles) form the black background. Table III shows that the water contents did not change significantly. The images in Figure 22 show that in the top

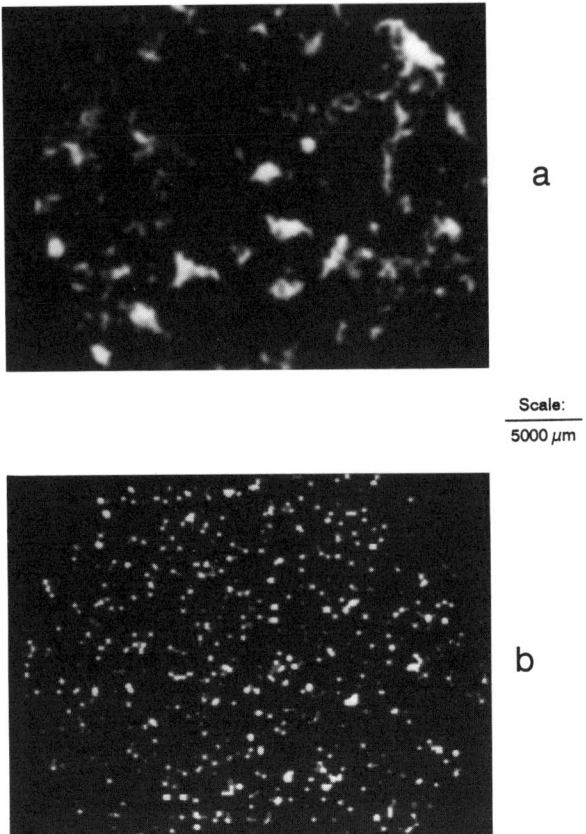

Figure 21. NMR images of froth sampled from the bottom of the froth layer in the continuous pilot plant flotation vessel. The images have been reconstructed to highlight the water droplets (a) and the gas bubbles (b).

Table III. Phase Saturations and Mean Droplet and Bubble Diameters for Froth Samples as Determined by Magnetic Resonance Imaging

Location in Froth Layer	Froth Composition (volume%)			Composition after Aging 26 Days (vol. % water)	Mean Water Droplet Diameters (μm)	Mean Gas Bubble Diameters (μm)
	Oil	Water	Gas			
Top	72.1	17.9	10.0	19.5	600–800	200–500
Middle	67.3	23.4	9.3	23.0	600–800	200–500
Bottom	68.2	23.6	8.2	28.5	600–800	200–500

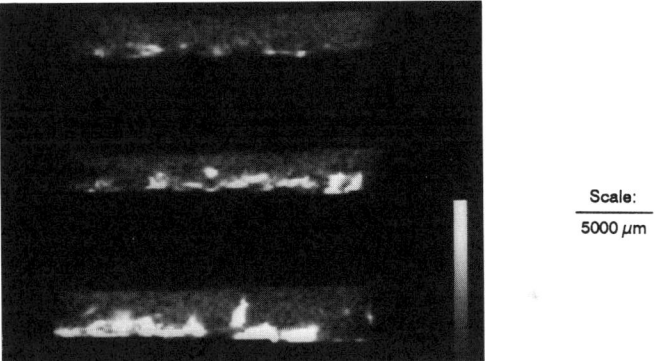

Figure 22. NMR images of froth sampled from the top, middle, and bottom of the froth layer in the continuous pilot plant flotation vessel. These images are from the same samples as in Figures 19–21, but after aging for 26 days at ambient temperature. The images have been reconstructed to highlight only the water droplets.

and middle samples some settling has occurred, but with little or no droplet size changes. In the bottom sample, there has been some settling and also aggregation of droplets, but again little or no coalescence of droplets has occurred. Separate determinations showed that over this period of time the gas contents of these samples did not significantly change either.

The total range of water droplet sizes observed in the froth samples was larger than suggested in Table III. Whereas the mean droplet sizes were consistently in the range 600–800 μm, some droplets as large as 1400 μm and as small as <100 μm were observed. That the smaller size droplets were observed qualitatively corresponds with the measurements made by Chung et al. (66) for "emulsified" water droplets that were up to about 18 μm. The larger droplets determined in the present work presumably correspond to the free water phase.

The range of gas bubble sizes observed in the froth samples also corresponds closely with the estimates made earlier based on Houlihan's measured rising droplet sizes, where froth gas bubble sizes of about 216 ± 140 μm were predicted. Table III shows that the mean bubble sizes fell in the range 200–500 μm. Here again, although the mean bubble sizes were consistently in that range, some bubbles as large as 1000 μm and as small as <100 μm were observed.

Froth Treatment

Deaerating. As shown, the primary and secondary froths comprise bitumen, air, water, and solids. The two most often cited properties of concern with regard to processing behavior are its high air content (compressible) and high viscosity (*50, 51*). To make them easier to pump the froths are deaerated as outlined previously. Despite the presence of appreciable amounts of water and solids, the deaeration process returns the viscosity of the froths to very nearly that of bitumen itself (*38*).

After they are deaerated, the primary and secondary froths are combined into a single stream for further treatment. The deaerated froth contains about 65% oil, 25% water, and 10% solids. It also contains emulsions. The microscopic studies of Swanson (*67*) showed that emulsified water droplets in froth persist through the deaeration process and were also found in naphtha-diluted deaerated froth. Emulsions of water in bitumen and of bitumen in water, both thought to be stabilized by asphaltenes and fine biwetted solids, have indeed been found in interface layer emulsions in enhanced gravity separators (*68*).

Dilution and Separating. The froth treatment process is used to remove water and fine solids from the deaerated froth. The froth is first diluted with heated naphtha in about a 1:1 volume ratio. The diluted froth is then centrifuged in scroll centrifuges (at about $350 \times g$) to remove coarse solids greater then 44 μm. The product from the scroll centrifuges is subsequently filtered and pumped to disc-nozzle centrifuges where higher G-force (about $2500 \times g$) is used to remove essentially all of the remaining solids and most of the water.

Centrifugation is quite efficient in separating hydrocarbon from water and solids, but it is an energy- and maintenance-intensive process. The application of lamella settlers has been advocated as a means of improving process performance (*69*). A lamella settler comprises a stack of parallel plates that are spaced apart from each other and inclined from the horizontal. The space between each set of plates forms a separate settling zone. The deaerated and diluted froth is pumped into these spaces at a point near the longitudinal middle of the plates. The relatively less dense diluted bitumen phase rises to the underside of the upper plates and flows to the tops of those plates. Meanwhile, the relatively more dense water and solid phases settle down to the upper side of the lower plates and flow to the bottoms of those plates. Thus, the diluted bitumen is collected at the tops of the plate stack, and the water and solids are collected at the bottom of the plate stack. Such an inclined lamella settling process is much more efficient than vertical gravity separation. However, with only the force of gravity available to drive the separation the rate of the separation process is quite sensitive to increases in fluid viscosity that reduce the rising–settling velocities and reduce the capacity of the continuous process vessels. Commercial plant performance tests have indicated that even for the higher G-force disc-nozzle centrifuges, oil losses to tailings are associated with higher viscosities in the feed.

The froth model described earlier, and shown in Figure 18, produces collapsed globules, composed of a water (and solids) droplet surrounded by a membranous layer made up of asphaltenes and biwetted solids. When such froth is contacted with naphtha, the time required to penetrate the bitumen membrane coating is on the order of 30 min, whereas in a commercial process the elapsed time between naphtha addition and introduction into a settling vessel is less than 1 min. Thus, the diluted froth process stream can contain these globules, probably in flocs, which would have a bulk density intermediate between diluted bitumen and water. Such flocs would then accumulate in the separation vessel and form an interface layer (sometimes called rag-layer) emulsion, and could potentially form an effective barrier to gravity separation (68).

In commercial practice, a staged series of lamella settlers is used in which the tailings stream from one vessel is treated and used to form the feed to the next vessel and so on (69). In field performance testing of such lamella settlers, highly emulsified samples have frequently been observed. Such emulsions are apparently composed of the water-in-oil globules, dispersed in water, that is, an emulsion of water in oil in water (W/O/W). These emulsions appear gel-like and exhibit extremely high viscosities at very low shear rates (as high as 200,000 mPa·s at 80 °C). However, the emulsions apparently invert so that their viscosities are dramatically reduced under moderate shear (at 80 °C to about 10–20 mPa·s). Figure 23 illustrates the effects of shear, including the shear-induced inversion from water-continuous (W/O/W) to oil-continuous (W/O). Depending upon how a separation vessel is operated, such emulsions could accumulate into an emulsified layer in the vessel and form an effective barrier to gravity separation.

According to the model advanced in this review, the emulsion structure-related problems encountered in the treatment of diluted froths have their origin in elements of the original froth structure.

List of Symbols

G	ratio of volume fraction of bitumen-rich component, V_R, to volume fraction of air, V_A, in froth
MRI	nuclear magnetic resonance imaging
V_A	volume fraction of air in froth
V_M	volume fraction of middlings phase in froth
V_B	volume fraction of bitumen-rich component in froth

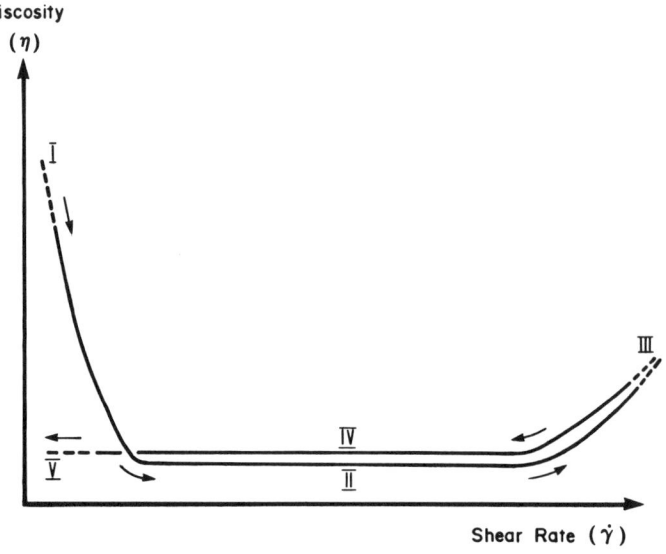

Figure 23. Illustration of changes taking place upon subjecting a diluted froth sample to progressively higher and then lower shear rates. The initial emulsion (Region I) is actually a water-in-oil-in-water multiple emulsion (68).

Acknowledgments

The authors thank Syncrude Canada Ltd. and the Petroleum Recovery Institute for permission to publish this work.

References

1. Martinez, A. R. In *The Future of Heavy Crude and Tar Sands;* Meyer, R. F.;

Wynn, J. C.; Olson, J. C., Eds.; United Nations Institute for Training and Research: New York, 1982; pp ixvii–ixviii.
2. Danyluk, M.; Galbraith, B.; Omana, R. In *The Future of Heavy Crude and Tar Sands;* Meyer, R. F.; Wynn, J. C.; Olson, J. C., Eds.; United Nations Institute for Training and Research: New York, 1982, pp 3–6.
3. Khayan, M. In *The Future of Heavy Crude and Tar Sands;* Meyer, R. F.; Wynn, J. C.; Olson, J. C., Eds.; United Nations Institute for Training and Research: New York, 1982; pp 7–11.
4. Walters, E. J. In *Oil Sands Fuel of the Future;* Hills, L. V., Ed.; Canadian Society of Petroleum Geologists: Calgary, Canada, 1974; Memoir 3, pp 240–263.
5. Demaison, G. J. In *The Oil Sands of Canada–Venezuela 1977;* Redford, D. A.; Winestock, A. G., Eds.; CIM Special Volume 17; Canadian Institute of Mining, Metallurgy, and Petroleum: Calgary, Canada, 1977; pp 9–16.
6. *Major Tar Sand and Heavy Oil Deposits of the United States;* Interstate Oil Compact Commission: Oklahoma City, OK, 1984.
7. *The Future of Heavy Crude Oils and Tar Sands;* Meyer, B. F. and Steele, C. T., Eds.; McGraw-Hill: New York, 1981.
8. Towson, D. In *Kirk–Othmer Encyclopedia of Chemical Technology,* 3rd Ed.; Wiley: New York, 1983; Vol. 22, pp 601–627.
9. Ruhl, W. *Tar (Extra Heavy) Sands and Oil Shales;* Ferdinand Enke: Stuttgart, Germany, 1982.
10. *Bitumens, Asphalts and Tar Sands;* Chilingarian, G. V.; Yen, T. F., Eds.; Elsevier: Amsterdam, The Netherlands, 1978.
11. *Oil Sands Fuel of the Future;* Hills, L. V., Ed.; Canadian Society of Petroleum Geologists: Calgary, Canada, 1974; Memoir 3.
12. *Guide to the Athabasca Oil Sands Area;* Carrigy, M. A.; Kramers, J. W., Eds.; Information Series 65; Research Council of Alberta: Edmonton, Canada, 1973.
13. Stewart, G. A.; MacCallum, G. T. *Athabasca Oil Sands Guide Book;* Canadian Society of Petroleum Geologists: Calgary, Canada, 1978.
14. Mossop, G. D. In *Facts and Principles of World Petroleum Occurrence;* Miall, A. D., Ed.; Canadian Society of Petroleum Geologists: Calgary, Canada, 1980; Memoir 6, pp 609–632.
15. Stewart, G. A. In *The K. A. Clark Volume: A Collection of Papers on the Athabasca Oil Sands;* Carrigy, M. A., Ed.; Alberta Research Council: Edmonton, Canada, 1963; pp 15–27.
16. Kendall, G. H. *Geology of the Heavy Oil Sands Symposium Proceedings— Alberta;* Geological and Mineralogical Associations of Canada: Department of Geology, University of Alberta, Edmonton, Canada, 1976.
17. O'Donnell, N. D.; Jodrey, J. M. *Proceedings of the SME of AIME Meeting;* Society of Mining Engineers of AIME: New York, 1984; preprint 84-421.
18. Mossop, G. D. *Science (Washington, D.C.)* **1980,** *207,* 145–152.
19. Vigrass, L. W. *Am. Assoc. Petrol. Geol. Bull.* **1968,** *52,* 1984.
20. Carrigy, M. A. *Proceedings of the 7th World Petroleum Congress;* Elsevier: Amsterdam, The Netherlands, 1967; Vol. 3, pp 573–581.
21. Carrigy, M. A. *Proceedings of the 5th World Petroleum Congress;* Fifth World Petroleum Congress, Inc.: New York, 1959; Vol. 1, p 575.

22. Camp, F. W. *Tar Sands of Alberta, Canada*, 3rd Ed.; Cameron Engineers Inc.: Denver, CO, 1976.
23. Carrigy, M. A. *J. Sediment. Petrol.* **1962**, *32*, 312.
24. Clark, K. A. *Trans. Can. Inst. Min. Metall.* **1944**, *47*, 257.
25. Ball. M. W. *Trans. Can. Inst. Min. Metall.* **1941**, *44*, 58.
26. Ball. M. W. *Bull. Am. Assoc. Petrol. Geol.* **1935**, *19*, 153.
27. Sobol, W. T.; Schreiner, L. J.; Miljkovic, L.; Marcondes-Helene, M. E.; Reeves, L. W.; Pintar, M. M. *Fuel* **1985**, *64*, 583.
28. Takamura, K. *Can. J. Chem. Eng.* **1982**, *60*, 538.
29. Zajic, J. E.; Cooper, D. G.; Marshall, J. A.; Gerson, D. F. *Fuel* **1981**, *60*, 619.
30. Schutte, R.; Ashworth, R. W. In *Ullmans Encyklopadie der Technischen Chemie;* Verlag Chemie GmbH: Weinheim, Germany, 1979; Vol. 17.
31. Innes, E. D.; Fear, J. V. D. *Proceedings of the 7th World Petroleum Congress;* Elsevier: Amsterdam, The Netherlands, 1967; Vol. 3, pp 633–650.
32. Schramm, L. L.; Smith, R. G. Can. Patent 1 188 644, June 11, 1985; U.S. Patent 4 462 892, July 31, 1984.
33. Perrini, E. M. *Oil from Shale and Tar Sands;* Noyes Data Corp.: Park Ridge, NJ, 1975.
34. Ranney, M. W. *Oil Shale and Tar Sands Technology;* Noyes Data Corp.: Park Ridge, NJ, 1979.
35. Erskine, H. L. In *Handbook of Synfuels Technology;* Meyers, R. A., Ed.; McGraw-Hill: New York, 1984; pp 5-1–5-79.
36. Schramm, L. L. *J. Can. Petrol. Technol.* **1989**, *28*, 73–80.
37. Schramm, L. L. Can. Patent 1 232 854, February 16, 1988; U.S. Patent 4 637 417, January 20, 1987.
38. Schramm, L. L.; Kwak, J. C. T. *J. Can. Petrol. Technol.* **1988**, *27*, 26–35.
39. Schramm, L. L.; Smith, R. G. *Colloids Surf.* **1985**, *14*, 67–85.
40. Schramm, L. L.; Smith, R. G., Syncrude Canada Ltd., Edmonton, unpublished results.
41. Ward, S. H.; Clark, K. A. *Determination of the Viscosities and Specific Gravities of the Oils in Samples of Athabasca Bituminous Sand;* Research Council of Alberta: Edmonton, Canada, 1950; Report No. 57.
42. Farouq Ali, S. M. In *Oil Sands Fuel of the Future;* Hills, L. V., Ed.; Canadian Society of Petroleum Geologists: Calgary, Canada, 1974, pp 199–211.
43. Dealy, J. M. *Can. J. Chem. Eng.* **1979**, *57*, 677.
44. Jacobs, F. A.; Donnelly, J. K.; Stanislav, J.; Svrcek, W. Y. *J. Can. Petrol. Technol.* **1980**, *19*, 46–50.
45. Svrcek, W. Y.; Mehrotra, A. K. *J. Can. Petrol. Technol.* **1982**, *21*, 31–38.
46. Jacobs, F. A. M.S. Thesis, University of Calgary, 1978.
47. Christensen, R. J.; Lindberg, W. R.; Dorrence, S. M. *Fuel* **1984**, *63*, 1312.
48. Misra, M.; Miller, J. D. *Min. Eng.* **1980**, *32*, 302.
49. Hupka, J.; Miller, J. D.; Cortez, A. *Min. Eng.* **1983**, *35*, 1635.
50. Wood, R. Can. Patent 1 137 906, December 21, 1982.
51. Kizior, T. E. Can. Patent 1 072 474, February 26, 1980.
52. Schramm, L. L.; Smith, R. G. *AOSTRA J. Res.* **1989**, *5*, 87–107.
53. Sanford, E. C.; Seyer, F. A. *CIM Bull.* **1979**, *72*, 164.
54. Sanford, E. C. *Can. J. Chem. Eng.* **1983**, *61*, 554–567.
55. Seitzer, W. H. *Proceedings of the ACS Division of Petroleum Chemistry;* San Francisco, CA, April 2–5, 1968; pp F19–F24.

56. Malmberg, E. W.; Bean, R. M. *Proceedings of the ACS Division of Petroleum Chemistry;* San Francisco, CA, April 2–5, 1968; pp F25–F37.
57. Malmberg, E. W.; Bean, R. M. *Proceedings of the ACS Division of Petroleum Chemistry;* San Francisco, CA, April 2–5, 1968; pp F38–F49.
58. Cymbalisty, L. M. *Proceedings of the 30th Canadian Chemical Engineering Conference;* Edmonton, Canada, 1980; Vol. 4, pp 1168–1182.
59. *Syncrude Analytical Methods for Oil Sand and Bitumen Processing;* Bulmer, J. T.; Starr, J., Eds.; Alberta Oil Sands Technology and Research Authority: Edmonton, Canada, 1979.
60. Schramm, L. L. In *Emulsions: Fundamentals and Applications in the Petroleum Industry;* Schramm, L. L., Ed.; ACS Advances in Chemistry 231; American Chemical Society: Washington, DC, 1992; pp 1–49.
61. Schutte, R., Syncrude Canada Ltd., Edmonton, unpublished results.
62. Houlihan, R. N., Syncrude Canada Ltd., Edmonton, unpublished results.
63. Iwata, H., Homma, T.; *Powder Technol.* **1974**, *10*, 79.
64. Danielson, L., Syncrude Canada Ltd., Edmonton, unpublished results.
65. Crickmore, P. J.; Schutte, R., Syncrude Canada Ltd., Edmonton, unpublished results.
66. Chung, K. H.; Ng, S.; Sanford, E. C. *AOSTRA J. Res.* **1991**, *7*, 183–193.
67. Swanson, W. D., Syncrude Canada Ltd., Edmonton, unpublished results.
68. Schramm, L. L.; Hackman, L. P., Syncrude Canada Ltd., Edmonton, unpublished results.
69. Shelfantook, W. E.; Hyndman, A. W.; Hackman, L. P. U.S. Patent 4 859 317, August 22, 1989.

RECEIVED for review October 14, 1992. ACCEPTED revised manuscript April 27, 1993.

12

Antifoaming and Defoaming in Refineries

V. E. Lewis and W. F. Minyard

Refinery Process Chemicals, Nalco Chemical Company, P.O. Box 87, 7701 Highway 90-A, Sugarland, TX 77487

> *This chapter is intended as an introduction to defoaming and antifoaming in refinery and petrochemical applications. A brief look at defoaming theory is presented along with chemistry of antifoams. This is followed by examination of antifoam applications in refineries and petrochemical plants. The causes of foam formation, from demulsifying surfactants to dispersants, are dealt with starting with the desalter and proceeding to the distillation towers and to the coker. Unit diagrams are provided, and a description of each process and how antifoams are used in the particular process is presented.*

FOAMING IN REFINERY PROCESSES leads to a number of operating problems. Foaming can occur in both aqueous and nonaqueous systems and in a variety of process types such as distillation, extraction, gas and liquid scrubbing, and other separations. The economic consequences of uncontrolled foaming can be significant. Foaming can cause serious losses in throughput and require expensive and impractical operating changes.

When crude oil enters a refinery, the first vessel in which it is treated is the desalter. Here it is mixed with water to remove any salts that may be present. In this process, an emulsion is formed, and it must be broken prior to further treatment. The majority of the emulsion is broken by an electric grid; however, chemicals called emulsion breakers (organic salts) are also used to aid this process.

The desalted crude is then fractionated by using a series of distillation towers at pressures from atmospheric to vacuum. In each of these towers, potential impediments to processing exist. These take the form of corrosion and fouling (deposit formation). Corrosion inhibitors are used to combat corrosion on the metal surfaces in the refinery or petrochemical plant. There are two classes: filmers and neutralizers. Filmers prevent corrosion by spreading a thin film on the metal surface, and neutralizers actually react with any acidic species that may cause corrosion. The bottoms (highest boiling material) from the vacuum distillation tower are sent to the coker, where the heavy, viscous residual oil is cracked to produce coke, light naptha, and distillate. Corrosion can still occur in this vessel and is treated with corrosion inhibitors. In addition, each of the chemicals described can end up in the coker because they are very high in molecular weight and boiling point.

Corrosion inhibitors, emulsion breakers, and dispersants, which are substances that prevent the deposit of polymers or other insoluble particles on vessel surfaces, all serve a valuable purpose in that the refinery could not be operated effectively, or in many cases at all, without them. However, they do have one drawback. They all have surface activity and can stabilize foam formation in any of the vessels in which they are present. In addition, the crude oil itself may contain elements that would stabilize foam formation.

Although sometimes petrochemical processes differ vastly from refinery processes, they also have conditions that can favor foam stabilization. Foam formation can be as costly as any of the other types of problems such as corrosion, fouling, or emulsions. A distillation tower, whether a refinery atmospheric tower or a petrochemical debutanizer (a distillation tower that removes hydrocarbons that boil in the range of butanes, butenes, or butynes) that is filled with foam can no longer provide the desired separation of compounds. Quite simply, in refinery or petrochemical operations, foam costs money; either in reduced capacity or contamination of products. To combat problems in these and other vessels, a group of compounds known as antifoams or defoamers have been developed. Although these terms are used interchangeably, they actually have specific definitions that apply to the method by which they are used. A defoamer is a compound added to a system that is already foaming. Its effectiveness is often expressed in terms of its "initial knockdown" capability, which is its ability to minimize foam production. An antifoam, on the other hand, is added to a system prior to foam formation and thus inhibits this process.

Figure 1 shows a graphic representation of defoamers and antifoams. The particular stream that was treated, styrene–butadiene latex, is outside the scope of this text, but the example is useful to help define defoamer and antifoamer. In this evaluation, a glass column with a steam inlet attached to the bottom is used. The column is graduated in centimeters; 90 cm is the top of the column. Latex was added, and steam was introduced into the bottom of the column, producing foam. Foam height was allowed to build up to 70 cm; then the defoamer was added. The initial knockdown time period for each defoamer was about 2 s (given by the second

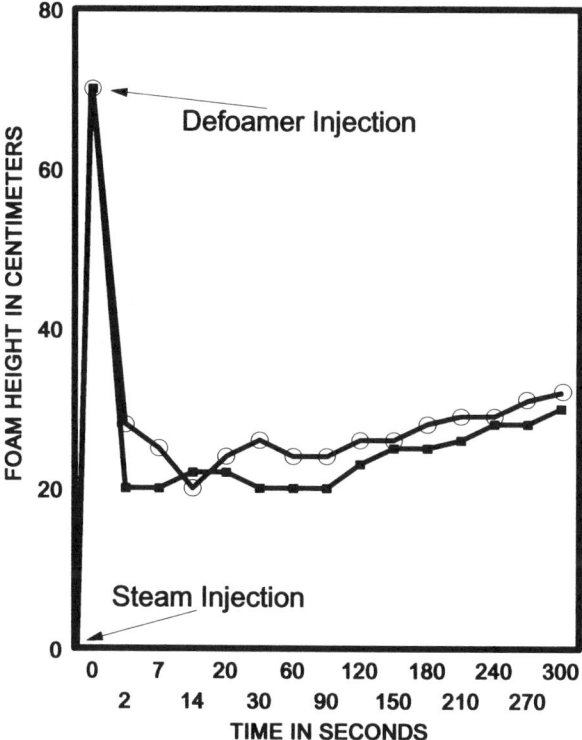

Figure 1. Comparison of hydrophobic silica defoamer action using a styrene–butadiene latex.

point on the graph for each compound). After this initial knockdown, the products then continued to inhibit foam production for the entire test period. At this point, they were acting as antifoams.

Before discussing specific classes of chemicals used as antifoams, a brief discussion of general antifoam characteristics is in order. Because foam formation is a surface phenomenon; that is, it occurs at a gas–liquid interface, an antifoam must concentrate at the surface. It must also be located at the gas–liquid interface.

If an antifoam is to be effective it must be able to enter the film that makes up the foam bubbles and spread across the film surface. Equations 1 and 2 define the entering coefficient, E, and the spreading coefficient, S, of an antifoam with respect to a particular foaming medium.

$$E = \sigma_m + \sigma_{ma} - \sigma_a \qquad (1)$$

$$S = \sigma_m - \sigma_{ma} - \sigma_a \qquad (2)$$

In these equations, σ_m represents the surface tension of the foaming medium, σ_a is the surface tension of the antifoam, and σ_{ma} is the interfacial tension between the two (*1*). For the antifoam to enter the film, $E >$ 0, and if the antifoam is to spread across the film bubble surface, then $S >$ 0. As the antifoam spreads, it takes large amounts of the film and surfactant with it. This spreading leads to a weakened film and eventually bubble rupture (*2, 3*). Although accurate, these two equations represent experiments performed on static systems. In a dynamic system, such as a delayed coker or other refinery vessel, the equations may not be completely representative.

Regardless of how well an antifoam penetrates a foam film or how well it spreads, one other criterion is necessary for good foam inhibition. The antifoam must be insoluble in the foaming medium. If a compound is soluble in a foaming system, it cannot act as an antifoam. In fact, in some cases, it may even increase foaming. Shearer and Akers (*4*) have shown this effect by using silicone oils as antifoams for different hydrocarbon oils. The silicone oil used was a poly(dimethylsiloxane) with a viscosity of 1000 centistokes (1 cSt = 1 mm^2/s). At low concentrations, this antifoam is soluble in No. 555 oil and actually increases foaming. After the hydrocarbon is saturated at about 300 ppm, it is insoluble and acts as an antifoam. The same silicone oil is insoluble in No. 702 oil at all concentrations and inhibits foam formation. As concentration increases, foam inhibition gets better.

Chemistry of Antifoams

Hydrophobic Silicas. Because foaming is a surface phenomenon, any antifoam used must concentrate at the surface (or gas–liquid interface). Hydrophobic silicas, which are silicas that have been treated with a compound that causes them to float on the top of water, have been used to fulfill this function for almost 30 years. U.S. Patent 3 408 306 (*5*) discloses the use of a hydrophobic silica dispersed in a hydrocarbon oil. Hydrophobic silica for this composition, which is still in use today, is made either by continuous ("dry roast") or batch process. In either process, precipitated silicas rather than silica gels or fumed silicas are typically used to make antifoams. During a continuous process, silicone oils, usually poly(dimethylsiloxane), are sprayed onto a bed of hydrophilic silica. The bed is heated to temperatures ranging up to 300 °C, and reaction times are up to 20 h. At these temperatures and reaction times, bond formation between the silica particle and silicone oil may occur in addition to simple coating of the particle.

After treatment, the silica particles may be blended with a hydrocarbon oil and any other additive as desired. Blending with hydrocarbon oil is critical for maximizing effectiveness of hydrophobic silica. Hydrophobic silica by itself or blended as an emulsion in water has no defoaming ability (6, 7). This phenomenon is depicted in Figure 2, which shows a comparison between two hydrophobic silicas when evaluated in a latex-foaming application (the blank is a sample of untreated latex). Product A is a silica–silicone blend that has been emulsified in water. Product B is a hydrophobic silica in a naphthenic oil. For this evaluation latex samples were dosed with Product A and B, separately, at levels of 250 ppm. As the graph indicates, A had almost no activity in this evaluation, and B kept the foam height at about 40 cm throughout the full 5-min test.

Hydrophobic silica production via batch processes is somewhat different. A slurry of hydrophilic silica in oil is treated with the hydrophobic agent and a catalyst. The hydrophobic agent may be either silicone polymer containing a reactive end group or a small reactive silicon-containing molecule such as trimethylchlorosilane. Typical silicone polymers used in this process are hydroxy-terminated poly(dimethylsiloxane)s. The catalyst

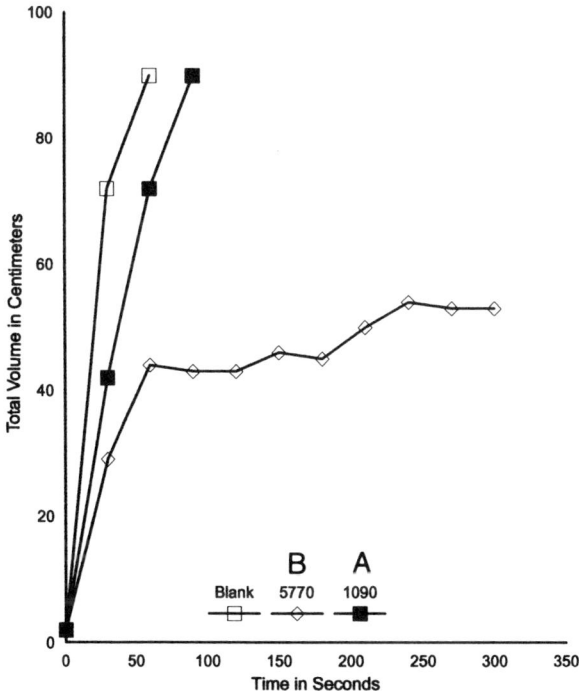

Figure 2. Comparison of hydrophobic silica antifoam properties. The blank is an untreated styrene–butadiene latex. Compound A is an aqueous blend. Compound B is a hydrocarbon blend.

is alkaline in nature. Organic polyethyleneamines are often used for this purpose.

Figure 3 shows an example of the reaction as carried out using a hydroxy-terminated poly(dimethylsiloxane). Extent of reaction is measured for this, and the continuous process, by hydrophobic index (HI). HI is measured by placing hydrophobic silica in water that contains varying percentages of an alcohol. The percent alcohol at which the silica starts to become wet (sink in the solution) is defined as its hydrophobic index.

Methanol and 2-propanol have both been used for this measurement. 2-propanol, which has two more carbons than methanol, is the less polar of the two alcohols. Owing to this fact, treated silicas become "hydrophilic" in 2-propanol–water solutions sooner than they do in methanol–water solutions. Thus, the same silica will have a higher HI in a methanol–water solution than in a 2-propanol–water solution. In Figure 3, the hydrophobic index was measured using a 2-propanol–water solution, and this hydrophobic silica just started to sink in a 20% solution of 2-propanol in water.

Small reactive silicon-containing molecules, such as chlorosilanes or silazanes, are also used to make hydrophobic silicas. Used in place of large silicone polymers, these silanes or silazanes react rapidly with the reactive sites on the silica particle. During this reaction, there is less steric crowding at the reactive sites on the silica particle, which allows for a greater portion of the silica to react with the hydrophobic agent. U.S. Patent 3 338 073 (8) discloses the use of dimethyldichlorosilane to form a unique bridging silicon ether. In a batch process, hydrophilic silica (typically containing ~0.5% moisture) is added to the carrier oil and dehydrated to a water content of <0.05%. Owing to the reactivity of chlorosilanes with water, the dehydration step must be carried out prior to chlorosilane

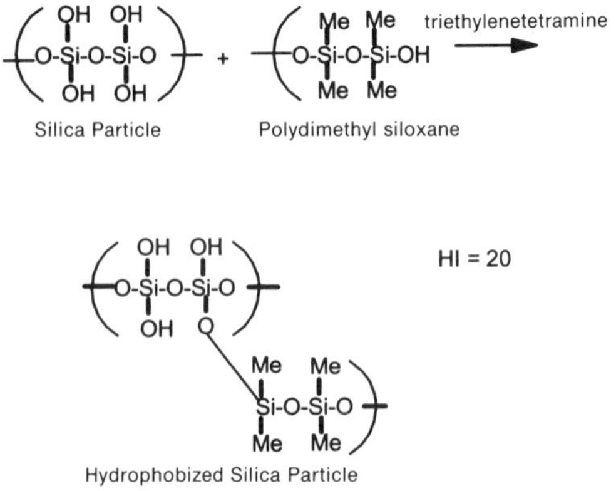

Figure 3. Reaction of silicone oil with hydrophilic silica particle.

addition. HCl is a by-product of this reaction and must be dealt with either by adding a scavenger or by using a glass-lined reactor. When the hydrophobic silica has been produced, it is blended with emulsifiers or spreading agents to enhance its performance in a particular application. The final product is often subjected to a grinding process to reduce the overall size of the silica particles. The reduction in size of the particles helps prevent settling during storage.

Silicone Oils. Silicone refers to a siloxane polymer. The name itself derives from early research in which it was thought that oxygen was bound to silicon via a double bond such as found in a ketone. The simplest of these polymers are the poly(dimethylsiloxane)s. Poly(dimethylsiloxane)s are made by hydrolyzing dimethyldichlorosilane with hydrochloric acid (9). Dimethyldichlorosilane is, in turn, made by the Rochow reaction. In this reaction, an alkyl halide (RX), methyl chloride in this instance, is treated with silicon metal in the presence of a catalyst (usually copper). The reaction results in a number of products that are separated by distillation (10).

$$RX + Si(Cu) \longrightarrow RSiX_3 + R_2SiX_2 + R_3SiX + R_4Si + \cdots$$

Poly(dimethylsiloxane)s may be linear or branched and are described using nomenclature shorthand developed by General Electric. The trimethylsilyloxy unit is given the designation of M. The repeating dimethylsilyloxy units are designated D, and branch points in the chain are either T or Q. T refers to a monoalkylated silyloxy group such as CH_3SiO_3, and Q is a tetraoxysilyl group, SiO_4 (10).

Linear poly(dimethylsiloxane)s are used in a number of refinery applications. They are normally trimethylsilyl end-capped polymers of the type MD_nM. They are made by reaction of dimethylsilicone fluid (D_n) and a source of chain terminator, $(CH_3)_3SiO$- (M). The ratio of M to D controls the molecular weight and viscosity of the product (9).

In refinery applications such as the delayed coker, linear poly(dimethylsiloxane)s of intermediate molecular weight are used. These products are normally described in terms of their viscosity in centistokes. Thus a 60,000- or 100,000-centistoke silicone fluid would be the typical recommendation for use in a delayed coker. More viscous and consequently higher molecular weight silicones have been used as described in U.S. Patent 3 700 587 (11).

Fluorinating dimethylsilicones provides a polymer with greater thermal stability. This type of poly(dimethylsiloxane) is made by equilibrating a fluorinated alcohol with a dimethylsilicone fluid as described in U.S. Patent 4 824 983 (*12*). As shown in this patent, fluorination is not present on carbons in the α or β position relative to silicon. The electropositive character of silicon renders compounds with fluorine in either of these positions thermally unstable (*9*).

Organic Compounds: Glycols. Poly(propylene glycol) (PPG) is made by treating propylene oxide (PO) with base; normally potassium hydroxide. In this reaction, an initiator such as dipropylene glycol may be used. KOH deprotonates dipropylene glycol, and the resulting nucleophilic attack of this oxide on PO causes an extended propylene glycol with an oxide terminus. Polymerization is propagated by reaction of this "living" oxide with PO.

Chain length and consequently molecular weight depends on reaction time. If the reaction is stopped at a point in which the molecular weight is around 400 (seven PO units), the polymer is soluble in water. However, if the reaction is allowed to digest until the molecular weight has built up to greater than 2000 (34 PO units), the polymer is insoluble in water. A polymer with molecular weight of greater than 2000 is an excellent antifoam in an ethylene plant caustic system's other applications (Figure 4).

Refinery Applications

This section will be limited to a discussion of antifoam applications commonly found in the refinery, a description of the processes, and the products used in those applications. For certain types of refinery processes, an antifoam is always required. In others, foaming is more symptomatic of an operational problem or contamination of the process fluids. In these cases, antifoams are used more as a "quick fix" until the causes of the foaming can be identified and eliminated.

For the refining industry, the main classes of antifoams used are silicones, polyglycols, and hydrophobic silicas. For hydrocarbon systems, the most common type of antifoam used is silicone, or poly(dimethylsiloxane), of various molecular weights and viscosities. In some cases, a combination of silicone and poly(propylene glycol) or silicone and hydrophobic silica is used.

For aqueous systems such as acid gas scrubbers and caustic scrubbers (where mercaptans are removed from gasoline), hydrophobic silicas or polyglycols are used. The specific chemistries and applications for both aqueous and nonaqueous foaming situations will be discussed.

Delayed Coking. By far, the most extensive use of antifoams is in the delayed coking unit. Coking is a high-temperature thermal operation

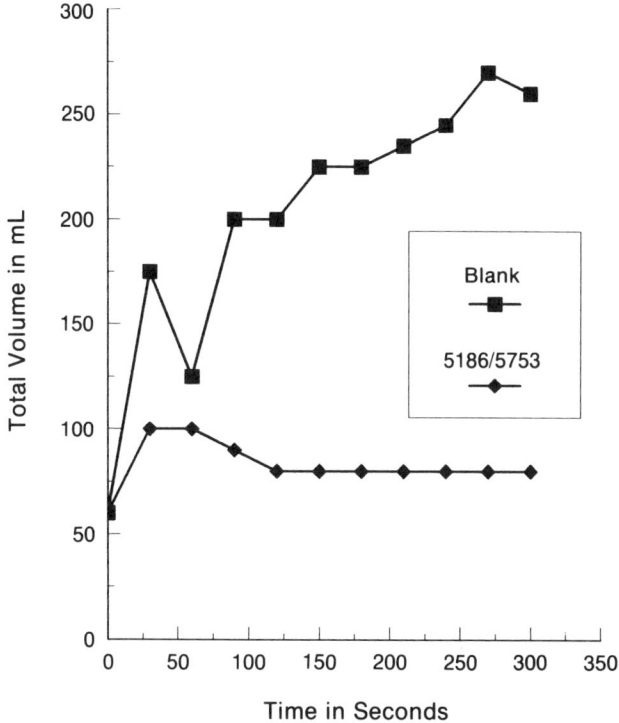

Figure 4. Antifoam evaluations on water taken from an ethylene plant dilution steam drum. The blank is an untreated sample of this water. The other sample was treated with a polyglycol antifoam.

that is used to convert vacuum "resid" (viscous, semiliquid bottoms products) to lighter products and coke. This process is used for conversion of the "bottom of the barrel" of crude oil. Because of the nature of the process, very few cokers can operate without using an antifoam.

Feedstocks to the coker generally consist of vacuum "resids"; along with some "slops" (refinery or petrochemical waste streams); heavy FCC (fluid catalytic cracking) slurry oils, which contain unconverted hydrocarbons and traces of catalyst fines entrained from the reactor; and asphalts. The coker is often the final disposition for many poor quality refinery process streams. As a result, the feedstock to the coker contains high levels of impurities.

Figure 5 is a diagram of a typical delayed coker, including the distillation tower and light ends section. The feedstock is preheated using some of the heavy and light gas oil product streams. It then enters the lower portion of the fractionation tower, above the point where the product vapors from the coke drums enter the tower. This introduction of the

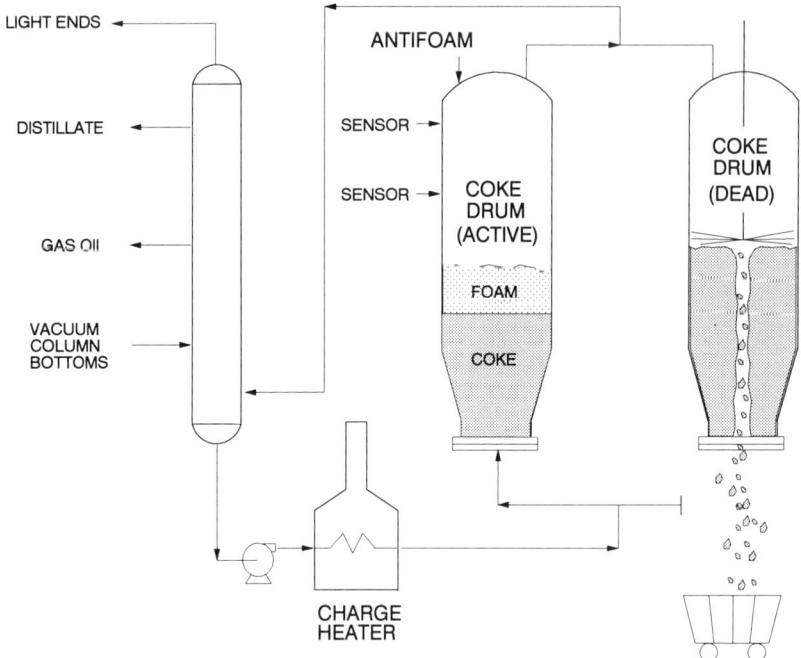

Figure 5. Schematic of a delayed coker showing the process flow.

feedstock serves to quench the product vapors and to further preheat the coker feed. The "resid" and the recycled coker bottoms are then fed to the furnace, where the "resid" is heated to 900–950 °F. The residence time in the furnace is very short, and the thermal-cracking reaction is delayed until the "resid" enters the coke drum. In the coke drum, the "resid" cracks and forms molten coke and hydrocarbon vapors. The vapors go overhead to the fractionation tower, where they are condensed and separated into their respective fractions. A minimum of two coke drums are used in this process. As one drum is filling, the coke in the other drum is being quenched and removed. Each cycle of filling and coke removal takes about 24 h.

In the coke drum, foaming occurs. As the bulk viscosity of the "resid" increases and solids form, a stable foam front forms, and it rises higher in the drum as the coke level rises. If left unchecked, the foam will reach near the top of the coke drum and potentially go overhead, or "foam-over". The consequences of a "foam-over" are immediate and disastrous. The transfer line from the coke drum to the fractionator becomes fouled with coke particles. As the solids reach the fractionation tower, the suction screens in the bottom of the tower become plugged. Finally, the finer particles of coke that pass through the suction screens will deposit in the furnace heater tubes and result in severe fouling. Foam carry-over results in an immediate unscheduled shutdown of the unit for cleaning and may result in throughput limitations for the whole refinery.

Although such scenarios are rare, they are a concern for the refiner, so proper application of the antifoam is critical. Another important aspect of the coking cycle is the amount of space remaining between the top of the coke bed and the top of the drum at the end of the cycle. This distance is called the outage. By minimizing the outage in each drum, higher throughput to the coker can be maintained.

High molecular weight, high-viscosity silicones are used to control foaming in the coking process. The silicones are usually characterized by their viscosity. Either 60,000- or 100,000-cSt silicones are most commonly used in the industry. Some refiners believe that 100,000-cSt silicone performs better than 60,000-cSt silicone. In either case, the silicone is diluted in kerosene or aromatic distillate to make the products.

Antifoam is normally applied using a light gas—oil slipstream. The slipstream—antifoam mixture is injected into the top of the coke drum that is being filled. The slipstream is used to keep the injection pipe clear and to aid in distributing the silicone over the foam. The typical dosage of product ranges from 2 to 20 ppm.

The antifoam can be added either on a continuous basis or when the foam front reaches a specified level. A series of level detectors is normally present on each drum, as shown in Figure 5. When the foam level is observed at the first level detector, the antifoam injection is started. An immediate drop in the level will be observed. At this time, the operators should be preparing the next drum to begin receiving feed. The antifoam injection will continue for some time after the drum switch occurs as the coke is being quenched with steam and water.

The amount of silicone added to the process is of great concern to the refiner, because silicon is a poison for the catalyst used to hydrotreat coker naphtha. Coker naptha is a liquid hydrocarbon fraction in the boiling range of 180–380 °F and is obtained from the cracking of very heavy "resids" in the delayed coking unit. Hydrotreating is a catalytic hydrogenation process used to remove sulfur-containing compounds from a refinery process stream. Because of the high temperatures in the coke drum, at least some of the silicone will decompose to a variety of volatile silicon compounds and end up in the fractionator with the liquid products. Several of the silicone decomposition products and their boiling ranges are shown in Table I. Most of these compounds are in the naphtha boiling range or lower. [To be considered naphtha, not less than 10% of the hydrocarbon stream should boil below 175 °C (345 °F), and not less than 95% of the material should boil below 240 °C (465 °F) under standard conditions (ASTM D-86) (*13*)]. Therefore, efforts are made to minimize

Table I. Silicone Decomposition Products and Their Boiling Points

Decomposition Product		Boiling Point (°F)
$(CH_3)_3-Si-O-Si-(CH_3)_3$		211
$(CH)_3-Si-(O-Si)_x-O-Si-(CH_3)_3$ $\hphantom{(CH)_3-Si-(O-Si)_x}\vert$ $\hphantom{(CH)_3-Si-(O-S}(CH_3)_2$	$x = 1$ $x = 2$	307 381
$(CH_3)_2-Si-O-Si-(CH_3)_2$ $\hphantom{(CH_3)_2-}\vert\hphantom{-Si-O-Si}\vert$ $\hphantom{(CH_3)_2-}O\hphantom{-Si-O-Si}O$ $\hphantom{(CH_3)_2-}\vert\hphantom{-Si-O-Si}\vert$ $(CH_3)_2-Si-O-Si-(CH_3)_2$		348
$(CH_3)_2-SiH_2$		−3
$(CH_3)_3-Si-OH$		209

the amount of silicon carry-over to the coker naphtha by minimizing the amount of silicone added to the coker.

Preflash, Atmospheric, and Vacuum Towers. The crude unit is designed to separate the various fractions of the crude oil on the basis of their volatility. The crude unit process equipment includes the desalter; the preheat exchanger train (a series of heat exchangers used to initially heat the crude oil and for heat recovery from refinery products); the atmospheric and vacuum furnaces; and the preflash, atmospheric, and vacuum distillation columns (towers). Figure 6 is a general diagram of a crude unit, showing the major pieces of equipment, feed and product streams, and typical operating conditions. Antifoam injection points are also shown.

The preflash column is used when the crude oil reaches a temperature of about 400–450 °F. At this point, some of the crude is above its boiling point, and vaporization occurs. The lightest portion of the crude is flashed (vaporized and separated from the bulk liquid) in the preflash column. The overhead is condensed and added into the atmospheric tower as a reflux, and the noncondensible vapors are added to the atmospheric tower overhead vapors. This step serves to conserve energy and unit investment for preheating the bulk of the crude.

Occasionally, foaming in the preflash column can limit throughput to the crude unit. Organic surfactants or water carry-over from the desalters can be the cause of the problem. Silicones, usually of 60,000-cSt viscosity, are normally used in this process. However, sometimes the normal silicones are too soluble in the crude to be effective in controlling foam. In

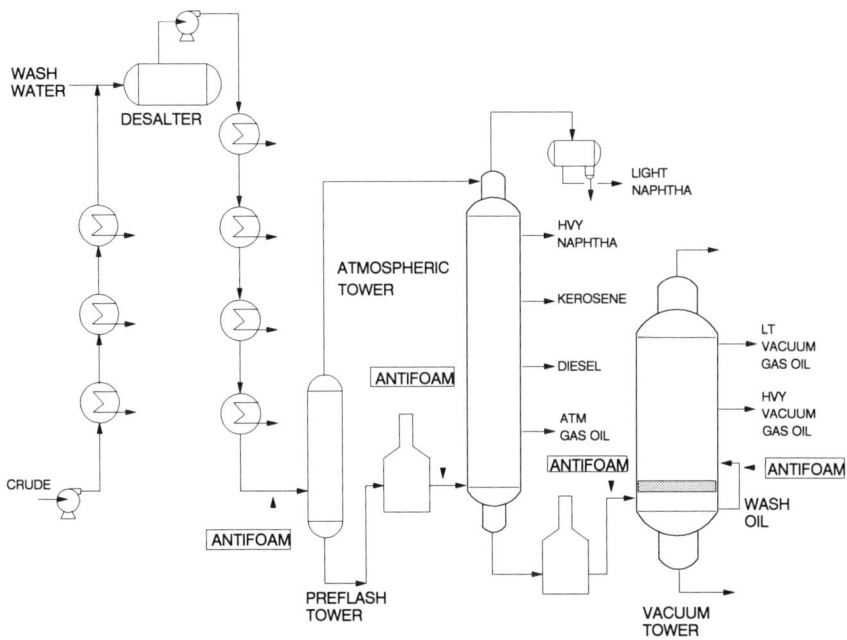

Figure 6. Refinery crude unit schematic showing the process flow.

this case, the use of a substituted silicone, such as polyfluorosilicone, is usually effective.

Foaming in the atmospheric column can sometimes occur but is not normally encountered. In the atmospheric column, the crude, at about 700 °F, enters a flash zone at the bottom of the tower, where the vapors and liquid are separated. The vapors travel up the tower, where the various fractions condense. When foaming does occur, difficulties may be encountered in meeting the specified cut points of the fractions. Again the 60,000-cSt silicones are used at dosages of 2–10 ppm.

In the vacuum tower, heavy liquids from the atmospheric tower, referred to as reduced crude, are fed to a vacuum furnace and heated to about 750–800 °F. To suppress coking in the furnace tubes, steam is added to increase the velocity of the hydrocarbon in the tubes. The reduced crude enters the vacuum tower flash zone, where the pressure is maintained at 20–30 mmHg absolute (2600–4000 Pa). When foaming occurs, fouling of the demister pads (sets of grids designed to minimize entrainment of heavier liquids into the upper sections of the tower) above the flash zone can occur, side stream gas oil products will be discolored, and gas oil end-point specification cannot be met. Again, silicone additives are used. The antifoam is normally injected into the feed to the tower, but can also be fed into the wash oil section above the demister pads.

Amine Units. Throughout the refinery, particularly in the cracking processes, light gases that are produced contain significant quantities of acid gases such as H_2S and CO_2. These acid gases are removed in a process called gas sweetening. This process is accomplished in an amine unit. Acid gas makes contact with a dilute alkanolamine solution to form a salt, and a clean fuel gas is produced. The salt is then thermally dissociated in a stripping section to regenerate the amine.

Typical amines used in the process are listed in Table II; the two most common amines are monoethanolamine (MEA) and diethanolamine (DEA). Figure 7 is a diagram of a typical amine unit. The feed gas containing H_2S and CO_2 enters the bottom of a scrubbing section. The lean amine enters the top of the scrubber and flows countercurrent to the gases.

The sweetened gas goes overhead and is sent to the fuel gas system. The rich amine exits the bottom of the scrubber and is heated in the lean–rich exchanger. It then enters the amine regenerator, where a steam reboiler is used to heat the amine to 225–250 °F. At these temperatures, the salts are thermally dissociated to regenerate the amine. The acid gas, composed primarily of H_2S, is sent to a sulfur recovery unit that generates elemental sulfur.

Foaming can occur in both the scrubber and in the regenerator. Foaming in either section can result in a reduced loading of acid gas on the system and can even result in throughput limitations for the refinery. In addition, losses of amine caused by foaming into the overheads cause an increase in fresh amine makeup to the system. Causes of foaming can vary. Because both MEA and DEA are corrosive, corrosion products can build up and create a stable foam. Carryover of light hydrocarbons or surfactants into the amine system can also create severe foaming problems.

Table II. Amines Used in Gas Sweetening

Chemical Name	Chemical Formula
Monoethanolamine (MEA)	$H_2NCH_2CH_2OH$
Diglycolamine (DGA)	$H_2NCH_2CH_2OCH_2CH_2OH$
Diethanolamine (DEA)	$(HOCH_2CH_2)_2NH$
Diisopropanolamine (DIPA)	$(HOCH(CH_3)CH_2)_2NH$
Triethanolamine (TEA)	$(HOCH_2CH_2)_3N$
Methyldiethanolamine (MDEA)	$(HOCH_2CH_2)_2NCH_3$

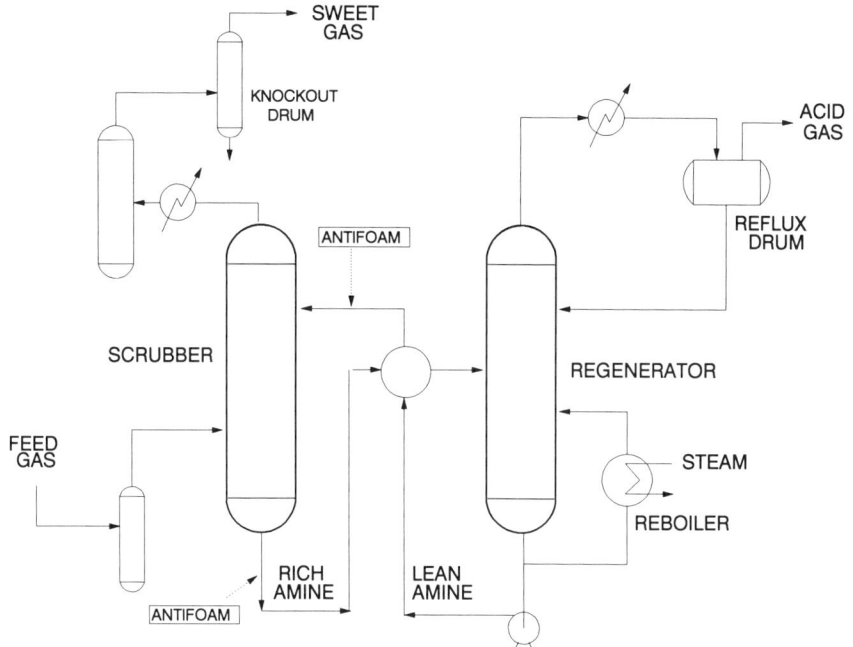

Figure 7. Schematic of an amine unit showing the process flow.

Finally, compounds can react with the amine to form heat-stable salts that are not dissociated at the temperatures in the regenerator. The concentration of heat-stable salts can build up to such an extent as to cause foaming problems.

The use of an antifoam in an amine unit can become a standard practice in a refinery or can be used only until the cause of foaming can be identified and corrected. Normally, the antifoam will be used only intermittently. Several products are used in amine systems. The selection of the best product can be determined in each case by performing a simple test, such as bubbling air or nitrogen into a sample of lean amine in a graduated column and determining foam height versus time.

Caustic Scrubbers and Sour-Water Strippers. Caustic scrubbers and sour-water strippers are used to remove sulfur from a variety of products, gases, and waste streams. Foaming is a fairly common occurrence in these systems because of contaminants found in the streams.

Caustic scrubbers use a solution of dilute NaOH to remove mercaptans and other sulfur compounds from gasoline streams. Feed is added to the bottom of the scrubber and makes contact with the caustic solution. The caustic-scrubbed product goes overhead. Spent caustic (mercaptan-contaminated NaOH solution) is removed for reclaiming, and fresh caustic is added. Figure 8 is a diagram of a caustic scrubber and the antifoam injection point.

The sour-water stripper removes contaminants from condensed water and wash water used in various process units. The water may then be reused, for example, as desalter wash water or to water-wash light gas streams. In this process, the contaminants such as H_2S are removed by stream stripping of the water. Figure 9 is a general diagram of a sour-water stripper.

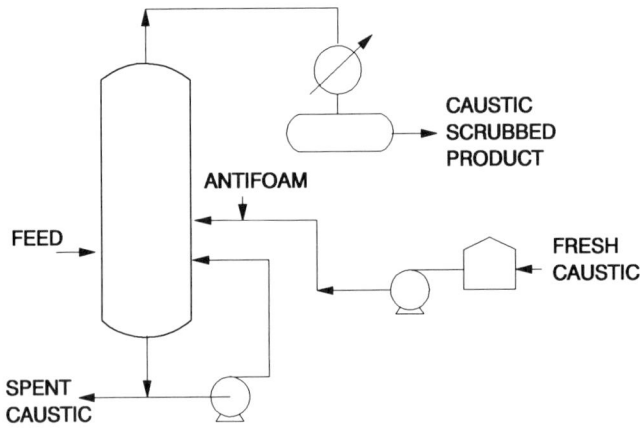

Figure 8. *Caustic scrubbing tower.*

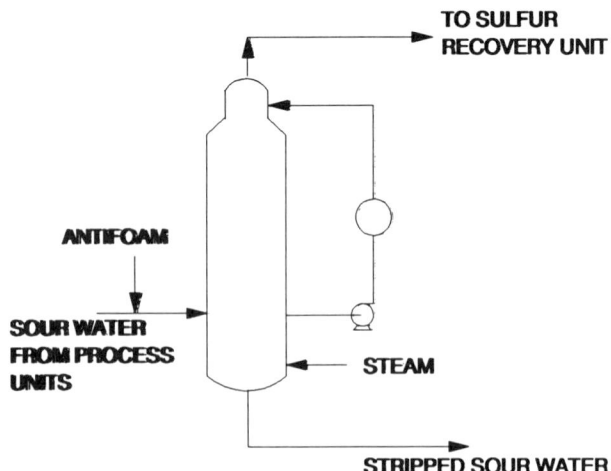

Figure 9. *Sour-water stripper.*

Removal of Aromatic Compounds. Because of the demand for high-purity aromatic compounds for petrochemical feedstocks, several processes have been developed for BTX (benzene, toluene, and xylenes) recovery from distillate streams. In these processes, aromatic compounds are separated from nonaromatic compounds by liquid–liquid extraction using polar solvents. The three major processes in use are the UOP–Dow UDEX process (di- or triethylene glycol solvent), the UOP sulfolane process (tetrahydrothiophene 1,1-dioxide), and the Union Carbide TETRA process (tetraethylene glycol).

Figure 10 is a diagram of a typical aromatic extraction process. When foaming becomes a problem, it normally occurs in the stripping column or the solvent recovery column. Foaming can lead to solvent losses and reduced throughput to the unit. Several antifoam products have been used in aromatic extraction units. A combination of silicone and polyglycol is normally effective in controlling foam. A combination of low molecular weight silicone and hydrophobic silica has also been used successfully. The antifoam is normally added to the reflux of the column at a dosage of 1–10 ppm.

Solvent Deasphalting. Solvent deasphalting (SDA) separates hydrocarbons on the basis of their solubility in a liquid solvent and is used

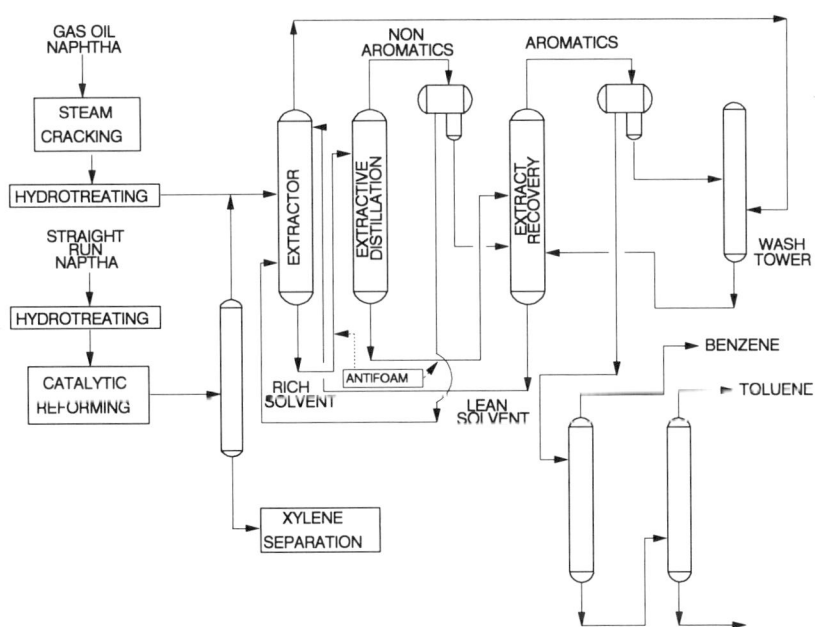

Figure 10. Extraction unit for aromatic compounds.

to upgrade feedstocks to other processes such as the FCC or hydrocracker. SDA produces a lower molecular weight, highly paraffinic deasphalted oil (DAO); and the asphalt raffinate, which is the portion of the treated liquid mixture that remains undissolved and is not removed by the selective solvent, is heavy and aromatic. Generally, propane is used as the solvent.

In the process, the feed makes contact with the solvent. A DAO–solvent mixture goes overhead, and an asphalt–solvent mixture exits the bottom. The DAO–solvent will go through a series of flash drums to recover most of the solvent. The final step is a DAO stripping section in which the remaining solvent is removed from the DAO by steam stripping. When foaming occurs, it normally does so in the DAO stripper shown in Figure 11. Foaming causes poor separation of the DAO–solvent mixture and increases the amount of makeup solvent to the unit. The antifoam is added to the DAO pumparound line at a dosage of 3–10 ppm, on the basis of the feed rate to the tower.

Petrochemical Applications

Ethylene Plants. In an ethylene plant, hydrocarbon feedstocks are fed to a furnace along with high-pressure steam where they are heated to about 800 °C. This temperature will vary depending on the type of feedstock. In the United States, most of the feed to the furnace is ethane,

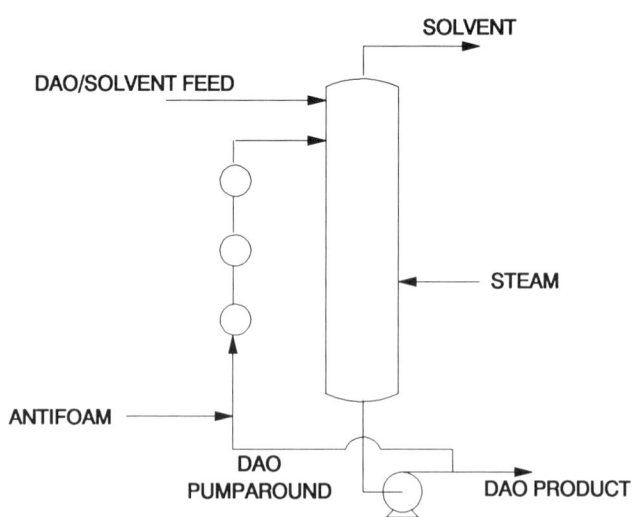

Figure 11. DAO–solvent stripper.

propane, or mixtures of the two. Because ethane and propane are smaller molecules, a higher temperature is required to crack them. On the other hand, many plants outside the United States use liquid feedstocks such as naphtha or gas oil to feed their furnaces, and consequently the furnaces do not have to be heated as strongly to crack these larger molecules.

As the plant name implies, the primary product of interest is ethylene; however, it is not the only compound produced. A large number of products are produced, ranging from methane to gasoline. When produced, the products must be cooled and the gaseous components compressed to a liquid for separation. After the furnace, the rest of the plant is devoted to this operation.

Figure 12 shows a simplified schematic of an ethylene plant from the furnace through the quench and compression sections. After the cracked gas stream leaves the furnace, the cooling process starts. If the plant is cracking heavier feedstocks such as naphthas, a gasoline fractionator or oil quench tower will be present. If the plant is cracking gas such as ethane or propane, the first quench tower will be the water quench tower. In Figure 12, the water from the quench settler is used to make steam for use in the furnace. If this water becomes contaminated, serious problems can result in the furnace. One potential source of contamination is foaming in the quench tower. Control of pH in this system is critical to prevent foaming. Alkaline conditions increase foaming in this unit.

Ideally this system would consist of three phases only. Two of these phases would be liquid, water and cooled hydrocarbons, and the third would be the cracked gas stream. Unfortunately, solids may accumulate in this vessel. Coke fines and small particles from the furnace may be entrained into this unit, and activated olefins such as butadiene or styrene may polymerize to form insoluble particles. All of this behavior promotes foaming. In this aqueous system, foam formation may be controlled by addition of a polyglycol.

From the quench tower, the cracked gas stream moves to the compression section. Compressors typically have from three to five stages. One stage of compression is defined as all the work performed before the gases are cooled. For example, the cracked gas stream may go through six impellers before exiting the compressor case to be cooled. A turbomachinery engineer will often refer to this as six stages; however, for this discussion, it will be considered only one stage.

Contained within Figure 12 is a schematic of a typical four-stage compressor. Between the third and fourth stage is the caustic tower. It is used to scrub acids gases, H_2S and CO_2, from the cracked gas stream. Polymer formation in this tower arises from self-condensation of acetaldehyde that may have been formed in the furnace by partial oxidation of ethylene. This polymer has been held responsible for inducing foam formation in this tower.

480 FOAMS: FUNDAMENTALS & APPLICATIONS IN THE PETROLEUM INDUSTRY

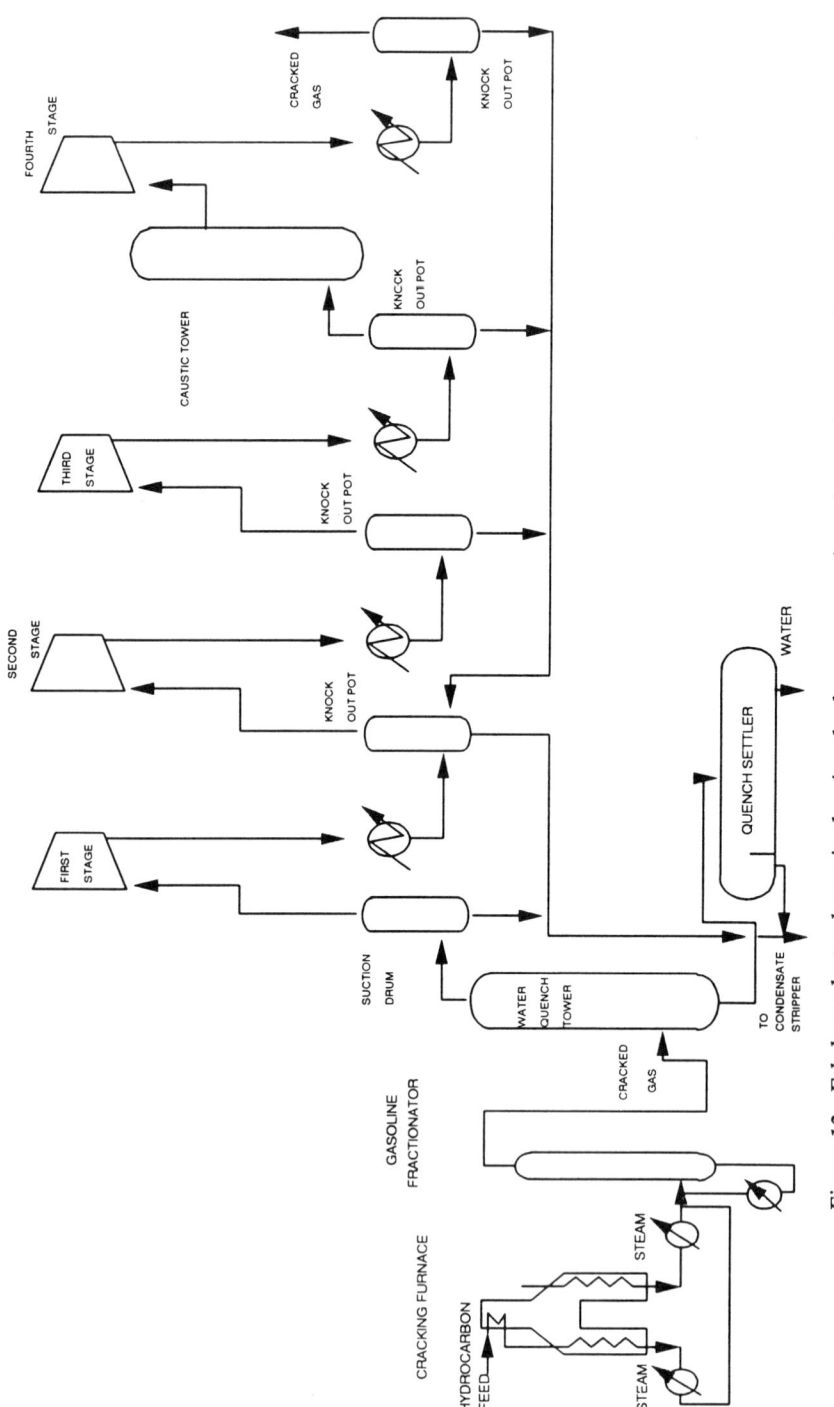

Figure 12. Ethylene plant schematic showing the furnace, quench section, and compression section.

About 20–30% of the aldehydes in the caustic tower are carried into the next stage of compression. If a foaming condition in the caustic tower is not treated, caustic can carry over into the next stage of compression and catalyze acetaldehyde polymer formation here. As in the quench tower, polyglycol antifoams are very effective at controlling foam, as shown in Figure 13.

Butadiene Extraction Units. The recovery section of an ethylene plant consists of a series of distillation towers. They remove components from the product stream in ascending order of molecular weight and consequently boiling point. The tower that removes C_4-chains from the product stream is called the debutanizer. Hydrocarbons in this molecular weight and boiling point range are carried overhead and sent to a dif-

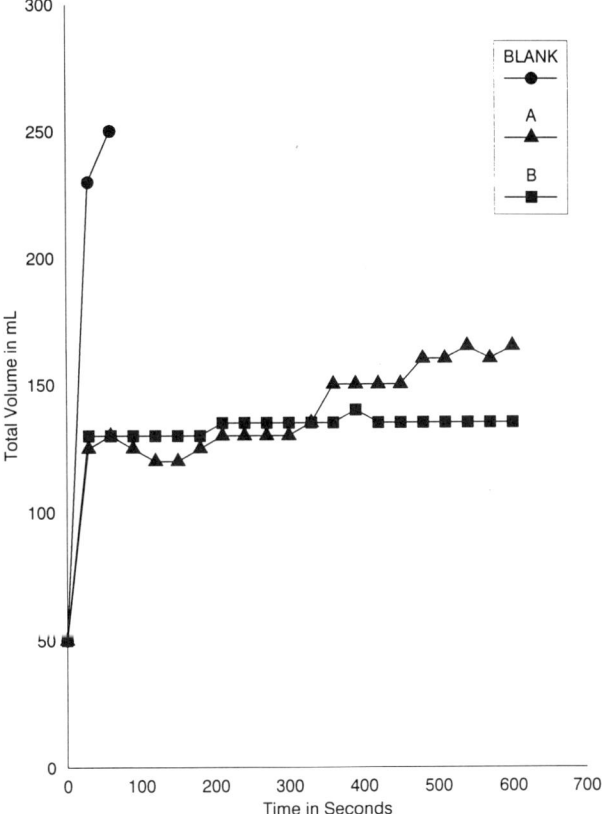

Figure 13. Ethylene plant caustic tower antifoam evaluation. The blank is an untreated sample of caustic water from this tower. A is a 2000 MW polyglycol antifoam and B is a 4000 MW polyglycol antifoam.

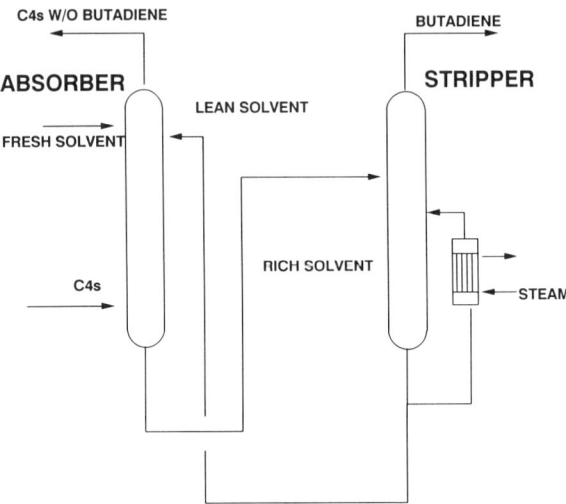

Figure 14. Schematic of a butadiene extraction unit.

ferent unit for further separation. This unit, called a butadiene extraction unit, consists of an absorber and a stripping tower (Figure 14). In the absorber, the mixed C_4 stream is washed with fresh and lean solvent (solvent that is free of extracted chemicals). Because butadiene is more polar than most of the other components, it is preferentially absorbed into the polar aprotic solvent (furfural, N-methylpyrrolidone, acetonitrile, and dimethylformamide are examples). The unabsorbed gases exit the top of this column, and the butadiene-enriched solvent exits from the bottom of this column.

In the stripper, the solvent is heated to a higher temperature to release the absorbed butadiene and other polar hydrocarbons such as methylacetylene. This lean solvent returns to the absorber, and the gases are carried overhead for further purification. The action of releasing gases from the rich solvent (the solvent that contains the extracted chemical) in the stripper induces foaming. Treatment of this condition is normally done by an emulsified silicone oil, although the antifoam itself may be a source of deposit formation (fouling). As Figure 14 indicated, a butadiene extraction unit operates in much the same fashion as the aromatic extraction unit.

Summary

In petroleum refining and petrochemical processing, the wide variety of processes provides conditions for foaming to occur. The types of systems in which foaming occurs can be classified as either aqueous or hydrocarbon. The causes of foaming are usually related to naturally occurring impurities found in the process streams, chemicals used in the producing oil fields, chemicals used in other parts of the plant, corrosion products, or changes in the process stream during processing.

Antifoam products are selected usually from a general class of chemicals including silicone oils, hydrophobic silicas, or glycols. The selection of the best product must be made after careful consideration of the process, the processing conditions, and the potential effect of the antifoam on product specifications or downstream processes.

References

1. Robinson, J. V.; Woods, N. W. *J. Soc. Chem. Ind.* **1948**, *67*, 361.
2. Ross, S. *J. Phys. Colloid Chem.* **1950**, *54*, 429.
3. Ewer, W. D.; Sutherland, K. L. *Aust. J. Sci. Res.* **1952**, *A5, 697.*
4. Shearer, L. T.; Akers, N. W. *J. Phys. Chem.* **1958**, *62*, 1264–1268.
5. Boylan, F. J. U.S. Patent 3 408 306, 1968.
6. Lichtman, I.; Gammon, T. In *Encyclopeadia of Chemical Technology;* John Wiley & Sons: New York, 1982; pp 440–441.
7. Lichtman, I.; Sinka, J.; Evans, D. *Assoc. Mex. Technol. Ind. Cellul. Pap.* **1975**, *15*, 26–32.
8. Domba, E. U.S. Patent 3 388 073, 1968.
9. Hardman, B. B.; Torkelson, A. In *Encyclopeadia of Chemical Technology;* John Wiley & Sons: New York, 1982; p 929.
10. Voorhoeve, R. J. H. *Organosilanes: Precursors to Silicones;* Elsevier: New York, 1967.
11. Hyde, J. A. U.S. Patent 3 700 587, 1972.
12. Fink, H.-F.; Berger, R.; Weitmeyer, C. U.S. Patent 4 824 983, 1989.
13. Speight, J. G. *The Chemistry and Technology of Petroleum,* 2nd ed.; Marcel Dekker: New York, 1991; p 692.

RECEIVED for review October 23, 1992. ACCEPTED revised manuscript May 26, 1993.

Glossary and Indexes

Glossary

Foam Terminology in the Petroleum Industry

Laurier L. Schramm

Petroleum Recovery Institute 100, 3512 33rd. St. N.W.,
Calgary, Alberta T2L 2A6, Canada

This glossary provides brief explanations for important terms in the science and engineering of foams in the petroleum industry. The petroleum operations in which foams may occur encompass aspects of so many different disciplines that there exists a large body of terminology. This selection of frequently encountered terms includes scientific terms related to the basic principles and properties of foams, as well as petroleum production and processing terms used to describe practical foams, their production, and their treatment. In addition, cross-references for the more important synonyms, abbreviations, and acronyms are included.

MANY TYPES AND OCCURENCES OF FOAMS are found in the petroleum industry. Also, many scientific and engineering disciplines are brought to bear in the study, formulation, and treatment of foams; each discipline brings with it its own special language. As a result, there exists a large body of terminology. This glossary presents some of the more common terms used in the petroleum science and engineering of foams. The selection of terms and their explanations matches the terminology and usage of the chapters of this book. As well, a considerable number of terms associated with related kinds of colloidal dispersions, especially emulsions, are included.

Some clearly nonfoam terms are included when they are used in a significant way in the chapters, and when they might be unfamiliar to many readers. Nonfoam terms include a number of petroleum industry terms such as the distinctions among different grades (gravities) of crude oils, which were made on the basis of United Nations Institute for Training and Research (UNITAR)-sponsored discussions (*1–3*). For terms drawn from colloid and interface science, I placed much reliance on the recommendations of the IUPAC Commission on Colloid and Surface Chemistry (*4*). For general petroleum industry terms, I tried to be consistent with the standard petroleum dictionaries (*5–7*).

In choosing the terms to be included, I assumed that the reader has some basic knowledge of the underlying fields of physical chemistry and chemical engineering. Although many named phenomena are included, only a few named equations and constants are included.

Absolute Viscosity   Viscosity measured by using a standard method with the results traceable to fundamental units. Absolute viscosities are distinguished from relative measurements made with instruments that measure viscous drag in a fluid, without known and uniform applied shear rates. *See also* Viscosity.

Absorbate   A substance that becomes absorbed into another material or absorbent. *See also* Absorption.

Absorbent   The substrate into which a substance is absorbed. *See also* Absorption.

Absorption   The increase in quantity (transfer) of one material into another or of material from one phase into another phase. Absorption may also denote the process of one material accumulating inside another.

Acid Number   *See* Total Acid Number.

ACN   Alkane carbon number, *see* Equivalent Alkane Carbon Number.

Activator   Any agent that may be used in froth flotation to selectively enhance the effectiveness of collectors for certain mineral components. *See also* Froth Flotation.

Adhesion   The attachment of one phase to another. *See also* Work of Adhesion.

Adsorbate   A substance that becomes adsorbed at the interface or into the interfacial layer of another material, or adsorbent. *See also* Adsorption.

Adsorbent   The substrate onto which a substance is adsorbed. *See also* Adsorption.

Adsorption   The increase in quantity of a component at an interface or in an interfacial layer. In most usage, it is positive, but it can be negative (depletion); in this sense, negative adsorption is a different process from desorption. Adsorption may also denote the process of components accumulating at an interface.

GLOSSARY

Adsorption Isotherm   The mathematical relationship between the equilibrium quantity of a material adsorbed and the composition of the bulk phase at constant temperature. *See also* Gibbs Isotherm, Langmuir Isotherm.

Aerated Emulsion   A foam in which the liquid consists of two phases in the form of an emulsion. Also termed foam emulsion. Example: whipped cream consists of air bubbles dispersed in cream, which is an emulsion. *See also* Foam.

Aerating Agent   *See* Foaming Agent.

Aerosol   Colloidal dispersion of liquids or solids in a gas. Distinctions may be made among aerosols of liquid droplets (e.g., fog) and aerosols of solid particles (e.g., smoke and dust). *See also* Mist.

Aging   The process by which the properties of many colloidal systems change with time in storage.

(Petroleum) In crude oils, changes in composition due to oxidation, precipitation of components, bacterial action, or evaporation of low-boiling components.

(Foams and Emulsions) In emulsions or foams, any aggregation, coalescence, creaming, or chemical changes. Aged emulsions and foams frequently have larger droplet sizes.

Agglomeration   The aggregation of particles, droplets, or bubbles in a dispersion. This term is sometimes used to include aggregation and coalescence.

Aggregate   A group of species, usually droplets, bubbles, or particles, that are held together in some way. A micelle can be considered to be an aggregate of surfactant molecules.

Aggregation   The process of forming a group of droplets, bubbles, or particles that are held together in some way. Sometimes referred to interchangeably as coagulation or flocculation The reverse process is termed deflocculation or peptization.

Aggregation Number   The number of surfactant molecules comprising a micelle. Example: The aggregation number for sodium dodecyl sulfate in water is about 70.

Alkane Carbon Number   (ACN) *See* Equivalent Alkane Carbon Number.

Amphipathic   Having both lyophilic and lyophobic groups (properties) in the same molecule, as in most surfactants. Also referred to as amphiphilic.

Amphiphilic   *See* Amphipathic.

Amphoteric Surfactant   A surfactant molecule for which the ionic character of the polar group depends on solution pH. For example, lauramidopropylbetaine, $C_{11}H_{23}CONH(CH_2)_3N^+(CH_3)_2CH_2COO^-$ is positively charged at low pH but is electrically neutral, having both positive and negative charges at intermediate pH. Other combinations are possible, and some amphoteric surfactants are negatively charged at high pH. *See also* Zwitterionic Surfactant.

Anionic Surfactant   A surfactant molecule whose polar group is negatively charged. Example: sodium dodecyl sulfate, $CH_3(CH_2)_{11}SO_4^-Na^+$.

Antibubbles   A dispersion of liquid-in-gas-in-liquid wherein a droplet of liquid is surrounded by a thin layer of gas that in turn is surrounded by bulk liquid (*8*). Example: In an air–aqueous surfactant solution system this dispersion would be designated as water-in-air-in-water, or W/A/W, in fluid film terminology. A liquid–liquid analogy can be drawn with the structures of multiple emulsions. *See also* Fluid Film.

Antifoaming Agent   Any substance that acts to reduce the stability of a foam; it may also act to prevent foam formation. In some usage, terms such as antifoamer or foam inhibitor are used to specify the prevention of foaming, and terms such as defoamer or foam breaker are used to specify the reduction or elimination of foam stability. Example: poly(dimethylsiloxane)s, $(CH_3)_3SiO[(CH_3)_2SiO]_xR$, where R represents any of a number of organic functional groups.

Aphrons   *See* Microgas Emulsions.

API Gravity   A measure of the relative density (specific gravity) of petroleum liquids. The API gravity in degrees is given by °API = (141.5/relative density) − 131.5, where the relative density at temperature T (°C) = (density at T)/(density of water at 15.6 °C).

Apparent Viscosity   Viscosity determined for a non-Newtonian fluid without reference to a particular shear rate for which it applies. Such viscosities are usually determined by a method strictly applicable to Newtonian fluids only.

Areal Elasticity   *See* Film Elasticity.

Asphalt   A naturally occurring hydrocarbon that is a solid at reservoir temperatures. An asphalt residue may also be prepared from heavy (asphaltic) crude oils or bitumen, from which lower-boiling fractions have been removed.

Asphaltene   A high molecular mass, polyaromatic component of some crude oils that also has high sulfur, nitrogen, oxygen, and metal contents. In practical work, asphaltenes are usually defined operationally by using a

GLOSSARY 491

standardized separation scheme. One such scheme defines asphaltenes as those components of a crude oil or bitumen that are soluble in toluene but insoluble in *n*-pentane.

Association Colloid   A dispersion of colloidal-sized aggregates of small molecules. These are lyophilic dispersions. Example: micelles of surfactant molecules in water.

Attractive Potential Energy   See Gibbs Energy of Attraction.

Autophobicity   See Spreading Coefficient.

A/O/W   An abbreviation for a fluid film of oil between air and water. Usually designated W/O/A. *See also* Fluid Film.

A/W/A   An abbreviation for a fluid film of water in air. *See also* Fluid Film.

A/W/O   An abbreviation for a fluid film of water between air and oil phases. Usually designated O/W/A. *See also* Fluid Film.

Basic Sediment and Water   That portion of solids and aqueous solution in an emulsion that separates out on standing or is separated by centrifuging in a standardized test method. Basic sediment may also contain emulsified oil. Also referred to as BS&W, BSW, bottom settlings and water, or bottom solids and water.

Batch Treating   In oil production or processing, the process in which a foam, froth, or emulsion is collected in a tank and then broken in a batch. This process is as opposed to continuous or flow-line treating.

Beaker Test   See Bottle Test.

Bilayer   See Bimolecular Film.

Biliquid Foam   A concentrated emulsion of one liquid dispersed in another.

Bimolecular Film   A membrane that separates two aqueous phases and is composed of polar organic molecules such as surfactants or lipids that are oriented with their hydrocarbon groups together and their polar groups facing the respective aqueous phases. *See also* Black Lipid Membrane.

Bingham Plastic Fluid   See Plastic Fluid.

Bitumen   A naturally occurring viscous hydrocarbon having a viscosity greater than 10,000 mPa·s at ambient deposit temperature and a density greater than 1000 kg/m^3 at 15.6 °C. In addition to high molecular mass hydrocarbons, bitumen contains appreciable quantities of sulfur, nitrogen, oxygen, and heavy metals.

Black Film Fluid films yield interference colors in reflected white light that are characteristic of their thickness. At a thickness of about 0.1 µm, the films appear white and are termed silver films. At reduced thicknesses, they first become grey and then black (black films). Among thin equilibrium (black) films, one may distinguish those that correspond to a primary minimum in interaction energy, typically at about 5-nm thickness (Newton black films) from those that correspond to a secondary minimum, typically at about 30-nm thickness (common black films).

Black Lipid Membrane A bimolecular film in which the molecules comprising the membrane film are lipid molecules. The term "black" refers to the fact that these films appear black when illuminated (no apparent interference colors). *See also* Bimolecular Film, Black Film.

Blender Test An empirical test in which an amount of potential foaming agent is added into a blender containing a specified volume of liquid to be foamed. After blending at a specified speed and for some specified time, the blending is halted, and the extent (volume) of foam produced is measured both immediately and after a period of time of quiescent standing. There are many variations of this test. *See also* Bottle Test.

BLM *See* Black Lipid Membrane.

Bottle Test (Foams) An empirical test in which an amount of potential foaming agent is added into a bottle containing a specified volume of liquid to be foamed. After shaking the bottle in a specific manner and for some specified time, the shaking is halted, and the extent (volume) of foam produced is measured both immediately and after a period of time of quiescent standing. There are many variations of this test. *See also* Blender Test.

(Emulsions) An empirical test in which varying amounts of a potential demulsifier or coagulant are added into a series of tubes or bottles containing subsamples of an emulsion or other dispersion that is to be broken or coagulated. After some specified time, the extent of phase separation and appearance of the interface separating the phases are noted. There are many variations of this test. For emulsions, in addition to the demulsifier, a diluent may be added to reduce viscosity. In the centrifuge test, centrifugal force may be added to speed up the phase separation. Again, there are many variations of this test. Other synonyms include jar test and beaker test.

Bottom Settlings and Water *See* Basic Sediment and Water.

Bottom Solids and Water *See* Basic Sediment and Water.

Boussinesq Number A measure of the ratio of interfacial and bulk viscous effects in a thinning foam film: $B_o = (\eta^s + \kappa^s)/(\eta R_f)$ where η^s is

GLOSSARY

the surface-shear viscosity, κ^s is the surface-dilational viscosity, η is the bulk liquid viscosity, and R_f is the thin film radius. *See also* Chapter 2.

Breaking The process in which a foam, froth, or emulsion separates. Usually coalescence causes the separation of a macrophase, and eventually the formerly dispersed phase becomes a continuous phase, separate from the original continuous phase.

Brownian Motion Random motion of small dispersed particles in a liquid caused by random fluctuations in density.

BS&W *See* Basic Sediment and Water.

Bubble Snap-Off *See* Snap-Off.

Bulk Foam Any foam for which the length scale of the confining space is greater than the length scale of the foam bubbles. The converse case categorizes some foams in porous media, distinguished by the term "lamellar foam". *See also* Foam, Foam Texture.

Capillarity A general term referring either to the general subject of or to the various phenomena attributable to the forces of surface or interfacial tension. The Young–Laplace equation is sometimes referred to as the equation of capillarity.

Capillary A tube having a very small internal diameter. Originally, the term referred to cylindrical tubes whose internal diameters were of similar dimension to hairs.

Capillary Constant A dimensionless number, a, is used in considerations of capillarity such as in capillary rise. For two phases in contact, $a^2 = 2\gamma/\Delta\rho g$ where γ is the surface (or interfacial) tension, g is the gravitational constant, and $\Delta\rho$ is the density difference between the phases.

Capillary Flow Liquid flow in response to a difference in pressures across curved interfaces. *See also* Capillary Pressure.

Capillary Forces The interfacial forces acting among oil, water, and solid in a porous medium. These determine the pressure difference (capillary pressure) across an oil–water interface in a pore. Capillary forces are largely responsible for oil entrapment under typical reservoir conditions.

Capillary Number (N_c) A dimensionless ratio of viscous forces to capillary forces. One form gives N_c as velocity times viscosity divided by interfacial tension. It is used to indicate how strongly trapped residual oil is in a porous medium.

Capillary Pressure The pressure difference across an interface between two phases. When the interface is contained in a capillary, it is sometimes referred to as the suction pressure. In petroleum reservoirs, it is the local pressure difference across the oil–water interface in a pore contained in a porous medium. Because one of the liquids usually preferentially wets the solid, the capillary pressure is normally taken as the pressure in the nonwetting fluid minus that in the wetting fluid.

Capillary Ripples   Surface or interfacial waves caused by perturbations of an interface. When the perturbations are caused by mechanical means (e.g., barrier motion), the transverse waves are known as capillary ripples or Laplace waves, and the longitudinal waves are known as Marangoni waves. The characteristics of these waves depend on the surface tension and the surface elasticity. This property forms the basis for the capillary wave method of determining surface or interfacial tension.

Capillary Rise   The tendency and process for a liquid to rise in a capillary. Example: Water rises in a partially immersed glass capillary. Negative capillary rise occurs when the liquid level in the capillary falls below the level of bulk liquid, as when a glass capillary is partially immersed in mercury. Capillary rise forms the basis for a method of determination of surface or interfacial tension.

Capillary Viscometer   An instrument used for the measurement of viscosity in which the rate of flow through a capillary under constant applied pressure difference is determined. This method is most suited to the determination of Newtonian viscosities. Various designs are used, such as the Ostwald and Ubbelohde types.

Capillary Waves   *See* Capillary Ripples.

Cationic Surfactant   A surfactant molecule whose polar group is positively charged. Example: cetyltrimethylammonium bromide, $CH_3(CH_2)_{15}N^+(CH_3)_3Br^-$.

CCC   *See* Critical Coagulation Concentration.

Centrifuge   An apparatus in which an applied centrifugal force is used to achieve a phase separation by sedimentation or creaming. For centrifuges operating at very high relative centrifugal forces (RCF or g-forces), the terms supercentrifuge (ca. tens of thousands of RCF, or gs) or ultracentrifuge (ca. hundreds of thousands of RCF, or gs). *See also* Relative Centrifugal Force.

Centrifuge Test   *See* Bottle Test.

Charge Density   In colloidal systems, the quantity of charge at an interface expressed per unit area.

Cloud Point   The temperature at which a nonionic surfactant starts to lose some of its water solubility and to become ineffective as a surfactant.

The originally transparent solution becomes cloudy because of the separation of a surfactant-rich phase. *See also* Coacervation.

CMC   *See* Critical Micelle Concentration.

Coacervate   *See* Coacervation.

Coacervation   When a lyophilic colloid looses stability, a separation into two liquid phases may occur. This process is termed coacervation. The phase that is more concentrated in the colloid is the coacervate, and the other phase is the equilibrium solution. *See also* Cloud Point.

Coagulation   *See* Aggregation.

Coalescence   The merging of two or more dispersed species into a single one. Coalescence reduces the total number of dispersed species and also the total interfacial area between phases. In emulsions and foams, coalescence can lead to the separation of a macrophase, in which case the emulsion or foam is said to break.

Co-ions   Ions that, compared to the colloidal ions in a system, have low molecular mass and the same charge sign. For example, in a suspension of negatively charged clay particles containing dissolved sodium chloride, the chloride ions are co-ions, and the sodium ions are counterions. *See also* Counterions.

Collector   A surfactant used in froth flotation to adsorb onto solid particles, make them hydrophobic, and thus facilitate their attachment to gas bubbles. *See also* Froth Flotation.

Colloid Stability   In colloid science, an indication that a specified process, such as aggregation, does not proceed at a significant rate, which is different from the definition of thermodynamic stability (4). The term colloid stability must be used with reference to a specific and clearly defined process, for example, a colloidally metastable emulsion may signify a system in which the droplets do not participate in aggregation, coalescence, or creaming at a significant rate.

Colloidal   A state of subdivision in which the particles, droplets, or bubbles, which are dispersed in another phase, have at least one dimension between about 1 and 1000 nm. A colloidal dispersion is a system in which colloidal species are dispersed in a continuous phase of different composition or state.

Colloidal Electrolyte   An electrolyte that dissociates to yield ions at least one of which is of colloidal or near-colloidal size. Example: ionic surfactants.

Common Black Film   *See* Black Film.

Compressibility   *See* Film Elasticity.

Compressional Modulus   *See* Film Elasticity.

Concentric Cylinder Rheometer   *See* Rheometer.

Condensate   Any light hydrocarbon liquid mixture obtained from the condensation of hydrocarbon gases. Condensate typically contains mostly propane, butane, and pentane.

Contact Angle   When two immiscible fluids (e.g., liquid–gas or oil–water) are both in contact with a solid, the angle formed between the solid surface and the surface of the more dense fluid phase is termed the contact angle. It is essential to state which interfaces are used to define the contact angle, although by convention, if one of the fluids is water, then the contact angle is the angle measured through the water phase. Distinctions may be made among advancing, receding, or equilibrium contact angles.

Continuous Phase   In a colloidal dispersion, the phase in which another phase of particles, droplets, or bubbles is dispersed. Sometimes referred to as the external phase. The opposite of Dispersed Phase.

Cosorption Lines   Contours of equal surface activity as measured by the Gibbs surface excess concentrations plotted on phase diagrams. *See* p 131 of reference 9.

Cosurfactant   Any chemical, whether surface active by itself or not, that may be added to a system to enhance the effectiveness of a surfactant.

Couette Flow   The flow of liquid in the annulus between two concentric cylinders that rotate at different speeds. In the Couette rheometer, one cylinder rotates, and torque is measured at the other. *See also* Rheometer.

Couette Rheometer   *See* Couette Flow.

Counterions   Ions that, compared to the colloidal ions in a system, have low molecular mass and opposite charge sign. For example, clay particles are usually negatively charged and are naturally associated with exchangeable counterions such as sodium and calcium. In the early literature, the term Gegenion was used to mean counterion. *See also* Co-ions.

Creaming   The process of droplets, or other species, floating upward under gravity or a centrifugal field to form a concentrated dispersed phase (cream) quite distinct from the underlying, more dilute dispersion. This is not the same as the breaking of an emulsion. *See also* Sedimentation.

Critical Coagulation Concentration   (CCC) The electrolyte concentration that marks the onset of coagulation. The CCC is very system-specific, although the variation in CCC with electrolyte composition has been empirically generalized. *See also* Schulze–Hardy Rule.

Critical Film Thickness   A fluid film may thin to a narrow range of film thicknesses within which it either becomes metastable to thickness changes (equilibrium film) or else ruptures. Persistent foams comprise fluid films at their critical film thickness.

Critical Micelle Concentration   (CMC) The surfactant concentration above which molecular aggregates, termed micelles, begin to be formed. In practice, a narrow range of surfactant concentrations represents the transition from a solution in which only single, unassociated surfactant molecules (monomers) are present to a solution containing micelles. *See also* Micelle.

Critical Thickness   *See* Critical Film Thickness.

Crude Oil   A naturally occurring hydrocarbon produced from an underground reservoir. *See also* Asphalt, Bitumen, Extra Heavy Crude Oil, Heavy Crude Oil, Light Crude Oil, Oil.

Darcy's Law   *See* Permeability.

Debye–Hückel Parameter   *See* Debye Length.

Debye Length   A parameter in the Debye–Hückel theory of electrolyte solutions, κ^{-1}. For aqueous solutions at 25 °C, $\kappa = 3.288\sqrt{I}$ in reciprocal nanometers, where I is the ionic strength of the solution. The Debye length is also used in the DLVO theory, where it is referred to as the electric double-layer thickness. *See also* Electric Double-Layer Thickness.

Deflocculation   The reverse of aggregation (or flocculation or coagulation). Peptization means the same thing.

Defoamer   *See* Foam Breaker, Antifoaming Agent.

Demulsifier   (Chemical) Any agent added to an emulsion that causes or enhances the rate of breaking of the emulsion (separation into its constituent liquid phases). Demulsifiers may act by any of a number of different mechanisms, which usually include enhancing the rate of droplet coalescence.
(Device) Any device that is used to break emulsions. Such devices may employ chemical, electrical, or mechanical means or a combination to break an emulsion and cause separation into its constituent liquid phases.

Depressant   Any agent that may be used in froth flotation to selectively reduce the effectiveness of collectors for certain mineral components. *See also* Froth Flotation.

Desorption   The process by which the amount of adsorbed material becomes reduced. That is, the converse of adsorption. Desorption is a different process from negative adsorption. *See also* Adsorption.

Detergent A detergent is a surfactant that has cleaning properties in dilute solutions. As a commercial cleaning product, detergents are actually formulations containing a number of chemical components, including surfactants, builders, bleaches, brighteners, enzymes, opacifiers, and fragrances. Detergent solutions may or may not have foaming properties.

Dewetting In antifoaming, where a droplet or particle of antifoaming agent enters the gas–liquid interface and displaces some of the original liquid from the interface. Because the liquid is usually an aqueous phase, the process is sometimes referred to as dewetting.

Differential Viscosity (η_D) The rate of change of shear stress (τ) with respect to shear rate ($\dot\gamma$), taken at a specific shear rate $\eta_D = d\tau/d\dot\gamma$.

Diffuse Double Layer *See* Electric Double Layer.

Dilatant A fluid for which viscosity increases as the shear rate increases. Also termed shear-thickening.

Dilatational Elasticity *See* Film Elasticity.

Discontinuous Phase *See* Dispersed Phase.

Disjoining Pressure The negative derivative with respect to distance of the Gibbs energy of interaction per unit area yields a force per unit area between colloidal species, termed the disjoining pressure. Example: In a thin liquid film, the disjoining pressure equals the pressure, beyond the external pressure, that has to be applied to the liquid in the film in order to maintain a given film thickness.

Dispersant Any species that may be used to aid in the formation of a colloidal dispersion. Often a surfactant.

Disperse *See* Dispersion.

Disperse Phase *See* Dispersed Phase.

Dispersed Phase In a colloidal dispersion, the phase that is distributed, in the form of particles, droplets, or bubbles, in a second, immiscible phase that is continuous. Also referred to as the disperse, discontinuous, or internal phase.

Dispersing Agent *See* Dispersant.

Dispersion (Colloids) A system in which finely divided droplets, particles, or bubbles are distributed in another phase. As it is usually used, dispersion implies a distribution without dissolution. An emulsion is an example of a colloidal dispersion. *See also* Colloidal.

GLOSSARY 499

(Oil Recovery) The mixing by convection of fluids flowing in a porous medium.

Dispersion Forces   The dipolar interaction forces between any two bodies of finite mass, including the Keesom forces of orientation among dipoles, Debye induction forces, and London forces between two induced dipoles. Also referred to as Lifshitz–van der Waals forces.

Dissolved Gas Flotation   See Froth Flotation.

Division   See Lamella Division.

DLVO Theory   An acronym for a theory of the stability of colloidal dispersions developed independently by B. Derjaguin and L. D. Landau in one laboratory (*10*), and by E. J. W. Verwey and J. Th. G. Overbeek in another (*11*). The theory was developed to account for the stability against aggregation of electrostatically charged particles in a dispersion.

Double Layer   See Electric Double Layer.

Dry Foam   See Foam.

du Nouy Ring Method   A method for determining surface or interfacial tension based on measuring the force needed to pull an inert ring through an interface. *See also* Wilhelmy Plate Method.

Duplex Film   Any film that is thick enough for each of its two interfaces to be independent of each other and exhibit their own interfacial tensions. A duplex film is thus thicker than a monomolecular film and covers an entire surface as opposed to a lens.

Dynamic Interfacial or Surface Tension   The surface or interfacial tensions that may change dramatically as a function of the age of the surface or interface. These preequilibrium tensions are distinguished from equilibrium (limiting) tensions. *See also* Equilibrium Interfacial or Surface Tension.

EACN   See Equivalent Alkane Carbon Number.

EDL   See Electric Double Layer.

Effective Foam Viscosity   For foam flowing in porous media, the foam's effective viscosity is that calculated from Darcy's law. This value is an approximation because foams are compressible and are also usually non-Newtonian.

Elasticity   The ability of a material to change its physical dimensions when a force is applied to it, and then restore its original dimensions when the force is removed. *See also* Surface Elasticity.

Elasticity Number   A dimensionless quantity (E_s) characterizing the surface tension gradient in a thinning foam film. For systems containing only one surfactant: $E_s = -(d\gamma/d\ln\rho^s)(R_f/[\eta D])$ where γ is the surface tension, ρ^s is the surface density of the surfactant, R_f is the thin-film radius, η is the bulk liquid viscosity, and D is the diffusivity of the surfactant. See also Chapter 2.

Electric Double Layer (EDL) An idealized description of the distribution of free charges in the neighborhood of an interface. Typically, the surface of a charged species is viewed as having a fixed charge of one sign (one layer), and oppositely charged ions are distributed diffusely in the adjacent liquid (the second layer). The liquid layer may be considered to be made up of a relatively less mobile Stern layer in close proximity to the surface, and a relatively more diffuse layer (Gouy layer or Gouy–Chapman layer) at greater distance.

Electric Double-Layer Thickness A measure of the decrease of potential with distance in an electric double layer. It is the distance over which the potential falls to 1/e, about one-third, of the value of the surface potential. Also termed the Debye length.

Electrokinetic A general adjective referring to the relative motions of charged species in an electric field. The motions may be of either charged dispersed species or of the continuous phase, and the electric field may be either an externally applied field or else created by the motions of the dispersed or continuous phases. Electrokinetic measurements are usually aimed at determining zeta potentials.

Electrokinetic Potential See Zeta Potential.

Electrophoresis The motion of colloidal species caused by an imposed electric field. The term replaces the older term cataphoresis. The species move with an electrophoretic velocity that depends on their electric charge and the electric-field gradient. The electrophoretic mobility is the electrophoretic velocity per unit electric-field gradient and is used to characterize specific systems.

Electrophoretic Mobility or Velocity See Electrophoresis.

Electrosteric Stabilization The stabilization of a dispersed species by a combination of electrostatic and steric repulsions.

Electroviscous Effect The increase in apparent viscosity of a dispersion of charged species caused by their mutual electrostatic repulsion.

Ellipsometry The subject concerned with the behavior of light when it passes through or is reflected by an interface. In ellipsometry, an incident beam of plane-polarized light is caused to be reflected from a coated surface, and the degree of elliptical polarization produced is measured. This measurement allows the thickness of the surface coating to be determined.

Emulsifier   Any agent that acts to stabilize an emulsion. The emulsifier may make it easier to form an emulsion, provide stability against aggregation, and provide stability against coalescence. Emulsifiers are frequently but not necessarily surfactants.

Emulsion   A dispersion of droplets of one liquid in another immiscible liquid in which the droplets are of colloidal or near-colloidal sizes. The term emulsion may also be used to refer to colloidal dispersions of liquid-crystals in a liquid. *See also* Macroemulsion, Microemulsion, Miniemulsion.

Enhanced Oil Recovery   The third phase of crude oil production, in which chemical, miscible fluid, or thermal methods are applied to restore production from a depleted reservoir. Also known as tertiary oil recovery. *See also* Primary Oil Recovery, Secondary Oil Recovery.

Equation of Capillarity   *See* Young–Laplace Equation.

Equilibrium Contact Angle   The contact angle that is measured when all contacting phases are in equilibrium with each other. The term arises because either or both of the advancing or receding contact angles may differ from the equilibrium value. It is essential to state which interfaces are used to define the contact angle. *See also* Contact Angle.

Equilibrium Film   *See* Fluid Film.

Equilibrium Interfacial or Surface Tension   The surface or interfacial tensions may change dynamically as a function of the age of the surface or interface. The equilibrium (or limiting) tensions are distinguished from the dynamic tensions. *See also* Surface Tension.

Equivalent Alkane Carbon Number   (EACN) Each surfactant, or surfactant mixture, in a reference series will produce a minimum interfacial tension (IFT) for a different n-alkane. For any crude oil or oil component, a minimum IFT will be observed against one of the reference surfactants. The EACN for the crude oil refers to the n-alkane that would yield minimum IFT against that reference surfactant. The EACN thus allows predictions to be made about the interfacial tension behavior of a crude oil in the presence of surfactant. *See* references 12 and 13.

Equivalent Film Thickness   Refers to an experimentally determined fluid film thickness for which some assumptions about the structure and properties of the film have been made. The experimental technique used should also be stated when using this term.

Evanescent Foam   A transient foam that has no thin-film persistence and therefore is very unstable. Such foams exist only where new bubbles can be created faster than existing bubbles can rupture. Examples: air bubbles blown rapidly into pure water; the foam created when a champagne bottle is opened.

Excess Quantities   *See* Gibbs Surface.

Expansion Factor   In foaming, the ratio of foam volume produced to the volume of liquid used to make the foam. Also termed the expansion ratio.

External Phase   *See* Continuous Phase.

Extra Heavy Crude Oil   A naturally occurring hydrocarbon having a viscosity less than 10,000 mPa·s at ambient deposit temperature and a density greater than 1000 kg/m^3 at 15.6 °C.

Fatty Acid Soaps   A class of surfactants comprising the salts of aliphatic carboxylic acids having hydrocarbon chains of between 6 and 20 carbon atoms. Fatty acid soaps are no longer restricted to molecules having their origin in natural fats and oils.

Fatty Alcohol Surfactants   The class of primary alcohol surfactants having hydrocarbon chains of between 6 and 20 carbon atoms. Fatty alcohol surfactants are no longer restricted to molecules having their origin in natural fats and oils.

Film   Any layer of material that covers a surface and is thin enough to not be significantly influenced by gravitational forces. *See also* Duplex Film, Monolayer Adsorption.

Film Balance   A shallow trough that is filled with a liquid on top of which is placed material that may form a monolayer. The surface area available can be adjusted by movable barriers, and any surface pressure thus created can be measured by means of a float. Also called Langmuir Film Balance, Langmuir Trough, and Pockels–Langmuir–Adam–Wilson–McBain Trough or PLAWM Trough.

Film Compressibility   The ratio of relative area change to differential change in surface tension. *See also* Film Elasticity.

Film Drainage   The drainage of liquid from a lamella of liquid separating droplets or bubbles of another phase (i.e., in a foam or emulsion). Also termed thin film drainage. *See also* Fluid Film.

Film Elasticity   The differential change in surface tension of a surface film with relative change in area. Also termed surface elasticity, dilatational elasticity, areal elasticity, compressional modulus, surface dilatational modulus, or modulus of surface elasticity. For fluid films, the surface tension of one surface is used. The Gibbs film (surface) elasticity is the equilibrium value. If the surface tension is dynamic (time-dependent) in character, then for nonequilibrium values, the term Marangoni film (surface) elasticity is used. The compressibility of a film is the inverse of the film elasticity.

Film Element   Any small homogeneous region of a thin film. The film element includes the interfaces.

Film Pressure   The pressure, in two dimensions, exerted by an adsorbed monolayer. It is formally equal to the difference between the surface tension of pure solvent and that of the solution of adsorbed solute. It can be measured by using the Film Balance.

Flocculation   *See* Aggregation. The products of the flocculation process are referred to as flocs.

Flotation   *See* Froth Flotation, Sedimentation.

Flow-Line Treating   In oil production or processing, the process in which a foam, froth, or emulsion is continuously broken and separated. *See also* Treater.

Fluid Film   A thin fluid phase, usually taken to be films of thickness of less than about 1 μm. Such films may be specified by abbreviations similar to those used for emulsions; for example, some common designations are A/W/A for a water film in air, W/O/W for an oil film in water, O/W/O for a water film in oil, W/O/A for an oil film between water and air. Fluid films are usually unstable to breakage because of thinning to the point of allowing contact of the separating phase(s), that is, to rupture. However, at some film thicknesses, a film may be stable or metastable to thickness changes. Films with this property are equilibrium films. Otherwise fluid films may be distinguished by rapid (mobile film) or slow (rigid film) thickness changes. *See also* Black Film.

Foam   A dispersion of gas bubbles in a liquid or solid in which at least one dimension falls within the colloidal size range. Thus a foam typically contains either very small bubble sizes or, more commonly, quite large gas bubbles separated by thin liquid films. The thin liquid films are called lamellae (or laminae). Sometimes distinctions are drawn between concentrated foams, in which liquid films are thinner than the bubble sizes and the gas bubbles are polyhedral (Dry Foam and Polyederschaum), and low concentration foams in which the liquid films have thicknesses on the same scale or larger than the bubble sizes and the bubbles are approximately spherical (Gas Emulsions, Gas Dispersions, Wet Foams, and Kugelschaum). Bulk foams may also be distinguished from lamellar foams. *See also* Aerated Emulsion, Foam Texture, Froth.

Foam Booster   *See* Foaming Agent.

Foam Breaker   Any agent that reduces or eliminates foam stability. Also termed defoamer, but a more general term is Antifoaming Agent.

Foam Drainage   The drainage of liquid from liquid lamellae separating bubbles in a foam. *See* Fluid Film.

Foam Emulsion   *See* Aerated Emulsion.

Foam Flooding   (Enhanced Oil Recovery) The process in which a foam is made to flow through an underground reservoir. The foam, which may either be generated on the surface and injected or generated in situ, is used to increase the drive fluid viscosity and improve its sweep efficiency.

(Petroleum Processing) In refinery distillation and fractionation towers, the occurrence of foams that can carry liquid into regions of the towers intended for vapor.

Foam Fractionation   A separation method in which a component of a liquid that is preferentially adsorbed at the liquid–gas interface is removed by foaming the liquid and collecting the foam that is produced. Foaming surfactants can be separated in this manner.

Foam Inhibitor   Any agent that prevents foaming. Also termed foam preventative, but a more general term is Antifoaming Agent.

Foam Number   A relative drainage rate test for foams in which a foam is formed in a vessel and thereafter the remaining foam volume determined as a function of time. The foam number is the volume of bulk liquid that has separated after a specified time interval, expressed as a percentage of the original volume of liquid foamed.

Foam-over   In an industrial process vessel, unwanted foam may occasionally build up to such an extent that it is carried out of the top of the vessel ("foam-over") and on to the next part of the process. This carryover of foam and any entrained material that comes with it is frequently detrimental to other parts of a processing operation. *See also* Chapter 13.

Foam Preventative   *See* Antifoaming Agent, Foam Inhibitor.

Foam Quality   The gas volume fraction in a foam. Expressed as a percentage, this is sometimes referred to as Mitchell foam quality. In three-phase systems, other measures are used. For example, when foams are formulated to contain solid particles as well, the slurry quality, Q_s, which gives the volume fraction of gas plus solid, may be used: $Q_s = (V_g + V_s)/(V_g + V_s + V_l)$ where V_g, V_s, and V_l denote the volumes of gas, solid, and liquid, respectively. *See also* Chapter 9.

Foam Stimulation Fluid   An aqueous or nonaqueous foam that is injected into a petroleum reservoir in order to improve the productivity of oil- or gas-producing wells. Some mechanisms of action for foam stimulation fluids include fracturing, acidizing to increase permeability, and diversion of flow.

Foam Texture   The bubble size distribution in a foam. For foams in porous media, foam texture may be expressed in terms of the length scale of foam bubbles as compared to that for the spaces confining the foam. When the length scale of the confining space is comparable to or less than the length scale of the foam bubbles, then the foam is sometimes termed lamellar foam in order to distinguish from the opposite case, termed bulk foam.

Foamer   See Foaming Agent.

Foaming Agent   Any agent that stabilizes a foam. The foaming agent may make it easier to form a foam and provide stability against coalescence. Foaming agents are usually surfactants. Also termed Foam Booster, Whipping Agent, and Aerating Agent.

Foaming Power   See Increase of Volume upon Foaming.

Free Water   The readily separated, nonemulsified, water that is coproduced with oil from a production well.

Froth   A type of foam in which solid particles are also dispersed in the liquid (in addition to the gas bubbles); the solid particles may be the stabilizing agent. The term froth is sometimes used to refer simply to a concentrated foam, which is not preferred usage. Example: Bituminous froth is a nonaqueous (oleic) foam that also contains water and solids (*See also* Chapter 11).

Froth Flotation   A separation process using flotation, in which particulate matter becomes attached to gas (foam) bubbles. The flotation process produces a product layer of concentrated particles in foam, termed froth. Variations include dissolved-gas flotation in which gas is dissolved in water that is added to a colloidal dispersion. As microbubbles come out of solution they attach to and float the colloidal species.

Frother   See Frothing Agent.

Frothing Agent   Any agent that acts to stabilize a froth. The frothing agent, usually a surfactant, may make it easier to form a froth and provide stability against coalescence. Analogous to Foaming Agent.

Gas Aphrons   See Microgas Emulsions.

Gas Dispersion   See Foam, Gas Emulsion, Solid Foam.

Gas Emulsion   "Wet" foams in which the liquid lamellae have thicknesses on the same scale or larger than the bubble sizes. Typically in these cases, the gas bubbles have a spherical rather than polyhedral shape. Other synonyms include Gas Dispersion and Kugelschaum. In petroleum production, gas emulsion specifies crude oil that contains a small volume fraction of dispersed gas. *See also* Foam.

Generalized Plastic Fluid   A fluid characterized by both of the following: the existence of a finite shear stress that must be applied before flow begins (yield stress), and pseudoplastic flow at higher shear stresses. *See also* Bingham Plastic Fluid.

Gibbs Adsorption   See Gibbs Isotherm, Gibbs Surface.

Gibbs Effect   The decrease in surface or interfacial tension that occurs as surfactant concentration increases toward the critical micelle concentration.

Gibbs Elasticity   See Film Elasticity.

Gibbs Energy of Attraction   When two dispersed-phase species approach, they may attract each other because of such forces as the London–van der Waals forces. The Gibbs energy of attraction is the difference between Gibbs attractive energies of the system at a specified separation distance and infinite separation. IUPAC (4) discourages the use of the synonyms Potential Energy of Attraction and Attractive Potential Energy, although they are still in common usage. *See also* Gibbs Energy of Interaction, Gibbs Energy of Repulsion.

Gibbs Energy of Interaction   When two dispersed-phase species approach, they experience repulsive and attractive forces as electrostatic repulsion and van der Waals attraction. The Gibbs energy of interaction is the difference between Gibbs energies of the system at a specified separation distance and infinite separation. An example of such a diagram is that calculated from DLVO theory. IUPAC (4) discourages the use of the synonyms Potential Energy of Interaction and Total Potential Energy of Interaction, although they are still in common usage. *See also* DLVO Theory, Gibbs Energy of Attraction, Gibbs Energy of Repulsion, Primary Minimum.

Gibbs Energy of Repulsion   When two dispersed-phase species approach, they may repel each other because of such forces as electrostatic repulsion. The Gibbs energy of repulsion is the difference between Gibbs repulsive energies of the system at a specified separation distance and infinite separation. IUPAC (4) discourages the use of the synonyms Potential Energy of Repulsion and Repulsive Potential Energy, although they are still in common usage. *See also* Gibbs Energy of Attraction, Gibbs Energy of Interaction.

Gibbs Film Elasticity   See Film Elasticity.

Gibbs Isotherm   An equation that relates the Gibbs surface excess (amount adsorbed) to the change in interfacial tension with activity of the adsorbing species.

Gibbs—Marangoni Effect The effect in thin liquid films and foams when stretching an interface causes the surface excess surfactant concentration to decrease, hence surface tension to increase (Gibbs effect); the surface tension gradient thus created causes liquid to flow toward the stretched region and provides both a "healing" force and also a resisting force against further thinning (Marangoni effect). Sometimes referred to simply as the Marangoni effect.

Gibbs—Plateau Border See Plateau Border.

Gibbs Ring See Plateau Border.

Gibbs Surface A geometrical surface chosen parallel to the interface and used to define surface excess properties such as the extent of adsorption. The surface excess amount of adsorption is the excess amount of a component actually present in a system over that present in a reference system of the same volume as the real system, and in which the bulk concentrations in the two phases remain uniform up to the Gibbs dividing surface. The terms surface excess concentration or surface excess have now replaced the earlier term superficial density.

Gibbs Surface Concentration The Gibbs surface excess adsorption amount divided by the area of the interface.

Gibbs Surface Elasticity See Film Elasticity.

Gouy—Chapman Layer See Electric Double Layer.

Gouy Layer See Electric Double Layer.

Hamaker Constant In the description of the London—van der Waals attractive energy between two dispersed bodies, such as particles or droplets, the Hamaker constant is a proportionality constant characteristic of the particle composition. It depends on the internal atomic packing and polarizability of the droplets. Also termed the van der Waals—Hamaker constant.

Head Group The lyophilic functional group in a surfactant molecule. In aqueous systems, it is the polar group of a surfactant. *See also* Surfactant, Surfactant Tail.

Heavy Crude Oil A naturally occurring hydrocarbon having a viscosity less than 10,000 mPa·s at ambient deposit temperature and a density between 934 and 1000 kg/m^3 at 15.6 °C.

Helmholtz Double Layer A simplistic description of the electric double layer as a condenser (the Helmholtz condenser) in which the condenser-plate separation distance is the Debye length. The Helmholtz double layer is divided into an inner Helmholtz plane (IHP) of adsorbed, dehydrated ions immediately next to a surface, and an outer Helmholtz plane (OHP), at the center of a next layer of hydrated, adsorbed ions just inside the imaginary boundary where the diffuse double layer begins. That is, both Helmholtz planes are within the Stern layer.

Helmholtz–Smoluchowski Equation   See Smoluchowski Equation.

Hemimicelle   An aggregate of adsorbed surfactant molecules that may form beyond monolayer coverage, the enhanced adsorption being due to hydrophobic interactions between surfactant tails. Hemimicelles have the form of surface aggregates or of a second adsorption layer with reversed orientation, somewhat like a bimolecular film.

Henry Equation   A relation expressing the proportionality between electrophoretic mobility and zeta potential for different values of the Debye length and size of the species. *See also* Electrophoresis, Hückel Equation, Smoluchowski Equation.

Heterodisperse   A colloidal dispersion in which the dispersed species (droplets, particles, etc.) do not all have the same size. Subcategories are paucidisperse (few sizes) and polydisperse (many sizes). *See also* Monodisperse.

HLB Scale   *See* Hydrophile–Lipophile Balance.

HLB Temperature   *See* Phase Inversion Temperature.

Homogenizer   Any machine for preparing colloidal systems, including foams, by dispersion. Examples: colloid mill, blender, and ultrasonic probe.

Hückel Equation   A relation expressing the proportionality between electrophoretic mobility and zeta potential for the limiting case of a species that can be considered to be small and that has a thick electric double layer. *See also* Electrophoresis, Henry Equation, Smoluchowski Equation.

Hydrophile–Lipophile Balance   (HLB scale) An empirical scale categorizing surfactants in terms of their tendencies to be mostly oil soluble or water soluble, hence their tendencies to promote W/O or O/W emulsions, respectively. *See also* Phase Inversion Temperature.

Hydrophilic   A qualitative term referring to the water-preferring nature of a species (atom, molecule, droplet, particle, etc.). In emulsions, hydrophilic usually means that a species prefers the aqueous phase over the oil phase. In this example, hydrophilic has the same meaning as oleophobic, but such is not always the case.

Hydrophobic A qualitative term referring to the water-avoiding nature of a species (atom, molecule, droplet, particle, etc.). In emulsions, hydrophobic usually means that a species prefers the oil phase over the aqueous phase. In this example, hydrophobic has the same meaning as oleophilic, but such is not always the case. A functional group of a molecule that is not very water soluble is referred to as a hydrophobe.

Hydrophobic Bonding The attraction between hydrophobic species in water that arises from the fact that the solvent–solvent interactions are more favorable than the solvent–solute interactions.

Hydrophobic Index An empirical measure of the relative wetting preference of very small solid particles. In one test method, solid particles to be tested are placed in samples of water containing increasing concentrations of alcohol. The percentage alcohol solution at which the particles just begin to become hydrophilic and sink is taken as the hydrophobic index. *See also* Chapter 12.

Imbibition The uptake of liquid by a porous medium or a gel.

Increase of Volume upon Foaming In foaming, 100 times the ratio of gas volume to liquid volume in a foam. Also termed the foaming power.

Indifferent Electrolyte An electrolyte whose ions have no significant effect on the electric potential of a surface or interface, as opposed to potential-determining ions, which have a direct influence on surface charge. This distinction is most valid for low electrolyte concentrations. Example: For the AgI surface in water, $NaNO_3$ would be an indifferent electrolyte, and both Ag^+ and I^- would be potential-determining ions.

Induced Gas Flotation *See* Froth Flotation.

Inherent Viscosity (η_{inh}) In solutions and colloidal dispersions, the natural logarithm of the relative viscosity (η/η_o), all divided by the solute or dispersed-phase concentration (C). $\eta_{inh} = C^{-1} \ln(\eta/\eta_o)$. In the limit of vanishing concentration, it reduces to the intrinsic viscosity. Also termed the logarithmic viscosity number.

Initial Knockdown Capability *See* Knockdown Capability.

Ink Bottle Pore A description of one kind of shape of pore in a porous medium, in which a narrow throat is connected to a larger pore body. *See also* Porous Medium.

Inner Potential (Stern) In the diffuse electric double layer extending outward from a charged interface, the electrical potential at the boundary between the Stern and the diffuse layer is termed the inner electrical potential. Synonyms include the Stern layer potential or Stern potential. *See also* Electric Double Layer, Zeta Potential.

(Galvani) Alternatively, the electric potential inside a phase. In this case also termed the Galvani Potential.

Interface The boundary between two immiscible phases, sometimes including a thin layer at the boundary within which the properties of one bulk phase change over to become the properties of the other bulk phase. An interfacial layer of finite specified thickness may be defined. When one of the phases is a gas, the term surface is frequently used.

Interface Emulsion An emulsion occurring between oil and water phases in a process separation or treatment apparatus. Such emulsions may have a high solids content and are frequently very viscous. In this case, the term interface is used in a macroscopic sense. Other terms are cuff layer, pad layer, or rag layer emulsions.

Interfacial Film A thin layer of material positioned between two immiscible phases, usually liquids, in which the composition of the layer is different from either of the bulk phases.

Interfacial Layer The plane or layer at an interface that contains adsorbed species. Also termed the surface layer.

Interfacial Potential   *See* Surface Potential.

Interfacial Rheology   *See* Surface Viscosity.

Interfacial Tension   *See* Surface Tension.

Interfacial Viscosity   *See* Surface Viscosity.

Interferometry An experimental technique in which a beam of light is reflected from a film. Light reflected from the front and back surfaces of the film travel different distances and yield interference phenomena, a study of which allows calculation of the film thickness.

Internal Phase The phase in an emulsion that is dispersed into droplets; the dispersed phase. *See also* Dispersed Phase.

Intrinsic Viscosity The specific increase in viscosity, η_{SP}, divided by the dispersed-phase concentration in the limits of both the dispersed-phase concentration approaching infinite dilution, and of shear rate, γ, approaching zero:

$$[\eta] = \lim_{c \to 0} \lim_{\dot{\gamma} \to 0} \eta_{SP}/C$$

Also termed limiting viscosity number.

Inverse Micelle A micelle that is formed in a nonaqueous medium, thus having the surfactants' hydrophilic groups oriented inward away from the surrounding medium.

GLOSSARY

Inversion The process by which one type of emulsion is converted to another, as when an O/W emulsion is transformed into a W/O emulsion, and vice versa.

Invert Emulsion A water-in-oil emulsion. This is different from the term Reverse Emulsion used in the petroleum field.

Ion Exchange A special kind of adsorption in which the adsorption of an ionic species is accompanied by the simultaneous desorption of an equivalent charge quantity of other ionic species.

Ionic Strength A measure of electrolyte concentration given by $I = 1/2 \Sigma c_i z_i^2$ where c_i are the concentrations, in moles per liter of the individual ions, i, and z_i are the ion charge numbers.

Iridescent Layers See Schiller Layers.

Isaphroic Lines Contours of equal foam stability plotted on foam phase diagrams. Example: p 312 of reference 9.

Isoelectric An ionic macromolecule that exhibits no electrophoretic or electroosmotic motion.

Isoelectric Point The solution pH or condition for which the electrokinetic or zeta potential is zero. Under this condition, a colloidal system will exhibit no electrophoretic or electroosmotic motions. *See also* Point of Zero Charge.

Isotherm The mathematical representation of a phenomenon occurring at constant temperature. *See also* Adsorption Isotherm.

Jar Test *See* Bottle Test.

Kelvin Equation An expression for the vapor pressure of a droplet of liquid, $RT \ln(p/p_o) = 2\gamma V/r$ where R is the gas constant, T is the absolute temperature, p is the vapor pressure of the liquid in bulk, p_o is that of the droplet, γ is the surface tension, V is the molar volume, and r is the radius of the liquid. *See also* Young–Laplace Equation.

Kinematic Viscosity Kinematic viscosity is the absolute viscosity of a fluid divided by the density.

Knockdown Capability A measure of the effectiveness of a defoamer. First, a column of foam is generated in a foam-stability apparatus and the foam height is recorded. A measured amount of defoamer is added, and the reduction in foam height over a specified time period, For example, 2 sec, is noted. The knockdown capability is the reduction in foam height. There are many variations of this test. Sometimes referred to as initial knockdown capability. *See also* Chapter 12.

Krafft Point The temperature above which the solubility of a surfactant increases sharply (micelles begin to form). In practice, a narrow range of temperatures represents the transition from a solution in which only single, unassociated surfactant molecules (monomers) can be present to a solution containing micelles. Numerous tabulations are given in references 14 and 15.

Kugelschaum *See* Gas Emulsion.

Lamella *See* Foam.

Lamella Division A mechanism for foam lamella generation in porous media. Typically, when a foam lamella reaches a branch point in a flow channel, then the lamella may divide into two lamellae (bubbles) rather than simply follow one of the two available pathways. *See also* Lamella Leave-Behind, Snap-Off.

Lamella Leave-Behind A mechanism for foam lamella generation in porous media. When gas invades a liquid-saturated region of a porous medium, it may not displace all of the liquid, but rather leave behind liquid lamellae that will be oriented parallel to the direction of the flow. A foam generated entirely by the lamella leave-behind mechanism will be gas-continuous. *See also* Lamella Division, Snap-Off.

Lamellar Foam Although all foams contain lamellae, this term is sometimes used to distinguish a certain kind of foam in porous media. If the length scale of the confining space is comparable to or less than the length scale of the foam bubbles, then the foam is termed "lamellar foam" to distinguish from the opposite case, termed bulk foam. *See also* Foam, Foam Texture.

Lamina *See* Foam.

Laminar Flow A condition of flow in which all elements of a fluid passing a certain point follow the same path, or streamline; there is no turbulence. Also referred to as streamline flow.

Langmuir Adsorption *See* Adsorption Isotherm.

Langmuir Film Balance *See* Film Balance.

Langmuir Isotherm An adsorption isotherm equation that assumes monolayer adsorption and constant enthalpy of adsorption. The amount adsorbed per mass of adsorbent is proportional to equilibrium solute concentration at low concentrations and exhibits a plateau or limiting adsorption at high concentrations. *See also* Adsorption Isotherm.

Langmuir Trough *See* Film Balance.

Laplace Flow *See* Capillary Flow.

GLOSSARY 513

Laplace Waves   *See* Capillary Ripples.

Lather   A foam produced by mechanical agitation on a solid surface. Example: the mechanical generation of shaving foam (lather) on a wet bar of soap.

Leave-Behind   *See* Lamella Leave-Behind.

Lens   (Colloid) Formed by a nonspreading droplet of liquid at an interface. The lens will be thick enough for its shape to be significantly influenced by gravitational forces.

Light Crude Oil   A naturally occurring hydrocarbon having a viscosity less than 10,000 mPa·s at ambient deposit temperature and a density less than 934 kg/m^3 at 15.6 °C.

Limiting Capillary Pressure   For foam flow in porous media, the maximum capillary pressure that can be attained by simply increasing the fraction of gas flow. Foams flowing at steady state do so at or near this limiting capillary pressure. In the limiting capillary pressure regime, the steady-state saturations remain essentially constant.

Limiting Viscosity Number   *See* Intrinsic Viscosity.

Lipophile   The organic liquid-preferring nature of a part of a molecule, the lipophile.

Lipophilic   The (usually fatty) organic liquid-preferring nature of a species. Depending on the circumstances, it may also be a synonym for oleophilic. *See also* Hydrophile–Lipophile Balance.

Lipophobe   The organic liquid-avoiding nature of a part of a molecule.

Lipophobic   The (usually fatty) organic liquid-avoiding nature of a species. Depending on the circumstances, it may also be a synonym for oleophobic.

Liquid Aerosol   *See* Aerosol.

Liquid Crystals   *See* Mesomorphic Phase.

Logarithmic Viscosity Number   *See* Inherent Viscosity.

Lyophilic   General term referring to the continuous-medium- (or solvent)-preferring nature of a species. *See also* Hydrophilic.

Lyophilic Colloid   An older term used to refer to single-phase colloidal dispersions. Examples: polymer and micellar solutions.

Lyophobic   General term referring to the continuous-medium- (or solvent)-avoiding nature of a species. *See also* Hydrophobic.

Lyophobic Colloid   An older term used to refer to two-phase colloidal dispersions. Examples: suspensions, foams, and emulsions.

Macroemulsion   In enhanced oil recovery terminology, the term macroemulsion is sometimes employed to identify emulsions having droplet sizes greater than some specified value, and sometimes simply to distinguish an emulsion from the microemulsion or micellar emulsion types. See also Emulsion.

Macroion   A charged colloidal species whose electric charge is attributable to the presence at the surface of ionic functionalities.

Macromolecule   A large molecule composed of many simple units bonded together. Macromolecules may be naturally occurring, such as humic substances, or synthetic, such as many polymers.

Macropore   See Pore.

Magnetic Resonance Imaging   (MRI Imaging, Nuclear Magnetic Resonance Imaging, NMR Imaging) A technique for imaging and quantifying the distribution of phases in multiphase systems, including dispersions in porous media. The technique employs a homogeneous, static, high magnetic field with a superimposed, time-dependent, linear gradient magnetic field so that the total magnetic field strength depends on position in the sample. Resonance is induced with radio-frequency energy. In the imaging system, the position-dependent resonance frequencies and signal intensities allow the determination of the concentration, chemical environment, and position of any NMR-active nuclei in the sample.

Marangoni Effect   In surfactant-stabilized fluid films, any stretching in the film causes a local decrease in the interfacial concentration of adsorbed surfactant. This lowered concentration causes the local interfacial tension to increase (Gibbs effect), which in turn acts in opposition to the original stretching force. With time, the original interfacial concentration of surfactant is restored. The time-dependent restoring force is referred to as the Marangoni effect and is a mechanism for foam and emulsion stabilization. The combination of Gibbs and Marangoni effects is properly referred to as the Gibbs–Marangoni effect, but is frequently referred to simply as the Marangoni effect.

Marangoni Elasticity   See Film Elasticity, Marangoni Effect.

Marangoni Flow   Liquid flow in response to a gradient in surface (or interfacial) tension. See also Marangoni Effect.

Marangoni Surface Elasticity   See Film Elasticity, Marangoni Effect.

Marangoni Waves   See Capillary Ripples.

Maximum Bubble Pressure Method   A method for the determination of surface tension in which bubbles of gas are formed and allowed to dislodge from a capillary tube immersed in a liquid. The maximum bubble pressure achieved during the growth cycle of the bubbles is used to calculate the surface tension based on the pendant drop shape analysis method. Variations include the differential maximum bubble pressure method, in which two capillaries are used and the difference in maximum bubble pressures is determined.

Meniscus   The uppermost surface of a column of a liquid. The meniscus may be either convex or concave depending on the balance of gravitational and surface or interfacial tension forces acting on the liquid.

Mesomorphic Phase   A phase consisting of anisometric molecules or particles that are aligned in one or two directions but randomly arranged in other directions. Such a phase is also commonly referred to as a liquid-crystalline phase or simply a liquid crystal. *See also* Neat Soap.

Mesopore   *See* Pore.

Metastable   *See* Thermodynamic Stability.

Micellar Charge   The net charge of surfactant ions in a micelle including any counterions bound to the micelle.

Micellar Emulsion   An emulsion that forms spontaneously and has extremely small droplet sizes (<10 nm). Such emulsions are thermodynamically stable and are sometimes referred to as microemulsions.

Micelle   An aggregate of surfactant molecules in solution. Such aggregates form spontaneously at sufficiently high surfactant concentration, the critical micelle concentration. The micelles typically contain from tens to hundreds of molecules and are of colloidal dimensions. If more than one kind of surfactant forms the micelles, they are referred to as mixed micelles. If a micelle becomes larger than usual because of either the incorporation of solubilized molecules or the formation of a mixed micelle, then the term swollen micelle is applied. *See also* Critical Micelle Concentration, Inverse Micelle.

Microelectrophoresis   *See* Electrophoresis.

Microemulsion   A special kind of stabilized emulsion in which the dispersed droplets are extremely small (<100 nm) and the emulsion is thermodynamically stable. These emulsions are transparent and may form spontaneously. In some usage, a lower size limit of about 10 nm is implied in addition to the upper limit. *See also* Micellar Emulsion.

Microgas Emulsions   A kind of foam in which the gas bubbles exist in clusters as opposed to either separated, nearly spherical bubbles or the more concentrated, system-filling polyhedral bubbles. Also termed aphrons or gas aphrons.

Microtome Method   A means of determining the surface concentration of species by cutting away a thin surface layer with a knife (microtome), physically separating the layer and analyzing it.

Middle Soap   *See* Neat Soap.

Miniemulsion   A term sometimes used to distinguish an emulsion from the microemulsion or micellar emulsion types. Thus a miniemulsion would contain droplet sizes greater than 100 nm and less than 1000 nm, or some other specified upper size limit. *See also* Emulsion.

Mist   A dispersion of a liquid in a gas (aerosol of liquid droplets) in which the droplets have diameters less than a specified size.

Mitchell Foam Quality   *See* Foam Quality.

Mixed Micelles   *See* Micelle.

Mobile Film   *See* Fluid Film.

Mobility Reduction Factor   (MRF) A dimensionless measure of the effectiveness of a foam at reducing gas mobility when flowing in porous media. The mobility reduction factor is equal to the mobility (or pressure drop) measured for foam flowing through porous media, divided by the mobility (or pressure drop) measured for surfactant-free solution and gas flowing at the same volumetric flow rates.

Modulus of Surface Elasticity   *See* Film Elasticity.

Monodisperse   A colloidal dispersion in which all the dispersed species (droplets, particles, etc.) have the same size. Otherwise, the system is heterodisperse (paucidisperse or polydisperse).

Monolayer Adsorption   Adsorption in which a first or only layer of molecules becomes adsorbed at an interface. In monolayer adsorption, all of the adsorbed molecules are in contact with the surface of the adsorbent. The adsorbed layer is termed a monolayer or monomolecular film.

Monolayer Capacity   In chemisorption, the monolayer capacity refers to the amount of adsorbate needed to satisfy all available adsorption sites. For physisorption, the monolayer capacity refers to the amount of adsorbate needed to cover the surface of the adsorbent with a complete monolayer.

Monomolecular Film   *See* Monolayer Adsorption.

Monomolecular Layer   *See* Monolayer Adsorption.

GLOSSARY

MRF   *See* Mobility Reduction Factor.

MRI   *See* Magnetic Resonance Imaging.

Multilayer Adsorption   Adsorption in which the adsorption space contains more than a single layer of molecules; therefore, not all adsorbed molecules will be in contact with the surface of the adsorbent. *See also* Monolayer Adsorption.

Multiple Emulsion   An emulsion in which the dispersed droplets themselves contain even more finely dispersed droplets of a separate phase. Thus, there may occur oil-dispersed-in-water-dispersed-in-oil (O/W/O) and water-dispersed-in-oil-dispersed-in-water (W/O/W) multiple emulsions. More complicated multiple emulsions such as O/W/O/W and W/O/W/O are also possible.

Naphtha   A petroleum fraction that is operationally defined in terms of the distillation process by which it is separated. A given naphtha is thus defined by a specific range of boiling points of its components. Naphtha is sometimes used as a diluent for W/O emulsions.

Neat Soap   A mesomorphic (liquid-crystal) phase of soap micelles that are oriented in a lamellar structure. Neat soap contains more soap and less water, in contrast to middle soap, which contains less soap than water, and is also a mesomorphic phase but with a hexagonal array of cylinders rather than a lamellar structure (*4*).

Negative Adsorption   *See* Adsorption.

Newton Black Film   *See* Black Film.

Newtonian Flow   Fluid flow that obeys Newton's law of viscosity. Non-Newtonian fluids may exhibit Newtonian flow in certain shear rate or shear stress regimes. *See also* Newtonian Fluid.

Newtonian Fluid   A fluid whose rheological behavior is described by Newton's law of viscosity. Here shear stress is set proportional to the shear rate. The proportionality constant is the coefficient of viscosity, or simply viscosity. The viscosity of a Newtonian fluid is a constant for all shear rates.

Nitrified Foam   A slang term used in the petroleum industry to denote foams in which nitrogen is the gas phase. Example: nitrified fracturing foams.

Nonionic Surfactant   A surfactant molecule whose polar group is not electrically charged. Example: poly(oxyethylene) alcohol, C_nH_{2n+1}-$(OCH_2CH_2)_mOH$.

Non-Newtonian Fluid   A fluid whose viscosity varies with applied shear rate (flow rate). Non-Newtonian flow refers to fluid flow that does not obey Newton's law of viscosity. Non-Newtonian fluids sometimes exhibit non-Newtonian flow only in certain shear rate or shear stress regimes. *See also* Newtonian Fluid.

Nonwetting   *See* Wetting.

Oil   Liquid petroleum (sometimes including dissolved gas) that is produced from a well. In this sense, oil is equivalent to crude oil. The term oil is, however, frequently more broadly used and may include, for example, synthetic hydrocarbon liquids, bitumen from oil (tar) sands, and fractions obtained from crude oil.

Oleophilic   The oil-preferring nature of a species. A synonym for lipophilic. *See also* Hydrophobic.

Oleophobic   The oil-avoiding nature of a species. A synonym for lipophobic. *See also* Hydrophilic.

Optimum Salinity   In microemulsions, the salinity for which the mixing of oil with a surfactant solution produces a middle-phase microemulsion containing an oil-to-water ratio of 1. In micellar enhanced oil recovery processes, extremely low interfacial tensions result, and oil recovery is maximized when this condition is satisfied.

Osmosis   The process in which solvent will flow through a semipermeable membrane from a solution of lower dissolved solute activity (concentration) to that of higher activity (concentration), when the membrane is not permeable to that solute).

Ostwald Viscometer   *See* Capillary Viscometer.

Outer Potential   The potential just outside the interface bounding a specified phase. Also termed the Volta potential. The difference in outer (Volta) potentials between two phases in contact is equal to the surface or interfacial potential between them. *See also* Inner Potential.

O/W   Abbreviation for an oil-dispersed-in-water emulsion.

O/W/A   An abbreviation for a fluid film of water between oil and air. *See also* Fluid Film.

O/W/O   (In multiple emulsions) Abbreviation for an oil-dispersed-in-water-dispersed-in-oil multiple emulsion. Here the water droplets have oil droplets dispersed within them, and the water droplets themselves are dispersed in oil, which forms the continuous phase.

(In fluid films) Abbreviation for a thin fluid film of water in an oil phase. *Note* the possibility of confusion with the multiple emulsion convention. *See also* Fluid Film.

GLOSSARY

Paucidisperse   A colloidal dispersion in which the dispersed species (droplets, particles, etc.) have a few different sizes. Paucidisperse is a category of heterodisperse systems. *See also* Monodisperse.

Pendant Drop Method   A method for determining surface or interfacial tension based on measuring the shape of a droplet hanging from the tip of a capillary (in interfacial tension the droplet may alternatively hang upward from the tip of an inverted capillary). Also termed the hanging drop (or bubble) method.

Peptization   The dispersion of an aggregated (coagulated or flocculated) system. Deflocculation means the same thing.

Peptizing Ions   *See* Potential Determining Ions.

Permeability   A measure of the ease with which a fluid can flow (fluid conductivity) through a porous medium. Permeability is defined by Darcy's law. For linear, horizontal, isothermal flow, permeability is the constant of proportionality between flow rate times viscosity and the product of cross-sectional area of the medium and pressure gradient along the medium.

Perrin Black Film   A Newton black film. *See also* Black Film.

Petroleum   A general term that may refer to any hydrocarbons or hydrocarbon mixtures, usually liquid, but sometimes solid or gaseous.

Phase Inversion Temperature   (PIT) Temperature at which the hydrophilic and oleophilic natures of a surfactant are in balance. As temperature is increased through the PIT, a surfactant will change from promoting one kind of emulsion, such as O/W, to another, such as W/O. Also termed the HLB Temperature.

PIT   *See* Phase Inversion Temperature.

Plastic Flow   The deformation or flow of a solid under the influence of an applied shear stress.

Plastic Fluid   A fluid characterized by both of the following: The existence of a finite shear stress that must be applied before flow begins (yield stress), and Newtonian flow at higher shear stresses. May be referred to as Bingham Plastic. *See also* Generalized Plastic Fluid.

Plateau Border   The region of transition at which thin fluid films are connected to other thin films or mechanical supports such as solid surfaces. For example, in foams Plateau borders form the regions of liquid situated at the junction of liquid lamellae. Sometimes referred to as a Gibbs ring or Gibbs–Plateau Border.

PLAWM Trough   (Pockels–Langmuir–Adam–Wilson–McBain Trough) *See* Film Balance.

Pockels–Langmuir–Adam–Wilson–McBain Trough   *See* Film Balance.

Point of Zero Charge   The condition, usually the solution pH, at which a particle or interface is electrically neutral. This condition is not always the same as the isoelectric point, which refers to zero charge at the shear plane, that exists a small distance away from the interface.

Polar Group   *See* Head Group.

Polydisperse   A colloidal dispersion in which the dispersed species (droplets, particles, etc.) have a wide range of sizes. Polydisperse is a category of heterodisperse systems. *See also* Monodisperse.

Polyederschaum   *See* Foam.

Polyelectrolyte   A kind of colloidal electrolyte comprising a macromolecule that, when dissolved, dissociates to yield a polyionic parent macromolecule and its corresponding counterions. Also termed a polyion, polycation, or polyanion. Similarly, a polyelectrolyte may be referred to in certain circumstances as a polyacid, polybase, polysalt, or polyampholyte (*4*). Example: carboxymethylcellulose.

Pore   In porous media, the interconnecting channels forming a continuous passage through the medium are made up of pores or openings that may be of different sizes. Macropores have diameters greater than about 50 nm. Mesopores have diameters of between about 2.0 and 50 nm. Micropores have diameters of less than about 2.0 nm.

Porosity   The ratio of the volume of all void spaces to total volume in a porous medium.

Porous Medium   A solid containing voids or pore spaces. Normally, such pores are quite small compared to the size of the solid and well distributed throughout the solid.

Potential Determining Ions   Ions whose equilibrium between two phases, frequently between an aqueous solution and a surface or interface, determines the difference in electrical potential between the phases, or at the surface. Example: For the AgI surface in water, both Ag^+ and I^- would be potential-determining ions. If such ions are responsible for the stabilization of a colloidal dispersion, then they are sometimes referred to as peptizing ions. *See also* Indifferent Electrolyte.

Potential Energy Diagram   *See* Gibbs Energy of Interaction.

Potential Energy of Attraction–Interaction–Repulsion   *See* Gibbs Energy of Attraction, Gibbs Energy of Interaction, Gibbs Energy of Repulsion.

Pour Point   The lowest temperature at which a material will flow under a standardized set of test conditions.

Power-Law Fluid   A fluid whose rheological behavior is reasonably well described by the power-law equation. Here shear stress is set proportional to the shear rate raised to an exponent n, where n is the power-law index. The fluid is pseudoplastic for $n < 1$, Newtonian for $n = 1$, and dilatant for $n > 1$.

Primary Minimum   In a plot of Gibbs energy of interaction versus separation distance, there may occur two minima. The minimum occurring at the shortest distance of separation is referred to as the primary minimum, and the minimum occurring at larger separation distance is termed the secondary minimum.

Primary Oil Recovery   The first phase of crude oil production, in which oil flows naturally to the wellbore. *See also* Enhanced Oil Recovery, Secondary Oil Recovery.

Protected Lyophobic Colloids   *See* Sensitization.

Protection   The process in which a material adsorbs onto droplet surfaces and thereby makes an emulsion less sensitive to aggregation and coalescence by any of a number of mechanisms. *See also* Sensitization.

Protective Colloid   A colloidal species that acts to protect the stability of another colloidal system. This term thus refers specifically to a protecting colloid and only indirectly to a second, protected colloid. Example: when a lyophilic colloid such as gelatin acts to protect another colloid in a dispersion by conferring steric stabilization. *See also* Protection.

Pseudoplastic   A fluid whose viscosity decreases as the applied shear rate increases. Also termed shear-thinning. Pseudoplastic behavior may occur in the absence of a yield stress and also after the yield stress in a system has been exceeded (i.e., when flow begins).

RCF   *See* Relative Centrifugal Force.

Reduced Adsorption   The relative Gibbs surface concentration of a component with respect to the total Gibbs surface concentration of all components. *See also* reference 4 for the defining equations.

Reduced Viscosity   (η_{red}) For solutions or colloidal dispersions, the specific increase in viscosity divided by the solute or dispersed-phase concentration, respectively ($\eta_{red} = \eta_{SP}/C$). Also termed the viscosity number.

Relative Adsorption   The relative Gibbs surface concentration of a component with respect to that of another specified component. *See also* reference 4 for the defining equations.

Relative Centrifugal Force   (RCF) When a centrifuge is used to enhance sedimentation or creaming, the centrifugal force is equal to mass times the square of the angular velocity times the distance of the dispersed species from the axis of rotation. The square of the angular velocity times the distance of the dispersed species from the axis of rotation, when divided by the gravitational constant, g, yields the relative centrifugal force or RCF. RCF is not strictly a force but rather the proportionality constant. It is substituted for g in Stokes' law to yield an expression for centrifuges and is used to compare the relative sedimentation forces achievable in different centrifuges. Because RCF is expressed in multiples of g, it is also termed g-force or simply gs.

Relative Molecular Mass The mass of a mole of species (actually it is the mass of a mole of species divided by the mass of 1/12 mole of ^{12}C). This has replaced the older term molecular weight.

Relative Viscosity (η_{rel}) In solutions and colloidal dispersions, the viscosity of the solution or dispersion divided by the viscosity of the solvent or continuous phase, respectively ($\eta_{rel} = \eta/\eta_o$). Also termed the viscosity ratio.

Repeptization Peptization, usually by dilution, of a once-stable dispersion that was aggregated (coagulated or flocculated) by the addition of electrolyte.

Repulsive Potential Energy See Gibbs Energy of Repulsion.

Reverse Emulsion A petroleum industry term used to denote an oil-in-water emulsion (most wellhead emulsions are W/O). The opposite of Invert Emulsion.

Rheology Strictly, the science of deformation and flow of matter. Rheological descriptions usually refer to the property of viscosity and departures from Newton's law of viscosity. *See also* Rheometer.

Rheomalaxis A special case of time-dependent rheological behavior in which shear rate changes cause irreversible changes in viscosity. The change can be negative, as when structural linkages are broken, or positive, as when structural elements become entangled (like work-hardening).

Rheometer Any instrument designed for the measurement of non-Newtonian as well as Newtonian viscosities. The principal class of rheometer consists of the rotational instruments in which shear stresses are measured, and a test fluid is sheared between rotating cylinders, plates, or cones. Various types of rotational rheometers are concentric cylinder, cone–cone, cone–plate, double cone–plate, plate–plate, and disc (*16*).

Rheopexy Dilatant flow that is time-dependent. At constant applied shear rate, viscosity increases, and a flow-curve hysteresis occurs (opposite of the thixotropic case).

Rigid Film   *See* Fluid Film.

Ross Foam   Foam produced from a binary or ternary solution under conditions in which its temperature and composition approach (but do not reach) the point of phase separation into separate immiscible liquid phases.

Rupture   *See* Fluid Film.

Salinity Requirement   *See* Optimum Salinity.

Salting In   *See* Salting Out.

Salting Out   (Surfactant) In salting out, the addition of electrolyte to a solution of nonionic surfactant causes the critical micelle concentration to decrease. Salting in refers to the reverse.

(Emulsions) In emulsions, salting out refers to the process of demulsification by the addition of electrolyte.

Schiller Layers   The layers of particles that may be formed during sedimentation such that the distances between layers are on the order of the wavelength of light, leading to iridescent, or Schiller layers.

Schulze–Hardy Rule   An empirical rule summarizing the general tendency of the critical coagulation concentration (CCC) of an emulsion, or other dispersion, to vary inversely with the sixth power of the counterion charge number of added electrolyte. *See also* Critical Coagulation Concentration.

Secondary Minimum   *See* Primary Minimum.

Secondary Oil Recovery   The second phase of crude oil production in which water or an immiscible gas is injected to restore production from a depleted reservoir. *See also* Enhanced Oil Recovery, Primary Oil Recovery.

Sediment   The process of sedimentation in a dilute dispersion generally produces a discernible, more concentrated dispersion termed the sediment, and has a volume termed the sediment volume.

Sediment Volume   *See* Sediment.

Sedimentation   The settling of suspended particles or droplets due to gravity or an applied centrifugal field. The sedimentation rate (or velocity) divided by acceleration is termed the sedimentation coefficient. The sedimentation coefficient extrapolated to zero concentration of sedimenting species is termed the limiting sedimentation coefficient. The sedimentation coefficient reduced to standard temperature and solvent is termed the reduced sedimentation coefficient. If the sedimentation coefficient is extrapolated to zero concentration of sedimenting species, it is termed the reduced limiting sedimentation coefficient. Negative sedimentation is also called flotation. Flotation, that is, when droplets rise upward, is also called creaming. Flotation in which particulate matter becomes attached to gas bubbles is also referred to as froth flotation.

Sedimentation Equilibrium   The state of a colloidal system in which sedimentation and diffusion are in equilibrium.

Sedimentation Potential   The potential difference, at zero current, caused by the sedimentation of dispersed species. The sedimentation may occur under gravitational or centrifugal fields. The potential difference per unit length in a sedimentation potential cell is the sedimentation field strength.

Sensitization   The process in which small amounts of added hydrophilic colloidal material make a hydrophobic colloid more sensitive to coagulation by electrolyte. Example: the addition of polyelectrolyte to an oil-in-water emulsion to promote demulsification by salting out. Higher additions of the same material usually make the emulsion less sensitive to coagulation; this is termed protective action or protection. The protected, colloidally stable dispersions that result in the emulsions that are less sensitive to coagulation are termed protected lyophobic colloids.

Separator   (Petroleum) A vessel designed to separate the oil phase in a petroleum fluid from some or all of the other three constituent phases: gas, solids, and water. Free-water knockouts fall under this category, but so do separators capable of breaking and removing water and solids from emulsions. The separators range from gravity to impingement (coalescence) to centrifugal separators.

Septum   The dividing wall between bubbles in a foam. That is, the thin liquid films or lamellae.

Sessile Bubble Method   See Sessile Drop Method.

Sessile Drop Method   A method for determining surface or interfacial tension based on measuring the shape of a droplet at rest on the surface of a solid substrate (in liquid–liquid systems, the droplet may alternatively rest upside down, that is underneath a solid substrate). This technique may also be used to determine the contact angle and contact diameter of the droplet against the solid.

Shear   The rate of deformation of a fluid when subjected to a mechanical shearing stress. In simple fluid shear, successive layers of fluid move relative to each other such that the displacement of any one layer is proportional to its distance from a reference layer. The relative displacement of any two layers divided by their distance of separation from each other is termed the shear, or shear strain. The rate of change with time of the shear is termed the shear rate, or strain rate.

Shear Plane   Any species undergoing electrophoretic motion moves with a certain immobile part of the electric double layer that is commonly assumed to be distinguished from the mobile part by a sharp plane, the shear plane. The zeta potential is the potential at the shear plane.

Shear Stress   A certain applied force per unit area is needed to produce deformation in a fluid. For a plane area around some point in the fluid, and in the limit of decreasing area, the component of deforming force per unit area that acts parallel to the plane is the shear stress.

Shear Thickening   *See* Dilatant.

Shear Thinning   *See* Pseudoplastic.

Silicone Oil   Any of a variety of silicon-containing polymer solutions. An example is a linear poly(dimethylsiloxane): $HO[(CH_3)_2SiO]_nH$. *See also* Chapter 12.

Silver Film   *See* Black Film.

Slurry Quality   *See* Foam Quality.

Smoluchowski Equation   (Electrophoresis) A relation expressing the proportionality between electrophoretic mobility and zeta potential for the limiting case of a species that can be considered to be large and having a thin electric double layer. Also termed Helmholtz–Smoluchowski Equation. *See also* Electrophoresis, Henry Equation, Hückel Equation.

Snap-Off   A mechanism for foam lamella generation in porous media. When gas enters and passes through a constriction in a pore, a capillary pressure gradient is created and causes liquid to flow toward the region of the constriction, where it accumulates and may cause the gas to pinch-off or snap-off to create a new gas bubble separated from the original gas by a liquid lamella. *See also* Lamella Division, Lamella Leave-Behind.

Soap   A surface-active, fatty acid salt containing at least eight carbon atoms. The term is no longer restricted to fatty acid salts originating from natural fats and oils. *See also* Surfactant.

Soap Film   A thin film of water in air in which the film is stabilized by surfactant. The term is used even though the film is not a film of soap and even when the surfactant is not a soap. *See also* Fluid Film.

Sol   A colloidal dispersion. In some usage, the term sol is used to distinguish dispersions in which the dispersed-phase species are of very small size so that the dispersion appears transparent.

Solid Aerosol   *See* Aerosol.

Solid Emulsion   A colloidal dispersion of a liquid in a solid. Examples: opal and pearl.

Solid Foam   A colloidal dispersion of a gas in a solid. Example: polystyrene foam.

Solubility Parameter   A means of estimating solubility. The square root of the ratio of energy of vaporization to molar volume. The ratio of energy of vaporization to molar volume is also known as the cohesive energy density. Pairs of substances having very similar solubility parameters are mutually soluble.

Solubilization   The process by which the solubility of a solute is increased by the presence of another solute. Micellar solubilization refers to the incorporation of a solute (solubilizate) into or on micelles of another solute, thereby increasing the solubility of the first solute.

Sorbate   A substance that becomes sorbed into an interface, another material, or both. *See also* Sorption.

Sorbent   The substrate into which or onto which a substance is sorbed, or both. *See also* Sorption.

Sorption   In a general sense, either or both of the processes of adsorption and absorption.

Specific Increase in Viscosity   The relative viscosity minus unity. Also referred to as specific viscosity.

Specific Surface Area   *See* Surface Area.

Specific Viscosity   *See* Specific Increase in Viscosity.

Spinning Drop Method   A method for determining surface or, more commonly, interfacial tension based on measuring the shape of a droplet (or bubble) suspended in the center of a horizontal, cylindrical tube filled with a liquid, as the tube is spinning about its long axis. This method is particularly suited to the determination of very low interfacial tensions.

Spread Layer   The interfacial layer formed by an adsorbate when it becomes essentially completely adsorbed out of the bulk phase(s). If the layer is known to be one molecule thick, then the term spread monolayer is used.

Spreading   The tendency of a liquid to flow and form a film coating an interface, usually a solid or immiscible liquid surface, in an attempt to minimize interfacial free energy. Such a liquid forms a zero contact angle as measured through itself.

Spreading Coefficient   A measure of the tendency for a liquid to spread over a surface (usually of another liquid). It is −1 times the Gibbs free energy change for this process (the work of adhesion between the two phases minus the work of cohesion of the spreading liquid) so that spreading is thermodynamically favored if the spreading coefficient is greater than zero. Early usage of the concept involved terms such as the spreading parameter or wetting power (*17*). When equilibria at the interfaces are not achieved instantaneously, then reference is frequently made to the initial spreading coefficient and final (equilibrium) spreading coefficient. If the initial spreading coefficient is positive and the final spreading coefficient is negative, then the system exhibits autophobicity.

Spreading Parameter   See Spreading Coefficient.

Spreading Pressure   See Surface Pressure.

Spreading Tension   A synonym for Spreading Coefficient.

Stability   See Colloid Stability, Thermodynamic Stability.

Static Interfacial Tension   See Static Surface Tension.

Static Surface Tension   The static surface or interfacial tension is a synonym for the equilibrium surface or interfacial tension.

Steric Stabilization   The stabilization of dispersed species induced by the interaction (steric stabilization) of adsorbed polymer chains. Example: Adsorbed proteins stabilize the emulsified oil (fat) droplets in milk by steric stabilization. Also termed Depletion Stabilization. *See also* Protection.

Stern Layer   A layer of specifically adsorbed ions immediately adjacent to a surface. The less mobile part of an electric double layer. *See also* Helmholtz Double Layer.

Stern Potential   See Inner Potential.

Stimulation Fluid   See Foam Stimulation Fluid.

Stokes' Law   The relation describing the terminal rising or settling velocity of an uncharged spherical body in a medium having a different density. The terminal velocity is given as $2r^2 \Delta \rho g/9\eta$, where r is the sphere radius, $\Delta \rho$ is the density difference between the phases, g is the gravitational constant, and η is the medium viscosity.

Strain, Strain Rate   See Shear.

Stratified Film   A fluid film in which several thicknesses can exist simultaneously and can persist for a significant amount of time. *See also* Fluid Film.

Streaming Potential   The potential difference, at zero current, created when liquid is made to flow through a porous medium.

Streamline Flow   See Laminar Flow.

Stress   *See* Shear Stress.

Substrate   A material that provides a surface or interface at which adsorption takes place.

Supercentrifuge   *See* Centrifuge.

Superficial Density   An older term now replaced by the Gibbs surface concentration, or simply, the surface excess.

Surface   *See* Interface.

Surface Active Agent   *See* Surfactant.

Surface Area   The area of a surface or interface, especially that between a dispersed and a continuous phase. The specific surface area is the total surface area divided by the mass of the appropriate phase.

Surface Charge   The fixed charge that is attached to, or part of, a colloidal species' surface and forms one layer in an electric double layer. Thus, a surface charge density is associated with the surface.

Surface Concentration   *See* Gibbs Surface.

Surface Conductivity   The excess conductivity, relative to the bulk solution, in a surface or interfacial layer per unit length. Also termed the surface excess conductivity.

Surface Coverage   The ratio of the amount of adsorbed material to the monolayer capacity. The definition is the same for either of monolayer and multilayer adsorption.

Surface Dilatational Modulus   *See* Film Elasticity.

Surface Dilatational Viscosity   *See* Surface Viscosity.

Surface Elasticity   *See* Film Elasticity.

Surface Excess (Concentration)   *See* Gibbs Surface.

Surface Excess Conductivity   *See* Surface Conductivity.

Surface Excess Isotherm   A function relating, at constant temperature and pressure, the relative adsorption, reduced adsorption, or similar surface excess quantity to the concentration of component in the equilibrium bulk phase.

Surface Fluidity   The inverse of the surface shear viscosity.

Surface Layer   *See* Interfacial Layer.

Surface of Tension   An imaginary boundary having no thickness at which surface or interfacial tension acts.

Surface Potential   The potential at the interface bounding two phases, that is, the difference in outer (Volta) potentials between the two phases. *See also* Inner Potential and Outer Potential.

Surface Pressure   Actually an analogue of pressure, refers to the force per unit length exerted on a real or imaginary barrier separating an area of liquid or solid that is covered by a spreading substance from a clean area on the same liquid or solid. Also referred to as spreading pressure.

Surface Rheology   *See* Surface Viscosity.

Surface-Shear Viscosity   *See* Surface Viscosity.

Surface Tension   The contracting force per unit length around the perimeter of a surface. Usually referred to as surface tension if the surface separates gas from liquid or solid phases, and interfacial tension if the surface separates two nongaseous phases. Although not strictly defined the same way, surface tension can be expressed in units of energy per unit surface area. For practical purposes, surface tension is frequently taken to reflect the change in surface free energy per unit increase in surface area. *See also* Surface Work.

Surface Viscosity   Surface or interfacial viscosity is usually thought of as simply the two-dimensional analog of viscosity acting along the interface between two immiscible fluids. In fact, there are two kinds of surface viscosity: surface-shear viscosity and surface-dilatational viscosity. Surface-shear viscosity is the component that is analogous to three-dimensional shear viscosity: the rate of yielding of a layer of fluid due to an applied stress. Surface-dilatational viscosity relates to the rate of area expansion and is expressed as the local gradient in surface tension per change in relative area per unit time. Any shear-rate dependence (non-Newtonian behavior) falls under the subject of surface rheology. Although usually termed surface viscosity or rheology, especially where one fluid is a gas, the more general terminology is surface or interfacial rheology. *See also* Viscosity.

Surface Work   The work required to increase the area of the surface of tension. Under reversible, isothermal conditions, the surface work (per unit surface area) equals the equilibrium or static surface tension.

Surfactant   Any substance that lowers the surface or interfacial tension of the medium in which it is dissolved. The substance does not have to be completely soluble and may lower surface or interfacial tension by spreading over the interface. Soaps (fatty acid salts containing at least eight carbon atoms) are surfactants. Detergents are surfactants or surfactant mixtures whose solutions have cleaning properties. Also referred to as surface-active agents or tensides.

Surfactant Effectiveness   The surface excess concentration of surfactant corresponding to saturation of the surface or interface. Example: One indicator of effectiveness is the maximum reduction in surface or interfacial tension achievable by a surfactant. This term has a different meaning from surfactant efficiency. *See also* references 14 and 15.

Surfactant Efficiency   The equilibrium solution surfactant concentration needed to achieve a specified level of adsorption at an interface. Example: One such measure of efficiency is the surfactant concentration needed to reduce the surface or interfacial tension by 20 mN/m from the value of the pure solvent(s). This term has a different meaning from surfactant effectiveness. *See also* references 14 and 15.

Surfactant Tail   The lyophobic portion of a surfactant molecule, which is usually a hydrocarbon chain having at least eight carbon atoms. *See also* Head Group.

Suspension   A system of solid particles dispersed in a liquid. A typical operational definition is any dispersed matter that can be removed by a 0.45-μm nominal pore size filter.

Swamping Electrolyte   An excess of indifferent electrolyte that severely compresses electric double layers and minimizes the influence of electric charges borne by large molecules or dispersed colloidal species.

Swelling   The increase in volume associated with the uptake of liquid or gas by a solid or a gel.

Swelling Pressure   The pressure difference between a swelling material and the bulk of fluid being imbibed that is needed to prevent additional swelling. *See also* Swelling.

Swollen Micelle   *See* Micelle.

Syneresis   The spontaneous shrinking of a colloidal dispersion due to the release and exudation of some liquid. Frequently occurs to gels and foams but also occurs in flocculated suspensions. Mechanical syneresis refers to enhancing syneresis by the application of mechanical forces.

TAN   *See* Total Acid Number.

Tenside   *See* Surfactant.

Tensiometer   A general term applied to any instrument that may be used to measure surface and interfacial tension.

Tertiary Oil Recovery   *See* Enhanced Oil Recovery.

Thermodynamic Stability   In colloid science, the terms thermodynamically stable and metastable mean that a system is in a state of equilibrium

GLOSSARY

corresponding to a local minimum of free energy (4). If several states of energy are accessible, then the lowest is referred to as the stable state, others are referred to as metastable states, and unstable states are not at a local minimum. Most colloidal systems are metastable or unstable with respect to the separate bulk phases. *See also* Colloid Stability.

Thickness of the Electric Double Layer   *See* Electric Double-Layer Thickness.

Thin Film Drainage   *See* Film Drainage, Fluid Film.

Thixotropic   Pseudoplastic flow that is time-dependent. At constant applied shear rate, viscosity gradually decreases, and a flow curve hysteresis occurs. That is, after a given shear rate is applied and then reduced, it takes some time for the original dispersed species' alignments to be restored.

Total Acid Number   (TAN) A property of crude oil. The acid number expresses the amount of base (potassium hydroxide) that will react with a given amount of crude oil in a standardized titration procedure. A large acid number indicates a high concentration of acids in the oil, usually including natural surfactant precursors.

Total Potential Energy of Interaction   *See* Gibbs Energy of Interaction.

Treater   A vessel used for the breaking of foams, froths, or emulsions, and the consequent removal of phases such as solids and water (BS&W). Foam or emulsion breaking may be accomplished through some combination of thermal, electrical, chemical, or mechanical methods.

Turbidity   A coefficient related to the amount of scattering of light when it passes through a dispersion (including emulsions). Turbidity is a function of the size and concentration of droplets or particles.

Turbulent Flow   A condition of flow in which all components of a fluid passing a certain point do not follow the same path. Turbulent flow refers to flow that is not laminar or streamline.

Ubbelohde Viscometer   *See* Capillary Viscometer.

Ultracentrifuge   *See* Centrifuge.

Ultrasonic Dispersion   The use of ultrasound waves to achieve or aid in the dispersion of particles or droplets.

van der Waals Forces   *See* Dispersion Forces.

Viscoelastic   A liquid (or solid) with both viscous and elastic properties. A viscoelastic liquid will deform and flow under the influence of an applied shear stress, but when the stress is removed the liquid will slowly recover from some of the deformation.

Viscometer   Any instrument employed in the measurement of viscosity. In most cases, the term is applied to instruments capable of measuring only Newtonian viscosity, and not capable of non-Newtonian measurements. *See also* Rheometer.

Viscosity   A measure of the resistance of a liquid to flow. It is the coefficient of viscosity and expresses the proportionality between shear stress and shear rate in Newton's law of viscosity.

Viscosity-Average Molecular Mass   Molecular mass determined on the basis of viscosity measurements coupled with an empirical equation such as the Staudinger–Mark–Houwink equation.

Viscosity Number   *See* Reduced Viscosity.

Viscosity Ratio   *See* Relative Viscosity.

Wet Foam   *See* Foam.

Wet Oil   An oil containing free water and emulsified water.

Wettability   A qualitative term referring to the water- or oil-preferring nature of surfaces, such as mineral surfaces. Example: The flow of emulsions in porous media is influenced by the wetting state of the walls of pores and throats through which the emulsion must travel. Wettability may be determined by direct measurement of contact angles, or inferred from measurements of fluid imbibition or relative permeabilities. Several conventions for describing wettability values exist. *See also* Contact Angle, Wetting.

Wetting   A general term referring to one or more of the following specific kinds of wetting: adhesional wetting, spreading wetting, and immersional wetting. Frequently, the term wetting is used to denote that the contact angle between a liquid and a solid is essentially zero and there is spontaneous spreading of the liquid over the solid. Nonwetting on the other hand is frequently used to denote the case in which the contact angle is greater than 90° so that the liquid rolls up into droplets. *See also* Contact Angle, Wettability.

Wetting Coefficient   In the equilibrium of a system of solid, gas, and liquid the wetting coefficient is given as $k = (\gamma_{sg} - \gamma_{sl})/\gamma_{lg}$, where γ is the interfacial tension, and the subscripts g, l, and s refer to gas, liquid, and solid phases at the interfaces, respectively. Complete wetting occurs for $k \geq 1$ and nonwetting for $k \leq 1$.

Wetting Power   *See* Spreading Coefficient.

GLOSSARY

Whipping Agent   *See* Foaming Agent.

Wilhelmy Plate Method   A method for determining surface or interfacial tension on the basis of measuring the force needed to pull an inert plate, held vertically, through an interface. Also termed the Wilhelmy slide method. *See also* du Nouy Ring Method.

W/O   Abbreviation for a water-dispersed-in-oil emulsion.

Work of Adhesion   The energy of attraction between molecules in a phase. Defined as the work per unit area done on a system when two phases meeting at an interface of unit area are separated reversibly to form unit areas of new interfaces of each with a third phase.

Work of Cohesion   The work per unit area done on a system when a body of a phase is separated reversibly to form two bodies of the phase each forming unit areas of new interfaces with a third phase.

Work of Separation   A synonym for the Work of Adhesion.

Work of Spreading   The work of spreading expressed per unit area is the same as the Spreading Coefficient.

W/O/A   Abbreviation for a thin fluid film of oil between water and air phases. *See also* Fluid Film.

W/O/W   (Multiple Emulsions) Abbreviation for a water-dispersed-in-oil-dispersed-in-water multiple emulsion. Here the oil droplets have water droplets dispersed within them, and the oil droplets themselves are dispersed in water, forming the continuous phase.

(In fluid films) Abbreviation for a thin fluid film of oil in a water phase. *Note* the possibility of confusion with the multiple emulsion convention. *See also* Fluid Film.

Yield Stress   For some fluids, the shear rate (flow) remains at zero until a threshold shear stress is reached, termed the yield stress. Beyond the yield stress, flow begins. Also termed the yield value.

Young–Laplace Equation   The fundamental relationship giving the pressure difference across a curved interface in terms of the surface or interfacial tension and the principal radii of curvature. In the special case of a spherical interface, the pressure difference is equal to twice the surface (or interfacial) tension divided by the radius of curvature. Also referred to as the equation of capillarity.

Young's Equation   A fundamental relationship giving the balance of forces at a point of three-phase contact. For a gas–liquid–solid system Young's equation is $\gamma_{SL} + \gamma_{LG} \cos \theta = \gamma_{SG}$, where γ_{SL}, γ_{LG}, and γ_{SG} are interfacial tensions between solid–liquid, liquid–gas, and solid–gas, respectively, and θ is the contact angle of the liquid with the solid, measured through the liquid.

Zero Point of Charge   *See* Point of Zero Charge.

Zeta Potential   Strictly called the electrokinetic potential, the zeta potential refers to the potential drop across the mobile part of the electric double layer. Any species undergoing electrokinetic motion, such as electrophoresis, moves with a certain immobile part of the electric double layer that is assumed to be distinguished from the mobile part by a distinct plane, the shear plane. The zeta potential is the potential at that plane.

Zwitterionic Surfactant   A surfactant molecule whose polar group contains both negatively and positively charged groups. Example: lauramidopropylbetaine $C_{11}H_{23}CONH(CH_2)_3N^+(CH_3)_2CH_2COO^-$ at neutral and alkaline solution pH. *See also* Amphoteric surfactant.

References

1. Martinez, A. R. In *The Future of Heavy Crude and Tar Sands;* Meyer, R. F.; Wynn, J. C.; Olson, J. C., Eds.; United Nations Institute for Training and Research: New York, 1982; pp ixvii–ixviii.
2. Danyluk, M.; Galbraith, B.; Omana, R. In *The Future of Heavy Crude and Tar Sands;* Meyer, R. F.; Wynn, J. C.; Olson, J. C., Eds.; United Nations Institute for Training and Research: New York, 1982; pp 3–6.
3. Khayan, M. In *The Future of Heavy Crude and Tar Sands;* Meyer, R. F.; Wynn, J. C.; Olson, J. C., Eds.; United Nations Institute for Training and Research: New York, 1982; pp 7–11.
4. *Manual of Symbols and Terminology for Physicochemical Quantities and Units, Appendix II;* Prepared by IUPAC Commission on Colloid and Surface Chemistry; Butterworths: London, 1972.
5. Williams, H. R.; Meyers, C. J. *Oil and Gas Terms, 6th ed.;* Matthew Bender: New York, 1984.
6. *A Dictionary of Petroleum Terms, 2nd ed.;* Petroleum Extension Service, The University of Texas at Austin: Austin, TX, 1979.
7. *The Illustrated Petroleum Reference Dictionary, 2nd ed.;* Langenkamp, R. D.; PennWell Books: Tulsa, OK, 1982.
8. Isenberg, C. *The Science of Soap Films and Soap Bubbles;* Tieto Ltd.: Clevedon, United Kingdom, 1978; p 101.
9. Ross, S.; Morrison, I. D. *Colloidal Systems and Interfaces;* Wiley: New York, 1988.
10. Derjaguin, B. V.; Churaev, N. V.; Miller, V. M. *Surface Forces;* Consultants Bureau: New York, 1987.
11. Verwey, E. J. W.; Overbeek, J. Th. G. *Theory of the Stability of Lyophobic Colloids;* Elsevier: New York, 1948.
12. Cash, L.; Cayias, J. L.; Fournier, G.; Macallister, D.; Schares, T.; Schechter, R. S.; Wade, W. H. *J. Colloid Interface Sci.* **1977,** *59,* 39–44.

13. Cayias, J. L.; Schechter, R. S.; Wade, W. H. *Soc. Petrol. Eng. J.* **1976**, *16*, 351–357.
14. Rosen, M. J. *Surfactants and Interfacial Phenomena;* Wiley: New York, 1978.
15. Myers, D. *Surfactant Science and Technology;* VCH: New York, 1988.
16. Whorlow, R. W. *Rheological Techniques;* Ellis Horwood: Chichester, England, 1980.
17. Ross, S.; Becher, P. *J. Colloid Interface Sci.* **1992**, *149*, 575–579.

RECEIVED for review October 14, 1992. ACCEPTED revised manuscript March 2, 1993.

Author Index

Axelson, Dave, 423
Chambers, David J., 355
Czarnecki, Jan, 423
Dalland, Mariann, 319
Green, M. K., 235
Hanssen, Jan Erik, 319
Heller, John P., 201
Isaacs, E. E., 235
Ivory, J., 235
Koczo, K., 47
Kovscek, A. R., 115
Lewis, V. E., 461
Maini, Brij B., 405
Mannhardt, Karin, 259
Minyard, W. F., 461
Nikolov, A. D., 47
Novosad, Jerry J., 259
Radke, C. J., 115
Sarma, Hemanta, 405
Schramm, Laurier L., 3, 165, 423
Shaw, Robert C., 423
Wasan, D. T., 47
Wassmuth, Fred, 3

Affiliation Index

Alberta Research Council, 235
Canadian Fracmaster Ltd., 355
Illinois Institute of Technology, 47
Nalco Chemical Company, 461
New Mexico Institute of Mining and Technology, 201
Petroleum Recovery Institute, 3, 165, 259, 405, 423
RF–Rogaland Research, 319
Syncrude Canada Ltd., 423
University of California, Berkeley, 115

Subject Index

A

Absolute viscosity, 488
Absorbate, 488
Absorbent, 488
Absorption, 488
Acid fracturing, mechanism, 356
Acid number, 488
Acid spending, dolomite versus limestone formations, 379
ACN, *see* Equivalent alkane carbon number
Activator, 488
Adhesion, 488
Adsorbate, 488
Adsorbed surfactant layer, role in foam stability, 407–408
Adsorbent, 488
Adsorption
  anionic and amphoteric surfactants
    effect of solid surface
    mechanisms, 292–293
  foam-forming surfactants
    difficulty in comparison, 284
    effluent profiles, 284f
    experiments, 284–298
    for hydrocarbon-miscible flooding at high salinities, 259–309
    isotherm, 284f
    methods of determination, 284
  measurement
    apparatus for straight-through flow, 228f
    typical results for straight-through flow, 229f
  onto pore walls of reservoir rock, 226
  surface-active chemical
    definition, 407

INDEX

effect on liquid surface elasticity, 407
mechanism, 407
theory, 91–92
Adsorption isotherm
  definition, 488–489
  surfactant mixture versus pure surfactant, 303
Adsorption level distribution, surfactants and solids of like and opposite charge, 301f
Adsorption trends with increasing salt concentration
  in amphoteric surfactants, 287f, 288
  in anionic surfactants, effects, 287f, 288
Aerated bitumen droplet density
  relationship to bulk density, 444f
  relationship to froth quality, 443f
Aerated emulsion, 489
Aerating agent, 489
Aerosol, 489
Agglomeration, 489
Aggregate, 489
Aggregation, definition, 489
Aggregation number, 489
Aging, 489
Air/oil/water (A/O/W), 491
Air/water/air (A/W/A), 491
Air/water/oil (A/W/O), 491
Alkane carbon number (ACN), see Equivalent alkane carbon number
Amine units, 474–475
Amount of noncondensible gas required to stabilize steam-foams
  dependence on pore radius, 239, 241f
  dependence on temperature, 239, 241f
Amphipathic, definition, 12, 489
Amphiphilic, definition, 12
Amphoteric surfactant, 490
Anionic surfactant
  definition, 490
  methods of loss, 236
Anionic surfactant precipitation, effect of temperature, 236–237
Anionic surfactant retention in porous media, 236–237
Antibubbles, 490
Antifoam(s)
  definition, 462
  effect of immersion depth, 96–97
  effect of position, 96–97
  effectiveness, 463–464
  polymers with molecular weight greater than 2000, 469f
  refinery applications, 468–478

Antifoam chemistry
  definition, 464–468
  hydrophobic silicas, 464–467
Antifoaming
  definition, 88
  hydrophobic particles, 90–91, 92f
  mechanisms, 38, 89–92, 101f
  mechanisms with and without oil, 47–111
  mixture of particles and oil, 90–91
  oil bridge formation, 90–91f, 92f
  refineries, 461–483
  role of the pseudoemulsion film, 92–97
  spreading mechanism, 89–90
  suggested mechanism, 97–105
Antifoaming agent
  definition, 490
  efficiency 90f
  types, 89
Antifoaming and defoaming action, influence of unstable pseudoemulsion film, 88–105
Antifoaming efficiency, dependence on stability of the pseudoemulsion film, 92–93
Aphrons, see Microgas emulsions
API gravity, 490
Apparent viscosity
  CO_2 foam, 220–221
  combined effects of bubble radius and capillary radius, 207f
  definition, 490
Aqueous foam(s)
  stability mechanisms with and without oil, 47–111
  use in petroleum industry, 47
Aqueous-phase saturation profiles
  experimental, 151f
  model transient, 151f
Areal elasticity, see Film elasticity
Aromatic compounds, removal, 477
Asphalt, 490
Asphalt raffinate, definition, 478
Asphaltene, 490–491
Association colloid, 491
Atmospheric column, 473
Attractive potential energy, see Gibbs energy of attraction
Autophobicity, see Spreading coefficient

B

Backbone channels, 123
Basic sediment and water, 491
Batch extraction froths, conclusions, 435

Batch extraction test, basic steps, 436f
Batch treating, 491
Beaker test, 492
Bilayer, 491
Biliquid foam, 491
Bimolecular film, 491
Binary foam, *see* combined N_2- and CO_2-foamed fluids
Bingham plastic fluid, 519
Bitumen
  composition and density, 432
  definition, 423, 491
  factors resulting in, 424–425
  UNITAR-sponsored established definitions, 424t
Bitumen droplet size distribution
  average quality oil sand versus poor quality oil sand, 441f
  effect of alkali addition levels, 440–441
Bituminous froths in hot-water flotation process, 423–455
Black film, 492
Black lipid membrane (BLM), 492
Blender test, 492
Blowdown experiments, 418–419
Born repulsion, definition, 26
Bottle test, 492
Bottom settlings and water, 491
Bottom solids and water, 491
Bottomhole pressure, measurement, 396
Bottomhole treating pressure, calculation, 396
Boussinesq number, 56, 492–493
Breakers, definition, 38
Breaking, 493
Breakthrough behavior, foam-generating displacement versus gas–water displacement, 329
Brownian motion, 493
BS&W, *see* Basic sediment and water
Bubble coalescence
  complex interactions, 239
  definition, 239
  effect of Na^+ concentration, 240f
Bubble division
  definition, 130
  dependence on occupancy of surrounding pores, 130
Bubble electrophoresis, description, 25
Bubble generation, 141–142
Bubble microelectrophoresis, description, 26
Bubble sizes, effect on foam stability, 37
Bubble snap-off, *see* Snap-off

Bubble trains, description, 126–127
Bubble trapping
  factors affecting, 123–124
  relationship to pore size, 126
Buildup tests, description, 249–250
Bulk density, relationship to aerated bitumen droplet density, 444f
Bulk foam
  causes of viscosity, 61
  definition, 118, 493
  effect of crude oil, 100
  effect of emulsified crude oil content on lifetime, 100, 102f
  tests, 169
Bulk foam–oil sensitivity tests, categories, 168–169
Bulk foams formed from foam-flooding surfactant solutions, presence of crude, 100
Bulk viscosity terms, 35t
Butadiene extraction unit, schematic, 482f

C

Calculated flowing foam texture, 152, 153f
Capillarity, 493
Capillary, 493
Capillary constant, 493
Capillary flow, 493
Capillary forces, 493
Capillary number, 493
Capillary pressure
  definition, 493–494
  dependence on wetting-liquid saturation, 133
  effect of gas fractional flow, 215
Capillary ripples, 494
Capillary rise, 494
Capillary-suction coalescence
  basic premise, 135–136
  primary mechanism for lamellae breakage, 132–137
  relationship to surfactant formulation, 132
Capillary viscometer, 494
Capillary waves, 494
Cationic surfactant, 494
Caustic scrubbers, 475–476f
CCC, *see* Critical coagulation concentration
Centrifuge, 494
Centrifuge test, 492
Charge density, 494

INDEX

Chemical stability of surfactant, description, 274
Cloud point, 16, 494–495
CO_2
  advantages as energizer, 370
  characteristics, 368
  density
    relationship to temperature, 202f
    relationship to viscosity, 203f
  solubility, 369f
  three-phase diagram, 368f
  unique properties, 202
CO_2 flood
  corrosion, 231
  factors affecting use, 204–205
CO_2 foam fracturing treatment, description, 394–395f
CO_2 foam in enhanced oil recovery, operational considerations, 229–231
CO_2 foam mobility, with various surfactants at different rock permeability, 221–222f
CO_2 foam versus everyday foam, difference, 209
CO_2-foamed fluids, 365, 368–372
Coacervate, 495
Coacervation, 495
Coagulation, 489
Coalescence
  aqueous phase saturation 143–144f
  concentration, 143
  definition, 8, 143–145, 495
  dependence on wetting-liquid saturation, 143–144
  illustration, 78, 79f, 80f
  limiting capillary pressure, 134
  stages, 51
  surfactant formulation, 143
Coalescence pressure, definition, 239
Coefficient of viscosity, definition, 33
Co-ions, 495
Coker, definition, 462
Coking, definition, 468
Collector, 495
Colloid crystal-like model, description, 67
Colloid stability, 495
Colloidal, 495
Colloidal electrolyte, 495
Combined N_2- and CO_2-foamed fluids
  advantages, 373–374
  description, 373
  disadvantages, 374–375
Commercial oil-sands mining, diagram, 429f

Common black film, definition, 27, 36, 492
Composite fluid leak-off coefficient, definition, 376
Compressibility, see Film elasticity
Compressibility factor for N_2, effect of temperature and pressure, 364f
Compressional modulus, see Film elasticity
Concentric cylinder rheometer, 522
Condensate, 496
Condensing gas drive, definition, 260
Conditions for obtaining gas blockage
  critical parameters in order of importance, 340
  summary, 340–341
Confined foam
  definition, 118
  low gas fractional flow versus high gas fractional flow, 118
Conservation equations, 140–141
Constant ΔP experiment, superiority over constant flow rate experiment, 244
Constant flow rate experiment, concerns, 243–244
Contact angle, 496
Continuous-gas foam
  description, 119, 120
  formation, 120
  illustration, 119, 121f
Continuous phase, 496
Continuous shear strain, motion, 211
Control fluid leak-off, mechanisms, 375–376
Control of fluid loss, best techniques, 377–378
Conversion table, 399
Core-flood experiments
  apparatus, 173, 174f
  foam preparation, 173
  measurements, 173–174
  proposed mechanisms, 186–189
Core-flood performed with mixture of two surface-active components, effluent profiles, 304f
Core hole, typical cross-section, 128f
Corrosion inhibitors, classes, 461–462
Cosorption lines, 496
Cosurfactant, 496
Couette flow, 496
Couette rheometer, 496
Counterions, 496
Creaming, 496
Critical capillary pressure for rupture, definition, 134
Critical coagulation concentration, 496–497

Critical film thickness, 497
Critical foaming oil saturation, 176
Critical micelle concentration
  definition, 14, 497
  determination of surfactant–monomer concentration, 14
  determination with plot of interfacial tension and concentration of surfactant, 224f
  factors affecting, 14
Critical parameters for foam study
  differential pressure, 334
  gas type, 339–340
  general, 333–340
  oil, 335–336
  permeability, 336
  porous medium length, 337–339
  pressure level, 339–340
  temperature–concentration–salinity, 334–335
Critical thickness, 497
Crude oil, 497
Crude unit
  process equipment, 472
  schematic, 473
  use, 472
Cyclic processes, 250

D

Darcy's law, 519
Data envelope, example, 446f
Deasphalted oil (DAO)–solvent stripper, schematic, 478f
Debye–Hückel parameter, 497
Debye length, 497
Decomposition of alkylarylsulfonate surfactant, effect of pH, 236–237f
Deflocculation, 497
Defoamer
  comparison of action, 463f
  definition, 38, 462
Defoaming
  correlation to entering coefficient (E), 78
  definition, 88
  mechanism, 88–89
  refineries, 461–483
Delayed coker
  process, 469–471
  schematic, 470f
Delayed coking, 468–472
Demulsifier, 497

Dendritic channels, definition, 123
Depressant, 497
Derjaguin–Landau–Verwey–Overbeek (DLVO) theory, 27, 499
Desalter, description, 461
Design considerations of foams, 382–384
Desorption, 497–498
Destabilizer
  definition, 39
  mechanisms, 39
Detergent, 498
Dewetting, 498
Differential interferometry (DI), 76
Differential viscosity, 498
Diffuse double layer, see Electric double layer
Dilatant, 498
Dilatational elasticity, 57
Dilatational modulus, 57
Diluted (flooded) slurry, contents, 427
Dimple, definition, 52
Dipolar configurations, types, 26
Discontinuous-gas foam
  formation, 120
  illustration, 119, 120f
Discontinuous phase, 498
Disjoining pressure
  apparatus, 29
  approach to pseudoemulsion film stability, 182
  definition, 28–29, 52, 180, 498
  determination, 28–30
  effect of film thickness, 29
Dispersant, 498
Disperse, 498–499
Dispersed phase, 498
Dispersing agent, 498
Dispersion, 498–499
Dispersion forces
  approximation, 27
  definition, 499
Displacing fluid, CO_2, 202–206
Dissolved gas flotation, 505
Diverters, types, 380
Division, see Lamella division
Double layer, see Electric double layer
Drive processes, 251, 252t, 253t
Drop size
  effect of foam lifetime, 87
  relationship to film energy, 84
Droplet diameters and size distribution of bitumen, relation to slurry conditioning variables, 440

INDEX

Dry foams
  definition, 7
  estimation of viscosity, 35
du Nouy ring method, 499
Duplex film, 499
Dynamic foam test
  apparatus, 31f
  description, 30
Dynamic interfacial or surface tension, 499

E

EACN, see Equivalent alkane carbon number
EDL, see Electric double layer
Effective foam viscosity, 499
Effective permeability
  effect of foam, 121–124
  method of measurement, 121
Effective permeability of porous medium, effect of presence of foam, 377
Effective viscosity
  definition, 214
  flow resistance of transporting bubble-trains, 124
  measurement, 124
  measurement in confined foam, 124
  relationship to tube size, 206–207
Elasticity
  correlation to gas blocking performance, 63
  definition, 499
  relationship to hydrocarbon chain length, 19
Elasticity number, 499–500
Electric double layer
  composition, 24
  definition, 24, 500
Electric double-layer repulsion, differences in importance in aqueous and nonpolar foams, 408
Electric double-layer thickness, 500
Electrical properties, specific conductivity, 37–38
Electrokinetic, 500
Electrokinetic potential, 534
Electrophoresis, 500
Electrophoretic mobility
  definition, 25
  of rock particles as function of pH and brine ionic content, 282–283t
Electrosteric stabilization, 500
Electroviscous effect, 500

Ellipsometry, 500
Emulsification
  model, 182–184
  process, 178f
Emulsified oil (macroemulsion) within the foam system, behavior 78–80
Emulsifier, 501
Emulsion, 501
Enhanced oil recovery (EOR)
  advantages of dense CO_2, 202–203
  advantages of using foam, 166
  applications of foam, 165–166
  definition, 501
  disadvantages, 203–204
  statistics, 260–261t
  types of techniques, 165–166
  use of CO_2 foams, 201–232
Entering, 178
Entering coefficient
  correlation to defoaming action, 78
  definition, 176, 178–179
Equation of capillarity, see Young–Laplace equation
Equilibrium contact angle, 501
Equilibrium film, 503
Equilibrium interfacial or surface tension, 501
Equipment and material for study of foamy-oil flow
  experimental apparatus, 410–411
  heavy oils, 411–412
Equivalent alkane carbon number, 501
Equivalent film thickness, 501
Ethylene plant
  caustic tower antifoam evaluation, 481f
  process flow, 479
  schematic, 480f
Evaluation of surfactant adsorption, importance, 270
Evanescent foams, definition, 20, 501–502
Everyday foam
  description, 206
  methods of generation, 208
Expansion factor, 502
External phase, 502
Extra heavy crude oil, 502

F

Fatty acid soaps, 502
Fatty alcohol surfactants, 502
Field examples of gas-blocking foams, 348–349

Field experience with steam-foams
  oil recovery rates, 246–251
  sweep efficiency, 246–251
Field foam tests, numerical modeling, 251–255
Field monitoring of foam fracturing, 395–397
Field safety during foam fracture, 397–398
Film, 502
Film balance, 502
Film compressibility, 502
Film-destabilizing effect of particles, dependence on location, 93–94
Film drainage, 502
Film drainage time
  interfacial viscosity, 58–59
  surface tension gradient, 58–59
Film elasticity
  definition, 502–503
  surface-chemical explanation, 17
Film element, 503
Film energy, relationship to drop size, 84
Film pressure, 503
Film rupturing, definition, 8
Film stability
  electrolytes, 71–72
  foam film area, 71
  formation of colloid crystal structure, 73–74
  solubilized oil, 74
  temperature, 72–74
Film stratification
  micellar concentration, 69–71
  size, 69–71
Film tension versus surface tension, 10
Film thinning, definition, 8
Film-thinning phenomena, effect of mobility of surfaces, 59
Filmers, definition, 461–462
Fingers versus channels, 205
First-contact miscible, definition, 204, 259
Flocculation, 489
Flotation applied to oil sands, 426–431
Flow-line treating, 503
Flow achieved for given pressure drop, definition, 211–212
Flow experiments, limitations, 418
Flow resistance of foams in porous medium, 240–243
Flowing foam relative permeability
  dependence on saturation of flowing gas, 146
  gas saturation of flowing bubbles, 126

Fluid and formation compatibility, 387–388
Fluid diversion, description, 379–380
Fluid film, 503
Fluid flow, models used to characterize, 388
Fluid loss, 387
Foam(s)
  aqueous, see Aqueous foam(s)
  basic information, 5
  bulk properties, 30–38
  characteristics, 358
  classification, 6–7
  colloidal systems, 6–8
  cross-sectional diagram, 359f
  definition, 6–7, 116, 358, 503,
  desirable versus undesirable, 4–5
  determination of structure, 51
  examples in everyday experience, 4
  for hydrocarbon-miscible flooding in reservoirs containing high-salinity brines, 262–270
  for well stimulation, 355–399
  formation, 3, 7
  importance, 3–5
  making, 7
  manner in which it undergoes shear, 207
  microstructure, 116–121f
  mobility-controlling agent, 270
  petroleum industry, 3–4, 5t
  population of lamellae in pores of reservoir rock, 206–209
  principles, 3–43
  structure, 6–7
Foam application
  requirements, 246–247
  temperature surveys, 247–248, 249f
Foam application foam injection system, 247f
Foam aqueous stability, mechanisms with and without oil, 47–111
Foam behavior
  after oil contact, 171f
  effect of entering coefficient, 179
  effect of foam-flooding additive, 102, 103f, 104f
  effect of spreading coefficient, 179
  types, 179–180
Foam breakage frequencies, lamella number, 184–186
Foam breaker, 503
Foam coarsening, bulk foam versus confined foam, 137
Foam collapse, two-stage process, 406
Foam decay by evaporation
  factors influencing rate, 347

INDEX

mechanism, 346–347
Foam decay rate, causes, 209
Foam density
 calculation, 31
 variance, 33
Foam destabilization
 by oil, 193–194
 effect of nature of oil, 190–192
 higher molecular weight alkanes versus lower molecular weight hydrocarbons, 190
 mechanisms, 167
 methods, 88
Foam destabilization study by single-component pure hydrocarbon oils, danger, 191
Foam destruction, mechanisms, 132–137
Foam-dilatational viscosity, 62
Foam drainage, 504
Foam durability test
 alternate methods of testing, 225
 apparatus, 223–224
 procedure, 224
Foam effectiveness
 effect of noncondensible gas, 239
 in porous media, 265–269
Foam emulsion, 489
Foam-enhanced oil recovery, 105–106f
Foam film
 formed from micellar surfactant solutions, 64–66
 interference pattern, 84
 stability, 51–75
 stability at high surfactant concentrations, 63–75
Foam flooding, 504
Foam flow
 in limiting capillary-pressure regime, 137–139
 in porous media, 116, 139
 not in limiting capillary-pressure regime, 138
 pressure difference, 16
 shearing strain, 210
Foam formation
 alternative approach to minimizating the effect of residual oil, 246
 bubble division, 144–145
 foam decay, 132–137
 mechanisms, 128–132
 requirements, 20
 reservoir applications, 116–117
 surfactants or surfactant mixtures selection, 166

Foam-forming surfactant adsorption
 dependence on brine salinity, 285, 287–290
 dependence on divalent ion content, 285, 287–290
 dependence on temperature, 285–286f
Foam fractionation, 504
Foam fracturing design criteria, choice of fluid, 386–387
Foam friction, 385–386
Foam gas-blockage performance
 apparatus, 326–327
 combined effects of temperature and concentration, 335f
 effect of oil, 335
 measurement, 325–332
 procedure, 326–327
Foam generation
 critical velocity for onset, 142
 dissolution and expansion process definition, 208
 old-fashioned way, 208
Foam generation rate, dependence on pore sizes and complexity, 209
Foam half-life
 definition, 383
 function of quality, 384f
Foam in porous media, body to throat size aspect ratio, 117
Foam inhibitors
 definition, 38, 504
 uses in petroleum industry, 88
Foam instability, causes, 383
Foam lamella destruction, relationship to flux into termination sites 134, 143
Foam lamella rupture pressure and bead pack pressure gradient, comparison, 136f
Foam lifetime
 effect of drop size, 87
 effect of oil viscosity, 97, 98f
Foam mobilization, 125
Foam mobilization and trapping, percolation models, 123–124
Foam number, 504
Foam-oil interaction
 behaviors, 176, 177f
 experimental studies, 168–176
 summary, 109f
 theories, 176–184
Foam-over
 consequences, 470–471
 definition, 504
 description, 470
Foam phase-behavior diagrams, effect of oil, 181f

Foam preventatives, definition, 38
Foam propagation
  elevated temperatures, 238–243
  limits, 270–274
  surfactant retention during flow, 270
Foam propagation rate, effect of prefoam slug injected prior to foam formation, 245-246
Foam quality
  definition, 358, 504
  versus foam viscosity, 360f
Foam rheology, 60–63
Foam rupture model, spreading-entering theory, 77–78
Foam sensitivity to oil
  emulsification and imbibition, 172
  manifestations, 168
Foam stabilities to different oils, comparison, 191f
Foam stability
  definition, 383–384
  dilatational elasticity, 60
  elevated temperatures, 238–243
  emulsification-imbibition of oil into foam, 187
  experimental assessment, 30–31, 32f
  factors determining, 8
  formation of colloid crystal structure, 73–74
  influence of additional phases, 38
  influence of emulsion film stability, 107–108f
  influence of macrophobe oil versus oil as emulsified droplet, 183
  influence of stable pseudoemulsion film, 85–88
  in nonpolar media, 406–409
  interfacial properties, 8
  lifetime of the pseudoemulsion film, 106
  mechanism, 103
  mechanisms of enhancement, 40
  method to quantify, 383
  methods of improvement, 360
  methods of testing, 30
  oil presence, 167
  oleophilic surfaces versus hydrophilic surfaces, 192–193
  presence of mixed surfactant, 23
  pressure increase, 104–105
  properties that determine, 20
  second liquid phase, 39–40
  solids, 40–41
  stabilizing mechanisms, 63–64
  stratification, 69–71
  temperature, 23, 102
  water-wet cores versus oil-wet cores, 192–193
  wettability of rock, 192–193
  with drop size, 87
Foam stability at low surfactant concentration, role of surface rheological properties, 54
Foam stimulation fluid, 504–505
Foam texture
  definition, 382, 505
  effect of surfactant type and concentration, 383
  factors affecting, 382
  relationship to time, 152
Foam transport
  comparison of theory and experiment, 149–155
  definition, 121–128
  experimental details, 148
Foam viscosity, effects of quality, 384–385
Foamability, effect of presence of liquid-crystalline phase, 408
Foamed acid fluids, 375–382
Foamed acid fracturing
  difficulties, 375
  goal, 375
Foamed diversion
  advantages, 380
  purpose, 380
  theory, 379
  use, 379
Foamed fluids, definition, 361
Foamed matrix acidizing
  benefit, 379
  definition, 378
  process, 378
Foamed stimulating fluids, advantages, 357
Foamer, 505
Foaminess
  effect of dilatational elasticity, 60
  relationship to phase change, 409
Foaming, in refinery processes, 461
Foaming agent
  composition, 8
  definition, 505
  use, 8
Foaming capacity, fresh versus aged surfactant solution, 265, 266f
Foaming power, 509
Foamy-oil behavior, description, 405–406

INDEX

Foamy-oil flow in porous media, experimental investigation, 409–413
Foamy-oil flow study, procedure, 412
Fouling, definition, 461
Fracture acidizing treatments, factors affecting success, 378
Fracture stimulation fluids, desirable features, 356–357
Free water, definition, 447, 505
Fresh primary froth
 typical rheogram, 438f
 viscosities, densities, and compositions, 434t
Friction pressures of foams, parameters affecting, 386
Froth
 constituents, 445
 definition, 505
 density, 432
 generation, 445
Froth density, effect of low-shear froth viscosity, 439f
Froth flotation, 505
Froth formation models, 438–447
Froth, production, 431
Froth quality
 effect of aerated bitumen droplet density, 445f
 effect of packing density, 445f
 relationship to aerated bitumen droplet density, 443f
Froth rheometer
 mounting location on primary separation vessel, 437f
 sensor system, 437f
Froth structures, 438–453
Froth treatment
 deaerating, 453–454
 dilution and separating, 454–455
 process, 430–431
Froth viscosities, commercial-scale froth production, 434
Frother, 505
Frothing agent, 505
Froths, physical properties, 432–438

G

Gas aphrons, 515–516
Gas blockage
 characterization, 341
 gas transport, 342–346
 saturation, 341–342
Gas-blockage data
 experimental, 345f
 repeatability, 330f
Gas-blockage foam, production history, 329f
Gas-blockage performance
 correlation to elasticity, 63
 effect of absolute permeability, 337f
 evaporation, 347f
 gas type, 339f
 porous medium length, 338f
 pressure level, 339f
Gas-blocking ability, 334
Gas-blocking and steady states of foam in porous media, conditions, 341t
Gas-blocking foam generation, typical displacement fronts, 328f
Gas-blocking state, network model, 344f
Gas-blocking versus flowing foams, main difference, 343
Gas–oil ratio (GOR) reduction in production wells
 background, 320
 use of foams, 320–325
Gas coning
 description, 321
 ways to model, 321–325
Gas cusping, definition, 348
Gas diffusion, definition, 137
Gas emulsions, 7, 505–506
Gas mobility
 correlation of dilatational modulus, 62
 definition, 145–147
 percolation models, 147
Gas sweetening, 474
Generalized plastic fluid, 506
Generation, dependence on wetting-liquid saturation, 143–144
Gibbs adsorption equation, definition, 13
Gibbs effect, 506
Gibbs elasticity, effects of surfactant concentration, 20
Gibbs energy of attraction, 506
Gibbs energy of interaction, 506
Gibbs energy of repulsion, 506
Gibbs film elasticity, see Film elasticity
Gibbs isotherm, 506–507
Gibbs–Marangoni effect
 definition, 55, 507
 function, 17–18
 measurement, 18
Gibbs–Plateau border
 aqueous films formed, 48, 49f
 definition, 6–8, 519

Gibbs ring, 6–8, 519
Gibbs surface, 507
Gibbs surface concentration, 507
Gibbs surface elasticity, definition, 18, 19f
Gilwood and Beaverhill Lakes formations, characteristics of reservoirs, 262, 263t
Gregoire Lake in situ steam pilot (GLISP) well configuration, 248f
Gouy–Chapman layer, see Electric double layer
Gouy layer, see Electric double layer
Grading, concept, 205
Gravity, effect on foam drainage, 16

H

Half-life, definition, 358
Hamaker constant, 507
Head group, 507
Heavy crude oil, 507
Heavy oils, types, 411–412
Helmholtz double layer, 507–508
Helmholtz–Smoluchowski equation, 525
Hemimicelle, 508
Henry equation, 508
Herschel–Bulkley model, definition, 388
Heterodisperse, 508
High gas–oil ratio (GOR) production, requirements of foams, 320
HLB scale, see Hydrophile–lipophile balance
HLB temperature, see Phase inversion temperature
Homogenizer, 508
Hot-water flotation process
 description, 426–431
 diagram, 429f
 two of steps, 432f
 use, 426
Hückel equation, 508
Hydraulic fracturing, mechanism, 356
Hydrocarbon-miscible flooding
 at high salinities, 259–309
 description, 259
 projects in Canada, 262
 significance, 259–261
Hydrophile–lipophile balance, 508
Hydrophilic, 508
Hydrophilic silica, reaction of silicone oil, 466f
Hydrophobic, 509
Hydrophobic bonding, 509

Hydrophobic index
 definition, 509
 method of measurement, 466
Hydrophobic silica
 comparison of antifoam properties, 465f
 definition, 464
 lending with hydrocarbon oil, 464–465
 preparation, 464
Hydrostatic pressure calculation, example, 365, 366–367t
Hydrotreating, definition, 471

I

Imbibition, 509
Implementation of foams for well stimulation, field equipment setup, 393–395
Implicit-pressure, explicit-saturation (IMPES), definition, 149
Improved oil recovery, correlation of pseudoemulsion film stability, 105–107
Incipient phase change, mechanism, 409
Increase of volume upon foaming, 509
Incremental oil recovered by different types of foams, 190f
Indifferent electrolyte, 509
Individual segment pressure, effect of pressure drop, 415–417
Induced gas flotation, see Froth flotation
Inherent viscosity, 509
Inhibition of ordering of micelles
 mechanism, 72
 presence of electrolytes, 71–72
Initial knockdown capability, 511
Injection of CO_2
 continuous CO_2 injection, 229
 water alternated with gas (WAG) method, 229–230
Injection profile survey, description, 248
Ink bottle pore, 509
Inner potential, 509–510
Insoluble surface films, mechanisms of foam stability, 408
Instability growth, mechanism, 205
Interface, definition, 9, 510
Interface emulsion, 510
Interfacial film, 510
Interfacial layer, 510
Interfacial mobility versus dimensionless film thickness, 56f
Interfacial potential, see surface potential

INDEX

Interfacial properties, importance in foams, 9
Interfacial rheology, 529
Interfacial tension
  behavior, 237–238f
  definition, 9–10f
  effect of temperature, 238
  method of measurement, 10
  see surface tension
Interfacial viscosity, *see* surface viscosity
Interferometry, 510
Internal phase, 510
Intrinsic viscosity, 510
Inverse micelle, 510
Inversion, 511
Invert emulsion, 511
Ion exchange, 511
Ionic strength, 511
Iridescent layers, 523
Isaphroic lines, 511
Isoelectric, 511
Isoelectric point, 511
Isotherm, 511

J

Jar test, 492

K

Kelvin equation, 511
Kinematic viscosity, 511
Kinetics of adsorption and desorption, illustration, 306f
Knockdown capability, 511
Krafft Point, definition, 14, 512
Kugelschaum, definition, 7, 118

L

Lamella
  contents, 10
  definition, 6
Lamella division
  definition, 512
  description, 130
  mechanism, 130f
Lamella leave-behind, 512
Lamella settler, description, 454
Lamellar foam, 512
Lamina, *see* Foam

Laminar flow, 512
Langmuir adsorption, *see* Adsorption isotherm
Langmuir film balance, 502
Langmuir isotherm, 512
Langmuir trough, 502
Laplace capillary suction, effect on foam drainage, 16
Laplace flow, 493
Laplace waves, 494
Lather, 513
Leave-behind
  description, 130–131
  mechanism, 131f
Lens, 513
Leverett J-function, control of critical capillary pressure, 134
Light crude oil, 513
Limiting capillary pressure
  behavior, 139
  definition, 135, 513
Limiting viscosity number, *see* Intrinsic viscosity
Lipophile, 513
Lipophilic, 513
Lipophobe, 513
Lipophobic, 513
Liquid aerosol, 489
Liquid crystals, 515
Liquid drainage, effect of temperature, 103–104, 105t
Liquid-phase viscosity, role in foam stability, 406–407
Live oil, *see* Recombined oil
Logarithmic viscosity number, 509
Loss of gas blockage
  diffusion, 348
  evaporation, 346–348
Low recovery efficiency, reasons, 204
Lyophilic, 513
Lyophilic colloid, 513
Lyophobic, 513
Lyophobic colloid, 514

M

Macroemulsion, 514
Macroion, 514
Macromolecule, 514
Macropore, *see* Pore
Magnetic resonance imaging
  definition, 514
  studies, 447–453

theory, 449
procedure in froths, 449–450
Making and breaking, definition, 147–148
Marangoni effect, 514
Marangoni flow, 514
Marangoni surface elasticity, definition, 18–19f
Marangoni waves, 494
Mass balance equation
 gaseous and aqueous phases, 140–141
 surfactant, 140
Matching surfactants to specific reservoir solids, 300–302
Matrix acidizing treatment
 ideal, 378–379
 versus acid fracturing treatments, 378
 versus hydraulic acid fracturing, 356
Maximum bubble-pressure method, definition, 20, 514–515
Measurement of adsorption of surfactant onto rock core
 apparatus for recirculation method, 227f
 methods, 226–227
 typical results for recirculation methods, 228f
Mechanical shear, sources, 183–184
Mechanisms
 aqueous foam stability and antifoaming action with and without oil, 47–111
 foam stability in stable pseudoemulsion film, foam drainage, 85–87
 surfactant adsorption, 276–279
Membranous emulsified water formation, mechanism, 448f
Meniscus, 515
Mesomorphic phase, 515
Mesopore, 117, 520
Metastable, 530–531
Micellar charge, 515
Micellar emulsion, 515
Micelle, 14, 515
Micelle ordering, relationship to concentration, 64
Micelle polydispersity, effect of oil solubilization, 75
Micelle size, effect of oil solubilization, 75
Micellization, effect of temperature, 14, 16f
Microelectrophoresis, 500
Microemulsion, 515
Microgas emulsions, 515–516
Microscopic displacement efficiency, effect of emulsification, imbibition, and transportation of oil, 189
Microscopic studies, 447–453
Microstructure formation (stratification), definition, 63
Microstructure regimes, types, 118
Microtome method, 516
Microvisual experiments, proposed mechanisms, 184–186
Microvisual simulations
 apparatus, 169, 170f
 procedures, 169
Middle soap, 517
Middlings, definition, 430
Miniemulsion, 516
Minifrac, definition, 387
Miscibility, factors affecting, 259
Mist, 516
Mitchell foam quality, definition, 358–359
Mixed micelles, 14, 515
Mobile film, 503
Mobility
 effect of volumetric CO_2 fraction, 219–220
 influence of surfactant concentration, 218–219
Mobility control
 of steam-flooding, 100
 requirements, 205–206
Mobility of CO_2 foam
 apparatus for measurement, 216–217
 definition, 213
Mobility of fluid through porous media, factors, 211–212
Mobility of gases in porous media, effect of aqueous surfactant-stabilized foam, 116
Mobility reduction
 evaluation, 210–222
 relationship to lamella number, 188–189
Mobility reduction factor
 calculation, 188
 definition, 174, 214, 265–266, 516
 dependence on brine salinity, 268
 dependence on permeability, 268–269f
 measured in oil-free Berea cores containing
 measured with nitrogen versus with light hydrocarbon solvent mixtures, 268–269
 versus breakage frequencies of foam lamellae, 187f
Mobility reduction factor measurement
 accuracy, dependence on several factors, 266

INDEX

Model parameters
  applicable to standard, two-phase flow, 149–150
  foam displacement, 149–150
Models of fluid flow, mathematical, 388–392
Modulus of surface elasticity, *see* Film elasticity
Monodisperse, 516
Monolayer adsorption, 516
Monolayer capacity, 516
Monomolecular film, 516
Monomolecular layer, 516
Movement of the crude oil in porous medium, stability of the pseudoemulsion film, 106
MRF, *see* Mobility reduction factor
MRI, *see* Magnetic resonance imaging
Multilayer adsorption, 517
Multiple-contact miscibility
  definition, 204, 259–260
  processes for development, 260
Multiple emulsion, 517

N

N_2, methods of isolation, 363
N_2 foam fracture equipment setup, diagram, 394f
N_2 foam versus CO_2 foam, primary difference, 368
N_2-foamed fluids
  history of use, 361
  method of formation, 361, 362f
  procedure for calculating density, 363–365
N_2 pumper, schematic, 362f
N_2 space factor, definition, 364–365
Naphtha, 517
Neat soap, 517
Neck snap-off, 129
Negative adsorption, *see* Adsorption
Neutralizers, definition, 462
Neutron logs, 248
Newton black film, definition, 27–28, 36
Newtonian external foams, relationship between viscosity, quality, and shear rate, 385
Newtonian flow, 517
Newtonian fluid, 517
Newtonian versus non-Newtonian foams, 385, 386f
Nitrified foam, definition, 517
Nitrified foam fracture treatment, description, 393–394
Nonaqueous foams in porous media, uses, 117
Nonionic surfactant, 517
Non-Newtonian fluid, 518
Nonpolar foams in production of heavy oils, role, 405–419
Nonwetting, 532
Nuclear magnetic resonance images of froth
  after aging, 453f
  sampled from bottom of froth layer, 452f
  sampled from middle of froth layer, 451f
  sampled from top of froth layer, 450f
Nucleation, definition, 51

O

Oil
  configurations
    at gas-aqueous interface, 75–76
    in foams, 75–80
  definition, 518
  effect on aqueous foam stability and antifoaming action, 47–111
  emulsification, 47–48
Oil production, down-dip versus up-dip wells, 247
Oil-production performance, effect of gas-blocking foam, 323, 324f
Oil-production rate, effect of pressure drawdown, 413–414
Oil recovery reduction, reasons, 165–166
Oil recovery through hydrocarbon-miscible flooding, mechanisms, 259
Oil sand
  arrangement of phases, 425, 427f
  definition, 423
  structure, 424–426
Oil transport, degree of emulsification of the crude oil, 106
Oil/water (O/W), 518
Oil/water/air (O/W/A), 518
Oil/water/oil (O/W/O), 518
Oleophilic, 518
Oleophobic, 518
One-dimensional flow models, 324–325
Optical properties
  determination of the film thickness, 36
  measurement of foam decay, 37
Optimum salinity, 518
Organic compounds in antifoaming and defoaming, glycols, 468

Osmosis, 518
Ostwald viscometer, 494
Outage, definition, 471
Outer potential, 518
Overall surfactant adsorption, parameters determining, 276–277

P

Packing density of aerated bitumen droplets, function of droplet size distribution, 439–440
Particle electrophoretic mobility, change with adsorbed anionic, cationic, or amphoteric surfactant, 294–295
Partitioning of anionic surfactants, effect of temperature, 236–237
Paucidisperse, 519
Pendant drop method, 519
Peptization, definition, 519
Peptizing ions, 520
Permeability, 519
Permeability of bead pack to gas, with and without foam, 123
Permeability reduction factors (PRF)
 calculation, 188
 definition, 174
Perrin black film, 519
Petrochemical applications, ethylene plants, 478–481
Petrochemical applications for antifoaming and defoaming, butadiene extraction units, 481–482
Petrochemical debutanizer, definition, 462
Petroleum, 519
Petroleum reservoirs wettability, effect of crude oil, 277
Phase-behavior diagram, representation of behavior, 180, 181f
Phase inversion temperature (PIT), 519
Phases, arrangement, 7–8
Plastic flow, 519
Plastic fluid, 519
Plateau border, definition, 6–8, 519
PLAWM trough, *see* Film balance
Pockels–Langmuir–Adam–Wilson–McBain trough, 512
Point of zero charge, 520
Polar group, 520
Polydisperse, 520
Polyederschaum
 definition, 118
 structure, 6f, 7–8
Polyelectrolyte, 520
Population-balance method, description, 140
Population-balance modeling of foam flow in porous media, 139–148
Pore
 definition, 520
 description, 117
 two-phase occupancy, 117
Pore-level microstructure of foam, summary, 126–128
Porosity, 520
Porous medium
 definition, 520
 in foamy-oil flow studies, 411t, 412
Potential determining ions, 520
Potential energy diagram, 506
Potential energy of attraction–interaction–repulsion, 520
Pour point, 520
Power-law fluid, 521
Preflash column, 472–473
Prefoam slug technology, 245–246
Preheat exchanger train, definition, 472
Preneck snap-off, description, 129
Preparation of wettability-altered cores, 295
Presence of ionic materials, effect of solubility of surfactant, 236
Pressure drop, calculation, 391–392
Pressure distributions in sand-pack, pressure profiles, 414f
Pressure drawdown, definition, 405, 412
Pressure-driven foam generation
 model, 330–332
 pressure profiles, 332f
 typical front behavior, 331f
 water-saturation profiles, 332f
Pressure drop (ΔP) behavior during drawdown and low-pressure phases, steam-only versus steam-surfactant, 245f
Pressure falloff test, description, 249–250
Pressure gradient
 factors affecting, 123–124
 in individual segments, 415
 required to drive lamellae through porous media, 125
Pressure variation between the Plateau borders, effect of radius curvature, 11f
Primary channels, definition, 123
Primary froth
 contents, 438
 mechanisms of water–solid interaction, 438–439
 quality, 439

INDEX

Primary minimum, 521
Primary oil recovery, 521
Primary viscous stresses, difference between wet and dry foams, 62
Proppant effects, example of viscosity calculation, 392–393
Proppants, definition, 356
Protected lyophobic colloids, see Sensitization
Protection, 521
Protective colloid, 521
Pseudoemulsion film
  definition, 180
  interference pattern, 84
  mechanisms of behavior, 81
  model, 180–182
Pseudoemulsion film stability
  correlation of improved oil recovery, 105–107
  experiment, 93, 94f,t
Pseudophase-separation model for surfactant solutions, description, 302
Pseudoplastic foams, definition, 33, 521

R

Radial flow models gas coning, description, 321
Radial surfactant propagation
  in homogeneous reservoir, 272–273
  in layered reservoir, 273–274
Range of foam sensitivities to crude oil, qualitative types of behavior, 169
RCF, see Relative centrifugal force
Recombined oil
  preparation, 411, 412
  properties, 412t
Rectilinear snap-off, description, 129
Reduced adsorption, 521
Reduced viscosity, 521
Reduction of effective permeability of gas in powers medium, mechanism, 123
Relative adsorption, 521
Relative apparent viscosities (RAV), definition, 174
Relative bitumen layer thicknesses of aerated bitumen droplets, illustration, 442f
Relative centrifugal force, 521–522
Relative gas mobility as function of gas velocity, 343f
Relative mobility, definition, 213–214

Relative molecular mass, 522
Relative permeability to gas (k_{rg})
  effect of gas velocity, 240–241, 242f
  effect of liquid velocity, 241, 242f
  effect of simultaneous change in gas and liquid velocities, 241–243
Relative permeability to water, in clean and wettability-modified Berea cores containing oils, 296, 297f, 298
Relative phase saturations, froth samples, 452t
Relative viscosity, 522
Repeptization, 522
Repulsive potential energy, 522
Reservoir solids, isoelectric points, 279
Resid, definition, 469
Resistance tests, 381t
Restoration of surfactant adsorption equilibrium
  description, 17
  relationship to film thickness, 17
Reverse emulsion, 522
Rheological behavior of foam, dependence on liquid–gas ratio, 61
Rheological properties, 33–36
Rheological response of foam, foam models, 61
Rheology
  bitumen, 433–438
  definition, 388, 522
  viscosity, 433–434
Rheomalaxis, 522
Rheometer, 522
Rheopexy, 522
Rigid film, 503
Rochow reaction, 467
Role in foam stability
  incipient phase change, 409
  insoluble surface films, 408–409
Roof snap-off, description, 129
Ross foam, 523
Rupture, 503
Rupture of pseudoemulsion film, based on contact angles, 95–96

S

Sacrificial adsorbates
  basic assumptions on effectiveness, 305
  description, 305
  problem, 305
Salinity requirement, 518

Salting in, 523
Salting out, 523
Schiller layers, 523
Schulze–Hardy rule, 523
Secondary channels, definition, 123
Secondary minimum, 521
Secondary oil recovery, 523
Sector model experiments
  comparison, 322f
  generation of gas-blocking foam, 323
  placement of foaming agent solution, 322–323f
  procedure, 322–323
Sediment, 523
Sediment volume, 524
Sedimentation, 523–524
Sedimentation equilibrium, 524
Sedimentation potential, 524
Selection process for foam-forming surfactants, criteria, 262, 264
Sensitivity tests, dependence on penetration depth, 323–324
Sensitization, 524
Separate mobilities of fluids, concept, 212
Separator, 524
Septum, 524
Sessile bubble method, 524
Sessile drop method, 524
Shear, 524
Shear plane, 525
Shear stress
  calculation, 34
  definition, 525
Shear thickening, 498
Shear thinning, 33, 521
Silicone oils, 467–468, 525
Silver film, 492
Single-foam film stability, mechanisms, 51–54
Size scale characteristic of foam in porous media, 118–119
Slops, definition, 469
Slurry quality, definition, 359
Smoluchowski equation, 525
Snap-off
  description, 128–129, 525
  dominant foam generation mechanism, 131
  gas bubbles size, 129
  lamella arrival, 142
  mechanism, 128–129
  relationship of collar growth, 129
  relationship to coalescence, 129
Soap, 525

Soap film, 525
Sol, 525
Solid aerosol, 489
Solid emulsion, 525
Solid foam, 526
Solubility of CO_2
  effect of pressure, 371t
  versus N_2, 373f
Solubility parameter, 526
Solubilization, 526
Solvent deasphalting, description, 477–478
Sorbate, 526
Sorbent, 526
Sorption, 526
Sour-water stripper
  schematic, 476f
  use, 475–476
Specific increase in viscosity, 526
Specific surface area
  definition, 528
  reservoir rocks and clays, 292t
Specific viscosity, see Specific increase in viscosity
Spinning cylinder method, description, 26
Spinning drop method, 526
Spontaneous filming of bitumen around gas bubble, photographs, 433f
Spread layer, 526
Spreading
  definition, 526
  process, 177, 178f
Spreading coefficient, definition, 176–178f, 526–527
Spreading parameter, 176–178f, 526–527
Spreading pressure, 529
Spreading tension, 527
Stabilization of foams in nonpolar liquids, other methods, 408
Stabilization of thin film, surface rheological properties, 59
Stabilizing effect of foams, strong electric double-layer repulsion between oil drops and aqueous–gas foam interfaces, 186
Stabilizing effect on foams, presence of emulsified oil droplets, 186
Stable foams, factors that produce, 383
Static foam test
  apparatus, 32f
  description, 30, 168
Static interfacial tension, 527
Static stability, definition, 358
Static surface tension, 527
Steady-state behavior, 153–155

INDEX

Steady-state experimental mode, description, 319
Steady-state flow, definition, 413
Steady-state pressure drop versus gas velocity
  experimental, 153–155
  model, 153–155
Steam diversion, 249
Steam-foam for heavy oil and bitumen recovery, 235–254
Steam-foam performance in laboratory cores
  constant ΔP, 244
  cyclic steam, 244–245
Steam-foam success, residual oil saturation, 166
Stepwise thinning, in vertical films and foams, 66–67
Stepwise transitions
  effect of electrolytes, 71–72
  effect of surfactant chain length, 71
Steric stabilization, 527
Stern layer, 527
Stern model, description, 25
Stern potential, 509–510
Stimulation fluid, 504–505
Stimulation treatments
  potential hazards, 397–398
  types, 355–356
Stokes' law, 527
Stratification
  definition, 65–66
  effect of temperature, 73
  explanation, 67
Stratification rate, temperature, 73
Stratified film, 527
Streaming potential, 527
Streamline flow, 512
Stress, 525
Structure and stability of the pseudoemulsion film
  dependence on capillary pressure, 84–85
  dependence on electrolyte concentration, 81–82
  dependence on film size 84–85
  dependence on surfactant concentration, 81–83
Structure of primary froth, role of variable G, 448f
Substrate, 528
Successful acid fracture stimulation, key, 375
Supercentrifuge, 494
Superficial density, 528
Surface, 9, 510
Surface-shear viscosity, 529
Surface active agent, see Surfactant(s)
Surface area
  definition, 528
  minimization, 9
Surface charge
  definition, 528
  kaolinite, 279, 281
  reservoir solids, 279–283
Surface concentration, 507
Surface conductivity, 528
Surface coverage, 528
Surface dilatational elasticity, definition, 18
Surface dilatational modulus, see Film elasticity
Surface dilatational viscosity
  concepts, 21f
  defintion, 529
Surface elasticity
  defintion, 17–20
  differences in importance in aqueous and nonpolar foams, 408
  effect of surface tension gradient on, 17
  measurements, 20
  origin, 17f
  relationship to compressibility of surface film, 18
Surface excess conductivity, see Surface conductivity
Surface excess isotherm, 528
Surface fluidity, 528
Surface layer, 528
Surface of tension, 528
Surface potential
  defintion, 529
  dispersion forces, 26
  electric double layer, 24–26
  repulsive forces, 26
  total interaction energy, 27
Surface pressure, 529
Surface rheology, definition, 21–24, 529
Surface shear, concepts, 21f
Surface stress tensor
  general form, 21
  simplified surface geometry example, 21f,22
Surface tension
  defintion, 9–10f, 529
  effect of surfactants, 9–16
  method of measurement, 10

Surface tension gradient in the thinning film, creation, 55
Surface viscosity, 529
Surface work, 529
Surfactant(s)
  association behavior, 13f
  classification, 14–16
  definition, 12–13, 529
  effect on interfacial tension, 13
  effect on interfacial viscosity, 13
  for CO_2 floods, 223–228
  molecular mass range, 14
  most favorable orientation, 12f
  passing solubility criteria, 265t
  properties at elevated temperature, 235–238
  role in bubble transport, 125
  solubility–micellization behavior, 16
  tested for solubility, 264
  used in gas-blocking foam generation experiments, 328t
Surfactant adsorption
  at solid–liquid interface, 276–298
  brine salinity, 278
  characterization of systems in which it was measured, 295–296
  charge of surfactant versus change of solid, 278
  dependence on rock type, 290–295
  dependence on sacrificial adsorbate injection sequence, 306–308
  divalent cations, 289–290
  ionic content, 278
  method of measurement, 295
  on hydrophobic surfaces, 277–278
  on polar insoluble solids, 277
  on saltlike minerals, 277
  relationship to surfactant solubility, 278–279
  rock type, 290–292
  salinity, 270
  surfactant-type, 279
  wettability and residual oil, 295–298
Surfactant adsorption equilibrium, reestablishment, 17
Surfactant adsorption levels, in clean and wettability-modified Berea cores containing different oils, 296, 297f, 298
Surfactant adsorption measurements, mineral composition of rocks, 300
Surfactant adsorption minimization
  approaches, 298
  importance, 298–299

Surfactant alternated with gas (SAG) floods, definition, 230
Surfactant behavior, effect of concentrations, 14
Surfactant chemical stability, 265
Surfactant chemistry, effect on surface tension, 9
Surfactant degradation assessment, approaches, 265
Surfactant effectiveness, 530
Surfactant efficiency, 530
Surfactant loss
  mechanisms, 274–276
  when oil is present, 275–276
Surfactant mixtures, mechanism of adsorption reduction, 302–305
Surfactant precipitation, description, 274
Surfactant retention in reservoir, mechanisms, 270
Surfactant selection
  for producing hydrocarbon foams, 409
  level of surfactant adsorption, 270
  requirements, 227
Surfactant solubility, 264–265
Surfactant tail, 530
Suspension, 530
Swamping electrolyte, 530
Swelling, 530
Swelling pressure, 530
Swollen micelle, 14, 515
Syneresis, 530

T

Tailings oil recovery (TOR) flotation circuit, description, 430
TAN, *see* Total acid number
Tenside, *see* Surfactant(s)
Tensiometer, 530
Tertiary oil recovery, *see* Enhanced oil recovery (EOR)
Thermal stability
  sulfonates, 236
  surfactants, 235–236
Thermodynamic stability, 530–531
Thin-film drainage process, Gibbs–Marangoni effect, 54–55
Thin films, making, 7
Thin lamella stability, factors governing, 52
Thin lamella thinning rate, factors governing, 52

INDEX

Thin liquid film formation, stages, 52
Thixotropic, definition, 33, 531
Three-phase foam stability, effect of interfacial tension gradient, 81, 82f
Total acid number, 531
Total potential energy of interaction, see Gibbs energy of interaction
Tracers, 249
Transient displacement, 151–153
Transient pressure profiles
 experimental, 152f
 model, 152f
Trapped gas saturation measurement, gas-phase tracer experiments, 122
Turbidity, 531
Turbulent flow, 531
Turbulent flow of N_2 foams, scale-up equation, 391t
Type A foams, description, 169, 171f, 172f
Type B foams, description, 169, 171, 173f
Type C foams, description, 171–172

U

Ubbelohde viscometer, see Capillary viscometer
Ultracentrifuge, 494
Ultrasonic dispersion, 531
Unfulfilled adsorption equilibrium, consequence, 226

V

Vacancy mechanism, description, 70
Vacuum tower
 foaming, 473–474
 use, 473
van der Waals forces, see Dispersion forces
Vaporizing gas drive, definition, 260
Velocity shot survey, description, 248
Viscoelastic, 531–532
Viscometer, 532
Viscosities of oil with CO_2, effect of temperature, 371t
Viscosities of various Athabasca bitumens, effect of temperature, 435f
Viscosity, 532
Viscosity-average molecular mass, 532
Viscosity number, see Reduced viscosity
Viscosity of CO_2, relationship to density, 203f
Viscosity ratio, see Relative viscosity
Volumetric sweep efficiency, different experimental procedures, 174–175

W

Water relative permeability, effect of foam, 122
Water saturation, effect of gas fractional flow, 215
Waterflood recoveries, clean and wettability-modified Berea cores containing different oils, 296, 297f, 298
Water/oil (W/O), 533
Water/oil/air (W/O/A), 533
Water/oil/water (W/O/W), 533
Well-site operation, safety equipment, 398
Well stimulation
 candidates, 355
 definition, 355
 objective, 397
Wet foam
 definition, 7
 estimation of viscosity, 35
Wet oil, 532
Wettability, 532
Wettability assessment, criteria, 296
Wetting, 532
Wetting coefficient, 532
Wetting power, 176–178f, 526–527
Whipping agent, see Foaming agent
Wilhelmy plate method, 533
Work
 of adhesion, 533
 of cohesion, 533
 of separation, 533
 of spreading, 533

Y

Yield stress, definition, 210–211, 533
Young–Laplace equation
 basis for some important methods for measuring surface and interfacial tensions, 12
 definition, 10–12, 533
Young's equation, 40–41, 533

Z

Zero point of charge, 520
Zeta potential, definition, 25f, 534
Zwitterionic surfactant, 534

Copy editing and indexing: Stephanie M. Patton
Production: Margaret J. Brown
Acquisition: Cheryl Shanks
Cover design: Eileen Hoff

Printed and bound by Maple Press, York, PA